APPLICATIONS OF STATISTICS

APPLICATIONS OF STATISTICS

PROCEEDINGS OF THE SYMPOSIUM HELD AT
WRIGHT STATE UNIVERSITY, DAYTON, OHIO,
14-18 JUNE 1976

Edited by

PARUCHURI R. KRISHNAIAH

Department of Mathematics and Statistics
University of Pittsburgh, Pittsburgh, Pa., U.S.A.

N·H
P&C

1977

NORTH-HOLLAND PUBLISHING COMPANY–AMSTERDAM • NEW YORK • OXFORD

North-Holland ISBN 0444 85034 1

Published by:
NORTH-HOLLAND PUBLISHING COMPANY
AMSTERDAM • NEW YORK • OXFORD

Distributors for the U.S.A. and Canada:
Elsevier North-Holland, Inc.
52 Vanderbilt Avenue
New York, N.Y. 10017

Library of Congress Cataloging in Publication Data

Symposium on Applications of Statistics, Wright State University, 1976.
Applications of statistics.

Includes bibliographical references.
1. Mathematical statistics. I. Krishnaiah, Paruchuri R. II. Title.
QA276.A1S86 1976 501'.82 77-14481
ISBN 0-444-85034-1

PRINTED IN THE NETHERLANDS

PREFACE

Statistical techniques play an important role in drawing meaningful conclusions from non-deterministic data in many disciplines. Therefore, there is a great need to improve the interaction between statisticians and scientists in other disciplines who are interested in applying statistical methodology. A Symposium on Applications of Statistics was organized under the sponsorship of the Air Force Flight Dynamics Laboratory to increase such interaction and to stimulate research in statistical sciences. This symposium was held at Wright State University, Dayton, Ohio, during the period June 14-18, 1976. Professor J. Neyman, who has made many pioneering contributions to the field of statistics, delivered the Inaugural Address entitled "Experimentation with Weather Control and Statistical Problems Generated by It". The program consisted of sessions with invited papers and contributed papers as well as clinical sessions. In the clinical sessions scientists and engineers in the U. S. Air Force presented practical problems that require solution.

This book consists of the invited papers presented at the symposium. In these papers, distinguished scientists discuss the current state-of-the-art on a very broad spectrum of topics in the field of applications of statistics. Some of the areas of applications covered include acoustics, cancer research, cluster analysis, communication, econometrics, hydrology, meteorology, model building, pattern recognition, pharmacokinetics, psychometrics, reduction of dimensionality, reliability, stability of structures, and turbulence. Unfortunately, the contributed papers and the presentations made at the clinical sessions are not included in this book due to the limitation of space, but their titles are listed at the end of the book.

I wish to thank Dr. R. Kahal (Office of the Assistant Secretary, Research and Development, U. S. Air Force) for making the Opening Remarks. Also, I wish to express my appreciation to Dr. D. Zonars (Air Force Flight Dynamics Laboratory) and Dr. J. V. Murray (Wright State University) for welcoming the delegates. Thanks are due to Professors R. E. Bargmann, A. P. Basu, C. B. Bell, R. Elashoff, K. S. Fu, S. S. Gupta, T. Jayachandran, R. H. Jones, M. Kabrisky, G. Kallianpur, F. Kozin, C. Maneri, M. R. Paruchuri, M. L. Puri, A. Ramakrishnan, C. R. Rao, J. R. Van Ryzin, D. G. Shankland, N. Timm, C. P. Tsokos, V. M. R. Tummala and Drs. J. S. Bendat, S. Dubey, I. I. Gringorten, H. Hughes, R. Launer, W. R. Riggan and E. A. Robinson for presiding over different sessions. Thanks are also due to Professors R. E. Bargmann, L. Kanal, C. G. Khatri, S. Kotz, K. V. Mardia, F. Proschan, J. S. Rustagi, S. Saunders, S. Sclove, B. K. Shah and V. R. R. Uppuluri for refereeing the papers. I am grateful to Drs. H. L. Harter and I. N. Shimi for their help in the organization of the symposium and to Professors J. C. Lee and C. Maneri for their help in making the local arrangements. I wish to thank Colonel M. Duggins for his encouragement and Mrs. R. Kish for typing the manuscript patiently.

I am indebted to Professor J. Neyman for delivering the Inaugural Address. Special thanks are due to the contributors to this volume and to the North-Holland Publishing Company for their excellent cooperation. I wish to express my appreciation to the Department of Statistics at Carnegie-Mellon University, the Department of Mathematics and Statistics at University of Pittsburgh and the Air Force Flight Dynamics Laboratory for providing the necessary facilities to edit this volume. Last but not least, I wish to thank my wife, Indira, for her constant encouragement in the organization of the symposium.

P. R. Krishnaiah

CONTENTS

PREFACE v

EXPERIMENTATION WITH WEATHER CONTROL AND STATISTICAL PROBLEMS GENERATED BY IT*

Jerzy Neyman

I. Introduction 1

II. First Decade, 1951 - 1960 2
1. Randomized Experiments vs. Non-Randomized Cloud Seeding Operations 2
2. Efforts to Learn the New Field of Study 3
3. Biblical "Seven Fat Years" Followed by "Seven Lean Years" 4
4. Convincing Domestic Freezer Experiment 6
5. First Disenchantment: "Report of the Advisory Committee on Weather Control" 7
6. Some Findings in the Theory of Statistics 8

III. Second Period of my Involvement, 1964 to Present 9
7. Arrival of Seven Annual Reports on Grossversuch III 9
8. Special Weather Modification Conference at the Fifth Berkeley Symposium on Mathematical Statistics and Probability 11
9. Our Use of Grossversuch III Data 11
10. Difficulties in Using Old Experiments to Verify New Hypotheses 14
11. Verification of Generality of the Findings at Grossversuch III 17
12. Hypothetical Mechanism of Far Away Losses of Rain at the Arizona Experiment 20

Acknowledgements 22
References 23

ON ENTROPY MAXIMIZATION PRINCIPLE

Hirotugu Akaike

1. Introduction 27
2. A Critical View of Conventional Procedures 27
3. Entropy Maximization Principle 29
4. Classical Procedures and the Entropy Maximization Principle 30
5. Some Realizations of the Principle 31
6. Conclusion 39
Appendix 40
References 40

STOCHASTIC PROBLEMS IN THE STABILITY OF STRUCTURES

S. T. Ariaratnam and D. C. F. Tam

1. Introduction 43
2. Formulation 43
3. Approximation to Markov Process and Stability Equations 46
4. Application 48
5. Conclusions 53
Acknowledgement 53
References 53

*Inaugural Address

A MINIMAX APPROACH TO COMPONENT ANALYSIS
Rolf E. Bargmann and Fred D. Baker
1. Introduction 55
2. Rank Reduction to Control Maximum Error 57
3. Numerical Procedures 59
4. Representation 62
5. Applications 65
References 69

COHERENCE ANALYSIS OF MULTIPLE INPUT/OUTPUT DATA
J. S. Bendat
1. Models for Arbitrary Inputs 71
2. Multiple and Partial Coherence Functions 71
3. Ordered Conditioned Records 74
4. Models for Conditioned Inputs 74
5. Acoustics and Vibration Applications 74
References 78

MULTIVARIATE RELIABILITY CLASSES
Henry W. Block
1. Introduction 79
2. Multivariate Exponential Distributions and Extensions 80
3. Multivariate Increasing Failure Rate 81
4. Multivariate Increasing Failure Rate Average 86
5. Other Multivariate Classes and Estimation Procedures 87
References 87

CHARACTERIZING A LATENT TRAIT DISTRIBUTION
R. Darrell Bock and William A. Thrash
1. Background 89
2. The Error Function of the Spatial Ability Test 91
3. The Test Score Distribution in the Presence of Major-Gene Effects and Measurement Error 92
4. Maximum Likelihood Estimation of p, μ_1, μ_2, and σ 93
5. Example 95
References 96

APPROXIMATIONS TO THE DISTRIBUTIONS OF THE LIKELIHOOD RATIO STATISTICS FOR TESTING THE HYPOTHESES ON COVARIANCE MATRICES AND MEAN VECTORS SIMULTANEOUSLY
T. C. Chang, P. R. Krishnaiah and J. C. Lee
1. Introduction 97
2. Likelihood Ratio Test Statistic for the Homogeneity of Real Multivariate Normal Populations 97
3. Likelihood Ratio Statistic for Testing the Hypotheses that the Mean Vector and Covariance Matrix are Equal to Specified Values 102
4. Likelihood Ratio Test Statistic for the Homogeneity of Complex Multivariate Normal Populations 103
5. Likelihood Ratio Statistic for Testing the Hypothesis that the Covariance Matrix and Mean Vector of Complex Multivariate Normal are Equal to Specified Values 107
References 108

AN ALTERNATING RENEWAL PROCESS WITH AN ABSORBING STATE

C. L. Chiang and J. P. Hsu

1. Introduction 109
2. Identities and Exact Formulas 112
3. Probability Generating Functions 115
4. Properized Probability Distributions 116
5. Asymptotic Behavior 117
6. An Example 120

References 121

STATISTICAL PROBLEMS ARISING IN THE THEORY OF AGGREGATION

John S. Chipman

1. Introduction 123
2. Aggregation as a Means of Improving Estimation 123
3. Limitations of James-Stein Estimation as Applied to Aggregative Models 127
4. Application to a Study of the Relation Between Internal and External Prices in Germany, 1963 - 1973 133

Appendix 136

References 139

METHODOLOGICAL DEVELOPMENTS IN SOME APPLICATIONS OF CLUSTERING

A. Cohen, R. Gnanadesikan, J. R. Kettenring and J. M. Landwehr

1. Introduction 141
2. Descriptions of the Four Examples 142
3. Problems and Methodological Developments at the Analytical Methods Stage 143
4. Problems and Methodological Developments at the Output Stage 149
5. Concluding Remarks 157

References 161

SOURCE CORRELATION EFFECTS ON STRUCTURAL RESPONSE

Stephen H. Crandall and Alexander P. Kulvets

1. Introduction 163
2. One-Dimensional Structures 164
3. Two-Dimensional Structures 173

Acknowledgement 182

References 182

SOME APPLICATIONS OF STOCHASTIC LANGUAGES

K. S. Fu

1. Introduction 183
2. Review of Stochastic Languages 183
3. Application to Communication and Coding 189
4. Application to Syntactic Pattern Recognition 194
5. Application to Error-Correcting Parsing 198
6. Conclusions and Remarks 199

References 199

STATISTICAL PROBLEMS IN THE SIZE EFFECT ON MATERIAL STRENGTH

H. Leon Harter

1. Introduction 201
2. Effect of Length 202
3. Effect of Cross-Section 204
4. Effect of Volume 206
5. Strength of Fiber Bundles 208
6. Strength of Composites 209
7. Size Effect in Fatigue 213
8. Unsolved Problems 215
References 216

BISPECTRA OF ATMOSPHERIC AND WIND TUNNEL TURBULENCE

K. N. Helland, K. S. Lii and M. Rosenblatt

1. Introduction 223
2. Definition of Bispectrum 223
3. Data Acquisition 226
4. Computation and Statistical Analysis 227
5. Results of the Computation 229
6. Concluding Remarks 247
Acknowledgement 248
References 248

URN MODELS: A USEFUL TOOL IN APPLIED STATISTICS

N. L. Johnson and Samuel Kotz

1. Introduction 249
2. Biological Models 250
3. Learning Models 254
4. Sampling Schemes 257
5. Capture-Recapture Problems 258
6. Occupancy Distributions 260
7. Some Other Applications 262
References 263

STRUCTURAL EQUATION MODELS IN THE SOCIAL SCIENCES: SPECIFICATION, ESTIMATION AND TESTING

Karl G. Jöreskog

1. Introduction 265
2. General Theory 266
3. Applications 273
4. Conclusion 284
References 286

THE FLABBINESS AND INSTABILITY OF REGRESSION ANALYSIS AND COMPUTATIONAL METHODS FOR IMPROVEMENT

Harriet Kagiwada, R. Kalaba and K. Mease

1. Introduction 289
2. Dynamic Programming Solution for Fixed Values of λ and C 292
3. Successive Approximations 294
4. Analytic Continuation 294
5. Numerical Experiments 295
6. Discussion 297
References 298

ON HIERARCHICAL CLASSIFIER THEORY AND INTERACTIVE DESIGN

Laveen N. Kanal

1. Introduction 301
2. Multistage Decision Schemes and Hierarchical Classifiers 302
3. Automated Optimal Tree Design 305
4. MIPACS 306
5. Cloud Pattern Classification with MIPACS 315

References 321

TRANSFORMATIONS YIELDING RELIABILITY MODELS BASED ON INDEPENDENT RANDOM VARIABLES: A SURVEY

N. Langberg, F. Proschan and A. J. Quinzi

1. Introduction 323
2. Distributions with Exponential and Proportional Hazard Minima 323
3. Additive Families of Distributions 327
4. Preserving System Life Length and Failure Patterns 329
5. Applications to the Theory of Competing Risks, Life Testing, and Censored Data Problems 332

References 336

STATISTICAL PROBLEMS IN MENTAL TESTING

Frederic M. Lord

Text 339

Appendix 347

References 348

SELF-SELECTIVITY PROBLEMS IN ECONOMETRIC MODELS

G. S. Maddala

1. Introduction 351
2. A Simple Model with no Explanatory Variables 352
3. Models with Explanatory Variables 359
4. Conclusion 365

References 366

SOME APPLICATIONS OF COMPLEX DISTRIBUTIONS TO THE TESTING OF STATISTICAL HYPOTHESIS

Kenneth S. Miller

1. Introduction 367
2. Distributions Associated with the Complex Gaussian 368
3. Some Simple Applications to Hypothesis Testing 372
4. The Multiple Cell Problem 376

References 380

THE WEIGHTED DISTRIBUTIONS: A SURVEY OF THEIR APPLICATIONS

G. P. Patil and C. Radhakrishna Rao

1. Introduction 383
2. Notation and Terminology 383

PART I: Discrete Models and Size-Biased Probability Sampling 385

3. The Analysis of Family Data 385
4. Alcoholism and Family Size 388
5. Aerial Survey and Visibility Bias 389

6. Forest Disease and Line Transect Sampling 390
7. Discrete Waiting Time Paradox 392

PART II. Continuous Models and Size-Biased Probability Sampling 393
8. Renewal Theory and Its Applications 393
9. Waiting Time Paradox 394
10. Cell Cycle Analysis and Pulse Labeling 395
11. Efficacy of Early Screening for Disease 396
12. Forest Products Research 398
13. Particle Size Distributions by Thin Section Methods 398
14. Aerial Survey in Traffic Research 399
15. Rejection Technique in Random Variables Generation 400

PART III. Weighted Distributions as Mass-Size Distributions in Geology and Economics 401
16. Weight-Size Distributions in a Breakage Process 401
17. Moment Distributions in Economics 402
References 403

AN INVESTIGATION OF THE PROPERTIES OF SOME TWO SAMPLE TWO-STAGE DESIGNS

Dale Preston and Robert M. Elashoff

1. Introduction 407
2. The Basic Statistical Model 407
3. An Alternative Statistical Model 412
4. The Minimization Algorithm 413
5. Results 416
6. Discussion 427
7. Summary 431
8. Acknowledgements 432
9. References 432

PATTERN COMPARISON PROBLEMS

J. M. Richardson

1. Introduction 433
2. Preliminaries 433
3. General Statement of the Problem 435
4. Estimation of the Rate of Transformation of an Image 436
5. More Complex Comparison Problems 441
6. Discussion 445
References 445

DECONVOLUTION OF TIME SERIES AS APPLIED TO SPEECH

Enders A. Robinson

1. Introduction 447
2. Acoustic Tube Model 449
3. Deep Source Model 454
4. Surface-Source Model 455
5. Normal Equations 456
6. Deconvolution of Speech 458
7. Conclusion 460
References 460

STATISTICAL ANALYSIS OF COMPARTMENTAL MODELS WITH APPLICATIONS TO PHARMACOKINETICS AND BIOAVAILABILITY

Jagdish S. Rustagi and Umed Singh

1. Introduction 461
2. Compartmental Models 462
3. Difference Equation Approach 466
4. Statistical Considerations 472
5. Illustrations 473
6. Discussion 476

Acknowledgements 477
References 478

THE RELIABILITY OF STRUCTURES SUBJECT TO FATIGUE UNDER INSPECTION AND REPAIR

Sam C. Saunders

1. Introduction 481
2. Residual Strength for a Component 482
3. The Structural Overload Model for a Component 482
4. The Crack Growth Model for a Component 483
5. Crack Detection and Inspection 486
6. Component Failure under Inspection and Repair 489
7. Generalization to Structures with Jointly Distributed Loads 492

References 496

SYSTEM FAILURES AND STOCHASTIC HAZARD RATE

I. N. Shimi

1. Introduction 497
2. The Model 498
3. The Main Result 501
4. Examples 503

References 505

A SURVEY OF SOME SMOOTHING PROBLEMS AND THE METHOD OF GENERALIZED CROSS-VALIDATION FOR SOLVING THEM

Grace Wahba

1. Introduction 507
2. Ridge Regression 509
3. Spline Smoothing 515
4. Density Estimation 517
5. The Approximate Solution of Linear Operator Equations when the Data are Noisy 519

Notes Added in Proof 521
References 522

STATISTICAL PROBLEMS IN WATER RESOURCES

Vujica Yevjevich

1. Character of Stochastic Processes 525
2. Sources of Stochasticity 526
3. Complexities of Short-Interval Runoff Time Series 529
4. Problem of Stationarity of Annual Hydrologic and Climatologic Time Series 534
5. Combination of Statistical and Physical Information 535

6. Discrimination of Changes in Climatologic and Hydrologic Time Processes 535
7. New Methods for Statistical Inference 537
8. Design of Experiments 537
9. Properties of Derived Variables 538
10. Multivariate Analysis 539
References 540

TITLES OF CONTRIBUTED PAPERS 541

TITLES OF PRESENTATIONS AT THE CLINICAL SESSIONS 543

P.R. Krishnaiah, ed., APPLICATIONS OF STATISTICS

EXPERIMENTATION WITH WEATHER CONTROL AND STATISTICAL PROBLEMS GENERATED BY IT*

JERZY NEYMAN

Statistical Laboratory, University of California
Berkeley, California

CONTENTS

I. INTRODUCTION

II. FIRST DECADE, 1951 - 1960

1. Randomized Experiments vs. Non-Randomized Cloud Seeding Operations
2. Efforts to Learn the New Field of Study
3. Biblical "Seven Fat Years" Followed by "Seven Lean Years"
4. Convincing Domestic Freezer Experiment
5. First Disenchantment: "Report of the Advisory Committee on Weather Control"
6. Some Findings in the Theory of Statistics

III. SECOND PERIOD OF MY INVOLVEMENT, 1964 TO PRESENT

7. Arrival of Seven Annual Reports on Grossversuch III
8. Special Weather Modification Conference at the Fifth Berkeley Symposium on Mathematical Statistics and Probability
9. Our Use of the Grossversuch III data
10. Difficulties in Using Old Experiments to Verify New Hypotheses
11. Verification of Generality of the Findings at Grossversuch III
12. Hypothetical Mechanisms of the Far Away Losses of Rain at the Arizona Experiment

I. INTRODUCTION

I feel honored by being asked to give this inaugural address at the present Symposium attended by so many distinguished scholars, some of them specialists in mathematical statistics, and others, specialists in the various domains of applications.

I must confess that the pleasure I experience by being so honored is augmented by a particular circumstance of personal character. While thinking of what I might present that would be likely to be interesting to the distinguished audience, I suddenly realized that the present year, 1976, marks the quarter of a century

*Inaugural Address

since the first report on weather modification has been issued by our Statistical Laboratory in Berkeley. In fact, this first report of ours [1] is dated December 15, 1951. The work we did was in response to the request of the Division of Water Resources, Department of Public Works of the State of California. Since that time, we worked on the subject of weather control almost continuously. Thus, to me personally, the present occasion is very much like a celebration of the silver jubilee of continuing efforts in an important domain of study!

During the past quarter of a century a large number of problems came to our attention. Quite a few of them were "substantive" in character, as was the question of the Division of Water Resources in Sacramento. This question was whether the enthusiastic claims of success voiced by the many commercial cloud seeding operators in California should be taken seriously. The specific question was whether reliable evidence exists that cloud seeding represents a realistic resource of water much needed in California, particularly in the southern part. Here, then, we had to deal with a "substantive" problem concerned with rainfall or snowfall.

However, as is almost invariably the case, the substantive problem relating to a natural phenomenon generated a number of theoretical statistical problems which we tried to solve. Then, there were repeated "feed-backs", etc. In this process, certain new statistical methodologies were developed and also certain substantive findings occurred. We believe, or should I say we have the illusion, that these substantive findings are quite important. To my personal regret, the cloud seeding community does not like our results, that is, apart from a few exceptional individuals. Just too bad!

It seems to me that a reasonable plan for the present paper might consist of a necessarily brief sketch of three distinct periods of our research, in a summary of our findings, and in pointing out certain regrettable phenomena occurring in the weather modification literature. Hopefully this plan would provide a reasonably comprehensive picture of the present state of the domain as seen from our Statistical Laboratory in Berkeley. Curiously, some of our substantive findings resulted not from some kind of systematic search, but unexpectedly. Then systematic work was needed to confirm the finding.

II. FIRST DECADE, 1951 - 1960

1. Randomized Experiments vs. Non-Randomized Cloud Seeding Operations. As I see it, the greatest achievement of R. A. Fisher consists in inventing, in developing and in propagandizing the idea that, when working with variable material, the randomization of experiments is a necessary condition of their reliability. A very interesting historical sketch of this subject was recently published by Cochran [2]. One of the interesting details is that some of the difficulties with non-randomized trials were noted more than two centuries ago. The individual in question is described by Cochran as Arthur Young, "the great English experimentalist." The year given is 1764! One of the difficulties pointed out is that the experimentalists tend to have personal preferences to the treatments they study. The unavoidable results are first self-deception and later deception of others. As Cochran states, the general remarks of Arthur Young are as much relevant now as they were in 1764.

2. *Efforts to Learn the New Field of Study.* In the early 1950's, our efforts were directed towards learning the field of study which at the time was new to us, and towards convincing the authorities in Sacramento that, because of the lack of randomization of the commercial cloud seeding operations, the successes claimed by the operators should be taken with a degree of skepticism [3, 4].

It will be noticed that the reports quoted were authored or coauthored by several individuals who, at the time, were in early stages of their respective careers, but now hold responsible positions in various institutions. In alphabetical order they are: Arnold Court (now Professor at California State University, Northridge), Terry A. Jeeves (now with the Westinghouse Company), Lucien M. LeCam (perhaps the "most mathematical" leader in the field of mathematical statistics, deeply interested in applied fields such as immunotherapy of cancer), Elizabeth L. Scott (our former Department Chairman and a respected colleague), George Steck (now with Sandia Corporation), B. V. Sukhatme (now Professor at Iowa State University), and several others.

With the kind help of Edward M. Vernon, then Chief Regional Forecaster of the U. S. Weather Bureau in San Francisco, it did not take us too long to learn some of the background, sufficient to initiate certain studies of our own. The point was that, while the idea of the necessity of randomization as a basic prerequisite to reliability of experimentation was clear to me, as well as to my colleagues, the prospects of success in our efforts to convince the authorities depended very much on producing some "tangible" evidence invalidating the evaluations of the commercial cloud seeders.

The methodology of evaluating non-randomized cloud seeding operations used at the time, and also now, is the so-called "historical regression" methodology. Consider two not very distant areas, one being the area in which the operator contracted to increase the precipitation, called "target", and the other called "control". The presumption was, and continues to be, that cloud seeding over the target cannot possibly affect the precipitation over the control.

Using precipitation data available for a few years before the beginning of the cloud seeding era, one establishes the linear regression line of the target precipitation, say $\underline{Y}$, on the control precipitation, say $\underline{X}$. (Frequently, this is done with $\underline{X}$ and $\underline{Y}$ not expressed in inches of actual precipitation, but in terms of some transformation, perhaps in terms of square roots of actual precipitation, etc.). Next, having the control precipitation for periods of actual cloud seeding, say $\underline{x}_1, \underline{x}_2, \ldots, \underline{x}_n$, this regression equation is used to compute the precipitation in the target, say $\underline{Y}(\underline{x}_i)$, to be expected without seeding. Finally, if $\underline{Y}_i$ stands for the target precipitation measured on the $\underline{i}$th seeding operation, the difference $\underline{Y}_i - \underline{Y}(\underline{x}_i)$ is treated as measuring the effect of seeding. At first sight this methodology may appear convincing. However, on closer examination of various circumstances, both meteorological and those connected with the way the methodology was applied, it appears valueless.

The meteorological aspect of the problem, worked out in consultation with Vernon [5], is connected with the existence of several "types" of storms, each with a different target-control regression line. It happens that the frequencies of these different types of storms vary considerably from one year to the next. Thus, the historical target-control regression line need not coincide with that appropriate for the period of cloud seeding operations. This fact was brought out by Brier and Enger [6], published in 1952, who performed two evaluations of the same cloud seeding operation in Arizona. One analysis was done using a 10-year historical period and a particular control, and the other using a 30-year historical period and a different control area. They obtained contradictory results.

Our own studies revealed the dependence of the target-control regression line for non-seeded storms on the types of these storms, thus partly explaining the results of Brier and Enger. Also, they revealed the following circumstance. It appeared that the commercial cloud seeding operators do not seed all the storms passing over the target but only about 60% of them, namely those that, in their opinion, are "seedable." Naturally, given some ability to forecast local precipitation, the storms actually seeded may well be those promising relatively more rain in the target than in the control area!

3. *Biblical "Seven Fat Years" Followed by "Seven Lean Years".* Another weighty argument against the reliability of the historical regression line is the climatological fact noticed in antiquity, as described in the title of the present subsection. It is illustrated in Figures 1 and 2.

Figure 1 shows the change in the yearly precipitation in Arizona over more than a quarter of a century, preceding the cloud seeding operations investigated by Brier and Enger [6]. It is seen that over a period of some two decades the annual precipitation in Arizona showed an increasing trend (twenty "fat" years). Subsequently, there was a period of five leaner and leaner years. This continued in the absence of any known human efforts to modify weather. Thus, any difference between rainfall in a subsequent period with cloud seeding, with that in some earlier period may well reflect a natural change in the precipitation patterns, and have nothing to do with seeding.

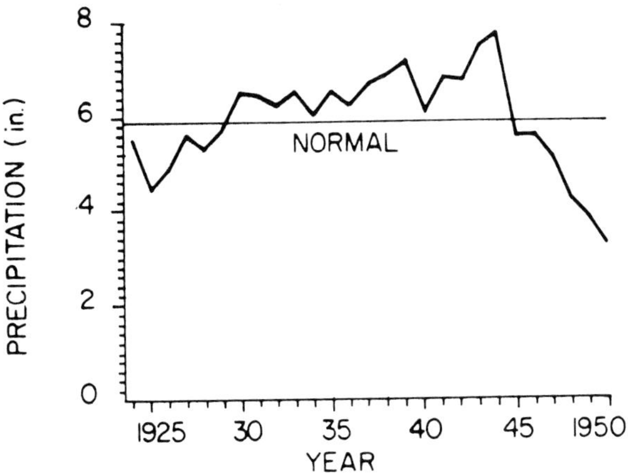

Fig. 1. Five year moving average of seasonal precipitation in Arizona.

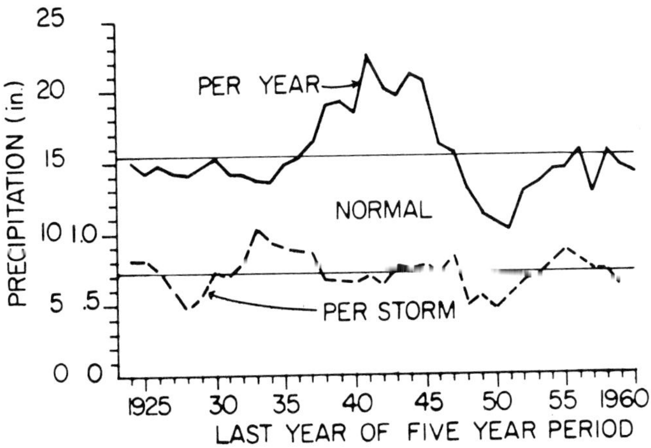

Fig. 2. Five year moving average of seasonal precipitation in Ventura County, California.

Figure 2, based on data from Ventura County, California, relating to the "pre-cloud seeding" era, illustrates the same phenomenon, but with more detail. It was constructed with the kind help of Mr. Byron D. Edde of the Department of Public Works of Ventura County, who was kind enough to separate the successions of "rainy" and "dry" days into separate storm periods. Here, then, the history of precipitation in the area could be characterized by averages of precipitation amounts per year, as well as by average amounts per storm. This latter detail ties up with the concept of "type of storm." As years went by, the frequencies of what may be called heavy storms varied in some kind of serial manner. This variation occurred naturally. Thus, the evaluation of any subsequent cloud seeding operations by comparing the seeded rainfall with that in some earlier period involves obvious hazards.

While we may perhaps claim the priority of pointing out the relevance of natural succession of "fat" and "lean" years to the evaluation of cloud seeding, the phenomenon itself is well known. Recently, it was a pleasure to witness a presentation at a conference concerned with cloud seeding which illustrated the same phenomenon, the "fat" and the "lean" successions of years [7]. Regretfully, the chart exhibited at the conference, very different in shape but comparable in its contents to Figure 1, is not reproduced in the publication quoted.

One might think that the findings of Brier and Enger, supported by the above results reported at various meetings, ought to have been fully convincing, not only to professional meteorologists, but also to the public at large. Regretfully, this is not so.

4. *Convincing Domestic Freezer Experiment.* While the findings just described made me a firm believer that the non-randomized cloud seeding operations have no scientific merit whatsoever and should not be discussed in scholarly literature, a simple experiment which a friendly cloud seeder made me perform in his office, convinced me that yes, seeding of clouds with silver iodide smoke CAN MODIFY these clouds and thus may be a tool to control the precipitation. Rather than to think in terms of "whether" one could stimulate rain, I began to think in terms of "how" one could do so.

The experiment is very simple. I was asked to approach a large domestic freezer; the door of the freezer was open and I was asked to blow into the facility. I did and saw a thin but noticeable "cloud" of what I was accustomed to call "vapor" from my breath. I was told that "vapor" is a transparent gas, and that the cloud I saw is composed of tiny liquid droplets which formed from water vapor I exhaled and were then "super-cooled". I was given an impressively large match, was requested to dip it into some liquid in a little glass, light it and move it through the "cloud" formed within the freezer. When I did so, some unexpected sparks became visible within the cloud. I was told that these "sparks" came from tiny ice crystals formed around silver iodide particles which were dispersed by my moving the lighted match through the "cloud" within the freezer. Of course, the light I saw as sparks was the light in the room reflected by the ice crystals.

Most of the above information about ice crystals formed around particles of silver iodide, etc., came from my host. However, I did see the "sparks" and what I heard appeared convincing. Since that time, many years ago, and after learning the essence of the Bergeron-Findeisen theory, I have been acting on the inspiring assumption that, by introducing certain chemicals into super-cooled clouds, with much invisible vapor within them, one can create ice nuclei. These can grow, using the surrounding vapor (while the existing super-cooled droplets evaporate) and eventually would fall, reaching the ground as some form of precipitation: rain or snow or hail. So--let's find out how one can do it, and in what identifiable conditions.

5. First Disenchantment: "Report of the Advisory Committee on Weather Control". The interest in cloud seeding as a possible way of controlling weather, affected not only our California Government but also the Federal Government. A special Advisory Committee to the President was appointed which published its Final Report in 1957. There were two volumes. Volume I was rather thin. It contained summaries and recommendations apparently for decision-making authorities who could not be expected to be interested in a variety of technicalities. Descriptions of a large number of studies formed the contents of Volume II. Because of the mass of essentially statistical work, the two volumes were sent to the Journal of the American Statistical Association for review. This review was written by K. Alexander Brownlee [8]. This is a good review. Brownlee, a statistician with a long experience in applications to science and technology, has a nice method of calling a spade a spade. However, his involvement with weather modification was just a little less than ours and, after reading his paper, we felt inspired to add a few details [9].

The main points of the two papers can be summarized as follows.

(i) Apart from the fact that Volume I of the Report is affected by, say, "cosmetic" treatment of the contents of Volume II, it contains statements which conflict with some of those in Volume II. One example is:

> "A(3). No evidence was found in the evaluation of any project [cloud seeding project] which was intended to increase precipitation that cloud seeding had produced a detectable negative effect on precipitation." [P. vii.]

Not true. A paper by Thom [10] in Volume II contains references to two cloud seeding operations with apparent negative results of cloud seeding found to be statistically significant. They are: a project in South Carolina and the Carrizo Plain Project, in California. We are aware of several others not mentioned by Thom.

(ii) Thom was aware of the dependence of the target-control regression line on types of storms, and claimed to do better "typing" than Vernon did for the Carrizo Plain Project. Also Thom reevaluated that project and found significant negative results. However, all the many other evaluations he performed were without any effort to "type" the storms.

(iii) Thom's paper contains a statement strengthening the findings of Brier and Enger. This statement relates to "freedom" of selecting rain gauges to represent the control area, as follows:

"If one takes the liberty of choosing among minimum distance controls, he can often find any result for seeding that suits his purposes, either positive or negative." [P. 33.]

This did not prevent him from using the historical regression line method to produce the substantial material in Volume II of the Final Report.

(iv) The Final Report of the Advisory Committee on Weather Control contains a paper by Robin R. Reynolds [11]. This article gives a description of a cooperative randomized cloud seeding project in which the Berkeley Statistical Laboratory participated. Mr. Reynolds served as the Chairman of the Board of Directors of the project. His article ends with the sentences:

"A rough estimate of a 23 percent increase can be made from the preliminary data. It should be noted, however, that the scarcity of data causes the results to be statistically nonsignificant."

Actually, our preliminary evaluation presented at the meeting of the Board of Directors of the Project has no resemblance to the figure of 23 percent. On the other hand, this figure, combined with the suggestion that it resulted from our studies, has been quoted in publicity material of the North American Weather Consultants who did the cloud seeding at the project. This circumstance forced us to withdraw from the cooperative enterprise and to give up the grant of the National Science Foundation. The complete results of our studies, including the description of the misrepresentation, have been published in the Bulletin of the American Meteorological Society [12]. The whole incident, combined with the discussion of some fancy methodology used in the Final Report was also described in the paper [9] already mentioned.

The circumstances connected with the Final Report of the Advisory Committee constituted my first experience of its kind. They resulted in the abandonment of the particular field of study which, at the time, appeared irrevocable. However, as will be seen below, due to certain novel experiences a few years later, my acute interest in the subject revived.

6. *Some Findings in the Theory of Statistics.* As I mentioned at the outset, practically all non-trivial problems of application generate novel problems of mathematical statistics. The studies of our early period of contacts with cloud

seeding generated several theoretical problems, of which two appear worth mentioning.

(i) Bias in the historical regression line method due to the use of transformations of the original precipitation data expressed in inches [13, 14]. When historical regressions are computed using transformed variables, the bias in question has always the same sign, favoring the conclusion that seeding increased the rainfall.

(ii) Problems of evaluating a randomized cloud seeding experiment contain a peculiarity. The observable random variable $\underline{X}$ has a positive probability, say θ, of being zero. Then, if $\underline{X} > 0$, it can vary continuously over a non-specified interval. In the latter case, the gamma density provides a satisfactory approximation to the conditional distribution of $\underline{X}$.

The cloud seeding can, and usually does, have two different effects. It can affect the value of θ and also it can affect the conditional expectation of $\underline{X}$, given $\underline{X} > 0$. Obviously, this is not a standard kind of situation. Several publication exemplified by [15, 16] provide an easy methodology to deal with the problem of testing the hypothesis that the combined effect of the two parts of the mechanism did not affect the unconditional expectation of $\underline{X}$. Of course, in this connection a variety of nonparametric tests may appear usable. However, as illustrated by Monte Carlo simulation, the new parametric procedure (known as $\underline{C}(\alpha)$ tests) offers a distinct advantage. This is particularly so in the case where the two parts of the mechanisms "work in opposite directions." The original theory has been variously generalized by Bartoo and Puri [17], and by Bühler and Puri [18]. An interesting paper concerned with summary evaluations of several experiments is due to Davies [19].

III. SECOND PERIOD OF MY INVOLVEMENT, 1964 TO PRESENT

7. *Arrival of Seven Annual Reports on Grossversuch III.* During several years following our liquidation of a variety of unfinished business connected with the earlier period of work on weather modification, we were concerned with problems of a different nature. Then, sometime in 1964, we received a large package of mimeographed material from Switzerland concerned with three successive summertime hail prevention experiments in Canton Ticino on the southern slopes of the Alps. The three experiments were conducted by the special commission of the Swiss government presided over by Professor R. Sänger. These experiments were supported by certain industries, including a special hail damage insurance company.

The first five-year-long experiment was not randomized. The method of hail suppression tried involved rockets which were used in Austria and Italy. The experiment in Ticino brought out two facts: (i) the rockets burst about one kilometer BELOW the clouds at which they were aimed, and (ii) the hail damage to crops during the experiment was about the same as in the preceding five years.

The second experiment of the series lasted four years. It was also not randomized, but the method of hail suppression tried was different, namely silver iodide seeding from ground-based generators. By that time, the Swiss authorities perceived that the evaluation of an experiment involves problems unfamiliar to meteorologists. Accordingly, a young Swiss statistician, Paul Schmid, became involved in the planning and conduct of the third experiment of the series, known as Grossversuch III.

Grossversuch III lasted seven years ending in 1963. The center of activities was the Osservatorio Ticinese, directed by J. C. Thams. The unit of experimentation was a twenty-four hour period. Seeding was conducted with pulses (five minutes on, ten minutes off) for fourteen hours, beginning at 7:30 a.m. The response variables were observed from 8:00 a.m. The decision whether a given calendar day was to be an "experimental day" was reached at the Osservatorio Ticinese on the preceding afternoon. This decision was based on the forecast--whether thunderstorms with hail were to be expected on the day in question. In the affirmative case a randomized decision was made whether to seed or not.

The seven annual reports on Grossversuch III we received in 1964 are impressive in their thoroughness and business-like spirit. While the main object of the experiment was hail suppression, the annual reports contain a wealth of information on many relevant observations: rainfall, including time and duration, temperatures, upper air soundings made in Milan (some twenty-five miles south of Ticino), etc., etc. The data on hail, including time of occurrence, duration, and a subjective appraisal of severity, consisted of reports from a number of observers.

The results of the first year, 1957, were somewhat regrettable: more thunderstorms, more hail reports, and substantially more rain on days with seeding than on experimental days without seeding. Yet, contrary to what one sees in reports on many other experiments, all the relevant information was presented in full detail, without any signs of attempts at "cover-ups" or "cosmetic treatment."

One detail of the annual reports deserves a special mention. This is the one-page protocols on each experimental day, beginning with the forecast on the preceding afternoon, the name of the forecaster, the name of the person who opened the envelope containing the random decision whether to seed or not and, of course, the random decision itself. The rest of each page contains a mass of qualitative information on happenings during the experimental day in question. Among other things, there is a list of silver iodide smoke generators which were found not operating at the time when they were supposed to be on. Also the reasons were given. On occasion the reasons indicated were "Bad weather" or "Violent electrical discharges"! The operational work at Osservatorio Ticinese was efficient. They checked whether the experiment was conducted as planned. During the first year of Grossversuch III there were quite a few days supposed to have been seeded on which some of the generators were found non-operative. In subsequent years the number of such days diminished markedly. The publication of cases of inactivity worked.

The above details of the annual reports on Grossversuch III convinced me that it was a real research enterprise fully reported, and my "post 1957" attitude changed completely. I became a believer in the possibility of learning how to modify weather, and I became emotionally involved in the study.

8. *Special Weather Modification Conference at the Fifth Berkeley Symposium on Mathematical Statistics and Probability.* The above change in attitude, combined with the awareness that for several years we had no contact with weather modification studies, dictated an effort to learn what was going on, both in the United States and abroad. Hence we organized a special conference on the subject. It was held in our Statistical Laboratory during the Christmas season, ending on January 7, 1967. We had a large number of American participants and also eight participants from abroad. Six of the latter were immediately connected with weather modification: (i) J. Bernier from France, (ii) K. R. Gabriel from Israel, (iii) H. G. Müller from Germany, (v) P. Schmid from Switzerland, (v) P. Siliceo from Mexico and (vi) E. J. Smith from Australia. The other two were statisticians: Robert B. Davies from New Zealand, the author of a novel method of simultaneous evaluation of several independent experiments [19], and Frank Yates from Britain, known for his many years of cooperation with R. A. Fisher on a variety of problems of randomized experimentation.

As of now, there are many conferences on the same subject with much larger international attendance. However, the *Proceedings* of our meeting [20] contain a section which has no counterpart in other similar publications known to me. This section contains day by day (or storm by storm) data on as many randomized experiments as we managed to assemble. After the ten years that elapsed, this section continues to be a source of important factual information, some of which will be used below.

Incidentally, the same volume contains reevaluations by our group of a number of early experiments in the United States, starting with SCUD, and also some papers given to methodology.

9. *Our Use of Grossversuch III Data.* After finding that the difference between average seeded and average not seeded precipitation at Grossversuch III can hardly be ascribed to chance variation, we proceeded to use the data in order to verify the various hypotheses underlying the efforts to modify the weather. In subsection 4 above, I described my impressions derived from the experiment with the domestic freezer. The experiment showed that a chemical introduced into a supercooled cloud can create ice nuclei. However, there was no direct evidence supporting the rest of the story I heard, about the crystals growing and then falling to the ground as rain or snow or hail. This was the first thing we tried to confirm

on the Swiss data.

Our reasoning went as follows. If the above description of the happenings is realistic, then the overall excess of rain on seeded days that we found must be concentrated on those days on which there were no stability layers in the atmosphere over Ticino, that is no stability layers (inversions or thick isothermals) occurring at temperatures warmer than some -5°C, at which the silver iodide smoke is supposed to be active. Accordingly, using the radiosonde data from Milan, we divided the close to 300 experimental days into two categories: days with uninhibited updrafts and those with warm stability layers. We confidently expected large positive effects of seeding on days with uninhibited updrafts and no noticeable effects on other days.

When the results came out of the computer [21], we got a shock. The results of the calculations were just the contrary to those expected: large significant excesses of seeded rain on days with stability layers, and marked but not significant losses of rain ascribable to seeding on days with uninhibited updrafts!

Professor Morris Neiburger of UCLA and, independently, Dr. Max Schüepp of Zürich provided a convincing explanation of the increases in rainfall on days with stability layers. Their explanation is that a warm stability layer over Milan does not reach the Alps which, on summer days, are quite warm and warm the surrounding air. Thus, the silver iodide smoke released over Ticino under a stability layer becomes thoroughly mixed with humid air, gradually drifts northward towards the Alps and then comes out from under the "lid" of the stability layer in an impressive cumulus cloud which spreads and rains. Actually, Dr. Schüepp provided a photograph taken from a high flying aircraft exhibiting the phenomenon.

So much for the increases in rain on days with stability layers. But what about the days with uninhibited updrafts? How to explain the absence of increases in rainfall on these days? For a time there was no answer to this question.

Obviously, further studies of Grossversuch III were needed using some more powerful methodology. The earlier evaluations were made through direct comparison of seeded vs. not seeded precipitation, without using any control areas. The next step was for us to secure rain data for some areas around Ticino which could serve as control areas. With the kind help of Dr. Schüepp we obtained data from two areas in Switzerland, one near Zürich and the other near Neuchâtel, with twenty reliable rain gauges in each. Similar, but slightly less abundant data were obtained for five areas in Italy. Before using these data it was necessary to check two things: (i) whether the precipitation in Ticino on non-seeded days is highly correlated with that in any of the seven areas contemplated as controls, and (ii) whether the seed-no seed difference in rain in an area contemplated as a control area shows any sign of being significant.

When the results came, we experienced another shock. The answer to question (i) was in the negative for every single contemplated control area. On the other hand, for days with stability layers, the seed-no seed differences in rainfall in the two areas in Switzerland proved large, positive and statistically significant. Also, similar results were found in one of the five areas in Italy; they were "almost significant." All these results are summarized in TABLE I. Here *P* stands for a two-tail significance probability.

TABLE I

Effects of Cloud Seeding in Ticino on Rainfall in 8 Areas
Averaged per Experimental Day

	Days with uninhibited updrafts 47 seeded, 50 not seeded				Days with stability layers 96 seeded, 94 not seeded			
Area	S	NS	% change	P	S	NS	% change	P
Zürich	4.54	7.04	-35	0.11	6.98	4.19	+67	0.012
Neuchâtel	4.43	5.86	-24	0.32	6.84	4.36	+57	0.037
Ticino	7.98	12.45	-36	0.15	14.47	8.82	+64	0.031
Brescia	4.18	6.65	-37	0.079	6.09	4.17	+46	0.066
Turin	4.24	4.07	+ 4	0.90	5.61	5.41	+ 4	0.89
Milan	2.93	3.78	-22	0.43	5.68	3.90	+45	0.13
Fidenza	3.76	3.94	- 5	0.87	5.26	3.68	+43	0.15
Genova	3.73	3.20	+16	0.65	4.24	3.65	+16	0.60

While the results for Zürich (about 80 miles north of Ticino) and for Neuchâtel (about 120 miles, north-northeast from Ticino), with the two-tail *P* = 0.012 and 0.037 appear quite convincing, the results for Brescia (*P* = 0.066) and similarly for Milan and for Fidenza are also impressive. The multiple column on the left, corresponding to days with uninhibited updrafts does not exhibit results on the customary level of significance. However, one feels impressed by the frequency of negative apparent effects of seeding.

What is the reasonable attitude to assume? Dr. Schüepp was incredulous: there must be some mistake somewhere. In the bright light of a summer day in Switzerland silver iodide is known to "disintegrate" within a very short period of time and the time needed for the seeding material to reach either Zürich or Neuchâtel must be of the order of some five to seven hours at least. Dr. Schüepp's suggestion was to make another stratification of days. The principal source of moisture in Switzerland is the Mediterranean, in the south. Thus, if one stratifies the experimental days according to wind direction in Milan with southerly velocity component and those with northerly, one might obtain interesting results. TABLE II summarizes the results of this stratification.

TABLE II

Effects of Cloud Seeding in Ticino on Rain in 8 Areas on Experimental Days Stratified According to Presence and Absence of Stability Layers and According to Wind Direction at 1,500 meters Above Sea Level

	Days with uninhibited updrafts				Days with stability layers			
	Days with low southerly winds							
	25 seeded, 22 not seeded				48 seeded, 46 not seeded			
Area	S	NS	% effect	P	S	NS	% effect	P
Zürich	5.22	7.20	-27	0.39	8.81	4.07	+116	0.004
Neuchâtel	5.51	7.22	-24	0.46	8.48	5.16	+ 64	0.060
Ticino	12.40	19.21	-35	0.23	17.78	8.61	+106	0.018
Brescia	5.38	9.85	-45	0.11	6.30	4.75	+ 33	0.33
Turin	5.65	4.87	+16	0.73	6.29	3.27	+ 93	0.063
Milan	3.96	5.60	-29	0.44	4.67	3.46	+ 35	0.42
Fidenza	3.43	6.01	-43	0.23	3.44	3.15	+ 9	0.81
Genova	3.96	3.91	+ 1	0.98	2.76	3.00	- 8	0.84
	Days with low northerly winds							
	19 seeded, 25 not seeded				42 seeded, 37 not seeded			
Zürich	4.32	6.90	-37	0.29	5.38	4.05	+ 33	0.41
Neuchâtel	3.55	4.19	-15	0.71	5.48	3.54	+ 55	0.27
Ticino	3.28	4.80	-32	0.42	11.03	8.76	+ 26	0.56
Brescia	3.14	3.71	-15	0.62	6.22	3.72	+ 67	0.11
Turin	2.98	3.27	- 9	0.86	5.02	9.18	+ 45	0.15
Milan	1.95	2.46	-21	0.60	7.26	5.07	+ 43	0.31
Fidenza	4.43	2.47	+79	0.11	7.92	5.11	+ 55	0.23
Genova	3.80	2.80	+36	0.51	6.31	5.28	+ 19	0.66

It is seen that, with southerly winds and stability layers, the precipitation in Zürich and in Ticino itself appears to have been doubled by seeding! After one more test Dr. Schüepp appeared convinced. This additional test consisted in examining hourly precipitation amounts near Zürich and near Bern. In both cases large average excesses of seeded days precipitation occurred some seven hours after the inception of seeding in Ticino.

10. Difficulties in Using Old Experiments to Verify New Hypotheses. One of the misconceptions one frequently hears at various discussions is the assertion that the validity of statistical analysis of experimental data is limited to those cases in which the question studied has been formulated BEFORE the experiment was performed. In this light, the findings exhibited in TABLES I and II would be labeled "post hoc" and, therefore, not valid. A little reflection should be convincing that any rule like that is based on a misunderstanding. On the contrary, due to the long period of time for a randomized cloud seeding experiment to accumulate enough data to be informative, a reasonably fast advance in the development of weather modification technology depends very much on using the already performed experiments in order to verify hypotheses that were not formulated at the time these experiments were planned and performed. Of course, with reference to the quotation from Thom given in subsection 5, one must be careful in selecting old experiments for further study. The dangers involved can be illustrated by the following incident.

In his recent article [22] Mielke asserts:

"The Israeli experiment is an example of a highly successful experiment employing the cross-over configuration (Gabriel, 1967)."

The reference is to Gabriel's article published in the *Proceedings* of our special weather modification conference [20] pp. 91-113. Also, the last section of the volume contains the day-by-day data supplied by Gabriel. Thus, on the surface, everything seems in order, including the final result of the experiment indicating 15% increase in rainfall ascribable to seeding. However, if Mielke looked at the paper a little more closely, he would have found good reasons for not quoting the Israeli experiment or, at least, for not quoting it in the manner he did.

The point is that the Israeli experiment was evaluated in what appears to be an unprecedented manner. There were three periods of the experiment in which the day's precipitation used in the evaluation was measured differently. During the first period, the day's precipitation was measured from 8:00 a.m. of the given day to 8:00 a.m. of the next. Then there was a substantial period when the day's precipitation used in the evaluation was measured from 8:00 p.m. to 8:00 p.m. and it is not quite clear whether this period began the day before that in question or in the evening of the day in question, extending to the evening of the next day.

The interesting thing is that, after a few years of this second method of measuring the day's precipitation, the experimentalists returned to the first method, 8:00 a.m. to 8:00 a.m. Why?

While thinking about it I feel interested in the details of the second method of measuring day's rainfall. There are two possibilities and both seem unexpected. The situation is illustrated in Figure 3. The quantity measured on the horizontal axis is time in hours. In the middle, there is symbolized a particular experimental day, from zero to twenty-four hours. Then there is indicated a hypothetical period of seeding and the "first" method of measuring this day's rainfall: from 8:00 a.m. of the given day to 8:00 a.m. of the next day. Various authors speak, and write, about the possible aftereffects of seeding and, therefore, the inclusion of the

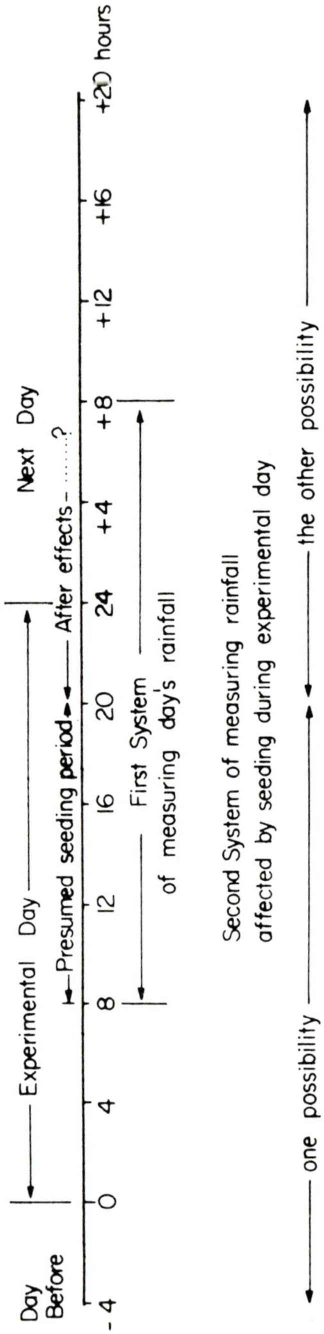

Fig. 3. Guesses about measures of day's precipitation in Israeli experiment.

night following a day with seeding does not seem objectionable. On the other hand, how to justify either of the two possibilities regarding the second method? If the rain measuring period starts in the evening of "preceding" day, then there seems to be a good chance of crediting the "given" day with precipitation affected by "yesterday's" seeding, this combined with the "loss" of rain of tonight likely to be affected by "today's" seeding. Again, if the rainfall of "today" is measured from 8:00 p.m. tonight and up to 8:00 p.m. tomorrow, similar questions arise.

However, the most fascinating question is about the changes in the method, from method 1 to method 2, and then back to method 1. If both methods have some advantages then is it not interesting to see two evaluations performed uniformly, one using method 1 and the other method 2?

While we are on this subject, I wish formulate a question regarding evaluations for which Dr. Mielke is at least partly responsible. As is well known, in company with Professor L. O. Grant, a cloud physicist, and C. F. Chappell, their former student, Dr. Mielke was involved in an important experiment, Climax I. This experiment is indeed important and, having visited Professor Grant's observatory high up on a mountain, and having heard Professor Grant speak at several meetings, I acquired a great respect for his work and for his opinions. On the other hand, I am uneasy about the statistical aspect of the situation.

Figure 4 summarizes the stratifications in the five consecutive papers concerned with Climax I, including the apparently unpublished M.S. thesis of Charles F. Chappell [23, 24, 25, 26, 27]. The puzzling thing about these five papers is the variation of the total number of experimental days used. My regret is that, up to recently at least, the complete set of all the observations at Climax I, including dates, etc., has not been published.

11. Verification of Generality of the Findings at Grossversuch III. With the risk of error characterized by significance probabilities given in TABLE II, I am acting on the assumption that the cloud seeding over Ticino did in fact cause the effects on rainfall near Zurich and Neuchatel exhibited in that table. However, Ticino is on the southern slopes of the Alps, a very special locality. While I find the results obtained unexpected and interesting by themselves, the important question is whether anything similar is to be expected in conditions in other lands with very different topography and different climate, say in the various parts of the United States, particularly in localities where efforts are made to modify the precipitation.

The important questions to answer are three:

Question 1. Are there any signs that cloud seeding over a target of conventional size affects the precipitation in localities some 100 miles away?

Question 2. Does the seeding under "warm" stability layers tend to increase the precipitation, at least in proper orographic conditions?

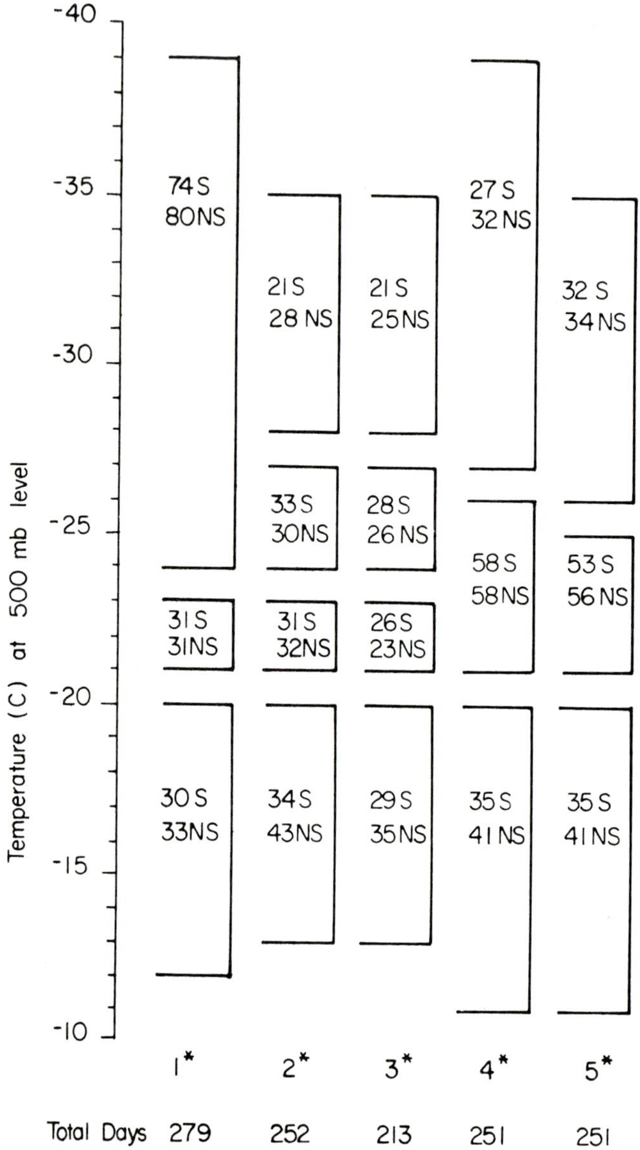

Fig. 4. Variability in the number of days used in evaluation of Climax I.

Question 3. Does the seeding of summer clouds, performed in conditions of uninhibited updrafts, tend to decrease the precipitation? If so, what is the plausible mechanism of this decrease?

Obviously, an affirmative answer to any of the three questions would imply a striking reorientation of the many cloud seeding experiments now planned or in progress, including the HIPLEX, the NHRE, and the contemplated experiment in connection with the possible increase of precipitation in the Sierra Nevada.

As mentioned in the preceding subsection, the only way to obtain the answer without waiting for the organization and completion of new appropriate experiments, is to look for reliable randomized experiments already completed, for which reliable data have been reliably published. The data needed are: (i) details of the randomization and of the method of seeding, (ii) dates of the experimental days (or other periods), (iii) location of the target and of the rain gauges in it, (iv) target precipitation for each experimental day, (v) winds aloft and, if at all possible, (vi) radiosonde data from a locality not very far from the target. Of course, in order to verify the far away effects of seeding, the precipitation data in far away localities must be available. Here, however, the network of the Weather Bureau gauges is the obvious source, even though in some cases this network is not very dense.

One might think that, with the possible exception of item (vi), all other data ought to be available in the reports on all the experiments performed with the support of a governmental agency. Regretfully, the actual situation is very different.

Our first thoughts and efforts were directed towards the Whitetop experiment. The data on the first four items were published. The data on winds aloft (pibal sounding) were not published but, with the kind help of the Office of Naval Research, we managed to get copies. However, the nearest radiosonde soundings were found to be quite far away and so we relied mainly on the pibal soundings. Besides, the cloud seeding was done from aircraft, likely to have been at levels above the possible "warm" stability layers.

Our studies of the Whitetop experiment are reflected in publications [28] to [35]. Regretfully, the last of these papers documents the conclusion that there must have been an important flaw in the implementation of randomized seeding. Therefore, no reliable appraisal of the effects of seeding at the Whitetop experiment is possible.

Accordingly, our subsequent efforts to answer the three questions above, concentrated on the Arizona experiment performed by L. J. Battan [36, 37]. Our studies of this experiment are reflected in publication [38] through [43]. Here we seem to have had much better luck than with the Whitetop experiment. Not only the Grossversuch findings appeared confirmed in Arizona, but also certain findings for Battan's experiment suggested a plausible mechanism of apparent losses of rain in Switzerland noticeable on days with uninhibited updrafts.

Battan's target of seeding was roughly of 15 x 25 miles dimension, a chain of Santa Catalina Mountains near Tucson. The experiment was conducted over the summers of seven years. Our first attempt to verify the far away effect of seeding was marked by finding a dense network of gauges in a very small area, at Walnut Gulch, established and maintained by the Soil and Water Conservation Division of the Agricultural Research Service, U. S. Department of Agriculture. The person in charge of the network is Dr. Herbert B. Osborn, with whom an effective cooperation was established. In our joint paper with Osborn [38] we established that on Battan's experimental days when Walnut Gulch was downwind from the Santa Catalinas, its 24-hour precipitation showed large and significant losses of rain ascribable to seeding, but no such losses on days when Walnut Gulch was upwind. This, then, was a result analogous to what was found in Switzerland, except that instead of large downwind gains in rain as near Zürich and Neuchâtel, there were losses.

Walnut Gulch is some 65 miles away from the Santa Catalinas. Compared to the dimensions of the Battan target, this distance is substantial, but something like one-half that between Ticino and Neuchatel. Thus, in other studies [39, 40] it was established that the average precipitation in localities in Arizona at distances from Battan's target of between 90 and 180 miles, in localities that on particular experimental days were downwind, there were losses of rain ascribable to seeding that averaged 34%, with two-tail significance probability $\underline{P} = 0.028$. Certain classifications of experimental days revealed 55% losses, with $\underline{P} = 0.005$! Also it was found that in the far away localities that on particular days were "on the right" of the day's wind direction there were GAINS of rain, averaging 72%, with $\underline{P} = 0.039$.

Thus, the unexpected far away effects of seeding in Switzerland appear to have taken place also in Arizona, with very different climatic and orographic conditions. In these circumstances it does seem prudent to expect such effects in other places too. If an experiment is to be performed with seeding of summertime convective clouds (such as HIPLEX and NHRE) the design ought to be appropriately adjusted. Rather than to think of happenings in particular clouds or, perhaps, in parallel with such thinking, one should visualize effects on a much larger scale. One obvious conclusion is that the cross-over design must be abandoned.

The implications of the various findings described above were discussed in some detail at two international meetings [41, 42].

12. Hypothetical Mechanism of Far Away Losses of Rain at the Arizona Experiment. Figure 5 illustrates the hypothetical mechanism of losses of rain ascribable to seeding, losses occurring at very large distances from the intended target. This mechanism was first proposed in [38]. It was proposed by statisticians with the hope of stimulating cloud physicists to do better!

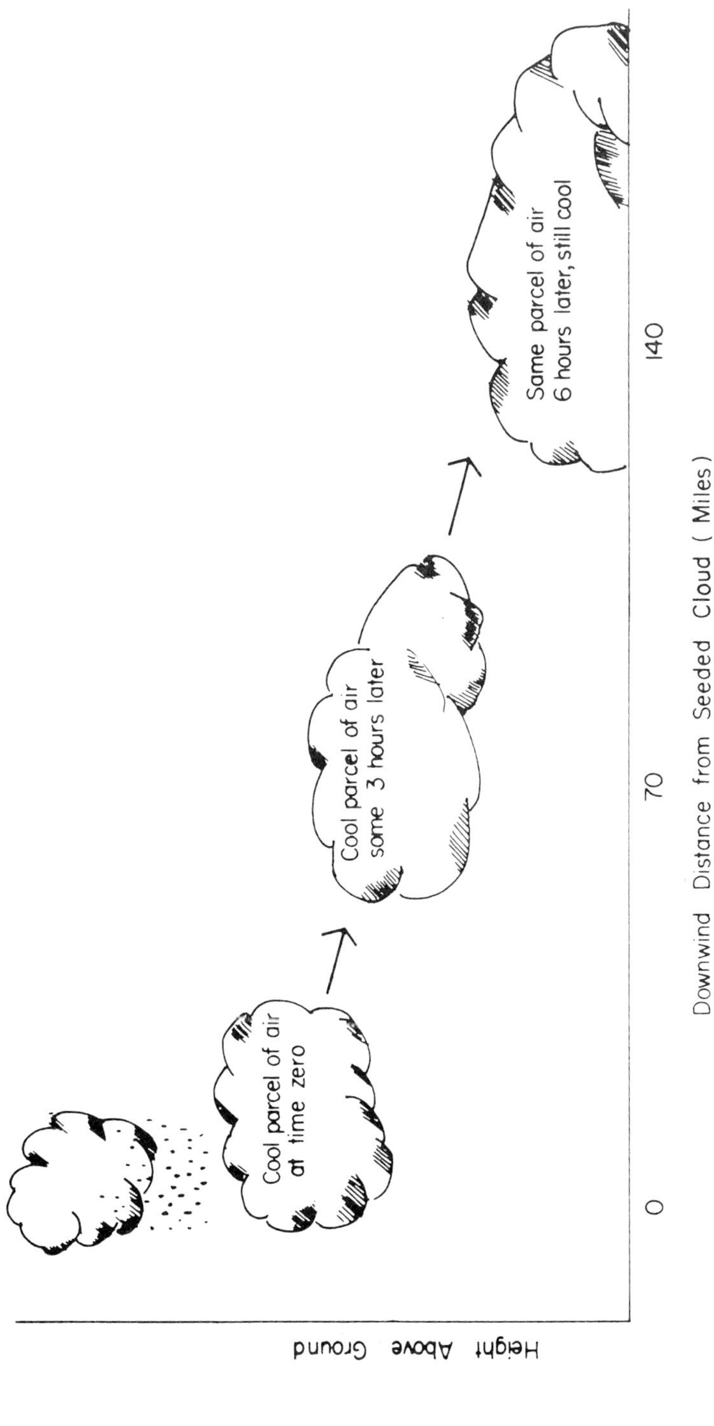

Fig. 5. Hypothetical mechanisms of downwind far away negative effects of seeding at the Arizona experiment.

Some further comments are now in order. It will be remembered that while describing our findings for the Arizona experiment my emphasis was on the presence of far away cloud seeding effects, similar to those found for Grossversuch III, but with a difference: rainfall increases in Switzerland and rainfall losses in Arizona. But was this restriction necessary? Let's remember that in Switzerland the far away gains in rainfall near Zürich and Neuchâtel were found for days with seeding from the ground under stability layers. Contrary to this, the Arizona seeding was performed from aircraft. During the first four years of experimentation the level of seeding was at -6°C, certainly above the "lid" of possible warm stability layers. During the subsequent three years Professor Battan's seeding was at the level of the base summertime cumuli, almost certainly above any warm stability layers that may have existed. Thus, if anything, the circumstances of cloud seeding in Arizona must have been comparable to those in Switzerland on days with uninhibited updrafts. A glance at TABLES I and II will show that on these particular days the prevailing results of seeding at Grossversuch III were predominantly negative, even though they were not significant at the conventional levels. Thus, to complete the "give and take", the Arizona experiment provides a plausible explanation of the findings for Grossversuch III which, by themselves, appeared puzzling.

What must have happened in Ticino is suggested by Figure 5. The bases of clouds on days with uninhibited updrafts must have been quite high. When the silver iodide smoke reached them from generators on the ground, it must have stimulated some rainfall. Falling through the air below the cloud, this rain must have cooled a parcel of air which, in turn, "killed" the updrafts and the rain reaching the ground.

One more detail of the empirical findings needs a tentative explanation. This is the finding for the Arizona experiment of a 72% increase in rainfall observed in localities 90 to 180 miles away from the center of seeding which on particular days were "on the right" of the day's wind direction. As mentioned, these average augmentations of rain were significant: the corresponding two-tail $\underline{P} = 0.039$. How to explain this? Here the discussions heard at various meetings bring to one's mind the so-called "squall lines" which develop as a result of approaching cold fronts. Could it be that the parcel of cool air in its third position exhibited in Figure 5 acted as an approaching cold front and created its own squall line, on the right of the wind direction as observed over the target some six hours earlier?

ACKNOWLEDGMENTS

It is a pleasure to acknowledge the help, not to say participation, of my colleagues at the Statistical Laboratory in the preparation of this paper, reviewing as it does a variety of our weather modification activities extending over a quarter of a century. In particular, I am indebted to Elizabeth L. Scott for correcting my memories, even of such details as when did we use the pibal and when the radiosonde data, etc. To Shyrl Dawkins I feel heartily indebted for the collection of our papers, both those published and also others, such as reports to the various agencies. To Terry Jue I am indebted for the preparation of figures, including the fanciful Figure 5. Finally, I feel indebted to the whole of our "cloud seeding crew", Shyrl Dawkins, Terry Jue, Jeanne Lovasich, Elizabeth L. Scott and Marcella Wells, for their decades-long interested cooperation in our joint studies and, especially, for putting finishing touches to the paper so it is legible and so that it tells the truth, as we see it, the whole truth compatible with the size allowed, and nothing but the truth.

This paper was prepared with support of the U. S. Office of Naval Research, contract number N00014-75-C-0159, at the Statistical Laboratory of the University of California, Berkeley.

REFERENCES

[1] NEYMAN, J. (1951). Methodology of statistical evaluation of rain making operations. Mimeographed report to the Division of Water Resources, Dept. of Public Works, State of Calif. Also see *Bulletin* 16, *State Water Resources Board, Weather Modification Operations in Calif.*, Preliminary Report, August 1954, C1-C24.

[2] COCHRAN, W. A. (1974). The vital role of randomization in experiments and surveys. In *The Heritage of Copernicus*, (J. Neyman, ed.) MIT Press 445-463.

[3] JEEVES, T. A., LECAM, L., NEYMAN, J. and SCOTT, E. L. (1953). On the methodology of evaluating cloud seeding operations. Mimeographed report to the Division of Water Resources, Dept. of Public Works, State of Calif. Also see *Bulletin* 16, *State Water Resources Board, Weather Modification Operations in Calif.*, 79-125.

[4] COURT, A. (1953). Evaluation of weather modification operations of the Southern Sierra Project. Mimeographed report to the Division of Water Resources, Dept. of Public Works, State of Calif. *Ibid.*, 126-145.

[5] VERNON, E. M. Report on classification of storms for statistical evaluation with special reference to Carrizo Plains and Santa Maria Coastal Area. *Ibid.*, 163-193.

[6] BRIER, G. W. and ENGER, I. (1952). An analysis of results of the 1951 cloud seeding operations in central Arizona. *Bull. Am. Meteor. Soc.*, **33**, 208-210.

[7] HANNAFORD, J. F. (1975). Hydrology and hydrologic applications. *Special Regional Weather Modification Conference, November* 11-13. American Meteorological Society , 158-164.

[8] BROWNLEE, K. A. (1960). Statistical evaluation of cloud seeding operations. *J. Amer. Statist. Assoc.*, **55**, 446-458.

[9] NEYMAN, J. and SCOTT, E. L. (1961). Further comments on the final report of the advisory committee on weather control. *J. Amer. Statist. Assoc.*, **56**, 580-600.

[10] THOM, H. C. S. (1957). An evaluation of a series of orographic cloud seeding operations. *Final Report of the Advisory Committee on Weather Control*, **II**, 25-50.

[11] REYNOLDS, R. R. Santa Barbara Project. *Ibid.*, 249-252.

[12] NEYMAN, J., SCOTT, E. L. and VASILEVSKIS, M. (1960). Statistical evaluation of the Santa Barbara randomized cloud seeding experiment. *Bull. Amer. Meteorol. Soc.*, 41, 531-547.

[13] NEYMAN, J. and SCOTT, E. L. (1960). Correction for bias introduced by a transformation of variables. *Ann. Math. Statist.*, 31, 643-655.

[14] NEYMAN, J. and SCOTT, E. L. (1960). Unbiased estimation based on transformed variables, with particular reference to cloud seeding experiments. *Proc. Fifth Conference on Design of Cloud Seeding Experiments, A.R.D.*, 353-372.

[15] NEYMAN, J. (1959). Optimal asymptotic tests of composite statistical hypotheses. *Probability and Statistics (The Harold Cramer Volume)*, (U. Grenander, ed.). Almquist and Wiksells, Stockholm, Sweden, 213-234.

[16] NEYMAN, J. and SCOTT, E. L. (1965). Asymptotically optimal tests of composite hypotheses for randomized experiments with non-controlled predictor variables. *J. Amer. Statist. Assoc.*, 60, 699-721.

[17] BARTOO, J. B. and PURI, P. S. (1967). On optimal asymptotic tests of composite statistical hypotheses. *Ann. Math. Statist.*, 38, 1845-1852.

[18] BUHLER, W. and PURI, P. S. (1966). On optimal asymptotic tests of composite hypotheses with several constraints. *Z. Wahrscheinlichkeitstheorie*, 5, 71-88.

[19] DAVIES, R. B. (1969). Beta-optimal tests and an application to the summary evaluation of experiments. *J. Roy. Statist. Soc. Ser. B* , 31, 524-538.

[20] LECAM, L. and NEYMAN, J. (1967). *Proc. Fifth Berkeley Symp. Math. Statist. Prob.* University of California Press, Berkeley.

[21] NEYMAN, J., SCOTT, E. L. and WELLS, M. A. (1968). Influence of atmospheric stability layers on the effect of ground-based cloud seeding. *Proc. Nat. Acad. Sci.*, 60, 416-423.

[22] MIELKE, P. W., JR. (1975). Some designs and evaluation considerations for the Sierra Nevada wintertime orographic cloud seeding project. *Special Regional Weather Modification Conference, November* 11-13. American Meteorological Society, 179-183.

[23] GRANT, L. O. and MIELKE, P. W. (1967). A randomized cloud seeding experiment at Climax, Colorado, 1960-65. *Proc. Fifth Berkeley Symp. Math. Statist. Prob.* (LeCam, L. and Neyman, J. eds.). University of California Press, Berkeley, 115-131.

[24] CHAPPELL, C. F. (1967). Cloud seeding opportunity recognition. Atmospheric Science Technical Paper 118, Dept. of Atmospheric Science, Colorado State University, Fort Collins, Colorado. M.S. thesis.

[25] GRANT, L. O., CHAPPELL, C. F. and MIELKE, P. W., JR. (1968). The recognition of cloud seeding opportunity. *Proc. First Nat'l Conf. on Weather Modification.* Albany, April 28-May 1.

[26] MIELKE, P. W., GRANT, L. O. and CHAPPELL, C. F. (1970). Elevation and spatial variation effects of wintertime orographic cloud seeding. *J. Appl. Meteor.*, 9, 475-488; corrigendum, 10, 842.

[27] MIELKE, P. W., GRANT, L. O. and CHAPPELL, C. F. (1970). Randomized orographic cloud seeding results for eight wintertime seasons at Climax, Colorado. *Proc. Second Nat'l Conf. on Weather Modification, Santa Barbara, California.* 66-69.

[28] NEYMAN, J. (1967). Experimentation with weather control. *J. Roy. Statist. Soc. Ser. A* , 130, 285-326.

[29] NEYMAN, J. and SCOTT, E. L. (1967). Rationale of statistical design of a rain stimulation experiment. *Proc. Skywater Conference II*, U. S. Department of Interior, Bureau of Reclamation, Denver, 195-250.

[30] NEYMAN, J., SCOTT, E. L. and WELLS, M. A. (1969). Statistics in meteorology. *Rev. of the Intern. Stat. Inst.*, **37**, 119-148.

[31] NEYMAN, J., SCOTT, E. L. and SMITH, J. A. (1969). Areal spread of the effect of cloud seeding at the Whitetop experiment. *Science*, **163**, 1445-1449.

[32] NEYMAN, J., LOVASICH, J. L., SCOTT, E. L. and SMITH, J. A. (1970). Statistical aspects of rain stimulation - problems and prospects. *Rev. Intern. Stat. Inst.*, **38**, 155-170.

[33] NEYMAN, J., LOVASICH, J. L., SCOTT, E. L. and SMITH, J. A. (1969). Wind directions aloft and effects of seeding on precipitation in the Whitetop experiment. *Proc. Nat. Acad. Sci.*, **64**, 810-817.

[34] NEYMAN, J., LOVASICH, J. L., SCOTT, E. L. and WELLS, M. A. (1971). Further studies of the Whitetop cloud seeding experiment. *Proc. Nat. Acad. Sci.*, **68**, 147-151.

[35] NEYMAN, J., LOVASICH, J. L., SCOTT, E. L. and WELLS, M. A. (1971). Hypothetical explanations of the negative apparent effects of cloud seeding in the Whitetop experiment. *Proc. Nat. Acad. Sci.*, **68**, 2643-2646.

[36] BATTAN, L. J. (1966). Silver iodide seeding and rainfall from convective clouds. *J. Appl. Meteorology*, **5**, 669-683.

[37] BATTAN, L. J. and KASSANDER, A. R., JR. (1967). Summary of results of a randomized cloud seeding experiment in Arizona. *Proc. Fifth Berkeley Symp. Math. Statist. Prob.* (LeCam and Neyman, eds.). University of California Press, 29-33.

[38] NEYMAN, J. and OSBORN, H. B. (1971). Evidence of widespread effects of cloud seeding at two Arizona experiments. *Proc. Nat. Acad. Sci.*, **68**, 649-652.

[39] NEYMAN, J., OSBORN, H. B., SCOTT, E. L. and WELLS, M. A. (1972). Reevaluation of the Arizona cloud seeding experiment. *Proc. Nat. Acad. Sci.*, **69**, 1348-1352.

[40] NEYMAN, J., SCOTT, E. L. and WELLS, M. A. (1973). Downwind and upwind effects in the Arizona cloud seeding experiment. *Proc. Nat. Acad. Sci.*, **70**, 357-360.

[41] NEYMAN, J. and SCOTT, E. L. (1973). Some current problems of rain stimulation research. *Proc. Intern. Symp. on Uncertainties in Hydrologic and Water Resource Systems.* (C. C. Kisiel and L. Duckstein, eds.). University of Arizona, Tucson, **III**, 1167-1244.

[42] NEYMAN, J. and SCOTT, E. L. (1974). Rain stimulation experiments: design and evaluation. *Proc. WMO/IAMAP Scientific Conf. on Weather Modification.* No. 399, World Meteorological Organization, Geneva, Switzerland, 449-457.

[43] NEYMAN, J. (1974). Problems of design and of evaluation of rain making experiments. *A Survey of Statistical Design and Linear Models.* (J. N. Srivastava, ed.). North-Holland Publishing Co., Amsterdam, 443-458.

P.R. Krishnaiah, ed., APPLICATIONS OF STATISTICS

ON ENTROPY MAXIMIZATION PRINCIPLE*

HIROTUGU AKAIKE
The Institute of Statistical Mathematics, Tokyo, Japan
and
Harvard University, Cambridge, Massachusetts

Under the principle of entropy maximization the object of statistical inference is defined as the estimation of the probability distribution of a set of observations. The entropy is a measure of the goodness of fit of an estimate to the true distribution and the principle demands the development of statistical procedures to maximize the expected entropy. The principle realizes a unification of conventional concepts and procedures of statistical inference. The discriminations between the test and estimation and the orthodox and Bayesian are eliminated and practical solutions to many important problems are obtained.

1. INTRODUCTION

In this paper we define the purpose of statistical inference as the estimation of the probability distribution of a given set of observations. The performance of an inference procedure is evaluated by the goodness of fit of the estimated distribution to the true distribution. The goodness of fit is measured by the probabilistic entropy which is defined in Section 2 as the difference of the expected log likelihoods of the estimated and true distributions. In the search for a good inferential procedure we follow the entropy maximization principle; that is, we try to maximize the expected entropy.

The principle is more fundamental than the classical maximum likelihood principle. It provides a unifying view of the conventional procedures of test and estimation. Some historically most successful statistical procedures may be viewed as realizations of this principle.

The implementation of the principle is carried out through the estimation of the expected log likelihoods of various models and the comparison of the estimates. Two significant examples of realization of the principle are discussed and their practical utilities are demonstrated by numerical examples.

2. A CRITICAL VIEW OF CONVENTIONAL PROCEDURES

Before going into the detailed discussion of the entropy maximization principle we will show that a combination of the conventional test and estimation pro-

*This work was done while the author was a Vinton Hayes Senior Fellow at the Division of Engineering and Applied Physics at Harvard University.

cedures leads to an essential extension of the classical theory of estimation. The theory of estimation as defined by R. A. Fisher is the theory of estimation of parameters within the function of known mathematical form which specifies the probability distribution of observations. In his Harvard Tercentenary Conference lecture Fisher admitted a wider type of inductive argument which would discuss methods of assigning the functional form of the population from the data but he denied the existence of such a theory ([14] p. 250).

The subject of fitting curved regression lines is discussed in Section 5 of the book by Fisher [16]. There the method of polynomial fitting is discussed using the numerical values of the differences of the yields from two plots of Broadbalk wheat field. The example is further discussed by Fisher and Yates [15] and the second order polynomial is chosen by the test of significance of the coefficients. This procedure of polynomial fitting is actually almost a realization of a practical procedure of determining the functional form of a population.

Consider the situation where a set of data $(x_1, x_2, \ldots, x_N)$ is given, where x_is are independently and identically distributed according to a distribution specified by the probability density function $f(x)$. Assume an initial guess $g(x)$ of $f(x)$. The choice of $g(x)$ corresponds to the choice of a null hypothesis. Transform x_i into y_i by the formula

$$y_i = \int_{-\infty}^{x_i} g(x)dx \qquad i=1,2,\ldots,N.$$

The distribution of y_i is now limited to the unit interval $[0,1]$. As a general system of alternative hypotheses assume the distributions H_k $(k=0,1,\ldots,K)$ which are specified by the density functions $h_k(y)$ of y_i such that

$$\log h_k(y) = \sum_{m=0}^{k} a_m \phi_m(y) , \tag{1}$$

where $\{\phi_m(y);\ m=0,1,\ldots,K\}$ is a system of functions linearly independent over $[0,1]$. By assuming $\phi_0(y) = 1$ we get H_0 = null hypothesis.

Now if we develop an estimation procedure of the coefficients a_m for $k = 0,1,\ldots,K$ and decide on the order k of the polynomial (1) this solves the problem of estimating the shape of the functional form of the population. Using the best choice of k and the corresponding estimate of a_m the estimate of the density function is given by

$$\exp \{ \sum_{m=0}^{k} a_m \phi_m(G(x)) \} g(x) ,$$

where $G(x)$ is defined by

$$G(x) = \int_{-\infty}^{x} g(t)dt .$$

If Fisher and Yates' use of the test of significance for the fitting of a polynomial regression is valid then its generalization to the present situation must provide a valid procedure of assigning a functional form of the population from the data. This observation shows that it was only a proper modeling of a general statistical hypothesis that was necessary for the development of what Fisher called a wider type of inductive argument. The selection of the basic functions $\phi_m(y)$ and the highest order K depends very much on the prior information of the problem. The Pearson's chi-square test of goodness of fit [24] corresponds to the case where 1-splines [27] are used as $\phi_m(y)$. If some free parameters are allowed in the specification of g(x) and the method of maximum likelihood is applied the log likelihood ratio test will realize a procedure which is exactly in the spirit of the original Neyman-Pearson formulation of the test of statistical hypotheses [23]. The best fit model suggests the most probable alternative and in many practical applications explicit specifications of this alternative would be very useful. It is the conventional discrimination between the test and estimation that prevented the development of this natural line of statistical thinking.

3. ENTROPY MAXIMIZATION PRINCIPLE

Denote by x the vector of observations $(x_1,x_2,\ldots,x_N)'$, where ' denotes transpose. Assume that the true distribution of x is specified by a density function $f(z) = f(z_1,z_2,\ldots,z_N)$ and denote by g(z:x) an estimate of f(z) which is a function of x. The situation may be viewed as the estimation of the distribution of the future observation z from the same distribution as that of x. We will use this interpretation whenever it is pertinent and g(z:x) will be called the predictive distribution. By an obvious analogy with the classical likelihood the value of g(z:x) at $z=z_o$ will be called the likelihood of the distribution at z_o.

As the measure of the goodness of fit of g(z:x) to f(z) we use the entropy of f(z) with respect to g(z:x) defined by

$$B(f;g) = -\int \left\{\frac{f(z)}{g(z:x)}\right\} \log \left\{\frac{f(z)}{g(z:x)}\right\} g(z:x)dz .$$

Boltzmann's probabilistic interpretation [12] of the thermodynamic entropy as the logarithm of the probability of a sample distribution which specifies the state of a system naturally extends to the present definition of entropy and B(f;g) can be considered to be proportionate to the logarithm of the probability of getting a sample distribution closely represented by f(z) by sampling from the population specified by g(z:x) [26]. By adopting B(f;g) as our criterion we are measuring the deviation of g(z:x) from f(z) through the process of conceptual sampling experiment from the distribution specified by g(z:x). In the inferential use of the entropy f(z) is factual and g(z:x) is hypothetical. The neg-entropy -B(f;g) is the Kullback information I(f;g) [21].

Our entropy maximization principle is stated as follows: *Formulate the object of statistical inference as the estimation of f(z) from the data x and try to find g(z:x) which will maximize the expected entropy* $E_x B(f;g) = \int B(f;g)f(x)dx$.

The most significant characteristic of the entropy for its use in statistical inference is demonstrated in the relation

$$B(f;g) = \int \log\{g(z:x)\}f(z)dz - \int \log\{f(z)\}f(z)dz. \qquad (2)$$

The second term in the right hand side of (2) is a constant which depends only on f(z). For the purpose of comparison of g(z:x) we have only to concentrate on the first term. This term is the expected log likelihood of g(z:x) with respect to the future observation z. Contrary to the thermodynamical use of the entropy, where the entropy of f(z) with respect to some fixed standard distribution is of main interest, here f(z) plays the role of the standard for the evaluation of the predictive distribution g(z:x) and the expected log likelihood carries the main interest.

A simple and natural estimate of the expected log likelihood is log{g(x:x)}, the value of log{g(z:x)} at z = x. g(x:x) is the likelihood at the present observation x. If we specify a family of possible g(z:x) the one with the maximum likelihood at x defines the maximum likelihood estimate of f(z). To maintain the natural correspondence with the classical theory the maximum likelihood at x will simply be called the maximum likelihood. Obviously the performance of the maximum likelihood estimate of f(z) depends on both the choice of the functional form, or the family, of predictive distributions g(z:x) and the statistical characteristic of log g(x:x) as an estimate of the expected log likelihood.

4. CLASSICAL PROCEDURES AND THE ENTROPY MAXIMIZATION PRINCIPLE

The method of maximum likelihood developed by Fisher may be viewed as an important but limited realization of our entropy maximization principle. Fisher recognized the striking similarity between his amount of information and the thermodynamic entropy ([13], p. 47). We already know that the expected log likelihood is almost the probabilistic definition of entropy itself.

With Fisher's correction of the degrees of freedom the Pearson's chi-square test of goodness of fit has been one of the most practically useful statistics. The relation between the chi-square and the log likelihood ratio was first discussed by Neymann and Pearson [23] and later by Wilks [28]. The chi-square test statistic is asymptotically equivalent to twice the increase of the logarithm of the maximum likelihood when some of the constraints of the parameters are removed.

It has often been lamented that in spite of its practical utility the chi-square test statistic is not an estimate of anything meaningful ([8] p. 25, [9], [25] p. 190). We now see clearly that being equivalent to twice the increase of the logarithm of the maximum likelihood the chi-square is measuring one essential quantity, the increase of entropy due to the elimination of constraints of the parameters within the model.

The above observations show that some of the most significant successes in the history of statistics were obtained when the statisticians were directly dealing with the entropy and its maximization.

5. SOME REALIZATIONS OF THE PRINCIPLE

First we review the definition of a statistic introduced in [1] which is an estimate of minus twice the expected log-likelihood. Consider the classical situation where $f(z) = g_1(z_1|\theta_o)g_1(z_2|\theta_o)\ldots g_1(z_N|\theta_o)$. It is assumed that $g_1(z_i|\theta)$ is specified by the vector of parameters $\theta = (\theta_1,\theta_2,\ldots,\theta_k)'$. By considering the family of $g(z:x) = g(z|\theta) = g_1(z_1|\theta)g_1(z_2|\theta)\ldots g_1(z_N|\theta)$ defined by a set of possible values of θ and applying the method of maximum likelihood we get the maximum likelihood estimate $\theta(x)$. Now we choose as our estimate $g(z:x) = g(z|\theta(x))$. Under proper regularity conditions of the family of $g(z|\theta)$, $-2B(f;g)$ is asymptotically distributed as a chi-square with k degrees of freedom. This shows that twice the expected decrease of entropy due to the replacement of θ_o by $\theta(x)$, the maximum likelihood estimate, is equal to k. Under the same circumstance the log likelihood ratio statistic $2 \log g(x:x)-2 \log f(x)$ is asymptotically distributed as a chi-square with k degrees of freedom. Thus if we use $(-2)\{\log g(x:x)-\log f(x)\}$ as an estimate of $-2B(f;g)$ it is asymptotically biased by the amount $-2k$. This bias was the main impediment for the direct extension of the maximum likelihood principle to cover those situations where log likelihood ratio tests were used instead. To compensate for this bias we add 2k to $(-2)\{\log g(x:x)-\log f(x)\}$ and since $2 \log f(x)$ is a common additive constant we simply ignore it and get a statistic $(-2)\log g(x:x)+2k$ which is an estimate of minus twice the expected log likelihood. This statistic which is called an information criterion (AIC) [2] is a measure of the badness of fit of $g(z:x)$ and can be described as follows:

$$\begin{aligned}\text{AIC} = {} & (-2)\log\ (\text{maximum likelihood})\\ & +2\ (\text{number of independently adjusted parameters}).\end{aligned}$$

When several different families of $g(z|\theta)$ are considered the estimate $g(z:x)$ which attains the minimum of AIC is the minimum AIC estimate (MAICE). MAICE can be considered as a realization of Occam's razor.

It is known that MAICE provides useful solutions to many practically important problems. It has been pointed out [3] that the final conclusions of the examples of Sections 20 and 21 of Fisher [16], obtained by somewhat subjective manipulations of chi-square test statistics, can be reached simply by adopting the MAICE's. In understanding the relation between MAICE and classical test procedures it is sometimes useful to note that the log likelihood ratio test statistic takes the form χ^2 = AIC(k)-AIC(K)+2(K-k), where AIC(k) denotes the AIC of a model with k free parameters. K is usually the highest possible value of k and χ^2 is tested as a chi-square with the degrees of freedom d.f.= K-k. Thus when χ^2 is given, χ^2-2d.f. can be used as AIC(k) for the purpose of comparison of models.

TABLE I gives the results of the factor analysis of the data given by Maxwell ([22] TABLE I).

TABLE I

Factor Analysis of Maxwell's Data

Number of factors	χ^2	d.f.	AIC(=χ^2-2d.f.)
1	193.46	35	123.46
2	48.35	26	- 3.65
3	18.15	18	-17.85[a]
4	7.32	11	-14.68

[a]Minimum AIC

By assuming the Guassian distribution and the linear factor model AIC attains its minimum at the three factors model which is identical to the Maxwell's choice. AIC for the two factors model is negative, which suggests that even this model will be a better choice than the non-structured model specified by the sample covariance matrix.

More manageable than the factor analysis is the analysis of covariance structure initiated by Bock and Bargmann [11]. TABLE II shows the results of the covariance structure analysis of the data discussed by Wiley, Schmidt and Bramble ([29] TABLE II).

In this example MAICE is Model (4) which coincides with the conclusion of the authors. Although the direct comparison of Models (2) and (3) is not considered by the classical log likelihood ratio test procedure the present results suggest that Model (2) would be a better choice than Model (3).

TABLE II

Covariance Structure Analysis of Wiley, Schmidt and Bramble's Data

Model	^{2}	d.f.	AIC(= ^{2}-2d.f.)
(1)	68.25	31	6.25
(2)	46.16	24	- 1.84
(3)	51.00	25	1.00
(4)	25.98	18	-10.02[a]

[a]Minimum AIC

Figure 1 shows a typical example of the estimation of a probability density function obtained by using the model of the general system of alternative hypothesis described in Section 2. The null hypothesis G(x) was chosen as a Gaussian distribution with unspecified mean and variance. In the notation of Section 2, the original distribution is defined by $f(x) = 0.6\phi(x)+0.4\phi(x-C)$ with $C = 5.0$, where $\phi(x)$ is the density function of the Gaussian distribution with mean 0 and variance 1. In Figure 1, M denotes the order, k, of (1). The m-th degree orthogonal polynomial over [0,1] was used as $\phi_m(y)$. Only crude approximations to the maximum likelihood estimates were obtained and a constant was added to AIC to make it equal to 200.00 at M = 0. The results show the potential of the present procedure as an automatic probability density function estimation procedure. At least there seems no reason for going back to histograms. With an improved method of the maximum likelihood computation and probably with the use of general k-splines [27] as $\phi_m(y)$ the produce will become very useful.

MAICE is extensively used in time series analysis ([2], [5], [6], [7]). Here only one example is illustrated in Figure 2 to show its use in the power spectral density function estimation.

Finishing the analysis of the statistical behavior of the log maximum likelihood as an estimate of the expected log likelihood, we now turn to the problem of the choice of the functional form of the predictive distribution g(z:x). In the classical estimation theory, when it is assumed that $f(z) = g_1(z_1|\theta_o)\ldots g_1(z_N|\theta_o)$, it has been customary to put $g(z:x) = g_1(z_1|\theta(x))g_1(z_2|\theta(x))\ldots g_1(z_N|\theta(x))$, where $\theta(x)$ is an estimate of θ_o. It is not at all obvious if the above choice of g(z:x) is in fact a good choice from the standpoint of the entropy maximization. The success of MAICE suggests that when the dimension of θ_o is high a better estimate might be obtained by using a model with a properly reduced number of parameters. In fact, the possibility of using a Bayesian type modeling for this purpose is already demonstrated in [4.]

As a typical example of practical importance, we consider the problem of polynomial fitting. Assume the situation where $f(z) = g_1(z_1|\theta_o)g_1(z_2|\theta_o)\ldots g_1(z_N|\theta_o)$ and $g_1(z_i|\theta_o)$ is Gaussian with mean $\theta_{0,1}\phi_0(i)+\theta_{0,2}\phi_1(i)+\ldots+\theta_{0,K}\phi_{K-1}(i)$ and variance $\sigma^2 = \theta_{0,K+1}$. It should be noticed that $g_1(z_i|\theta_o)$ now depends on i. $\theta_{0,m}$ denotes the m-th component of the (k+1)-dimensional vector θ_o and it is assumed that $\theta_{0,m} = 0$ $(k < m \leq K)$. k defines the order of the regression. $\phi_m(i)$ $(m=0,1,\ldots,K-1;\ i=1,2,\ldots,N)$ denotes the discretized version of the m-th degree orthogonal polynomial characterized by $\Sigma\{\phi_m(i)\}^2 = 1$ and $\Sigma\phi_m(i)\phi_n(i) = 0$ $(m \neq n)$, where the summation extends over $i=1,2,\ldots,N$. The problem is to get an estimate of the regression curve. For the successive values of $r = 0,1,\ldots,K$ we assume a Bayesian type model which is obtained by formally assuming that $\theta_1,\theta_2,\ldots,\theta_r$, the first r coefficients of regression, are distributed according to the Gaussian distribution with zero mean and variance $\tau^2 I_r$, where I_r is an rxr identity matrix. It is assumed that $\theta_{r+1} = \theta_{r+2} = \ldots = \theta_K = 0$. The likelihood for the original model is given by $g_o(x|\theta) = (2\pi\sigma^2)^{-N/2}\exp\{-||x-\Phi\theta_{-1}||^2(2\sigma^2)^{-1}\}$, where $||x||^2$ denotes the sum of squares of the components of a vector x, Φ is an NxK matrix of which (i,j) element is $\phi_{j-1}(i)$ and $\theta_{-1} = (\theta_1,\theta_2,\ldots,\theta_K)'$. The likelihood of the modified model is given by $g(x|r) = \int g_o(x|{}_r\theta)(2\pi\tau^2)^{-r/2}\exp\{-||{}_r\theta_{-1}||^2(2\tau^2)^{-1}\}d_r\theta_{-1}$, where ${}_r\theta = (\theta_1,\theta_2,\ldots,\theta_r,0,\ldots,0,\sigma^2)$ and $d_r\theta_{-1} = d\theta_1 d\theta_2\ldots d\theta_r$. We have

$$(-2)\log g(x|r) = (N-r)\log(2\pi\sigma^2)+||x-\Phi_r\hat{\theta}_{-1}||^2\sigma^{-2}$$
$$+ r\log\{2\pi(\sigma^2+\tau^2)\}+||{}_r\hat{\theta}_{-1}||^2(\sigma^2+\tau^2)^{-1},$$

where ${}_r\hat{\theta}_{-1}$ is the K-dimensional vector $({}_r\hat{\theta}_1,{}_r\hat{\theta}_2,\ldots,{}_r\hat{\theta}_r,0,\ldots,0)'$ defined by ${}_r\hat{\theta}_j = x'\Phi(\cdot j)$ $(j=1,2,\ldots,r)$, where $\Phi(\cdot j)$ denotes the j-th column of Φ. The present likelihood $g(x|r)$ contains, independently of the value of r, only three parameters; r, σ^2 and τ^2. For a fixed r the values of σ^2 and τ^2 which maximize $g(x|r)$ are given by $\sigma_r^2 = ||x-\Phi_r\hat{\theta}_{-1}||^2(N-r)^{-1}$ and $\tau_r^2 = ||{}_r\hat{\theta}_{-1}||^2 r^{-1}-\sigma_r^2$. τ_r^2 must be equal to zero when $||{}_r\hat{\theta}_{-1}||^2 r^{-1}$ is less than σ_r^2, but for simplicity we ignore this possibility and assume that $\sigma_r^2+\tau_r^2 = ||{}_r\hat{\theta}_{-1}||^2 r^{-1}$. Now we have

$$(-2)\log \mathrm{Max}\ g(x|r) = (N-r)\log\left\{\frac{||x-\Phi_r\hat{\theta}_{-1}||^2}{N-r}\right\} + r\log\left\{\frac{||{}_r\hat{\theta}_{-1}||^2}{r}\right\} + C,$$

where C is a constant independent of r. Since the number of the parameters

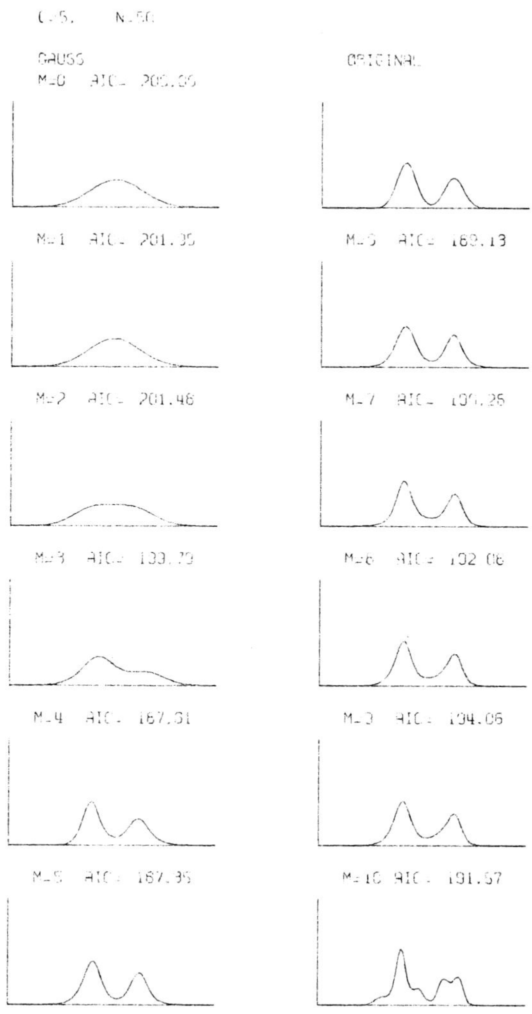

Fig. 1. Probability Density Function Estimation

MAICE is the estimate with M=5 which reproduces the original density function fairly well. Sample size N = 50. Original data are given in Appendix.

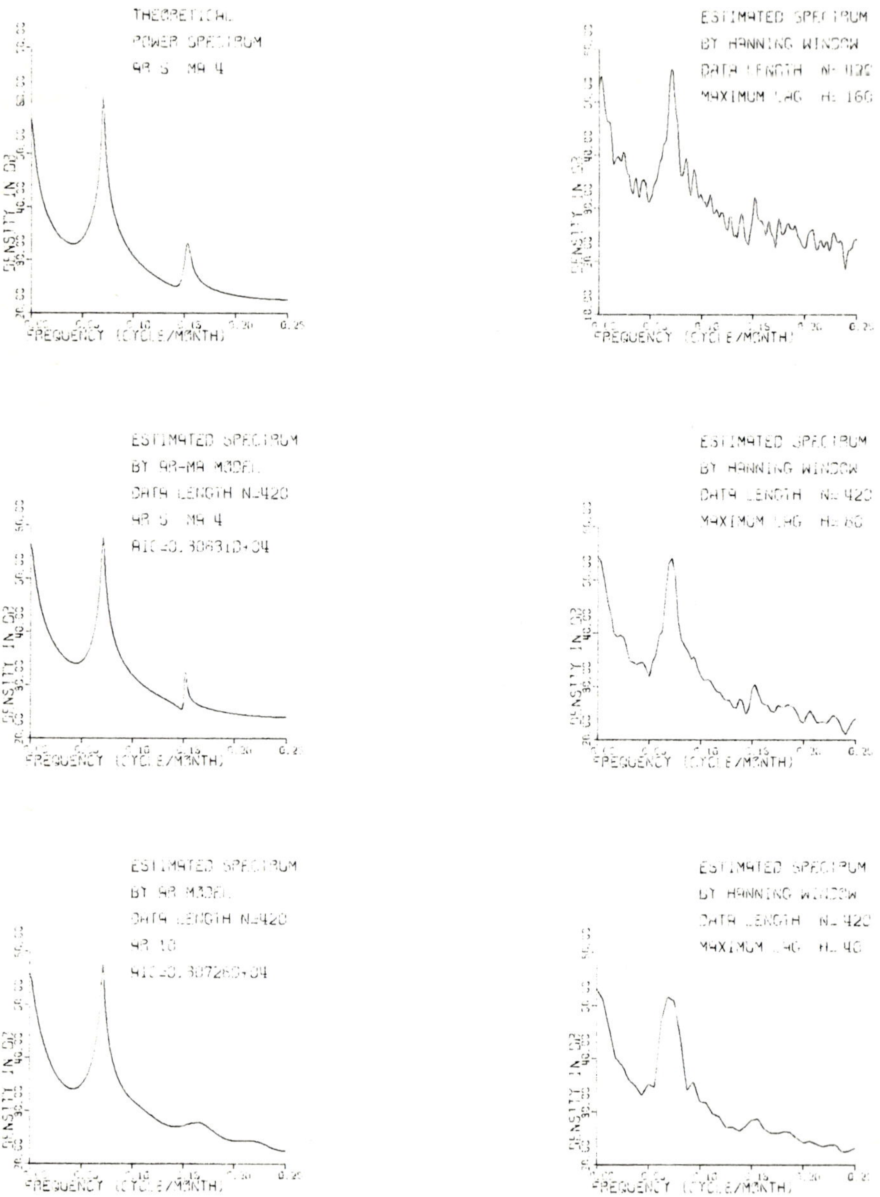

Fig. 2. Power Spectral Density Function Estimation

MAICEs within autoregressive moving average models and auto-regressive models are compared with the theoretical spectrum and other estimates obtained by windowing procedures.

within each model is equal to two, MAICE of k is identical to the maximum likelihood estimate. The maximum likelihood estimate of k is now obtained as the value of r which gives the minimum of

$$\mathrm{BIC} = (N-r)\log\left\{\frac{||x-\Phi_r\hat{\theta}_{-1}||^2}{N-r}\right\} + r\log\left\{\frac{||\Phi_r\hat{\theta}_{-1}||^2}{r}\right\} .$$

It should be noted that $||\Phi_r\hat{\theta}_{-1}||^2 = ||_r\hat{\theta}_{-1}||^2$.

The relation between BIC and AIC of the original model can be analyzed as follows. We have $\mathrm{BIC} = (N-r)\log S(r)-(N-r)\log(1-rN^{-1})+r\log\{P(r)r^{-1}\}+r\log N$, where $S(r) = ||x-\Phi_r\hat{\theta}_{-1}||^2N^{-1}$ and $P(r) = ||\Phi_r\hat{\theta}_{-1}||^2N^{-1}$. By using the approximate equality $-(N-r)\log(1-rN^{-1}) = r$ when rN^{-1} is small, we get an approximate equality $\mathrm{BIC} = N\log S(r) + r + r\log N + r\,\mathrm{Log}\{P(r)r^{-1}\} = \mathrm{AIC}+r\{\log N-1\}+r\log\{P(r)r^{-1}\}$. Although it was assumed that Φ is the matrix of the discretized orthogonal polynomials, Φ could be replaced by an arbitrary matrix. In this case $\Phi_r\hat{\theta}_{-1}$ should be taken as the projection of x onto the linear space spanned by the first r columns of Φ. Thus BIC can be applied to any regression analysis type situation, including the subset regression. When S(r) and P(r) converge stochastically to $S_\infty(r)$ and $P_\infty(r)$, respectively, as N is increased indefinitely and $S_\infty(r) = S$ and $P_\infty(r) = P$ for $r \geqq k$ and $S_\infty(r) > S_\infty$ for $r < k$, the value of r which gives the minimum of BIC, the minimum BIC estimate (MBICE) of k, is a consistent estimate of the order k. MAICE does not have this characteristic.

To see the practical utility of MBICE four experiments were performed each with one thousand repetitions of polynomial fitting to N = 180 data points. The results are given in TABLE III. The basic model was the fifth-degree polynomial regression treated by Guttman [17]. The same model has been treated by Hager and Antle [18] and Halpern [19]. In Experiments 1, 2 and 3 MBICE was giving better results than MAICE in terms of the average sum of squares of the deviations of the estimated regression from the true regression. In Experiment 4, where the absolute values of the coefficients of regression were reduced significantly, MAICE was giving better results. These results show that when some of the coefficients have significant values and others are close to zero MBICE would work quite effectively.

The condition for the consistency of order determination by MBICE holds in many models of regression type. For the case of an autoregressive moving average model of stationary time series, replacing r and S(r) of BIC by the number of coefficients and the maximum likelihood estimate of the innovation variance, respectively, and using the relation P(r) = S(0)-S(r), we can get a consistent estimate of the lowest order representation by MBICE. One example of order determination of an autoregressive model by MAICE and MBICE is given in TABLE IV.

TABLE III

Frequencies of the Degree of Polynomial Determined by MAICE and MBICE

Data	Criterion	Estimated degree[a] 3	4	5[b]	6	7
Experiment 1	AIC	0	0	777	139	84
	BIC	0	0	999	1	0
Experiment 2	AIC	0	0	776	139	85
	BIC	0	0	1000	0	0
Experiment 3[c]	AIC	0	0	759	155	86
	BIC	0	0	1000	0	0
Experiment 4[d]	AIC	0	11	760	152	77
	BIC	33	61	881	22	3

[a]Limited between 3 and 7.

[b]True degree.

[c]With the coefficients reduced by the factor 0.1.

[d]With the coefficients reduced by the factor 0.01.

TABLE IV

Frequencies of the Order of Autoregression Determined by MAICE and MBICE

Criterion	Estimated order[a] 14	15[b]	16	17	18	19	20	21	22	23	24	25
AIC	0	74	7	7	4	2	2	1	2	0	1	0
BIC	0	98	1	1	0	0	0	0	0	0	0	0

[a]Limited between 0 and 50.

[b]True order.

The basic model was the 15th order autoregressive model used by Bhansali [10] and reanalyzed by Jones [20]. The present results correspond to those given in Table 2A of Jones. The length of data of each realization was 1000. The results of TABLE IV clearly show the consistency property of MBICE. The use of AIC and BIC in time series analysis is discussed in [5] and [6].

It is interesting to note that MAICE may be viewed as the mode of the posterior distribution corresponding to the prior distribution defined by the product of the improper uniform distribution of the parameters within the model and the probability of the model defined to be proportionate to the exponential of minus the number of parameters. By analogy it may be expected that MBICE can be defined as the mode of the posterior distribution corresponding to the prior distribution defined by the product of the proper Gaussian distribution of the parameters and the probability of the model which is equal to a constant. If we literally followed this interpretation we would have adopted the estimates of the parameters which are shrinked towards zero like the Stein estimator [4].

Actually each of the Gaussian priors has a parameter τ^2 which is adjusted by the data. Thus they are not priors in the strict Bayesian sense. The results of an extensive Monte Carlo experiment of autoregressive model fitting have shown a significant reduction of the expected entropy by the use of the shrinked estimates instead of the maximum likelihood estimates of the parameters of the model chosen by MBICE. This shows that a proper definition of MBICE of the parameters is given by the mode of the "posterior" distribution corresponding to the "prior" distribution defined by the product of the improper uniform distribution over the parameters and the probability of the model defined to be proportionate to the ratio of the maximum likelihood of the modified model $g(x|r)$ to the maximum likelihood of the original model $g_o(x|_r\theta)$. Since $g(x|r)$ is a weighted average of $g_o(x|_r\theta)$ this choice of the "prior" seems to be a natural one. " " denotes that the related priors is adjusted by the observations.

Although the above interpretations may be interesting it is the power of the entropy maximization principle that has lead to these explicit specifications of the prior which are practically useful. Exploration of the possible use of the distributions of the weights $\exp\{(-1/2)\mathrm{AIC}\}$ or $\exp\{(-1/2)\mathrm{BIC}\}$ over the models might produce interesting results.

6. CONCLUSION

The principle of entropy maximization provides a very practical and versatile approach to statistical inference. It eliminates the dogmatic discriminations which existed among various conventional approaches and provides a unifying view of the subject of statistical inference. The importance of prior information in statistical model building is stressed and the way of its systematic use is explored. In this way the principle will continue to yield reasonable and useful solutions to practically important problems.

APPENDIX

The following is the list of the original data $x_1, x_2, \ldots x_N$ (N=50), rounded to two decimal places, which were used for the computation of the results illustrated in Figure 1.

1.38	2.20	0.07	1.45	-0.43	-1.12	1.24	-2.96	0.64	-0.25
-0.89	-0.53	-0.99	0.51	0.01	0.62	0.17	0.29	-1.29	0.52
-2.60	2.11	-0.49	-0.55	0.03	-0.01	-1.92	0.04	1.90	-0.29
6.24	5.35	5.29	3.79	6.27	5.12	3.37	6.63	3.74	5.50
6.19	5.06	5.14	4.20	6.20	4.42	3.65	4.89	4.45	5.22

REFERENCES

[1] AKAIKE, H. (1973). Information theory and an extension of the maximum likelihood principle. In *2nd International Symposium on Information Theory*, (B. N. Petrov and F. Csaki, Eds.) pp. 267-281. Akademiai Kiado, Budapest.

[2] AKAIKE, H. (1974). A new look at the statistical model identification. *IEEE Trans. Automat. Contr.* AC-19, 716-723.

[3] AKAIKE, H. (1975). What is the information criterion AIC? *Suri Kagaku (Mathematical Sciences)*, **153**, 5-11. (In Japanese).

[4] AKAIKE, H. (1976). An objective use of Bayesian models. Research Memo. No. 91, The Institute of Statistical Mathematics, Tokyo.

[5] AKAIKE, H. (1976). Time series analysis and control through parametric models. A paper presented at the Applied Time Series Analysis Symposium, University of Tulsa, Tulsa, Oklahoma, May 14-15, 1976.

[6] AKAIKE, H. (1976). Spectrum estimation through parametric model fitting. To be presented at the IUTAM Symposium on Stochastic Problems in Dynamics, Southampton, England, July 19-23, 1976.

[7] AKAIKE, H. (1976). Canonical correlation analysis of time series and the use of an information criterion. In *System Identification: Advances and Case Studies* (R. K. Mehra and D. G. Lainiotis, Eds.) (To appear). Academic Press, New York.

[8] ANSCOMBE, F. J. (1956). Contribution to the discussion of a paper by F. N. David and N. L. Johnson. *J. Roy. Statist. Soc. Ser. B*, **18**, 24-27.

[9] BERKSON, J. (1938). Some difficulties of interpretation encountered in the application of the chi-square test. *J. Amer. Statist. Assoc.*, **33**, 526-536.

[10] BHANSALI, R. J. (1973). A Monte Carlo comparison of the regression method and the spectral methods of prediction. *J. Amer. Statist. Assoc.*, **68**, 621-625.

[11] BOCK, R. D. and BARGMANN, R. E. (1966). Analysis of covariance structure. *Psychometrika*, **31**, 507-534.

[12] BOLTZMANN, L. (1877). Uber die Beziehung zwischen dem zweiten Hauptsatz der mechanischen Warmetheorie und der Wahrscheinlichkeitrechnung respective den Satzen uber das Warmegleichgewicht. *Wiener Berichte*, **76**, 373-435.

[13] FISHER, R. A. (1935). The logic of inductive inference. *J. Roy. Statist. Soc.*, XCVIII, 39-54.

[14] FISHER, R. A. (1936). Uncertain inference. *Proc. Am. Acad. Art. Sci.*, **71**, 245-258.

[15] FISHER, R. A. and YATES, F. (1949). *Statistical Tables for Biological, Agricultural and Medical Research.* Oliver and Boyd, London.

[16] FISHER, R. A. (1973). *Statistical Methods for Research Workers.* Hafner, New York.

[17] GUTTMAN, I. (1967). The use of the concept of a future observation in goodness-of-fit problems. *J. Roy. Statist. Soc. Ser. B*, **29**, 83-100.

[18] HAGER, H. and ANTLE, C. (1968). The choice of the degree of a polynomical. *J. Roy. Statist. Soc. Ser. B*, **30**, 469-471.

[19] HALPERN, E. F. (1973). Polynomial regression from a Bayesian approach. *J. Amer. Statist. Assoc.*, **68**, 137-143.

[20] JONES, R. H. (1975). Fitting autoregression. *J. Amer. Statist. Assoc.*, **70**, 590-592.

[21] KULLBACK. S. (1959). *Information Theory and Statistics.* Wiley, New York.

[22] MAXWELL, A. E. (1961). Recent trends in factor analysis. *J. Roy. Statist. Soc. Ser. A*, **124**, 49-59.

[23] NEYMAN, J. and PEARSON, E. S. (1928). On the use and interpretation of certain test criteria for purposes of statistical inference. Part II. *Biometrika*, **20A**, 263-294.

[24] PEARSON, K. (1900). On the criterion that a system of deviations from the probable in the case of a correlated system of variables is such that it can be reasonably supposed to have arisen from random sampling. *The London, Edinburgh and Dublin Philosophical Magazine and Journal of Science. Fifth Series*, **50**, 157-175.

[25] PRATT, J. W. (1965). Bayesian interpretation of standard inference statements. *J. Roy. Statist. Soc. Ser. B*, **27**, 169-203.

[26] SANOV, I. N. (1961). On the probability of large deviations of random variables. In *IMS and AMS Selected Translations in Mathematical Statistics and Probability*, **1**, 213-244.

[27] SCHOENBERG, I. J. (1964). On interpolation by spline functions and its minimal properties. In *On Approximation Theory*, (P. L. Butzer and J. Korevaar, Eds.). pp. 109-129. Birkhauser Verlag, Basel.

[28] WILKS, S. S. (1936). The likelihood test of independence in contingency tables. *Ann. Math. Statist.*, **6**, 190-196.

[29] WILEY, D. E., SCHMIDT, W. H. and BRAMBLE, W. J. (1973). Studies of a class of covariance structure models. *J. Amer. Statist. Assoc.*, **68**, 317-323.

P.R. Krishnaiah, ed., APPLICATIONS OF STATISTICS

STOCHASTIC PROBLEMS IN THE STABILITY OF STRUCTURES

S. T. ARIARATNAM AND D. S. F. TAM

Solid Mechanics Division, University of Waterloo
Waterloo, Ontario, Canada

The problem of parametric instability in elastic structures subjected to stochastically fluctuating loads is investigated. A formulation for determining the stability in the second norm of the response, applicable to both non-gyroscopic and gyroscopic systems, is presented. The theory is applied to the example of the flexural-torsional instability of a simply supported, thin, elastic beam under a stochastically varying transverse load.

1. INTRODUCTION

This paper deals with a class of problems in the stability of elastic structures in which the loss of stability is characterized by a bifurcation of the equilibrium configuration of the structure. The distinct mathematical feature of these problems is that the applied loads appear as a coefficient or parameter in the governing equations of motion. Thus, when the external loads are dynamic, this instability phenomenon is referred to as 'parametric' instability. Problems of parametric instability in both linear and non-linear mechanical systems have been fairly extensively studied in the case when the applied loads (or forcing functions) are deterministic periodic functions of time (see, e.g. [2], [5]). A common example of such loading is that arising from unbalanced forces in rotating machinery. However, in modern aerospace structures, propelled by jet and rocket engines, the external loads fluctuate in a random fashion over a wide range of frequencies, and hence cannot be adequately described solely in terms of deterministic time functions; they have to be regarded as stochastic processes. It is therefore necessary to investigate the stability behaviour of structures under the action of stochastically varying loads.

The present paper considers the problem of parametric instability in stochastically loaded structures that can be described mathematically as linear systems with a finite number of degrees of freedom. A general formulation is presented that includes the cases of both non-gyroscopic and gyroscopic systems. The theory is applied to an example of the flexural-torsional stability of a transversely loaded elastic beam.

2. FORMULATION

Consider elastic systems described by linear equations of motion of the form

$$M\ddot{\underline{q}} + G\dot{\underline{q}} + K\underline{q} + \varepsilon D^\circ\dot{\underline{q}} + \varepsilon^{1/2}f(t)\ P^\circ\underline{q} = 0 \tag{1}$$

where $\underline{q}$ is an n-vector of generalized coordinates, M the mass matrix, K the stiffness matrix, G the gyroscopic matrix, D° the damping matrix and P° the load transfer matrix. f(t) represents the external 'parametric' load, and is taken to be a stationary stochastic process with zero mean; ε is a small parameter. The matrices M, K, and D° are symmetric and positive-definite, while G is skew-symmetric.

In terms of the 2n-dimensional state-vector $\underline{z} = \{\underline{q}', \dot{\underline{q}}'\}$, where the prime indicates the transpose, Eq.(1) may be written in the form

$$A\ \dot{\underline{z}} + B\underline{z} = -\ \varepsilon D\underline{z} - \varepsilon^{1/2}f(t)\ P\underline{z} \tag{2}$$

where

$$A = \begin{bmatrix} K & 0 \\ 0 & M \end{bmatrix},\quad B = \begin{bmatrix} 0 & -K \\ K & G \end{bmatrix},\quad D = \begin{bmatrix} 0 & 0 \\ 0 & D^\circ \end{bmatrix},\quad P = \begin{bmatrix} 0 & 0 \\ P^\circ & 0 \end{bmatrix} \tag{3}$$

It may be noted that A = A', B = -B'. It will be convenient to further convert Eq.(2) to a set of equations in 'standard form'. Towards this end, we follow a procedure due to Meirovitch [4] and first consider Eq.(2) with ε set to zero, namely,

$$A\dot{\underline{z}} + B\underline{z} = \underline{o} \tag{4}$$

together with the associated eigenvalue problem

$$\lambda\ A\underline{z} + B\underline{z} = \underline{o} \tag{5}$$

Since the matrix A is real and symmetric and the matrix B is real and skew-symmetric, the eigenvalues λ are purely imaginary and of the form $\lambda_r = \pm\, i\omega_r$ (r = 1, 2, ...,n), where the ω_r are real. Let the corresponding eigenvectors be $\underline{z}_r = \underline{x}_r \pm i\underline{y}_r$. Then $\underline{x}_r$, $\underline{y}_r$ satisfy the equations

$$\begin{aligned} -\omega_r\ A\ \underline{y}_r + B\ \underline{x}_r &= \underline{o} \\ \omega_r\ A\ \underline{x}_r + B\ \underline{y}_r &= \underline{o} \end{aligned} \tag{6}$$

which can be reduced to

$$\omega_r^2\ A\underline{x}_r = C\ \underline{x}_r,\quad \omega_r^2\ A\ \underline{y}_r = C\ \underline{y}_r \tag{7}$$

where

$$C = B'\ A^{-1}B = C' \tag{8}$$

Furthermore, the vectors $\underline{x}_r$, $\underline{y}_r$ satisfy the orthogonality relations

$$\begin{aligned} \underline{x}_s' A \underline{x}_r &= \underline{x}_r' C \underline{x}_s = \underline{o} \\ \underline{y}_s' A \underline{y}_r &= \underline{y}_r' C \underline{y}_s = \underline{o}, \qquad (r \neq s) \end{aligned} \tag{9}$$

Let X, Y denote the 2nxn matrices defined by

$$X = [\underline{x}_1, \underline{x}_2, \ldots, \underline{x}_n], \; Y = [\underline{y}_1, \underline{y}_2, \ldots, \underline{y}_n]$$

and Ω, the nxn diagonal matrix

$$\Omega = \text{diag}\,\{\omega_1, \omega_2, \ldots, \omega_n\}$$

Then, we have the orthogonality properties

$$X'BX = Y'BY = X'AY = Y'AX = 0 \tag{10}$$

It is convenient to normalize the vectors $\underline{x}_r$, $\underline{y}_r$ by the relations

$$X'AX = Y'AY = I \tag{11}$$

where I is the nxn unit-matrix. Eq(6) may be written in the form

$$X'BY = -Y'BX = -\Omega \tag{12}$$

Using Eq.(12), it can be shown that the matrices X and Y are of the form

$$X = \begin{bmatrix} X_1 \\ -Y_1\Omega \end{bmatrix}, \quad Y = \begin{bmatrix} Y_1 \\ X_1\Omega \end{bmatrix}$$

where X_1, Y_1 are nxn matrices.

Returning to Eq. (2), let the solution vector be of the form

$$\underline{z}(t) = \sum_s \left[\xi_s(t)\, \underline{x}_s + \eta_s(t)\, \underline{y}_s\right] \tag{13}$$

where $\xi_s(t)$, $\eta_s(t)$ are scalar stochastic functions of t. Substituting (13) in Eq.(2) and premultiplying by X' and Y' respectively, and using Eqs. (10), (11) and (12), yields

$$\begin{aligned} \dot{\underline{\xi}} - \Omega\, \underline{\eta} &= -\varepsilon(D^{11}\underline{\xi} + D^{12}\underline{\eta}) - \varepsilon^{1/2} f(t)(P^{11}\underline{\xi} + P^{12}\underline{\eta}) \\ \dot{\underline{\eta}} + \Omega\underline{\xi} &= -\varepsilon(D^{21}\underline{\xi} + D^{22}\underline{\eta}) - \varepsilon^{1/2} f(t)(P^{21}\underline{\xi} + P^{22}\underline{\eta}) \end{aligned} \tag{14}$$

where

$$D^{11} = X'DX,\ D^{12}=X'DY,\ D^{21} = Y'DX,\ D^{22}=Y'DY,$$

$$P^{11} = X'PX,\ P^{12}=X'PY,\ P^{21} = Y'PX,\ D^{22}=Y'DY,$$

$$\underline{\xi} = \{\xi_1, \xi_2, .., \xi_n\},\ \underline{\eta} = \{\eta_1, \eta_2, ..., \eta_n)$$

We finally transform to amplitude and phase variables $a_r(t)$, $\phi_r(t)$ by the transformations

$$\xi_r(t) = a_r \sin\Phi_r, \quad \eta_r(t) = a_r \cos\Phi_r, \ \Phi_r = \omega_r t + \phi_r \tag{15}$$

Substituting Eq.(15) in Eqs. (14) leads to the following equations in 'standard form':

$$\begin{aligned}\dot{a}_r = & - \varepsilon\sum_s [F_{rs}^{(1)} \sin\Phi_r + F_{rs}^{(2)} \cos\Phi_r] a_s \\ & - \varepsilon^{1/2} f(t) \sum_s [Q_{rs}^{(1)} \sin\Phi_r + Q_{rs}^{(2)} \cos\Phi_r] a_s\end{aligned} \tag{16}$$

$$\begin{aligned}a_r\dot{\phi}_r = & - \varepsilon\sum_s [F_{rs}^{(1)} \cos\Phi_r - F_{rs}^{(2)} \sin \Phi_r] a_s \\ & - \varepsilon^{1/2} f(t) \sum_s [Q_{rs}^{(1)} \cos\Phi_r - Q_{rs}^{(2)} \sin\Phi_r] a_s\end{aligned} \tag{17}$$

where

$$\begin{aligned}F_{rs}^{(1)} &= d_{rs}^{11} \sin\Phi_s + d_{rs}^{12} \cos\Phi_s \\ F_{rs}^{(2)} &= d_{rs}^{21} \sin\Phi_s + d_{rs}^{22} \cos\Phi_s \\ Q_{rs}^{(1)} &= p_{rs}^{11} \sin\Phi_s + p_{rs}^{12} \cos\Phi_s \\ Q_{rs}^{(2)} &= p_{rs}^{21} \sin\Phi_s + p_{rs}^{22} \cos\Phi_s\end{aligned} \tag{18}$$

3. APPROXIMATION TO MARKOV PROCESS AND STABILITY EQUATIONS

Eqs. (16) and (17) are in a form suitable for approximate solution by the use of the limit theorem of Stratonovich [6] and Khas'minskii [3]. This theorem provides a technique by which the vector $(\underline{a}, \underline{\phi})$ may be approximated, in the weak sense, by a vector Markov diffusion process governed by a set of Itô equations. The method is an extension to stochastic differential equations of the well-known 'averaging method' of Bogoliubov and Mitropolsky [1]. In the case of Eqs. (16) and (17), it turns out that the amplitude vector $\underline{a}$ by itself approximates to a Markov process. Hence differential equations governing the various moments of the amplitudes may be set up using the Itô differential rule. Following this procedure, it is found that the second moments of the amplitudes, namely

$$m_r = E[a_r^2],$$

where E [ ] denotes the expectation, are governed by the equation

$$\dot{m}_r = \varepsilon \sum_s h_{rs} m_s, \tag{19}$$

where

$$\begin{aligned} h_{rr} = & -(d_{rr}^{11} + d_{rr}^{22}) \\ & + 1/4 \sum_s \Big[(p_{rs}^{11}p_{sr}^{11} + p_{rs}^{12}p_{sr}^{21} + p_{rs}^{21}p_{sr}^{12} + p_{rs}^{22}p_{sr}^{22})\, S^{+}(\omega_r,\omega_s) \\ & - (p_{rs}^{11}p_{sr}^{22} - p_{rs}^{12}p_{sr}^{12} - p_{rs}^{21}p_{sr}^{21} + p_{rs}^{22}p_{sr}^{11})\, S^{-}(\omega_r,\omega_s) \\ & - 2(p_{rs}^{11}p_{sr}^{12} + p_{rs}^{12}p_{sr}^{22} - p_{rs}^{21}p_{sr}^{11} - p_{rs}^{22}p_{sr}^{21})\, \psi^{+}(\omega_r,\omega_s) \\ & - 2(p_{rs}^{11}p_{sr}^{21} - p_{rs}^{12}p_{sr}^{11} + p_{rs}^{21}p_{sr}^{22} - p_{rs}^{22}p_{sr}^{12})\, \psi^{-}(\omega_r,\omega_s) \Big] \\ & + 1/4 \Big[(p_{rr}^{11} - p_{rr}^{22})^2 + (p_{rr}^{12} + p_{rr}^{21})^2 \Big]\, S(2\omega_r) \\ & + 1/4 \Big[(p_{rr}^{11} + p_{rr}^{22})^2 + (p_{rr}^{12} - p_{rr}^{21})^2 \Big]\, S(o) \end{aligned} \tag{20}$$

$$\begin{aligned} h_{rs} = & \; 1/4 \Big[(p_{rs}^{11})^2 + (p_{rs}^{12})^2 + (p_{rs}^{21})^2 + (p_{rs}^{22})^2 \Big]\, S^{+}(\omega_r,\omega_s) \\ & - 1/2 \Big[p_{rs}^{11}p_{rs}^{22} - p_{rs}^{12}p_{rs}^{21} \Big]\, S^{-}(\omega_r,\omega_s). \quad (r \neq s) \end{aligned} \tag{21}$$

In the above expressions, the functions $S^{\pm}(\omega_r,\omega_s)$, $\psi^{\pm}(\omega_r,\omega_s)$ are defined by

$$S^{\pm}(\omega_r,\omega_s) = S(\omega_r+\omega_s) \pm S(\omega_r-\omega_s)$$

$$\psi^{\pm}(\omega_r,\omega_s) = \psi(\omega_r+\omega_s) \pm \psi(\omega_r-\omega_s)$$

where

$$S(\omega) + i\,\psi(\omega) = 2\int_o^\infty E\,[f(t)f(t+\tau)]\; e^{-i\omega\tau}\, d\tau.$$

$S(\omega)$ is the spectral density function of the excitation process $f(t)$.

We define the norm of the response vector $\underline{z}$ by

$$\| \underline{z} \|^2 = \underline{z}'A\underline{z}$$

Using the relations (10), (11), and (13), it can be shown that

$$\| \underline{z} \|^2 = \sum_r a_r^2 = \| \underline{a} \|^2$$

where $\|\underline{a}\|$ is the Euclidean norm. Thus the stability in the second norm of the response is determined by the signs of the real parts of the eigenvalues of the matrix $H=[h_{rs}]$. This can be accomplished by applying the well-known Routh-Hurwitz criterion to H. It can be shown that stability in the second norm implies stability of the second moments of the response.

4. APPLICATION

As an application, the flexural-torsional instability of a thin simply-supported elastic beam subjected to a stochastically varying transverse load at mid-span is considered, Figure 1. Two forms of loading are investigated: (a) where the load maintains a constant vertical direction, and (b) where the load changes its direction so as to always remain in the plane of the cross-section at mid-span, i.e. a so-called 'follower' load.

Assuming deflection functions of the form $u(z,t) = q_1(t) \sin(\pi z/L)$, $\phi(z,t) = q_2(t) \sin(\pi z/L)$, which satisfy the boundary conditions of simple support, and using the Galerkin technique, it can be shown [2] that the equations of motion reduce to:

$$\begin{aligned} \ddot{q}_1 + \varepsilon d_1 \dot{q}_1 + \omega_1^2 q_1 + \varepsilon^{1/2} p_{12} f(t) q_2 &= 0 \\ \ddot{q}_2 + \varepsilon d_2 \dot{q}_2 + \omega_2^2 q_2 + \varepsilon^{1/2} p_{21} f(t) q_1 &= 0 \end{aligned} \tag{22}$$

where

$$\omega_1{}^2 = \pi^4 \, E \, I_y/(mL^4),$$

$$\omega_2^2 = \pi^2 \, (\pi^2 EK + GJL^2)/(m\rho^2 L^4),$$

$$\varepsilon^{1/2} f(t) = F(t)/P_{cr},$$

and EI_y, GJ, EK denote respectively, the flexural, torsional and warping rigidities of the beam; m is the mass per unit length, ρ the polar radius of gyration and d_1, d_2 are viscous damping coefficients. P_{cr} is the critical load for time independent

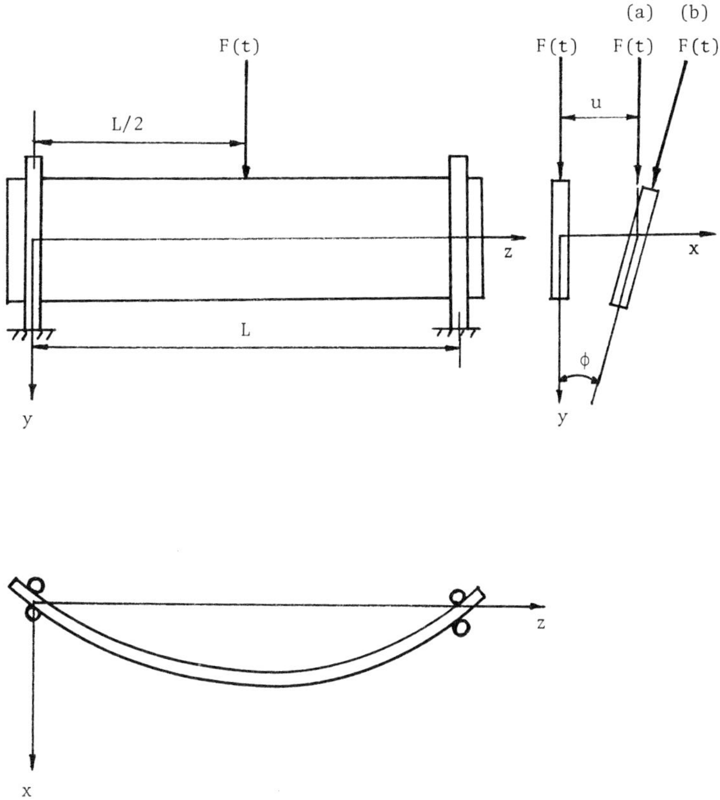

Fig. 1

loading.

For constant directional (vertical) loading:

$$P_{cr} = \frac{8\ mL\rho\omega_1\omega_2}{\pi^2+4},\ p_{12}=-\rho\omega_1\omega_2,\ p_{21}=-\omega_1\omega_2/\rho,$$

while, for follower-type loading:

$$P_{cr} = \frac{4mL\rho\left|\omega_1^2-\omega_2^2\right|}{\left[(12-\pi^2)(\pi^2+4)\right]^{1/2}},\quad p_{12}=\left[\frac{12-\pi^2}{4+\pi^2}\right]^{1/2}\frac{\rho\left|\omega_1^2-\omega_2^2\right|}{2\omega_1},$$

$$p_{21} = - \left[\frac{4+\pi^2}{12-\pi^2} \right]^{1/2} \frac{|\omega_1^2-\omega_2^2|}{2\rho\omega_2} .$$

It is worth noting that the product $p_{12}p_{21}$ is positive for case (a) and negative for case (b).

Eq. (22) may be written in the matrix form (1) where

$$M = I, \; G = 0, \; K = \mathrm{diag}\,\{\omega_1^2, \omega_2^2\} = \Omega^2,$$

$$D^{\circ} = \begin{bmatrix} d_1 & 0 \\ 0 & d_2 \end{bmatrix}, \qquad P^{\circ} = \begin{bmatrix} 0 & p_{12} \\ p_{21} & 0 \end{bmatrix}$$

Then, the matrices A, B, D, P become

$$A = \begin{bmatrix} \omega_1^2 & 0 & 0 & 0 \\ 0 & \omega_2^2 & 0 & 0 \\ 0 & 0 & 1 & 0 \\ 0 & 0 & 0 & 1 \end{bmatrix}, \quad B = \begin{bmatrix} 0 & 0 & -\omega_1^2 & 0 \\ 0 & 0 & 0 & -\omega_2^2 \\ \omega_1^2 & 0 & 0 & 0 \\ 0 & \omega_2^2 & 0 & 0 \end{bmatrix},$$

$$D = \begin{bmatrix} 0 & 0 & 0 & 0 \\ 0 & 0 & 0 & 0 \\ 0 & 0 & d_1 & 0 \\ 0 & 0 & 0 & d_2 \end{bmatrix}, \quad P = \begin{bmatrix} 0 & 0 & 0 & 0 \\ 0 & 0 & 0 & 0 \\ 0 & p_{12} & 0 & 0 \\ p_{12} & 0 & 0 & 0 \end{bmatrix}$$

The normalized matrices X, Y are found to be

$$X = \frac{1}{\sqrt{2}} \begin{bmatrix} \frac{1}{\omega_1} & 0 \\ 0 & \frac{1}{\omega_2} \\ -1 & 0 \\ 0 & -1 \end{bmatrix}, \quad Y = \frac{1}{\sqrt{2}} \begin{bmatrix} \frac{1}{\omega_1} & 0 \\ 0 & \frac{1}{\omega_2} \\ 1 & 0 \\ 0 & 1 \end{bmatrix}$$

so that

$$D^{11} = D^{22} = \frac{1}{2} \begin{bmatrix} d_1 & 0 \\ 0 & d_2 \end{bmatrix},$$

$$P^{11} = P^{12} = -P^{21} = -P^{22} = -\frac{1}{2}\begin{bmatrix} 0 & \frac{p_{12}}{\omega_2} \\ \frac{p_{21}}{\omega_1} & 0 \end{bmatrix}$$

Hence, from Eqs. (20) and (21),

$$h_{11} = -d_1 + \frac{p_{12}p_{21}}{4\omega_1\omega_2} S^-(\omega_1,\omega_2)$$

$$h_{12} = \frac{p_{12}^2}{4\omega_2^2} S^+(\omega_1,\omega_2)$$

$$h_{21} = \frac{p_{21}^2}{4\omega_1^2} S^+(\omega_1,\omega_2)$$

$$h_{22} = -d_2 + \frac{p_{12}p_{21}}{4\omega_1\omega_2} S^-(\omega_1,\omega_2)$$

The eigenvalues of H have negative real parts, if, and only if,

$$h_{11} < 0, \quad h_{22} < 0, \quad h_{11}h_{22} > h_{12}h_{21}$$

Thus the conditions for stability in the second norm are:

$$d_1,\ d_2 > \frac{p_{12}p_{21}}{4\omega_1\omega_2} S^-(\omega_1,\omega_2)$$

$$\left[d_1 - \frac{p_{12}p_{21}}{4\omega_1\omega_2} S^-(\omega_1,\omega_2)\right]\left[d_2 - \frac{p_{12}p_{21}}{4\omega_1\omega_2} S^-(\omega_1,\omega_2)\right] > \frac{p_{12}^2p_{21}^2}{16\omega_1^2\omega_2^2}\left[S^+(\omega_1,\omega_2)\right]^2$$

It may be noted that the stability conditions depend only on the values of the excitation spectral density at the frequencies $\omega_1+\omega_2$ and $\omega_1-\omega_2$. The regions of stability in the d_1,d_2 - plane are shown in Figure 2.

In the case when the spectrum of the load f(t) is of constant magnitude S_o over a wide frequency range, the stability conditions simplify to:

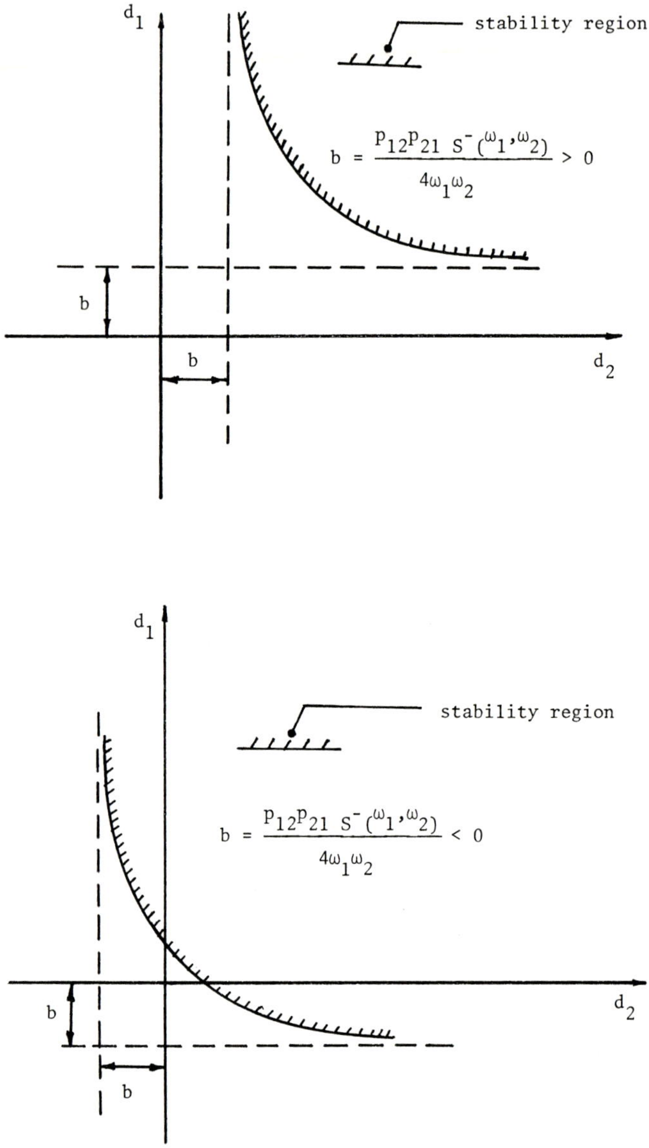

d_1
stability region
$b = \frac{p_{12}p_{21}\, S^-(\omega_1,\omega_2)}{4\omega_1\omega_2} > 0$
b
b
d_2
d_1
stability region
$b = \frac{p_{12}p_{21}\, S^-(\omega_1,\omega_2)}{4\omega_1\omega_2} < 0$
b
b
d_2

Fig. 2

$$d_1 > 0,\ d_2 > 0,\ d_1 d_2 > \left(\frac{P_{12} P_{21} S_o}{2\omega_1 \omega_2} \right)^2 .$$

5. CONCLUSIONS

A method for investigating parametric instability of stochastically loaded elastic structures is presented. The formulation includes structures whose linearized equations of motion also contain gyroscopic terms. As an application, the problem of the flexural-torsional instability of an elastic beam under a stochastic transverse load is investigated.

ACKNOWLEDGEMENT

The research for this paper was supported by the National Research Council of Canada under Grant No. A1815.

REFERENCES

[1] BOGOLIUBOV, N. N. and MITROPOLSKY, Y. A. (1961). *Asymptotic Methods in the Theory of Non-linear Oscillations*. Hindustan Publishing Corp., Delhi, India.

[2] BOLOTIN, V. V. (1964). *The Dynamic Stability of Elastic Systems*. Holden Day, San Francisco, California.

[3] KHAS'MINSKII, R. Z. (1966). A limit theorem for the solution of differential equations with random right-hand sides. *Theor. Probability Appl.*, **11**, 390-406.

[4] MEIROVITCH, L. (1975). A modal analysis for the response of linear gyroscopic systems. *J. Appl. Mech.*, **42**, 446-450.

[5] METTLER, E. (1967). Stability and vibration problems of mechanical systems under harmonic excitation. *Dynamic Stability of Structures* (G. Herrmann, ed.), 169-188. Pergamon Press, New York.

[6] STRATONOVICH, R. L. (1963). *Topics in the Theory of Random Noise*, **1**. Gordon and Breach, New York.

P.R. Krishnaiah, ed., APPLICATIONS OF STATISTICS

A MINIMAX APPROACH TO COMPONENT ANALYSIS

ROLF E. BARGMANN AND FRED D. BAKER

Department of Statistics and Computer Science
Boyd Graduate Studies Research Center
University of Georgia, Athens, Georgia

Component analysis is presented as a technique designed to represent a given data matrix by one of lesser rank which has the same informational content as the original data matrix. Traditional least squares approximations to this problem are quite inadequate, as is shown conspicuously in one of the applications. Numerical methods to move toward a minimax solution are discussed, and incorporated in a computer program. Preferred representations to facilitate interpretation of results are outlined. The method has been applied to three examples in food science.

1. INTRODUCTION

Statistical reasoning is concerned with inference from a sample of observations to a population or universe (mathematical model) from which this sample is assumed to have arisen. By contrast, many aspects of data analysis, and component analysis in particular, consider only the data at hand. One object of data analysis, and data pre-processing for subsequent statistical inference, is to replace a given data matrix by another one which has the same information and exhibits a certain structure. Consider the following data:

X:	60	61	62	63	64	65	66	67	68	69	70
Y:	16	16	17	17	18	18	19	19	20	21	21

The correlation between X and Y is estimated to be 0.986. The statistical hypothesis $\rho = 1$ (alternative $\rho < 1$) must be rejected. In fact, the *null hypothesis* $\rho = 1$ is not testable, for it implies *deterministic linear dependence* between the two sets, and hence every sample, no matter how large or small, must have $r = 1$.

The above data are selected points of the Fahrenheit and Celsius scales, rounded to the nearest integer. The problem of reducing the rank of the above matrix to one is not a problem in statistical inference. Rather, we wish to replace the above data matrix by one with the same informational content (Izenman [8]) in this case by adding and subtracting numbers the largest of which must be less than 0.5, in such a way that the resulting matrix have rank one. Least squares approximations to such data have been known for a long time. Legendre [10] is one of the earliest references which we were able to find. Karl Pearson [12] was probably the first to recognize that the rank reduction problem is associated with the problem of finding characteristic roots and vectors of scaled covariance matrices. Hotelling [7] and Eckart and Young [3] are the most frequently cited recent authors concerned with the problem of rank reduction by least squares. The rank one least squares approximation to the above data is:

$\underset{\sim}{X}$:	60.14	60.91	62.11	62.88	64.07	
$\underset{\sim}{Y}$:	15.74	16.16	16.80	17.22	17.86	
$\underset{\sim}{X}$:	64.85	66.04	66.81	68.01	69.20	69.97
$\underset{\sim}{Y}$:	18.28	18.93	19.34	19.99	20.63	21.05

It has a maximum error of approximation of 0.37, compared with the original data, and is thus well within rounding. The formula connecting these reproduced data is $X = 1.83Y + 31.28$; it is close to, though not identical with, the known formula $X = 1.8Y + 32$. A closer approach to a solution *minimizing the maximum error* is available by minimizing a power higher than the second, and balancing errors; it is:

$\underset{\sim}{X}$:	60.227	60.874	62.170	62.818	64.113	
$\underset{\sim}{Y}$:	15.773	16.125	16.830	17.182	17.887	
$\underset{\sim}{X}$:	64.761	66.057	66.704	68.000	69.296	69.943
$\underset{\sim}{Y}$:	18.239	18.943	19.296	20.000	20.705	21.057

The relation between X and Y is $X = 1.839Y + 31.220$.
To facilitate understanding of later notation, the last result will be stated formally as:

$$\hat{x}_{i1}\ (=\hat{x}_i) = \bar{x}_1 + c_1 + a_i b_1$$

$$\hat{x}_{i2}\ (=\hat{y}_i) = \bar{x}_2 + c_2 + a_i b_2$$

where

$$\bar{x}_1 = 65 \qquad c_1 = 0.222 \qquad \bar{x}_2 = 18.3636 \qquad c_2 = 0.1258$$

$$b_1 = .8811 \qquad b_2 = .4791, \text{ and the row vector}$$

$$\underset{\sim}{a}' = [-5.669 \quad -4.934 \quad -3.464 \quad -2.728 \quad -1.258 \quad -.523 \quad .948 \quad 1.683 \quad 3.153 \quad 4.623 \quad 5.359]$$

The maximum discrepancy between the raw and the reproduced data is under 0.30 (0.296). In fact, this is a solution which we would have obtained if we had specified the maximum permissible error to be 0.30. Attempts to reduce the rank of data matrices using a minimax procedure were first considered by Laplace [9], who abandoned the problem because of the need for excessive hand calculations.

Rank-reducing data analysis is especially valuable when data are merely recorded on an ordinal scale. The typical 5 or 9 point scales in opinion surveys or product testing are particularly well suited for this kind of reduction technique, for the neutral point on a five-point scale is not a realization of a random variable, but merely a convenient point chosen between 2.5 and 3.5; any value in this interval represents the same information as the number 3. It is suggested that this type of data analysis precede statistical inference. A computer program has been developed which attempts to reduce the rank of a given data matrix, keeping the maximum error within prescribed bounds; if, to this end, an approach to minimax is necessary, the program will proceed to such a solution. Preferred representations of the solution will be discussed. Three examples from taste testing experiments will be presented to illustrate the considerable gain that can be realized by not limiting approximation techniques to least squares.

2. RANK REDUCTION TO CONTROL MAXIMUM ERROR

2.1. Standardization

Let $\underset{\sim}{X}^*$ with typical elements x^*_{ij} be a *data matrix* with n rows and p columns, representing observations for n experimental units on p variables or "attributes". Of course, rank $(\underset{\sim}{X}^*) = r \leq p$. For convenience let the arithmetic mean of its column, $\bar{x}_j$, be subtracted from each element. Let $\underset{\sim}{X}$ be the resulting matrix. Now, if we approximate $\underset{\sim}{X}$ by a matrix of rank $k < r$ by least squares, then $\underset{\sim}{X} = \underset{\sim}{A}_1\underset{\sim}{B}'_1 + \underset{\sim}{E}_1$, where $\underset{\sim}{A}_1$ is of order n by k $(n > k)$, $\underset{\sim}{B}'_1$ is of order k by p $(k < p)$ and $\underset{\sim}{E}_1$ is an n by p matrix of errors of approximation. The maximum permissible error for each attribute (column) depends, of course, on the unit and precision of measurement of that attribute. If each column of X is divided by its precision constant p_j (absolute value of *maximum permissible error* in column j) a matrix $\underset{\sim}{X}$ is obtained, with elements $x_{ij} = (x^*_{ij} - \bar{x}_j)/p_j$. In the rank k approximation

$$\underset{\sim}{X} = \underset{\sim}{A}\underset{\sim}{B}' + \underset{\sim}{E}, \tag{1}$$

the elements of $\underset{\sim}{E}$ have equal importance. An approximation, satisfactory in the sense that the informational content of the original data matrix $\underset{\sim}{X}^*$ be preserved by the reduced-rank approximation, would need to have the largest element in $\underset{\sim}{E}$ to be less than 1 in absolute value. For approximations other than least squares a minor correction (to balance the sum of a high odd power of errors) c_j, permits a closer fit. Thus

$$\underset{\sim}{X} = \underset{\sim}{A}\underset{\sim}{B}' + \underset{\sim}{j}\underset{\sim}{c}' + \underset{\sim}{E} \tag{2}$$

where $\underset{\sim}{j}$ is a column vector of ones, and $\underset{\sim}{c}'$ is a row vector containing the elements c_j.

2.2. Initial Guess at Solution

The well-known least squares approximation of rank k to the standardized matrix $\underset{\sim}{X}$ in (2) is

$$\underset{\sim}{A} = \underset{\sim}{X}\underset{\sim}{Q}\underset{\sim}{G}^{-1} \tag{3}$$

$$\underset{\sim}{B} = \underset{\sim}{Q}\underset{\sim}{G}' \tag{4}$$

where $\underset{\sim}{Q}$ is a (p by k) matrix whose columns are the eigenvectors associated with the k largest characteristic roots of $\underset{\sim}{X}'\underset{\sim}{X}$, and $\underset{\sim}{G}$ is an arbitrary, real valued, non-singular matrix of order k. The correction vector $\underset{\sim}{c}'$ of formula (2) is zero in this case. This solution has been used as an initial guess for iterative procedures designed to obtain more satisfactory solutions.

2.3. Approach to Minimax

The principle of least squares, while convenient and optimum under certain assumptions (Gauss-Markov-Theorm) in statistical estimation, has no merit in data analysis or numerical analysis; in these types of analysis the maximum error, or error bound, must be guaranteed to remain below a prescribed value; this value is 1 for the standardized data matrix in formula (2). Now, if a least-squares approximation of rank k produces a maximum error less than 1, a satisfactory approximation in the data analytic sense has been found, and no further search is necessary for that particular rank; (recall the temperature example in the introduction, where the near-minimax solution was no better than the least-squares approximation to the formula which, in this case, was known). However, an approximation of lesser rank may be sought (quite successfully, see Section 5), even if for this smaller rank k the least squares approximation produces a maximum error greater than 1. Several techniques are available for rank 1 approximation to minimax solutions (Hoel [6], Arens, David [1]) in fact, the procedures used in numerical analysis (Tchebyshev [15], Maehli [11]) are but special cases of such solutions. For k > 1, however, no universally satisfactory approach is available.

The concept of a generalized mean of the errors lends itself readily for an approach to minimax solutions. Let

$$m_q = \left\{\frac{1}{N}\sum_{k=1}^{N} |e_k|^q\right\}^{1/q} \tag{5}$$

be the "q-mean" of errors e_k (k = 1, 2, ... N). For q = 1, m_q is the absolute deviation, for q = 2 the standard deviation, for q = -1 the harmonic mean, for $q \to 0$ the geometric mean, and for $q \to \infty$ the maximum $|e_k|$. Hamming [5] suggests that an iterative weighted least squares be used as an approach to a minimax solution; for the present problem this would require evaluating formulas (3) and (4) of Section 2.2, where the columns of $\underset{\sim}{Q}$ are now the eigenvectors associated with the k largest roots of $\underset{\sim}{X}'\underset{\sim}{D}_{w}\underset{\sim}{X}$, i.e.

$$\underset{\sim}{X}'\underset{\sim}{D}_{w}\underset{\sim}{X}\underset{\sim}{Q} = \underset{\sim}{Q}\underset{\sim}{D}_{\lambda} \tag{6}$$

where $\underset{\sim}{D}_w$ is a diagonal, n by n matrix, whose elements at iteration k are $e^q_{i,(k-1)}$, where $q = 2^{2k-2}$, and $e_{i,(k-1)}$ is the largest error of approximation in row i of the reproduced matrix $\underset{\sim}{A}\underset{\sim}{B}'$, as obtained in the k'th weighted least squares cycle. Unfortunately, the reduction of the maximum error by this sequence of weighted least squares iterations is slow, especially for larger (>3) values of k. For large data matrices, however, it may be the only method which is computationally feasible. Furthermore, addition of a constant c_j to the j'th column of $\underset{\sim}{A}\underset{\sim}{B}'$ yields a better balanced reduction. The approach to minimax solutions, as discussed in Section 3, is through a sequence of even powers of q, and minimization of m_q as defined in (5).

3. NUMERICAL PROCEDURES

3.1. Functions and Derivatives

The function to be minimized in order to reduce the maximum error is

$$f(\underset{\sim}{A}, \underset{\sim}{B}, \underset{\sim}{c}') = \sum_{i=1}^{n} \sum_{j=1}^{p} e^q_{ij}, \tag{7}$$

where q is a high even power (see Section 3.3), $e_{ij}=x_{ij} - \hat{x}_{ij}$, x_{ij} is an element in the data matrix $\underset{\sim}{X}$ of formula (2), i.e., standardized so as to make the maximum permissible error equal to 1, and

$$\hat{x}_{ij} = c_j + \sum_{z=1}^{k} a_{iz}b_{zj} \tag{8}$$

(see matrix notation in formula (2)).

The object is to find, for given k and q, the values c_j, a_{iz}, and b_{zj} which tend toward minimizing f in formula (7), where the iterative process can be terminated as soon as a set of values has been found for which max $|e_{ij}| < 1$. For the Newton-Raphson approach, the gradient vector and the Hessian matrix of second derivatives have elements:

$$\frac{\partial f}{\partial a_{rs}} = -q \sum_{j=1}^{p} b_{sj}e^{q-1}_{rj} \qquad r=1,2,\ldots,n; \quad s=1,2,\ldots,k: \tag{9}$$

$$\frac{\partial^2 f}{\partial a_{rs}\partial a_{vt}} = \begin{cases} 0, & r\neq v \\ q(q-1)\sum_{j=1}^{p} b_{sj}b_{tj}e_{rj}^{q-2}, & r=v \end{cases} \qquad v=1,2,\ldots,n;\; t=1,2,\ldots,k. \tag{10}$$

$$\frac{\partial f}{\partial b_{rs}} = -q\sum_{i=1}^{n} a_{ir}e_{is}^{q-1} \qquad r=1,2,\ldots,k;\; s=1,2,\ldots,p; \tag{11}$$

$$\frac{\partial^2 f}{\partial b_{rs}\partial b_{vt}} = \begin{cases} 0, & s\neq t \\ q(q-1)\sum_{i=1}^{n} a_{ir}a_{iv}e_{is}^{q-2} & s=t \end{cases} \qquad v=1,2,\ldots,k;\; t=1,2,\ldots,p. \tag{12}$$

$$\frac{\partial^2 f}{\partial a_{rs}\partial b_{vt}} = -qe_{rt}^{q-1}\,\delta_s^v + q(q-1)a_{rv}b_{st}e_{rt}^{q-2} \qquad r=1,2,\ldots,n;\; s=1,2,\ldots,k;\; v=1,2,\ldots,k;\; t=1,2,\ldots,p. \tag{13}$$

$$\frac{\partial f}{\partial c_s} = -q\sum_{i=1}^{n} e_{is}^{q-1} \qquad s=1,2,\ldots,p. \tag{14}$$

$$\frac{\partial^2 f}{\partial c_s^2} = q(q-1)\sum_{i=1}^{n} e_{is}^{q-2} \tag{15}$$

Second derivatives involving c_j have been ignored because of the smallness of the required correction.

The rapidity with which the Fletcher-Powell [4] approach converges depends on the availability of a very good guess of the minimum which the function will eventually reach. Since for a previously chosen power of q (say q=8) it is very difficult indeed to assess the absolute magnitude of the function for a higher q (say q=64), the function to be minimized by this method was chosen to be

$$f(\underset{\sim}{A}, \underset{\sim}{B}, \underset{\sim}{c}') = \left\{\sum_{i=1}^{n}\sum_{j=1}^{p} e_{ij}^{q}\right\}^{1/q}, \tag{16}$$

i.e., the q'th root of f defined in (17). The elements of the required gradient vector are, in this case:

$$\frac{\partial f}{\partial a_{rs}} = -\sum_{j=1}^{p} u_{sj}\,(e_{rj}/f)^{q-1} \qquad r=1,2,\ldots,n;\; s=1,2,\ldots,k. \tag{17}$$

and

$$\frac{\partial f}{\partial u_{rs}} = -\sum_{i=1}^{n} a_{ir} / \sqrt{\sum_{m=1}^{p} u_{im}^2} \; \{(e_{is}/f)^{q-1} + b_{rs} \frac{\partial f}{\partial a_{is}}\}$$

$$r=1,2,\ldots k; \quad s=1,2,\ldots,p. \tag{18}$$

where b_{sj} are related to the u_{sj} by the formula $b_{sj} = u_{sj}/\sqrt{\sum_{m=1}^{p} u_{sm}^2}$, i.e., the columns of the matrix B (rows of B' as defined in (2)) are chosen to have unit length, whereas the u's are allowed to vary, a technique well known to give greater power to steepest descent and modified steepest descent techniques. Since the c_j are obtained by Newton-Raphson only, the corresponding gradient vector for f as defined in (16) has been omitted.

3.2. Conditional Minimization

For a data matrix of order n by p, an approach to minimax of rank k requires evaluation of functions of nk + p + k(p-1) variables (with simple constraining of rows of $\underset{\sim}{B}'$), which can be slightly reduced (by k(k-1)/2 variables) if $\underset{\sim}{B}'$ is orthogonalized after each iteration. For all but the smallest data matrices this is a very large number of variables and some compromise seemed to be indicated especially since the actual minimum is rarely needed - merely a solution which keeps the largest error below 1.

After considerable experimentation, the following conditional minimization was found to be most economical: Using the $\underset{\sim}{A}$, $\underset{\sim}{B}'$, and $\underset{\sim}{c}'$ values from a lower q (initially q = 2) we treated (a) the elements in $\underset{\sim}{A}$ and $\underset{\sim}{c}'$ fixed and improved the columns of $\underset{\sim}{B}'$ by Newton-Raphson - it should be noted that the mixed second derivatives with respect to different columns of $\underset{\sim}{B}'$ are zero; (b) improved the elements of $\underset{\sim}{c}'$ treating $\underset{\sim}{A}$ and $\underset{\sim}{B}'$ as fixed; and (c) improved the rows of $\underset{\sim}{A}$ by Newton-Raphson, treating the elements of $\underset{\sim}{c}'$ and $\underset{\sim}{B}'$ as fixed. This procedure resulted in quick and efficient reduction of the maximum error except that, in the very first iteration for q=4 or q=8 or q=64, starting with least squares as initial guesses, the Newton-Raphson technique made a bad move which took several iterations to correct itself.

This unfortunate behavior of the conditional minimization process prompted us to precede the Newton-Raphson iterations by just two iterations using the Fletcher-Powell technique; these two steps (steepest descent followed by one cycle of modified gradient) brought the guesses sufficiently close to the minimizing values, so that Newton-Raphson iterations thereafter converged essentially uniformly. In the Fletcher-Powell cycles it was advantageous to iterate the entire matrix $\underset{\sim}{B}'$ given $\underset{\sim}{A}$ and $\underset{\sim}{c}'$, since in the required q'th root the columns of $\underset{\sim}{B}'$ were no longer independent.

3.3. The Choice of Exponents

In order to decide on an efficient procedure of choosing exponents q, we performed several experiments on 6 by 3 matrices, attempting a rank 1 modification and a 12 by 4 matrix attempting a rank 2 approximation (the latter was obtained by perturbing a rank 2 matrix by balanced errors of + 1 and -1). The following cases are illustrative:

Case I: Proceeding from least squares to q = 4, q = 64, and q = 256, the four largest errors in each instance were:

q = 4 (16 iterations)

$e_{23} = .9951 \quad e_{21} = -.9812 \quad e_{33} = -.8547 \quad e_{41} = .7676;$

q = 64 (18 iterations)

$e_{21} = -.9217 \quad e_{23} = .9167 \quad e_{31} = .9128 \quad e_{33} = -.9079;$

q = 256 (5 iterations)

$e_{21} = -.9189 \quad e_{31} = .9172 \quad e_{23} = .9166 \quad e_{33} = -.9091.$

Case II: Proceeding from least squares to q=8, q=64, and q=256 the four largest errors were

q = 8 (19 iterations)

$e_{21} = -.9578 \quad e_{23} = .9296 \quad e_{31} = .8675 \quad e_{33} = -.8575;$

The four largest errors for q=64 (15 iterations) and q=256 (5 iterations) were the same as in Case I.

Case III: Proceeding from least squares to q=64, and then to q=256, we needed 20 iterations at q=64 and 4 iterations at q=256 and obtained results slightly less satisfactory than Case I or Case II but clearly indicating a minimax solution which, for the given example, had the four largest errors as

$$e_{21} = -.9165 \quad e_{23} = .9165 \quad e_{31} = .9165 \quad e_{33} = -.9165$$

If 1 were chosen as the maximum permissible error, even the q=4 solution would be acceptable as a rank one matrix containing the same information as the data matrix.

The computer program RKRED (Baker [2]) (available at the University of Georgia) proceeds from least squares to q=64, then to q=128 and q=256; if thereafter the maximum error is ≤ 1.03 (but > 1) attempts are made at q equal to higher powers of 2 (to a maximum of 2^{19}). In fact, in the 12 by 4 matrix which was constructed so as to have the maximum error appear in every element, a successful rank 2 approximation was not attained until q was 8,192. The computer program makes provisions to handle such unusual cases.

4. REPRESENTATION

4.1. Simple Structure

Two objectives of data analysis are to facilitate interpretation and to suggest economies in further analyses. In component analysis, the elements in the original data matrix, X^*, are approximated by those of a rank k matrix differing from X^* by no more than a maximum permissible error of precision. If the reproduced matrix is standardized in such a way that the sum of elements in columns equals 0 and the sum of squares (corrected for means) is n, then the matrix of standardized observations can be expressed as

$$Y_{st} = AB', \tag{19}$$

where A is of order n by k, and B' of order k by p. The elements in A can be similarly standardized, i.e., $A_{st} = AD_{\alpha}$; the elements of A_{st} are called "*component scores*". Thus

$$Y_{st} = A_{st} D_{\alpha}^{-1} B' = A_{st} B_1' \tag{20}$$

where D_{α} is a diagonal matrix with elements equal to the scaling constants which standardize A. Even with this standardization, the decomposition is not unique, unless k=1. For, let T be any k by k matrix (non-singular), and let $B_1 T = V$. Then

$$Y_{st} = A^*_{st} V^{*'} \tag{21}$$

where $V^* = V D_{\alpha}^{-1}$

and $A^*_{st} = A_{st} (T')^{-1} D_{\alpha}$,

where the elements of the diagonal matrix D_{α} are chosen, again, so that A^*_{st} is standardized. In the terminology of factor analysis, the elements of V^* are called "loadings"; they are the regression coefficients by which the component scores A^*_{st} must be multiplied to produce the reproduced standard scores, i.e., to predict the standardized original variables within errors of measurement or recording. Conversely, $A^*_{st} = Y_{st}(V^{*'})^{-1} = Y_{st} B^*$ defines a p by k matrix which, if applied to the standardized reproduced scores, produces standardized component scores. The matrix $\frac{1}{n} Y'_{st} A^*_{st}$ is a "*correlation* matrix" whose elements express the relation between each attribute and the k components. For a different choice of T in formula (21), a different B^* and a different correlation matrix will result. Thurstone [13] suggested that the transformation T be chosen in such a way that, in a given column of V^*, as many as possible entries are near zero ("*zero loadings*"). The reasoning is that only the few attributes having *non-zero* loadings are needed to describe the component. This simple structure property can also be sought in the correlation matrix, $\frac{1}{n} Y'_{st} A^*_{st}$. The latter representation has a particularly important significance in component analysis. If, on the basis of a given simple structure "*plane*" (column of the correlation matrix) those attributes that have non-zero entries are eliminated, and the data matrix consisting of n rows and only those attributes which appeared as near zero entries is re-analyzed by minimax component analysis, the resulting reproduced matrix should have rank at least one less than the original data matrix. This verification of simple structure was always performed and, in our analyses, produced rank (k-1) even if all small entries (not just the customary $\pm$ 0.1) were declared as zero entries. The simple structure $\underset{\sim}{B}^*$ would be of interest for economization for, if a column in $\underset{\sim}{B}^*$ has only very few appreciably high entries, only these attributes need to be observed to estimate a component score.

It should be emphasized that all these decompositions are merely different *representations* of *the same solution*; this freedom of representations is a well known phenomenon in many mathematical models (e.g., a fitted polynomial trend can be represented equally well as a power series or as a series in terms of any sequence of orthogonal polynomials) and is not peculiar to structural analyses.

*4.2. Identification**)

In the attempt to identify "planes" of the simple structure, we followed a technique proposed by Thurstone [14], with considerable simplification, since we had a very efficient (Givens-Householder) program for obtaining all or selected characteristic roots of a symmetric matrix. After a trial $\underset{\sim}{t}_o$ (systematically, each row of $\underset{\sim}{B}$ standardized to unit length) weights were attached to each attribute depending on the values of $\underset{\sim}{B}\underset{\sim}{t}_o = \underset{\sim}{v}_o$; Thurstone's suggestion was to use

$$w_i = 10(.6 - |v_{oi}|) \quad \text{if} \quad |v_{oi}| \leq .6$$

$$\text{if} \quad |v_{oi}| > .6,$$

i.e., maximum weight when v_{oi} was near zero. The improved trial vector is then obtained by weighted least squares; in fact, $\underset{\sim}{t}_1$ is the eigenvector associated with the smallest root of $\underset{\sim}{B}'\underset{\sim}{D}_w\underset{\sim}{B}$, a small (k by k) matrix. Secondly, on the basis of $\underset{\sim}{B}\,\underset{\sim}{t}_1 = \underset{\sim}{v}_1$, weights of 1 and 0 were attached, depending on whether, after another weighted least squares iteration, the resulting element in $\underset{\sim}{v}$ was expected to become a "zero loading" (weight 1) or not (weight 0). With these new weights, the eigenvector associated with the smallest root of $\underset{\sim}{B}'\underset{\sim}{D}_w\underset{\sim}{B}$ identified the plane of the simple structure. The first author who has applied this identification method on hundreds of factor analyses over the past 20 years has always succeeded in finding well defined planes. The present application was no exception.

4.3. Verification

Each simple structure plane of interest, which was identified by the method

*) For a more detailed exposition, see Baker [2].

of section 4.2., was verified by a renewed minimax component analysis. If the original data matrix restricted to only those attributes (columns) which had near zero entries in the corresponding simple structure plane produced an acceptable rank (k-1) approximation, the component was considered to be verified. For details of the decision made by the computer program see Baker [2], Chapter 5.

5. APPLICATIONS

5.1. Potato Chip Data

The Food Science Department at the University of Georgia conducted an experiment to find the determining attributes in the assessment of preferences of potato chips. After considerable selection, 48 records were available (12 judges evaluating 4 products) on a general descriptor (Acceptability) and nine detailed attributes (dry, oily, greasy, overcooked, sharp, harsh, scorched, potatoey, and persistent aftertaste). In the first analysis, the maximum error ($\pm$ 1 on a 9-point scale) occurred entirely on one record; this record was removed as an "outlier". A subset of the computer printout of raw data (Records 1 to 10 and 41 to 47) is presented in TABLE V.I.

TABLE V.I.

Record	ACCP	DRY	OILY	GRSY	OVER	SHRP	HRSH	SCOR	PTAT	LAST
1	2.00	4.00	6.00	6.00	4.00	4.00	4.00	5.00	6.00	4.00
2	5.00	6.00	2.00	2.00	7.00	4.00	4.00	7.00	4.00	4.00
3	4.00	4.00	6.00	6.00	3.00	4.00	4.00	3.00	3.00	4.00
4	2.00	5.00	6.00	6.00	3.00	4.00	4.00	4.00	4.00	4.00
5	2.00	4.00	5.00	5.00	6.00	4.00	4.00	5.00	4.00	3.00
6	4.00	6.00	3.00	3.00	6.00	4.00	4.00	6.00	4.00	6.00
7	5.00	6.00	5.00	6.00	6.00	4.00	4.00	6.00	4.00	3.00
8	3.00	3.00	5.00	5.00	4.00	4.00	4.00	4.00	4.00	5.00
9	2.00	4.00	5.00	5.00	4.00	4.00	4.00	4.00	4.00	4.00
10	5.00	4.00	6.00	5.00	6.00	4.00	4.00	6.00	4.00	5.00
41	6.00	3.00	3.00	3.00	7.00	6.00	6.00	7.00	3.00	7.00
42	2.00	5.00	4.00	4.00	6.00	5.00	5.00	5.00	3.00	5.00
43	3.00	3.00	5.00	5.00	4.00	4.00	4.00	4.00	5.00	2.00
44	4.00	5.00	3.00	3.00	4.00	4.00	4.00	4.00	5.00	4.00
45	5.00	5.00	3.00	3.00	6.00	5.00	5.00	5.00	3.00	5.00
46	2.00	3.00	4.00	3.00	4.00	4.00	4.00	4.00	6.00	3.00
47	2.00	6.00	3.00	3.00	4.00	4.00	4.00	4.00	6.00	2.00

After least squares approximation, the largest error associated with each rank was

Rank	Max Error
1	2.264
2	1.989
3	1.932
4	1.994
5	1.966
6	1.832
7	0.811

Since this was based on the scaled data, 1 was the largest permissible error; for rank 7 an acceptable least squares approximation can be obtained.

However, with q = 64, after only one Fletcher-Powell (i.e., just steepest descent) iteration, a rank 6 approximation was found with maximum error .799; a feasible rank 5 approximation was obtained, again after just one Fletcher-Powell iteration, with a maximum error of .967. To study the feasibility of a rank 4 approximation, 2 Fletcher-Powell and 20 Newton-Raphson iterations were performed, at q=128, followed by two additional Fletcher-Powell iterations at q=512; the largest error of this rank 4 representation was 0.992, with the following distribution of the 5 largest errors.

Record	Attribute	Error
6	SHRP	.992
6	LAST	.988
5	GRSY	.985
42	ACCP	.982
5	OVER	.981

The computer printout contains the largest errors for np/10 (at least 4) entries. It is thus noted that, while least squares produced a matrix, with the same informational content as the data matrix, for rank 7, the approach to minimax reduced the rank to 4. This should illustrate the fact that, for problems in numerical analysis and data analysis, least squares is not a very efficient technique. The subset of reproduced records corresponding to TABLE V.I. is contained in TABLE V.II.

5.2. *Peaches*

The Department of Food Science at the University of Georgia evaluated four different brands of peaches, on five attributes: Acceptability, color, appearance, flavor, and texture. This is an instance where further rank reduction would be unlikely. In fact, by least squares, the rank 4 reduction had a maximum error (scaled data) of 1.412; after very considerable effort (exponent q = 65,536) a

TABLE V.II.

The Reduced Rank Representation of the Data Matrix for Rank 4 is given in Matrix Form

Record	ACCP	DRY	OILY	GRSY	OVER	SHRP	HRSH	SCOR	PTAT	LAST
1	2.83	3.36	5.78	5.65	4.76	3.06	3.90	4.47	5.79	3.10
2	4.31	6.22	2.29	2.22	6.57	4.71	4.16	6.38	4.83	4.19
3	3.28	4.08	5.94	6.27	3.25	4.15	3.93	3.91	2.65	3.44
4	2.54	4.11	5.56	5.79	3.23	3.52	3.79	3.67	3.64	3.06
5	2.97	4.53	4.08	4.02	5.02	3.75	4.00	4.82	4.92	3.66
6	3.80	5.02	2.62	2.45	6.15	4.99	4.41	5.72	4.00	5.01
7	5.35	6.17	5.23	5.64	5.16	4.47	3.80	6.05	3.93	2.78
8	2.78	2.97	5.09	4.95	4.56	4.12	4.28	4.20	3.92	4.48
9	2.40	3.79	4.86	4.89	3.91	3.70	4.00	3.89	3.98	3.71
10	5.18	4.07	5.49	5.47	6.08	4.90	4.34	6.04	4.10	4.35
41	5.65	3.22	3.46	2.98	7.91	6.67	5.21	6.88	2.99	7.20
42	2.98	4.42	3.13	3.03	5.03	4.76	4.35	4.74	3.33	4.93
43	2.62	3.96	5.34	5.36	4.24	3.15	3.81	4.22	5.16	2.97
44	3.03	4.81	3.67	3.68	4.82	4.17	4.07	4.76	4.08	3.99
45	4.32	4.95	3.12	3.09	5.81	5.52	4.49	5.70	2.86	5.27
46	1.68	3.35	4.02	3.68	4.90	3.10	4.06	4.10	5.84	3.86
47	1.93	5.67	2.89	2.93	4.36	3.13	3.68	4.35	5.29	2.91

feasible near-minimax rank 4 approximation was obtained. TABLE V.III. compares these values.

The computer printout showed the distribution of maximum errors in the scaled data (½ these numbers are the errors in the raw data) as follows:

Record	Attribute	Error
13	APPR	.9996
4	APPR	.998
21	APPR	.998
6	APPR	.997
5	APPR	.997
21	COLR	.991

The near equality of these largest errors indicates that the representation is near minimax.

TABLE V.III.

Rank 4 Minimax Representation of Peach Data

	ACCP		COLR		APPR.		FLAV.		TEXT	
Record	Raw	Repr.	Raw	Repr.	Raw	Repr.	Raw	Repr.	Raw	Repr.
1	2.97	3.00	3.82	4.00	3.27	3.00	3.04	3.00	2.87	3.00
2	4.95	5.00	3.88	4.00	4.21	4.00	4.95	5.00	4.00	4.00
3	2.99	3.00	1.72	2.00	2.29	2.00	4.01	4.00	3.97	4.00
4	2.08	2.00	2.44	2.00	1.50	2.00	1.98	2.00	2.17	2.00
5	1.11	1.00	3.46	3.00	2.50	3.00	2.00	2.00	2.24	2.00
6	3.13	3.00	2.46	2.00	1.50	2.00	2.09	2.00	2.20	2.00
7	3.00	3.00	4.31	4.00	3.78	4.00	1.97	2.00	3.06	3.00
8	3.96	4.00	2.72	3.00	2.31	2.00	4.04	4.00	2.85	3.00
9	3.80	4.00	4.62	5.00	4.49	4.00	2.91	3.00	3.62	4.00
10	2.98	3.00	2.79	3.00	3.26	3.00	3.99	4.00	4.00	4.00
11	2.04	2.00	1.05	1.00	.902	1.00	2.03	2.00	3.02	3.00
12	3.96	4.00	3.84	4.00	4.24	4.00	4.95	5.00	4.03	4.00
13	4.75	5.00	3.53	4.00	3.50	3.00	3.85	4.00	3.55	4.00
14	1.02	1.00	2.76	3.00	1.07	1.00	4.06	4.00	.933	1.00
15	3.01	3.00	3.17	3.00	2.84	3.00	2.97	3.00	3.07	3.00
16	1.01	1.00	1.57	2.00	1.39	1.00	3.08	3.00	2.88	3.00
17	4.08	4.00	3.36	3.00	2.51	3.00	4.97	5.00	2.27	2.00
18	2.02	2.00	3.00	3.00	1.94	2.00	3.02	3.00	1.99	2.00
19	2.03	2.00	1.85	2.00	1.00	1.00	4.03	4.00	2.00	2.00
20	2.98	2.00	4.07	4.00	1.93	2.00	2.07	2.00	.831	1.00
21	4.15	4.00	3.50	3.00	3.50	4.00	4.49	4.00	3.49	3.00

5.3. *White Wine*

The following data were extracted from a much larger set of experiments designed to seek the determiners of white wine flavor ratings,(Wu, et al [16]). Eight attributes were selected (from 28) which had been identified as two well overdetermined simple structure planes in a *factor analysis*. The intent was to show if *component* and *factor* analysis of such data would lead to the same interpretation. Forty-eight records were selected. The least squares approximation produced a feasible solution of rank 4. A reduction to rank 2 was obtained after very considerable effort (iteration at a power of q = 262,144). Nine of the largest discrepancies between raw and reproduced data were above .995, hence, in this case, a minimax solution was required to perform the rank reduction.

The eight attributes selected in this study were tart, bitter, biting, astringent, sharp (describing one factor in the previous factor analysis) and winey, fruity and grapey. A "rotated" representation of the rank 2 solution identifies the same two *factors* as *components;* however, *winey* is not a zero-loading on the astringency component, but is appreciably negative. In the (easily obtained) 3 component reduction, winey was a component by itself.

The computer printouts go into considerable detail, including standardized reproduced scores, standardized component scores, regression coefficients to predict one from the other, and correlation matrices.

REFERENCES

[1] ARENS, B. and DAVID, H. A. (1959). Optimal spacing in regression analysis. *Ann. Math. Statist.*, 30, 1072-1081.

[2] BAKER, F. D. (1976). Minimax approach to data analysis. Unpublished Ph.D. dissertation, University of Georgia.

[3] ECKART, C. and YOUNG, G. (1936). The approximation of one matrix by another of lower rank. *Psychometrika*, 1, 211-218.

[4] FLETCHER, R. and POWELL, M. J. D. (1963). A rapid descent method for minimization. *Comput. J.*, 6, 163-168.

[5] HAMMING, R. W. (1962). *Numerical Methods for Scientists and Engineers.* McGraw-Hill, New York.

[6] HOEL, P. G. (1965). Minimax designs in two dimensional regression. *Ann. Math. Statist.*, 36, 1097-1106.

[7] HOTELLING, H. (1933). Analysis of a complex of statistical variables into principal components. *Journal of Educational Psychology*, 24, 417-441, 498-500.

[8] IZENMAN, A. J. (1975). Reduced-rank regression for the multivariate linear model. *Journal of Multivariate Analysis*, 5, 248-264.

[9] LAPLACE, P. S. (1799). *Mecanique Celeste.* Tome III, No. 39.

[10] LEGENDRE, A. M. (1805). Nouvelles methodes pour la determination des orbites des cometes. Appendix. Paris.

[11] MAEHLI, H. (1956). Progress report on rational approximation. ONR, NR 044-196. Princeton University. Exposition in Ralston and Wilf, *Mathematical Methods for Digital Computers.* (1960). Wiley & Sons, New York.

[12] PEARSON, K. (1901). On lines and planes of closest fit to systems of points in space. *Philosophical Magazine*, 2, 559-572.

[13] THURSTONE, L. L. (1935). *The Vectors of Mind.* University of Chicago Press, Chicago.

[14] THURSTONE, L. L. (1954). An analytical method for simple structure. *Psychometrika*, 19, 173-182.

[15] TSHEBYSHEV, P. L. (1859). Sur les questions de minima qui se rattachent a la representation approximative des fonctions. *Oevres* I, 273-378.

[16] WU, L. S., BARGMANN and POWERS, J. J. (1975). Analysis of wine descriptors and their meaning. Abstracts, *35th Annual Meeting Institute of Food Technologists.* Chicago, Illinois. June. (To appear in Journal of Food Science).

P.R. Krishnaiah, ed., APPLICATIONS OF STATISTICS

COHERENCE ANALYSIS OF MULTIPLE INPUT/OUTPUT DATA

J. S. BENDAT

J. S. Bendat Company
Los Angeles, California

This paper describes coherence analysis techniques for decomposing and predicting various system identification frequency domain relationships when transient or stationary random data passes through physical systems. Procedures are indicated to obtain a set of ordered conditioned records from original records which yield optimum constant parameter linear systems connecting general multiple input/output data. Applications are noted to acoustics and vibration problems.

1. MODELS FOR ARBITRARY INPUTS

Figure 1 shows a multiple input/output model for arbitrary inputs where the inputs $\{x_i(t)\}$, $i = 1,2,\ldots,q$, and the output $y(t)$ are assumed to be representative members of transient or stationary random data passing through physical systems. These systems are constant parameter linear frequency response functions denoted by $\{H_i\}$, $i = 1,2,\ldots,q$, where the dependence upon frequency f has been omitted in each $H_i = H_i(f)$ for simplicity in notation. The input records are not assumed to be mutually independent and, in fact, may be partially (not fully) coherent with each other in any allowable way. In general system identification problems, it is required to determine the optimum form of the various $\{H_i\}$ by linear least-square prediction techniques, or some other prescribed fashion, from appropriate measurement and analysis of the $(q + 1)$ records $\{x_i(t)\}$ and $y(t)$.

This problem is treated in [1] and other books where solutions are obtained by involved matrix computations or similar operations which are difficult to carry out and interpret. New solutions for this problem have been published recently in [2,3] which are simpler to conduct and interpret using partial and multiple coherence functions. The purpose of this present paper is to indicate some special features in these recent developments. For further understanding of these points, these references should be studied.

2. MULTIPLE AND PARTIAL COHERENCE FUNCTIONS

Prior to computing the forms taken by the optimum systems, the first step should be to determine whether or not the model in Figure 1 is appropriate to fit the available data. This question can be answered quantitatively by computing the multiple coherence function $\gamma^2_{y\cdot x}(f)$ between the output y and all of the given inputs $\{x_i\}$. By definition, this is given by

$$\gamma^2_{y\cdot x}(f) = \frac{S_{yy}(f) - S_{nn}(f)}{S_{yy}(f)}$$

where $S_{yy}(f)$ is the total (measured) output autospectral density function and $S_{nn}(f)$ is the (unknown noise output autospectral density function due to non-linear operations, extraneous measurement noise and any other deviations from the ideal linear model.

Observe from Eq. (1) that

$$S_{yy}(f) - S_{nn}(f) = \gamma^2_{y\cdot x}(f)\ S_{yy}(f) \tag{2}$$

and

$$S_{nn}(f) = S_{yy}(f)\ [1 - \gamma^2_{y\ x}(f)] \tag{3}$$

Eq. (2) yields the "coherent output" autospectra due to the various $\{x_i\}$ passing through the optimum $\{H_i\}$, and Eq. (3) yields the unknown noise output autospectra.

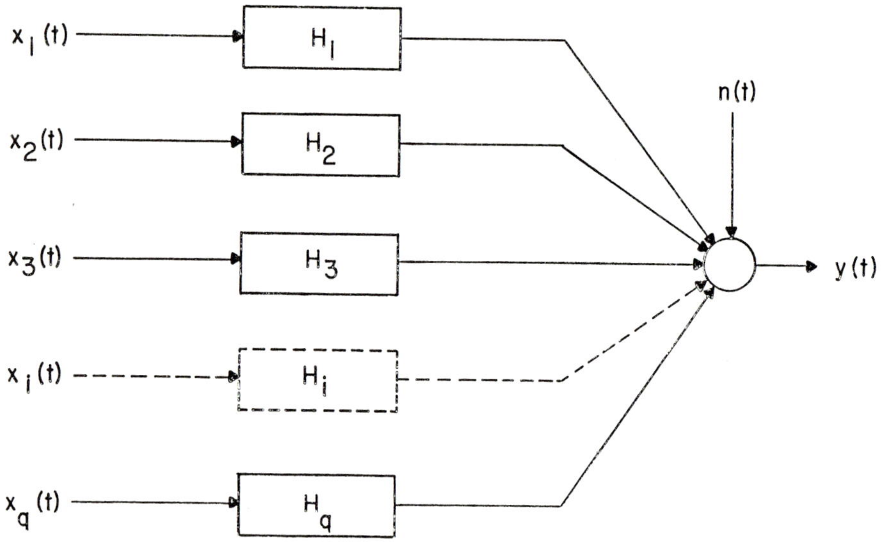

Fig. 1. Multiple input/output model for arbitrary inputs.

For all values of f, the multiple coherence function (similarly for ordinary coherence functions and partial coherence functions) will be bounded between zero and unity, namely,

$$0 \leq \gamma^2_{y \cdot x}(f) \leq 1 \tag{4}$$

As a practical guideline, the model in Figure 1 would be an appropriate model to use in physical problems if the value of $\gamma^2_{y \cdot x}(f)$ were, say, above 0.75 or 0.80 for this would then indicate that various theoretical assumptions are reasonably valid.

New general formulas to compute multiple coherence functions for any number of inputs are derived in [2]. The first three cases are shown in Figure 2. Note that the single input case gives the ordinary coherence function, while the cases of two and three inputs involve associated partial coherence functions. Definitions for these ordinary and partial coherence functions are shown in Figure 3 which can be extended to any number of records. Terms shown in the partial coherence functions are various conditioned (residual) autospectral and cross-spectral density functions which are obtained by removing linear effects of preceding records by optimum least-square prediction techniques, or the equivalent, as described in [2,3].

Single Input $\gamma^2_{y \cdot x} = 1 - \left(1 - \gamma^2_{1y}\right) = \gamma^2_{1y}$

Two Inputs $\gamma^2_{y \cdot x} = 1 - \left[\left(1 - \gamma^2_{1y}\right)\left(1 - \gamma^2_{2y \cdot 1}\right)\right]$

Three Inputs $\gamma^2_{y \cdot x} = 1 - \left[\left(1 - \gamma^2_{1y}\right)\left(1 - \gamma^2_{2y \cdot 1}\right)\left(1 - \gamma^2_{3y \cdot 12}\right)\right]$

For 3 uncoherent inputs, $\gamma^2_{y \cdot x} = \gamma^2_{1y} + \gamma^2_{2y} + \gamma^2_{3y}$

Fig. 2. Formulas for multiple coherence functions

$$\gamma^2_{1y} = \frac{|S_{1y}|^2}{S_{11}\, S_{yy}} = \gamma^2_{y1}$$

$$\gamma^2_{2y \cdot 1} = \frac{|S_{2y \cdot 1}|^2}{S_{22 \cdot 1}\, S_{yy \cdot 1}} = \gamma^2_{y2 \cdot 1}$$

$$\gamma^2_{3y \cdot 12} = \frac{|S_{3y \cdot 12}|^2}{S_{33 \cdot 12}\, S_{yy \cdot 12}} = \gamma^2_{y3 \cdot 12}$$

Fig. 3. Partial coherence functions.

3. ORDERED CONDITIONED RECORDS

The new recommended procedure for solving the general multiple input/output problem is to replace the original arbitrary input records by a set of ordered conditioned input records which are mutually uncoherent with each other. It then becomes a trivial matter to obtain the optimum systems for the ordered conditioned inputs since the problem is now essentially a set of single input/output problems where the other inputs merely result in extraneous noise at the output.

Figure 4 shows a procedure which comes from work in [2], for producing ordered conditioned records from arbitrary records. The constant parameter linear systems defined in the boxes contain terms which can be calculated from the given data by new computational algorithms derived in [2]. These results extend to any number of records.

4. MODELS FOR CONDITIONED INPUTS

Figure 5 shows a multiple input/output problem for ordered conditioned inputs $\{x_{i \cdot 12,\ldots,(i-1)}\}$, $i = 1,2,\ldots,q$. The first input is x_1; the second input is $x_{2\cdot 1}$ which is obtained by removing the linear effects of x_1 from x_2; the third input is $x_{3\cdot 12}$ which is obtained by removing the linear effects of both x_1 and x_2 from x_3, and so on to any number of input records. The optimum systems in Figure 5 have constant parameter linear frequency response functions denoted by $\{L_i\}$, $i = 1,2,\ldots,q$, where L_i is used instead of H_i to distinguish them from the different systems in Figure 1.

The optimum form of L_1, L_2 and L_3 for the three input cases are shown in Figure 6. These equations are simpler than the forms that are known and can be computed for H_1, H_2 and H_3 in Figure 1. General relations between systems $\{L_i\}$ and the corresponding $\{H_i\}$ have been derived by the author for any number of inputs. The author has also derived (but not yet published) new error analysis results for determining the statistical accuracy in estimates of $\{L_i\}$ from measured data.

5. ACOUSTICS AND VIBRATION APPLICATIONS

Applications of coherence analysis to multiple input/output data in acoustics and vibration problems have been very limited to date, except for those cases where the input noise sources are statistically independent (uncorrelated or uncoherent) with each other. This occurs when noise sources are not related and when mechanical systems are not structurally connected. For these cases, the ordinary coherence function and its associated coherent output spectra can provide many useful results as discussed in [4].

However, when noise sources are measured on a structurally connected systems

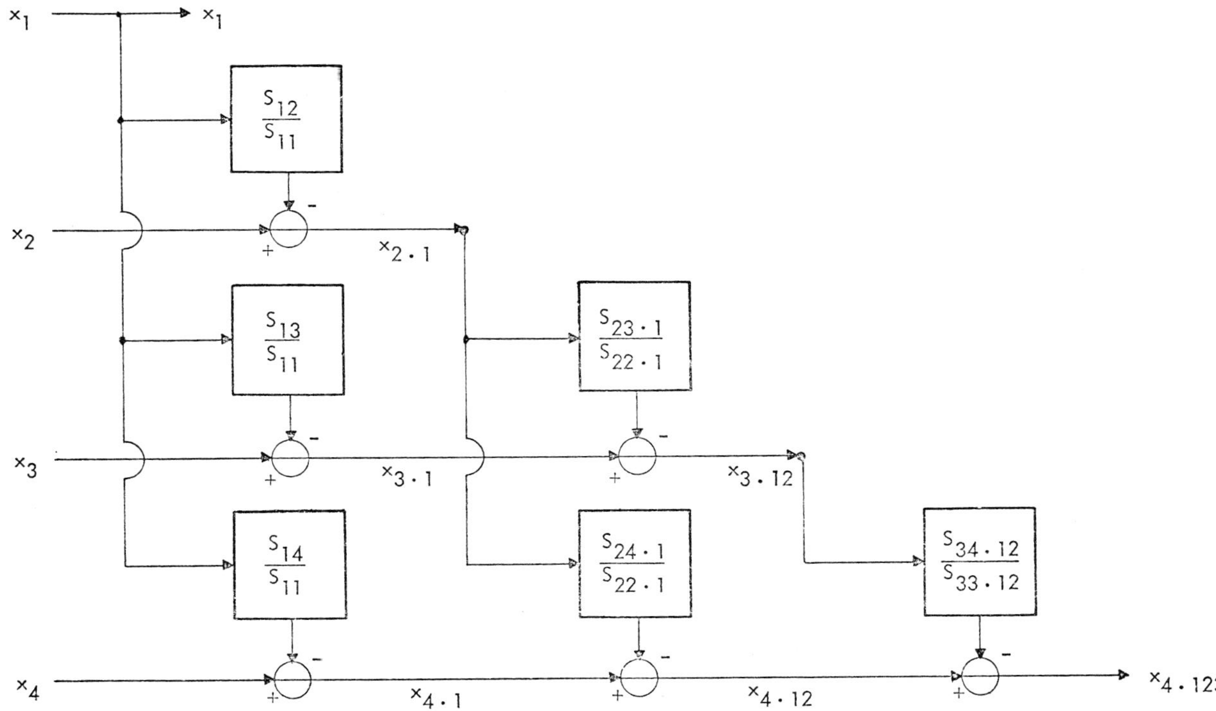

Fig. 4. Ordered conditioned records from arbitrary records. (These results extend to any number of records).

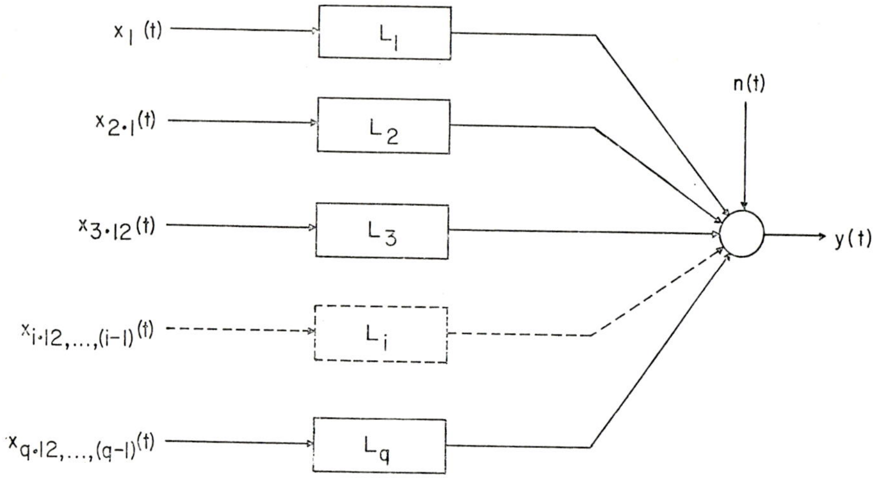

Fig. 5. Multiple input/output model for conditioned inputs.

Three Input Case

$$L_1 = \frac{S_{1y}}{S_{11}}$$

$$L_2 = \frac{S_{2y\cdot 1}}{S_{22\cdot 1}}$$

$$L_3 = \frac{S_{3y\cdot 12}}{S_{33\cdot 12}}$$

Fig. 6. Optimum frequency response functions for conditioned inputs.

such as a punch press or a diesel engine, the ordinary coherence function will not separate the various sources or distinguish between the possible transmission paths. Use of the ordinary coherence function by itself will give erroneous results and wrong interpretations. For these physical situations, correct results can be obtained by partial coherence functions and multiple coherence functions as discussed in [5,6]. Both of these advanced applications employ computational techniques of [1] which the author would now replace with the techniques described in this paper and in [2,3].

Figure 7 shows how a three input/output model can be analyzed by present recommended procedures. The original records x_1, x_2 and x_3 would first be replaced by ordered conditioned records x_1, $x_{2\cdot1}$ and $x_{3\cdot12}$ as calculated in Figure 4. Various autospectral and cross-spectral density functions would then be computed between these ordered conditioned records and the output record y. This would give the multiple coherence function from the formula in Figure 2 as well as all the terms shown in Figure 7.

The total output spectral density function for the case is given in Figure 8. This result states exactly what part of the output is due to each of the ordered conditioned inputs and it also clearly identifies the system between each of these ordered conditioned inputs and the output. In addition, a quantitative evaluation is given of possible nonlinear operations and extraneous noise effects from estimation of the noise quantity. This concludes the paper.

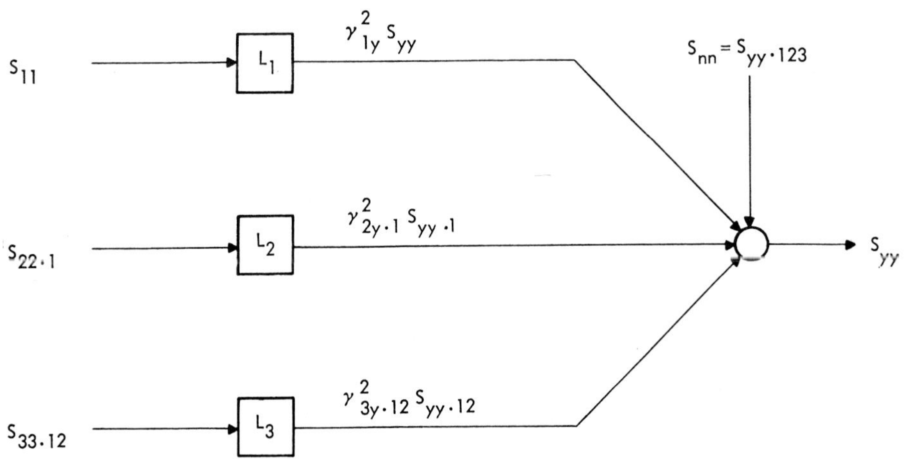

Fig. 7. Decomposition of three input/output model (Similar results follow for any number of inputs).

Three Input Case

$$S_{yy} = \gamma^2_{1y} S_{yy} + \gamma^2_{2y \cdot 1} S_{yy \cdot 1} + \gamma^2_{3y \cdot 12} S_{yy \cdot 12} + S_{yy \cdot 123}$$

$$\gamma^2_{1y} S_{yy} = |L_1|^2 S_{11}$$

$$\gamma^2_{2y \cdot 1} S_{yy \cdot 1} = |L_2|^2 S_{22 \cdot 1} \qquad S_{yy \cdot 123} = S_{nn}$$

$$\gamma^2_{3y \cdot 12} S_{yy \cdot 12} = |L_3|^2 S_{33 \cdot 12}$$

Fig. 8. Formulas for total output spectral density functions.

REFERENCES

[1] BENDAT, J. S. and PIERSOL, A. G. (1971). *RANDOM DATA: Analysis and Measurement Procedures*. John Wiley and Sons, Inc., New York.

[2] BENDAT, J. S. (1976). Solutions for the multiple input/output problem. *Journal of Sound and Vibration*, 44, 311-325.

[3] DODDS, C. J. and ROBSON, J. D. (1975). Partial coherence in multivariate random processes. *Journal of Sound and Vibration*, 42, 243-249.

[4] HALVORSEN, W. G. and BENDAT, J. S. (1975). Noise source identification using coherent output power spectra. *Sound and Vibration*, 9, 8, 15-24.

[5] KOSS, L. L. and ALFREDSON, R. J. (1974). Identification of transient sound sources on a punch press. *Journal of Sound and Vibration*, 34, 11-33.

[6] CHUNG, J. Y., CROCKER, J. J. and HAMILTON, J. F. (1975). Measurement of frequency responses and the multiple coherence function of the noise-generation system of a diesel engine. *J. Acoust. Soc. Am.*, 58, 635-642.

P.R. Krishnaiah, ed., APPLICATIONS OF STATISTICS

MULTIVARIATE RELIABILITY CLASSES*

HENRY W. BLOCK
University of Pittsburgh, Pittsburgh, Pennsylvania

Multivariate analogs of univariate failure rate functions are presented and using these, various classes of multivariate distributions with increasing failure rate are discussed and contrasted. Classes having multivariate increasing failure rate average (IFRA) are also discussed.

1. INTRODUCTION

The concepts of increasing and decreasing failure rates for univariate distributions have been found to be quite useful in reliability theory. The classes of distributions having these properties are designated the *IFR* and *DFR* distributions respectively and have been extensively studied. Other, more general classes, such as the class of distributions with increasing failure rate average *(IFRA)* or the classes of distributions which are now better than used *(NBU)* or now better than used in expectation *(NBUE)* have also been of great interest. See Barlow and Proschan [4] for a discussion of these classes.

All of these concepts apply only in the case where components are assumed to be independent. In many realistic situations, component lives are not independent and the above concepts are not applicable. In this paper, this latter situation, in which the component life times are not independent, will be discussed.

In Section 2, various multivariate exponential distributions are discussed along with the types of physical models from which they arise.

Several possible multivariate analogs of univariate failure rate functions are presented in Section 3. Using these functions, various definitions of multivariate increasing failure rate can be defined. These concepts are compared and some relations are discussed. The important concept of the hazard gradient and some of its properties are also presented.

Section 4 deals with multivariate versions of *IFRA*. These multivariate versions of *IFRA* are based on recent results of Block and Savits concerning univariate *IFRA* distributions.

Finally, in Section 5, multivariate analogs of the univariate *NBU*, *NBUE* and other classes are mentioned as well as recent results concerning estimation procedures related to the material discussed in the preceding sections.

*This research was supported by the Office of Naval Research under Contract N00014-76-C-0839.

2. MULTIVARIATE EXPONENTIAL DISTRIBUTIONS AND EXTENSIONS

Unlike the multivariate normal distributions, there is no single multivariate exponential distribution. Any distribution with exponential marginals will be said to be a multivariate exponential distribution. However all of the distributions discussed below, with the exception of the Gumbel distribution, arise from some physically motivated model. Distributional forms will be given only for distributions which are used in the succeeding sections.

2.1. *Bivariate Case*

Gumbel [18] introduced two bivariate exponential distributions. One of them has survival function

$$\bar{F}(x,y) = P\{X>x,\ Y>y\} = \exp(-x-y-\delta xy),\ x,y\geq 0,\ 0\leq\delta\leq 1. \tag{2.1}$$

This distribution does not appear to be connected with a particular model. A distribution which does arise from a variety of plausible models is the distribution of Marshall and Olkin [24]. This distribution has survival function

$$\bar{F}(x,y) = \exp(-\lambda_1 x-\lambda_2 y-\lambda_{12}\max(x,y)),\ x,y\geq 0,\ \lambda_1,\lambda_2,\lambda_{12}\geq 0 \tag{2.2}$$

and is a consequence of a fatal shock model, a nonfatal shock model and a bivariate lack of memory property. This distribution is at this time the most useful distribution in reliability theory. See Barlow and Proschan [4] for a detailed discussion of this distribution.

A closely related distribution to the Marshall and Olkin distribution, is the distribution of Block and Basu [11]. This distribution also arises from a bivariate lack of memory property. Furthermore it arises from a failure affected model which describes the situation where for a system of two components, the failure or one component tends to weaken the other component.

Downton [16] and Hawkes [20] have proposed distributions which are derived from a generalized property of the univariate exponential distribution. The distribution considered by Downton is a special case of a well-known bivariate gamma distribution. Paulson [25] has introduced a bivariate exponential distribution which arises from a compartment model.

Arnold [3] has introduced a hierarchy of exponential distributions. Block [10] and Block, Paulson and Kohberger [12] have discussed a related family of bivariate exponential distributions. The Arnold family is derived from a nonfatal shock model and the family of distributions of Block, Paulson and Kohberger is derived from a characteristic function equation. The two approaches turn out to be closely related and the families contain many of the distributions mentioned above.

2.2. Multivariate Case

Marshall and Olkin [24] have also given a multivariate analog of their bivariate exponential distribution. It arises from models similar to those in the bivariate situation. A multivariate generalization of the distribution of Block and Basu has been given by Block [7]. Furthermore a multivariate distribution which contains both of the above as special cases is also given in the same paper. Arnold's nonfatal shock model mentioned above is not restricted to the bivariate case and a multivariate family is given by this author.

3. MULTIVARIATE INCREASING FAILURE RATE

A single random component or system lifetime T with distribution function $F(t) = P\{T \leq t\}$ and $F(0) = 0$ is said to have *increasing failure rate (IFR)* if for $\bar{F}(t) = P\{T>t\}$

$$\frac{\bar{F}(x+t)}{\bar{F}(x)} \text{ decreases in } x \geq 0 \text{ for all } t \geq 0 \tag{3.1}$$

If T has a density f(x), this condition is equivalent to

$$\frac{f(x)}{\bar{F}(x)} \text{ increases in } x \geq 0. \tag{3.2}$$

In the multivariate context there are several possible ways of defining analogs of this univariate concept. To do this, multivariate versions of the quantities in (3.1) and (3.2) are required. Three such quantities are given below and in the following it will be shown how each of these can be used to define multivariate increasing failure rate concepts. Decreasing failure rate concepts can be similarly defined and will be mentioned only in certain cases.

Let $T_1, T_2, \ldots, T_n$ be the lifetimes of n components, not necessarily independent, with distribution function (d.f.) $F(\underset{\sim}{t})$ or survival function $\bar{F}(\underset{\sim}{t}) = P\{T_1>t_1, \ldots, T_n>t_n\}$, $\bar{F}(\underset{\sim}{0}) = 1$. Harris [19] introduced the quantity

$$\frac{\bar{F}(\underset{\sim}{x}+\underset{\sim}{t})}{\bar{F}(\underset{\sim}{x})} \tag{3.3}$$

which is seen to be a direct multivariate analog of the quantity in (3.1). This quantity was used by Brindley and Thompson [14] and Marshall [23] as will be described later. A second quantity

$$\frac{f(\underset{\sim}{x})}{\bar{F}(\underset{\sim}{x})} = \frac{\dfrac{\partial^n}{\partial x_1 \ldots \partial x_n} F(\underset{\sim}{x})}{\bar{F}(\underset{\sim}{x})} \tag{3.4}$$

was introduced by Basu [6] and used by Puri [26] and Puri and Rubin [27]. This also is seen to be a straighforward generalization of the quantity in (3.2). The n quantities

$$\frac{-\frac{\partial}{\partial x_i}\bar{F}(\underset{\sim}{x})}{\bar{F}(\underset{\sim}{x})} \quad \text{for } i = 1,2,\ldots, n \tag{3.5}$$

have been introduced independently by Block [8], Johnson and Kotz [21] and Marshall [22]. It will be seen that increasing failure rate concepts defined using (3.3) can be defined via the quantities in (3.5) and conversely.

The failure rate function given in (3.3) was utilized by Brindley and Thompson to define a multivariate increasing failure rate concept. This is given in the following definition

Definition 3.1. A d.f. F is said to be *MIFR* if

$$\frac{\bar{F}(\underset{\sim}{x}+t\underset{\sim}{1})}{\bar{F}(\underset{\sim}{x})} = \frac{\bar{F}(x_1+t,\ldots, x_n+t)}{\bar{F}(x_1,\ldots, x_n)} \downarrow \underset{\sim}{x}\underset{\sim}{\geq}\underset{\sim}{0} \text{ for all } t\geq 0 \tag{3.6}$$

and if any marginal distribution of F satisfies a similar condition. F is *MDFR* if ↓ is replaced by ↑ in the above.

Harris [19] also gave a definition of multivariate increasing failure rate in which he assumed (3.6) for F but not for marginal distributions and in addition assumed a type of dependence called right corner set increasing *(RCSI)*. See Barlow and Proschan [4] for a discussion of this concept and of *MIFR*.

Condition (3.6) can be interpreted as meaning that the life length of a series system formed with components of various ages decreases stochastically as the ages of the components increase.

Distributions which are *MIFR* have the following properties. The proofs of these can be found in Barlow and Proschan [4].

(a) A univariate *MIFR* distribution is *IFR*.

(b) The union of independent sets of *MIFR* distributions is *MIFR*.

(c) The marginals of *MIFR* distributions are *MIFR*.

(d) Minima of *MIFR* variables are *MIFR*.

(e) A distribution is *MIFR* and *MDFR* if and only it has the Marshall and Olkin multivariate exponential distribution.

Any reasonable definition of multivariate *IFR* should satisfy conditions (a), (b), (c), (d) and have as a boundary (in the sense of (e)) some meaningful multivariate exponential distribution. The definition of Harris [19] satisfies (a), (b), (c) and (d) but the boundary of this definition as shown by Brindley and Thompson [14] is the multivariate exponential with independent marginals, a kind of degenerate boundary. Furthermore, the appropriateness of the *RCSI* condition in Harris' definition is not easy to justify.

No useful multivariate monotonicity concepts have been defined using (3.4). See Block [8] for a discussion of this. The quantity given by (3.4) has been used in certain characterization results. Basu [6] gave the first of these.

Theorem 3.1. If

$$\frac{f(x_1,x_2)}{\bar{F}(x_1,x_2)} = c \quad \text{for all } x_1,x_2 \geq 0$$

and if the marginal distributions are exponential, then they are independent.

This result has been extended by Puri [26] to the case where only one of the marginals is assumed to be exponential. A general result of this type has also been given by Puri and Rubin [27] and is given in the theorem below.

Theorem 3.2. If

$$\frac{f(\underset{\sim}{x})}{\bar{F}(\underset{\sim}{x})} = c \quad \text{for all } \underset{\sim}{x} \underset{\sim}{\geq} \underset{\sim}{0}$$

then $f(\underset{\sim}{x})$ is a mixture of independent exponential distributions. (For the exact nature of this mixture see the original article).

The above theorem tends to show that if (3.4) is used to define a multivariate *IFR* concept, then the boundary distribution will not have a meaningful multivariate distribution since its marginals will be mixtures of exponential distributions which are *DFR*. Furthermore knowledge of (3.4) does not uniquely determine $\bar{F}(\underset{\sim}{x})$ as in the univariate case.

The n quantities introduced in (3.5) constitute a vector quantity which is given in the following definition.

Definition 3.2. Let $R(\underset{\sim}{t}) = -\log \bar{F}(\underset{\sim}{t})$ and

$$r_i(t) = \frac{\partial}{\partial t_i} R(\underset{\sim}{t}) = \frac{-\frac{\partial}{\partial t_i} \bar{F}(\underset{\sim}{t})}{\bar{F}(\underset{\sim}{t})}, \quad \underset{\sim}{t} \underset{\sim}{\geq} \underset{\sim}{0}, \; i=1,2,\ldots, n.$$

Then $\underset{\sim}{r}(\underset{\sim}{t}) = (r_1(\underset{\sim}{t}),\ldots, r_n(\underset{\sim}{t})$ is called the *hazard gradient*.

Notice that in the univariate case this reduces to the univariate failure rate function $r(x) = f(x)/\bar{F}(x)$ and recall that

$$\bar{F}(t) = \exp(-R(t)) = \exp(- \int_0^t r(x)dx)$$

so that r(x) completely determines $\bar{F}$. Keilson noticed that a similar relation holds for the hazard gradient, i.e.

$$\bar{F}(\underset{\sim}{t}) = \exp(-R(\underset{\sim}{t})) = \exp\left(-\int_{\underset{\sim}{0}}^{\underset{\sim}{t}} \underset{\sim}{r}(\underset{\sim}{x})\cdot d\underset{\sim}{x}\right) \tag{3.7}$$

where the integral is a line integral over a smooth path from $\underset{\sim}{0}$ to $\underset{\sim}{t}$ (see Marshall [22]) and so in the multivariate case the hazard gradient determines $\bar{F}$.

Block [8] used the hazard gradient to define multivariate *IFR* concepts and to show that these coincided with the concept of *MIFR* given in Definition 3.1. This is summarized in the next theorem.

Theorem 3.3. Let F have absolutely continuous univariate marginal distributions. Then F is *MIFR* if and only if

$$\underset{\sim}{r}(\underset{\sim}{x}+t\underset{\sim}{1}) \uparrow \text{ in } t \text{ for all } \underset{\sim}{x}\geq\underset{\sim}{0}$$

and the same condition holds for the hazard gradient of any marginal distribution.

It is also possible to characterize the multivariate exponential distribution of Marshall and Olkin [24] in terms of the hazard gradient. The proof of this and the previous theorem are given in Block [8].

Theorem 3.4. Let F have absolutely continuous univariate marginal distributions. The quantity r(x) is stationary (i.e. $\underset{\sim}{r}(\underset{\sim}{x}+t\underset{\sim}{1}) = \underset{\sim}{r}(\underset{\sim}{x})$ for all $t\geq 0$) if and only if F has the multivariate exponential distribution of Marshall and Olkin.

Block [8] also shows that the *RCSI* dependence condition can be expressed in terms of the hazard gradient and so the Harris multivariate *IFR* condition can be expressed in terms of r(t).

Another multivariate *IFR* condition defined via the hazard gradient was given by Johnson and Kotz [21].

Definition 3.3. A d.f. F(x) is said to have *vector multivariate IHR (DHR)* if

$$r_i(\underset{\sim}{x}) \uparrow (\downarrow) x_i \geq 0 \text{ for all } \underset{\sim}{x}\geq\underset{\sim}{0},\ i=1,2,\ldots,n.$$

This condition can be equivalently expressed in terms of the function (3.3) as follows:

$$\frac{\bar{F}(x_1,\ldots,x_i+t,\ldots,x_n)}{\bar{F}(x_1,\ldots,x_i,\ldots,x_n)} \downarrow (\uparrow) x_i \geq 0 \text{ for all } \underset{\sim}{x}\geq\underset{\sim}{0},\ i=1,2,\ldots,n. \tag{3.8}$$

The interpretation of this concept in the increasing case is that given that n components survive until time $\underset{\sim}{x}$, the probability that the ith component survives an additional t units (and the other n-1 components remain the same age) decreases in the age of the ith component. This gives that the univariate marginals are

IFR (taking $x_j = 0$ for $j \neq i$), but specifies a multivariate condition also.

As shown by Johnson and Kotz, the boundary distribution (i.e. the distribution which is both *IHR* and *DHR* simultaneously) in the bivariate case is the Gumbel distribution given in (2.1) and in the multivariate case is a multivariate form of the Gumbel distribution (see Johnson and Kotz [21] Erratum).

Other multivariate *IFR* conditions have been given by Marshall [23]. These are given in the following. All quantities and their arguments are assumed to be nonnegative. It is also assumed that the conditions hold for all marginal distributions.

(I) $$\frac{\bar{F}(\underset{\sim}{x}+\underset{\sim}{t})}{\bar{F}(\underset{\sim}{x})} \downarrow \underset{\sim}{x} \text{ for all } \underset{\sim}{t}.$$

(This condition implies that $\bar{F}(\underset{\sim}{t}) \leq \bar{F}_1(t_1)\bar{F}_2(t_2)\ldots\bar{F}_n(t_n)$ where $\bar{F}_i(t_i)$ is the ith univariate survival function).

(II) $$\frac{\bar{F}(x\underset{\sim}{1}+t\underset{\sim}{1})}{\bar{F}(x\underset{\sim}{1})} \downarrow x \text{ for all } t \ (\underset{\sim}{1} = (1,\ldots,1))$$

(i.e. the minimums are *IFR*).

(III) $$\frac{\bar{F}(\underset{\sim}{x}+t\underset{\sim}{1})}{\bar{F}(t\underset{\sim}{1})} \downarrow t \text{ for all } \underset{\sim}{x}$$

(i.e. the probability that components of the same age survive $\underset{\sim}{x}$ units longer decreases as the common ages of the components increase).

(IV) *MIFR*

(V) $\log \bar{F}(\underset{\sim}{t})$ concave

(VI) $\log \bar{F}(\underset{\sim}{t})$ concave on rays emanating from the origin (i.e. $\log \bar{F}(\alpha\underset{\sim}{t})$ concave in α).

(VII) $\log \bar{F}(\underset{\sim}{t})$ concave on diagonals from any point $\underset{\sim}{t}_0$ (i.e. $\log \bar{F}(\underset{\sim}{t}_0+\alpha\underset{\sim}{1})$ concave in α).

Marshall also gives *DFR* versions of I-VII and also versions of I-VII expressed in terms of the hazard gradient. The following two models are also given by Marshall. Let $X_1, X_2,\ldots, X_k$ be independent and be *IFR* and let $S_i \subset \{1,2,\ldots,k\}$ for $i = 1,2,\ldots, n$.

Model A: $(T_1,T_2,\ldots, T_n)$ where $T_i = \sum_{j\in S_i} X_j$, $i = 1,2,\ldots, n$

Model B: $(T_1, T_2, \ldots, T_n)$ where $T_i = \min_{j \varepsilon S_i} X_j$, $1 = 1,2,\ldots, n$.

The relationships among conditions I-VII and the models have been given in Marshall [23].

4. MULTIVARIATE INCREASING FAILURE RATE AVERAGE

A univariate distribution F(x) has *increasing failure rate average (IFRA)* if

$$\bar{F}(\alpha x) \geq \bar{F}^{\alpha}(x) \text{ for all } 0\leq\alpha\leq 1,\ x\geq 0. \tag{4.1}$$

If the distribution has density f(x) an equivalent form of (4.1) is given by

$$t^{-1}\left(\int_0^t \frac{f(x)}{\bar{F}(x)}\, dx\right) \uparrow t\geq 0. \tag{4.2}$$

The class of *IFRA* distributions is important in reliability since (1) it is the smallest class of distributions containing the exponential distributions which is closed under both formation of coherent systems and also limits in distribution (see Chapter 4, Section 1 of Barlow and Proschan [4]) and (2) it is a consequence of a natural nonfatal shock model (see Theorem 3.8, p. 94, Barlow and Proschan [4]).

A recent result obtained by Block [9] and Savits [28] is that the convolution of *IFRA* distributions is *IFRA*. Savits [28] has characterized the *IFRA* class of distributions as the distributions F(x) such that

$$\int_0^\infty h(x)dF(x) \leq \left\{\int_0^\infty h\ (x/\alpha)dF(x)\right\}^{\frac{1}{\alpha}} \tag{4.3}$$

for all $0\leq\alpha\leq 1$ and all nonnegative nondecreasing h(x). See also Block and Savits [13]. These results can be used to define multivariate versions of *IFRA*.

Several different versions of multivariate *IFRA* will now be defined. Let $F(\underset{\sim}{x})$ be the d.f. of the lifetimes $(T_1, T_2, \ldots, T_n)$. Three multivariate *IFRA* concepts are:

(I) $(X_1, X_2, \ldots, X_k)$ independent *IFRA*, $S_i \subset \{1,2,\ldots, k\}$, $i = 1,2,\ldots, n$ and $T_i = \sum_{j\varepsilon S_i} X_j$, $i = 1,2,\ldots, n$.

(II) $\int h(\underset{\sim}{x})dF(\underset{\sim}{x}) \leq \{\int h^{\alpha}(\underset{\sim}{x}/\alpha)dF(\underset{\sim}{x})\}$ for all nonnegative nondecreasing $h(\underset{\sim}{x})$ and $0\leq\alpha\leq 1$.

(III) $\bar{F}(\alpha\underset{\sim}{x}) \geq \bar{F}^{\alpha}(\underset{\sim}{x})$, $0<\alpha<1$.

Other definitions have been given by Esary and Marshall [17]. Results concerning these definitions will be discussed in a forthcoming paper.

5. OTHER MULTIVARIATE CLASSES AND ESTIMATION PROCEDURES

Multivariate analogs of other univariate reliability classes have been given by Buchanan and Singpurwalla [15]. These include multivariate versions of *NBU* (new better than used), *NBUE* (new better than used in expectation), and *DMRL* (decreasing mean residual life).

Estimates for the hazard gradient given in Definition 3.2 for the bivariate case have been proposed by Ahmad and Lin [1]. These are based on estimates of $-\frac{\partial}{\partial t_i}\bar{F}(\underset{\sim}{t})$, $i = 1, 2$ and of $\bar{F}(\underset{\sim}{t})$ due to Cacoullos. These estimates are shown to be consistent (Ahmad and Lin [1]) and jointly asymptotically normal (Ahmad and Lin [2]).

Barlow and Proschan [5] have recently attempted to analyze dependency in failure data from the transmission and the transmission pumps of caterpillar trailers using the hazard gradient.

REFERENCES

[1] AHMAD, I. A. and LIN, P.-E. (1975). Nonparametric estimation of a vector-valued bivariate failure rate I: consistency. FSU Statistics Report M363.

[2] AHMAD, I. A. and LIN, P.-E. (1976). Nonparametric estimation of a vector-valued bivariate failure rate II: asymptotic normality. FSU Statistics Report M372.

[3] ARNOLD, B. C. (1975). Multivariate exponential distributions based on hierarchial successive damage. *J. Appl. Probability*, **12**, 142-147.

[4] BARLOW, R. E. and PROSCHAN, F. (1975). *Statistical Theory of Reliability: Probability Models*. Holt, Rinehart and Winston, Inc., New York.

[5] BARLOW, R. E. and PROSCHAN, F. (1976). Techniques for analyzing multivariate failure data. Operations Research Center Report 76-11, University of California, Berkeley.

[6] BASU, A. P. (1971). Bivariate failure rate. *J. Amer. Statist. Assoc.*, **66**, 103-104.

[7] BLOCK, H. W. (1975). Continuous multivariate exponential extensions. *Reliability and Fault Tree Analysis*, (R. Barlow, J. Fussell and N. Singpurwalla, eds.). *SIAM*. 285-306.

[8] BLOCK, H. W. (1975). Monotone failure rates for multivariate distributions. *Naval Research Logistics Quarterly*. (To appear).

[9] BLOCK, H. W. (1975). Convolutions of *IFRA* distributions. Operations Research Center Report #75-22, University of California, Berkeley.

[10] BLOCK, H. W. (1976). A family of bivariate life distributions. *Theory and Applications of Reliability*, (C. Tsokos, ed.). (To appear).

[11] BLOCK, H. W. and BASU, A. P. (1974). A continuous bivariate exponential extension. *J. Amer. Statist. Assoc.*, **69**, 1031-1037.

[12] BLOCK, H. W., PAULSON, A. S. and KOHBERGER, R. D. (1976). A class of bivariate distributions. (To appear).

[13] BLOCK, H. W. and SAVITS, T. (1976). The *IFRA* closure problem. *Ann. Probability.*, **4** ,(to appear).

[14] BRINDLEY, E. C. and THOMPSON, W. A., JR. (1972). Dependence and aging aspects of multivariate survival. *J. Amer. Statist. Assoc.*, **67**, 822-830.

[15] BUCHANAN, W. B. and SINGPURWALLA, N. D. (1975). Some stochastic characterizations of multivariate survival. Serial T-319, George Washington University.

[16] DOWNTON, F. (1970). Bivariate exponential distributions in reliability theory. *J. Roy. Statist. Soc., Ser. B.*, **32**, 822-830.

[17] ESARY, J. D. and MARSHALL, A. W. (1975). Multivariate IHRA distributions. (To appear).

[18] GUMBEL, E. J. (1960). Bivariate exponential distributions. *J. Amer. Statist. Assoc.*, **56**, 335-349.

[19] HARRIS, R. (1970). A multivariate definition for increasing hazard rate distribution functions. *Ann. Math. Statist.*, **37**, 713-717.

[20] HAWKES, A. G. (1972). A bivariate exponential distribution with applications to reliability. *J. Roy. Statist. Soc. Ser. B* ,**34**, 129-131.

[21] JOHNSON, N. L. and KOTZ, S. (1975). A vector multivariate hazard rate. *Journal of Multivariate Analysis*, **5**, 53-66. Erratum, ibid. 498.

[22] MARSHALL, A. W. (1975). Some comments on the hazard gradient. *Stoch. Proc. Appl.*, **3**, 293-300.

[23] MARSHALL, A. W. (1975). Multivariate distributions with monotone hazard rate. *Reliability and Fault Tree Analysis*, (R. Barlow, J. Fussell and N. Singpurwalla, eds.). *SIAM*. 259-284.

[24] MARSHALL, A. W. and OLKIN, I. (1967). A multivariate exponetial distribution. *J. Amer. Statist. Assoc.*, **62**, 30-44.

[25] PAULSON, A. S. (1973). A characterization of the exponential distribution and a bivariate exponential distribution. *Sankhyā, Ser. A* ,**35**, 69-78.

[26] PURI, P. S. (1973). On a property of exponential and geometric distributions and its relevance to failure rate. *Sankhyā, Ser. A* ,**35**, 61-68.

[27] PURI, P. A. and RUBIN, H. (1974). On a characterization of the family of distributions with constant multivariate failure rates. *Ann. Probability* , **2**, 738-740.

[28] SAVITS, T. (1976). The *IFRA* closure problem. (To appear).

P.R. Krishnaiah, ed., APPLICATIONS OF STATISTICS

CHARACTERIZING A LATENT TRAIT DISTRIBUTION*

R. DARRELL BOCK AND WILLIAM A. THRASH
The University of Chicago, Chicago, Illinois

1. BACKGROUND

The statistical problem considered in this paper arose in connection with a behavior genetic study of spatial visualizing ability by Bock and Kolakowski (1973). Earlier investigators had found, among parents and offspring identified by sex, a pattern of familial correlations suggesting the influence of an X-linked major gene for spatial ability (Stafford, 1961; Hartlage, 1970). Bock and Kolakowski (1973) confirmed the existence of this pattern of correlation and also demonstrated the effect of the major-gene on spatial score distributions in a large sample of male and female subjects. The score distributions they reported are reproduced in Figure 1 with the superimposed Gaussian decompositions which Bock and Kolakowski fitted using Day's (1969) maximum likelihood method. For purposes of the decomposition, two distinguishable major phenotypes with respective effects μ_1 and μ_2 and population frequencies p and q=1-p were assumed; the polygenic background of major-gene effects was assumed normally distributed with constant variance σ^2.

From a genetic point of view, the main conclusion of the Bock and Kolakowski study was that two-component Gaussian decompositions fit the data in both the male and female samples (χ^2 = 11.7 and 21.2, each on 18 d.f.), whereas the fit of a single Gaussian component was unacceptable in both samples (χ^2 = 248.0 and 235.0, 20 d.f.). Moreover, the estimated proportions of high-scoring and low-scoring phenotypes (51:49, 80:20), estimated separately for the two sexes, were approximately in the ratios expected if high-scoring were due to an X-linked gene, recessive in females, with population frequency approximately 0.5.

In order to decompose the distributions in these proportions, however, it was necessary for Bock and Kolakowski to trim from the data the scores of those subjects whose frequencies are shown shaded in Figure 1. When these scores were included in the analysis, Day's procedure placed one component at the small cluster of very high-scoring subjects and the other over the remainder of the distribution.

The source of these outlying scores probably can be attributed to unreliability of the spatial test at the extremes of the range of scores. While measurement errors that displaced subjects toward the center of the distribution would have little effect on the decomposition, those that displaced scores toward the

*Supported in part by N.S.F. Grant BNS76-02849

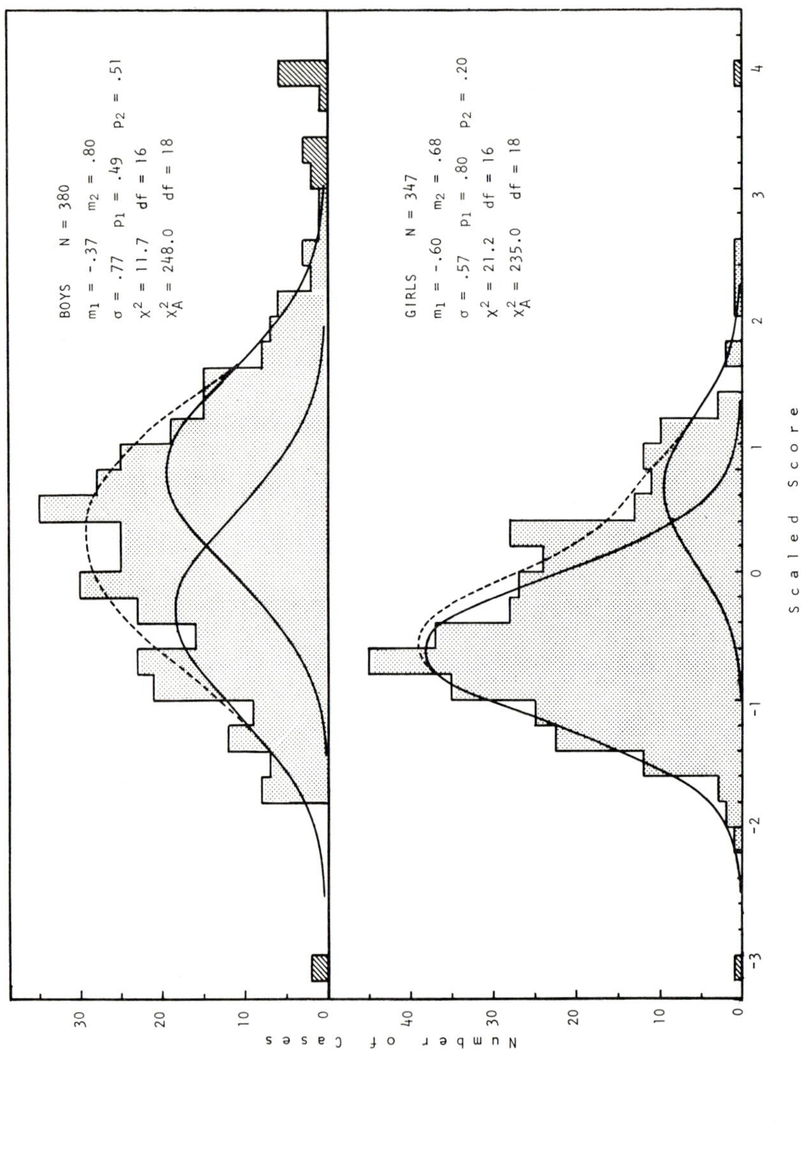
BOYS N = 380
m1 = -.37 m2 = .80
σ = .77 P1 = .49 P2 = .51
χ2 = 11.7 df = 16
χ2A = 248.0 df = 18
GIRLS N = 347
m1 = -.60 m2 = .68
σ = .57 P1 = .80 P2 = .20
χ2 = 21.2 df = 16
χ2A = 235.0 df = 18
Number of Cases
Scaled Score
-3
-2
-1
0
1
2
3
4
0
10
20
30
40

Fig. 1.

extremes could, and apparently did, generate spurious components in the decomposition. In the present paper, we attempt to handle in a more objective manner the effects of measurement error on this type of Gaussian decomposition of frequency distributions.

2. THE ERROR FUNCTION OF THE SPATIAL ABILITY TEST

In order to satisfy the metric requirements of the Gaussian decomposition, Bock and Kolakowski assumed the normal ogive model (Lord and Novick, 1968) for the item response probability as a function of the ability of the subject. Together with the assumption that item responses are independent, conditional on ability θ, this model leads to the likelihood equation (1) for estimating ability of subject i from his responses

$$\theta: \quad \sum_{j=1}^{n} a_j \left(\frac{\phi(y_{ij})}{\Phi(y_{ij})} \right)^{k_{ij}} \left(\frac{-\phi(y_{ij})}{1-\Phi(y_{ij})} \right)^{1-k_{ij}} = 0 \tag{1}$$

where $k_{ij} = \begin{cases} 1 \text{ if subject i responds correctly to item j,} \\ 0 \quad \text{otherwise,} \end{cases}$

and $\Phi(y_{ij})$ is the normal area and $\phi(y_{ij})$ the normal ordinate, corresponding to the deviate $y_{ij} = c_j + a_j\theta_i$. The c_j and a_j are item intercept and slope parameters estimated in a "calibration" sample drawn from the subject population in question (in this instance the item parameters were estimated by maximum likelihood simultaneously with the abilities). For given values of the item parameters, (1) may be solved efficiently by the Newton-Raphson method beginning from the initial value

$$\theta_i^{(1)} = \Phi^{-1} \; (\Sigma_j k_{ij}/n) \quad .$$

(Kolakowski and Bock, 1973).

According to standard results, the limiting distribution of the errors in estimating θ is normal with mean 0 and variance

$$\sigma_\varepsilon^2(\theta) = 1/ \sum_{j=1}^{n} \frac{a_j^2\phi^2 \; (y_j)}{\Phi(y_j)[1-\Phi(y_j)]} \tag{2}$$

where $y_j = c_j + a_j\theta$. The limit is, however, with respect to the number of items, which cannot be really large in a practical test. To gain some idea of the adequacy of (2) for a finite test, and to check on the bias of the maximum likelihood estimator, Kolakowski and Bock (1976), simulated responses of 750 subjects with normally distributed ability to a 38-item secondary school vocabulary test.

Although the results of this simulation showed some downward bias and some excess of scores beyond the $\pm 2\sigma$ bands, they indicated that (2) represents reasonably accurately the dependence of the measurement error on θ, especially the increasing instability of the estimates for extreme values of θ.

3. THE TEST SCORE DISTRIBUTION IN THE PRESENCE OF MAJOR-GENE EFFECTS AND MEASUREMENT ERROR

Lord (1969) has pointed out that the distribution of observed scores

$$y = \theta + \varepsilon \quad ,$$

where θ is a latent trait with probability density $f(\theta)$ and ε is a measurement error with probability density $g(\varepsilon)$, is given by the convolutional equation studied by Eddington,

$$h(y) = \int_{-\infty}^{\infty} g(y-\theta)\, f(\theta)\, d\theta \quad .$$

Eddington gave a solution $g(y-\theta)$ for normal $f(\theta)$ and known $h(y)$ (see Kurth, 1965). In the present application, we are assuming that $f(\theta)$ is the Gaussian mixture-distribution

$$f(\theta) = \frac{p}{\sqrt{2\pi}\sigma} \exp\left[-\frac{(\theta-\mu_1)^2}{2\sigma^2}\right] + \frac{q}{\sqrt{2\pi}\sigma} \exp\left[-\frac{(\theta-\mu_2)^2}{2\sigma^2}\right]$$

$$= p\phi\left[\frac{\theta-\mu_1}{\sigma}\right] + q\phi\left[\frac{\theta-\mu_2}{\sigma}\right] \quad , \qquad (4)$$

where p = the population frequency of the first phenotype (low spatial ability)

$q = (1-p)$ = the population frequency of the second phenotype (high spatial ability)

μ_1 = the effect of phenotype 1

μ_2 = the effect of phenotype 2

σ = the common standard deviation of the polygenic background.

Our problem is to estimate the four independent population parameters p, μ_1, μ_2, and σ, when the measurement error is distributed

$$g(\varepsilon) = \frac{1}{\sqrt{2\pi}\,\sigma_\varepsilon(\theta)} \exp\left[-\frac{(y-\theta)^2}{2\sigma_\varepsilon^2(\theta)}\right] = \phi\left[\frac{y-\theta}{\sigma_\varepsilon(\theta)}\right]$$

and estimated abilities are distributed as (4).

We do not attempt to find a general solution of (3) for this situation, but merely determine numerically the values of the population parameters that maximize the likelihood,

$$L = \prod_{i=1}^{N} h(y_i) \quad ,$$

or, equivalently,

$$\log L = \sum_{i=1}^{N} \log[h(y_i)] \quad . \tag{5}$$

In these computations, we employ the Gauss-Hermite quadrature

$$\int_{-\infty}^{\infty} g(x)e^{-x^2}\,dx \cong \sum_{k=1}^{s} A_k g(x_k) \quad ,$$

to approximate (3) as follows:

$$h(y) = \int_{-\infty}^{\infty} p\phi\left[\frac{y-\theta_1}{\sigma_\varepsilon(\theta_1)}\right]\phi\left[\frac{\theta_1-\mu_1}{\sigma}\right]d\theta_1 + \int_{-\infty}^{\infty} q\phi\left[\frac{y-\theta_2}{\sigma_\varepsilon(\theta_2)}\right]\phi\left[\frac{\theta_2-\mu_2}{\sigma}\right]d\theta_2$$

Substituting in the first and second integral, respectively,

$$\begin{aligned} x &= (\theta_1-\mu_1)/\sqrt{2}\sigma \\ \theta_1 &= \sqrt{2}\sigma x + \mu_1 \\ d\theta_1 &= \sqrt{2}\sigma dx \end{aligned} \qquad \text{and} \qquad \begin{aligned} x &= (\theta_2-\mu_2)/\sqrt{2}\sigma \\ \theta_2 &= \sqrt{2}\sigma x + \mu_2 \\ d\theta_2 &= \sqrt{2}\sigma dx \end{aligned}$$

we have

$$h(y_i) \cong \frac{1}{\sqrt{\pi}} \sum_{k=1}^{s} [p\phi(u_{ik}) + g\phi(v_{ik})]A_k \quad ,$$

where

$$u_{ik} = (y_i-\sqrt{2}\sigma x_k-\mu_1)/\sqrt{2}\sigma_\varepsilon(\theta_1)$$

and

$$v_{ik} = (y_i-\sqrt{2}\sigma x_k-\mu_2)/\sqrt{2}\sigma_\varepsilon(\theta_2)$$

4. MAXIMUM LIKELIHOOD ESTIMATION OF p, μ_1, μ_2, AND σ

The maximum likelihood equations for the population parameters are

$$p : \sum_{i=1}^{N} \frac{1}{h(y_i)} \sum_{k=1}^{s} [\phi(u_{ik}) - \phi(v_{ik})]A_k = 0$$

$$\mu_1 : \sum_{i=1}^{N} \frac{p}{h(y_i)} \sum_{k=1}^{s} \frac{(y_i-\sqrt{2}\sigma x_k-\mu_1)\ \phi(u_{ik})A_k}{\sqrt{2}\sigma_\varepsilon^2(\theta_1)} = 0$$

$$\mu_2 : \sum_{i=1}^{N} \frac{q}{h(y_i)} \sum_{k=1}^{s} \frac{(y_i-\sqrt{2}\sigma x_k-\mu_2)\ \phi(v_{ik})A_k}{\sqrt{2}\sigma_\varepsilon^2(\theta_2)} = 0$$

$$\sigma : \sum_{i=1}^{N} \frac{1}{h(yi)} \sum_{k=1}^{s} \left[p\frac{(y_i-\sqrt{2}\sigma x_k-\mu_1)x_k\ \phi(u_{ik})}{\sigma_\varepsilon^2\ (\theta_1)} + q\frac{(y_i-\sqrt{2}\sigma x_k-\mu_2)x_k\ \phi(v_{ik})}{\sigma_\varepsilon^2\ (\theta_2)} \right] A_k = 0\ .$$

Following the example of Day (1969), we put these equations in a form suitable for solution by relaxation:

$$p = \sum_{ik}\sum \phi(u_{ik})A_k / \sum_{ik}\sum [\phi(u_{ik}) + \phi(v_{ik})]A_k$$

(using the restriction $\sum_i h(y_i) = 1$);

$$\mu_1 = \sum_i \frac{1}{h(y_i)} \sum_k \frac{(y_i-\sqrt{2}\sigma x_k)\ \phi(u_{ik})}{\sigma_\varepsilon^2(\sqrt{2}\sigma x_k+\mu_1)} A_k / \sum_i \frac{1}{h(y_i)} \sum_k \frac{\phi(u_{ik})}{\sigma_\varepsilon^2(\sqrt{2}\sigma x_k+\mu_1)} A_k;$$

$$\mu_2 = \sum_i \frac{1}{h(y_i)} \sum_k \frac{(y_i-\sqrt{2}\sigma x_k)\ \phi(v_{ik})}{\sigma_\varepsilon^2(\sqrt{2}\sigma x_k+\mu_2)} A_k / \sum_i \frac{1}{h(y_i)} \sum_k \frac{\phi(v_{ik})}{\sigma_\varepsilon^2(\sqrt{2}\sigma x_k+\mu_2)} A_k;$$

$$\sigma = \frac{\sum_i \frac{1}{h(y_i)} \sum_k \left[p\frac{(y_i-\mu_1)u_{ik}\ \phi(u_{ik})}{\sigma_\varepsilon^2(\sqrt{2}\sigma x_k+\mu_1)} + q\frac{(y_i-\mu_2)v_{ik}\phi\ (v_{ik})}{\sigma_\varepsilon^2(\sqrt{2}\sigma x_k+\mu_2)} \right] A_k}{\sqrt{2} \sum_i \frac{1}{h(y_i)} \sum_k \left[p\frac{\phi(u_{ik})}{\sigma_\varepsilon^2(\sqrt{2}\sigma x_k+\mu_1)} + q\frac{\phi\ (v_{ik})}{\sigma_\varepsilon^2(\sqrt{2}\sigma x_k+\mu_2)} \right] X_k^2 A_k}\ .$$

These estimators are intuitively plausible. In the estimation of μ_1 and μ_2, the ϕ (u)'s and ϕ (v)'s have appreciable values only when u or v is near zero, that is, when the variable of integration is close to $(y_i-\mu_1)/\sqrt{2}\sigma$ or $(y_i-\mu_2)/\sqrt{2}\sigma$. The estimates of μ_1 and μ_2 are thus just means of the observations weighted in this special way and also weighted inversely as the theoretical error variance for each value of the variable of integration. Similarly, σ is estimated by the common regression of the manifest trait estimate on the latent variable, again in this special weighted sense. Finally, p and q are determined so that the probabilities of each observation arising from phenotype 1 or phenotype 2 add to 1.

One would expect these estimators to approach the Day estimators as the error goes to zero. This would happen, provided that the distances between adjacent quadrature points were small relative to σ_ε. This will, of course, not always be possible computationally; for a fixed number of quadrature points there will always be some smallest value for error below which this procedure would have to be abandoned in favor of Day's.

5. EXAMPLE

A large number of quadrature points must be used in this procedure to allow for reasonably small error variance; we somewhat arbitrarily chose to use 44 quadrature points ranging from -5.5 to +5.5 in our analysis of the spatial data.

Using this many quadrature points would result in a great deal of computation for a large sample, particularly since the convergence of the relaxation method is rather slow. As an economy we grouped the data as was shown in Figure 1, used the midpoints of the intervals as the y_i, and multiplied each term in the sums over i by the corresponding number of observations.

Although this procedure was slow to converge, requiring 125 relaxations for the females and over 300 for the males (starting with the respective manifest decompositions), it did successfully resolve both distributions without trimming the outliers. The results are shown in TABLE I.

TABLE I

Gaussian Decompositions of Score Distributions

Estimate	Male Sample		Female Sample	
	Manifest	Latent	Manifest	Latent
p	.49	.50	.80	.58
q	.51	.50	.20	.42
μ_1	-.37	-.01	-.60	-.78
μ_2	.80	.64	.68	.29
σ	.77	.96	.57	.42

The results for the male sample are fairly similar to those obtained using Day's procedure, although the separation between the means is smaller and the common standard deviation larger. Presumably, this is the result of including the more extreme values while weighting the central values more heavily.

The results for the female sample are quite different. The mean for phenotype 2 and the common standard deviation are appreciably reduced, and, worse, the phenotype proportions are now far from the Mendelian frequency of p^2 for an expression of an X-linked recessive gene in females.

If we examine the distribution of observed scores for females, we can understand what has happened. The latent resolution, which gives little weight to extreme values, has moved the second component down to a minor mode in the distribution at about y=.3.

This may be a reasonable result, but it does not agree with the supposition of X-linkage, nor with the results of Yen (1975), who in a large-scale study of spatial ability obtained the proportions p and p^2 in male and female subjects, respectively, using Day's method of Gaussian resolution. Nevertheless, the present study provides another demonstration, in addition to that of Lord (1969), that characteristics of a latent distribution can be inferred from manifest data in which measurement error variance is functionally dependent on the value of the latent variable.

REFERENCES

[1] BOCK, R. D. and KOLAKOWSKI, D. (1973). Further evidence of sex-linked major gene influence on human spatial visualizing ability. *American Journal of Human Genetics*, **25**, 1-14.

[2] DAY, N. E. (1969). Estimating the components of a mixture of normal distributions. *Biometrika*, **56**, 463-474.

[3] HARTLAGE, L. C. (1970). Sex-linked inheritance of spatial ability. *Perceptual Motor Skills*, **31**, 610.

[4] KOLAKOWSKI, D. and BOCK, R. D. (1976). Multivariate probit analysis. (In preparation).

[5] KURTH, R. (1965). On Eddington's solution of the convolutional integral equation. *Rend. Circ. Mat. Palermo*, **14**, 76-84.

[6] LORD, F. M. (1969). Estimating true-score distributions in psychological testing. (An empirical Bayes estimation problem.) *Psychometrika*, **34**, 259-299.

[7] LORD, F. M. and NOVICK, M. R. (1968). *Statistical Theories of Mental Test Scores.* Addison-Wesley, Reading, Massachusetts.

[8] STAFFORD, R. E. (1961). Sex differences in spatial visualization as evidence of sex-linked inheritance. *Perceptual Motor Skills*, **13**, 428.

[9] YEN, W. M. (1975). Sex-linked major gene influences on selected types of spatial performance. *Behavior Genetics*, **5**, 281-298.

P.R. Krishnaiah, ed., APPLICATIONS OF STATISTICS

APPROXIMATIONS TO THE DISTRIBUTIONS OF THE LIKELIHOOD RATIO STATISTICS FOR TESTING THE HYPOTHESES ON COVARIANCE MATRICES AND MEAN VECTORS SIMULTANEOUSLY

T. C. CHANG

University of Cincinnati, Cincinnati, Ohio

P. R. KRISHNAIAH

University of Pittsburgh, Pittsburgh, Pennsylvania

J. C. LEE

Wright State University, Dayton, Ohio

In this paper, the authors investigated approximations to distributions of the likelihood ratio statistics for testing the hypotheses on covariance matrices and mean vectors simultaneously.

1. INTRODUCTION

There are several situations where the experimenter is interested in testing the hypotheses that several real or complex multivariate normal populations are homogeneous. In these situations, it is of interest to test simultaneously for the equality of the mean vectors and covariance matrices of these populations. Similarly, it is of interest to test simultaneously that the mean vector and covariance matrix of the real or complex multivariate normal population are equal to specified values. In this paper, we approximated the distributions of the likelihood ratio test statistics for the hypotheses discussed above with suitable Pearson type distributions. The accuracy of these approximations is sufficient for practical purposes. Using these approximations, percentage points of the distributions of the likelihood ratio test statistics referred above are constructed. For a more complete set of tables, the reader is referred to a Technical Report by Chang, Krishnaiah and Lee (1975).

2. LIKELIHOOD RATIO TEST STATISTIC FOR THE HOMOGENEITY OF REAL MULTIVARIATE NORMAL POPULATIONS

Let $\underset{\sim}{X}_1,\ldots,\underset{\sim}{X}_k$ be random vectors of order px1 which are distributed independently as multivariate normal with mean vectors $\underset{\sim}{\mu}_1,\ldots,\underset{\sim}{\mu}_k$ and covariance matrices $\Sigma_1,\ldots,\Sigma_k$ respectively. Let H_1 denote the hypothesis that

$$H_1: \begin{cases} \Sigma_1 =\ldots=\Sigma_k = \Sigma \\ \underset{\sim}{\mu}_1 =\ldots=\underset{\sim}{\mu}_k = \underset{\sim}{\mu} \end{cases}$$

The likelihood ratio statistic for testing H_1 and the moments of this statistic were derived by Wilks (1932).

Now, let $\underset{\sim}{X}_{ij}$ $(j = 1, \ldots, N_i)$ denote j-th independent observation on $\underset{\sim}{X}_i$, and

$$S_i = \sum_{j=1}^{N_i} (\underset{\sim}{X}_{ij} - \overline{\underset{\sim}{X}}_{i.})(\underset{\sim}{X}_{ij} - \overline{\underset{\sim}{X}}_{i.})', \quad N = \sum_{j=1}^{k} N_j$$

$$\overline{\underset{\sim}{X}}_{i.} = \frac{1}{N_i} \sum_{j=1}^{N_i} \underset{\sim}{X}_{ij}, \quad \overline{\underset{\sim}{X}}_{..} = \frac{1}{N} \sum_i \sum_j \underset{\sim}{X}_{ij},$$

$$n_j = N_j - 1, n = \sum_{j=1}^{k} n_j, \quad S = \sum_{i=1}^{k} S_i.$$

The modified likelihood ratio statistic λ_1 and the moments of λ_1 are as given below (see Anderson (1958)):

$$\lambda_1 = \frac{n^{pn/2} \prod_{i=1}^{k} |S_i|^{n_i/2}}{\prod_{i=1}^{k} n_i^{pn_i/2} \left| S + \sum_{i=1}^{k} N_i (\overline{\underset{\sim}{X}}_{i.} - \overline{\underset{\sim}{X}}_{..})(\overline{\underset{\sim}{X}}_{i.} - \overline{\underset{\sim}{X}}_{..})' \right|^{n/2}} \tag{2.1}$$

$$E(\lambda_1^h) = \left(\frac{n^{phn/2}}{\prod_{i=1}^{k} n_i^{phn_i/2}} \right) \prod_{i=1}^{p} \left\{ \prod_{j=1}^{k} \frac{\Gamma[\frac{1}{2}(n_j + hn_j + 1 - i)]}{\Gamma[\frac{1}{2}(n_j + 1 - i)]} \right\} \frac{\Gamma[\frac{1}{2}(n + k - i)]}{\Gamma[\frac{1}{2}(n + hn + k - i)]}. \tag{2.2}$$

Using the first four moments, we approximated the distribution of $\lambda_1^{1/b}$ with Pearson Type I distribution whose density is given by

$$f(x) = \frac{1}{\beta(\alpha+1, \varepsilon+1) \ (\sigma - \sigma_o)^{\alpha+\varepsilon+1}} (x - \sigma_o)^{\alpha} (\sigma - x)^{\varepsilon}, \sigma_o \le x \le \sigma.$$

Here b is a suitably chosen integer.

The following table gives a comparison of the percentage points obtained by using Pearson type approximation with the corresponding values obtained using Box's asymptotic expression of order N^{-13} when $n_i = n_o$.

In the following table, the value of c_1 is defined by the equation

$$P[\tilde{\lambda}_1 \le c_1 | H_1] = (1 - \alpha) \tag{2.3}$$

TABLE I

Comparison of the Pearson Type Approximation with Asymptotic Expression

		p = 2			p = 3		
n_o	k	c_1	α_1	α_2	c_1	α_1	α_2
7	3	20.28	0.05	0.0501	34.75	0.05	0.0496
7	6	40.52	0.05	0.0502	71.35	0.05	0.0491
10	3	19.62	0.05	0.0500	32.60	0.05	0.0499
10	6	39.54	0.05	0.0501	67.79	0.05	0.0497
15	3	19.15	0.05	0.0500	31.20	0.05	0.0499
15	6	38.86	0.05	0.0500	65.46	0.05	0.0499
20	3	18.93	0.05	0.0500	30.56	0.05	0.0500
20	6	38.54	0.05	0.0500	64.42	0.05	0.0499

where $\tilde{\lambda}_1 = -2\log \lambda_1$. Also, α_1 is the value of α obtained by using Pearson type approximation whereas α_2 is the value of α obtained by using the asymptotic expression of order N^{-13}. TABLE I indicates that the accuracy of the Pearson type approximation is sufficient for practical purposes. Using this approximation, percentage points of the distribution of $\tilde{\lambda}_1$ are constructed and these are given in TABLE II. This table gives the values of c_1 where c_1 is defined by (2.3) for $p = 1(1)4$, $\alpha=0.05$ and $M = n_o-p=1(1)20,25,30$.

TABLE II

Upper 5% Points of $-2 \log \lambda_1$

p = 1

M	k = 2	k = 3	k = 4	k = 5	k = 6	k = 7	k = 8
1	6.96	10.39	13.42	16.26	19.00	21.66	24.26
2	6.68	10.13	13.18	16.03	18.78	21.45	24.06
3	6.52	9.99	13.04	15.91	18.66	21.34	23.95
4	6.42	9.89	12.96	15.83	18.59	21.27	23.89
5	6.36	9.83	12.90	15.78	18.54	21.23	23.85
6	6.31	9.78	12.85	15.74	18.51	21.19	23.83
7	6.27	9.75	12.82	15.71	18.48	21.17	23.80
8	6.24	9.72	12.80	15.68	18.46	21.15	23.79
9	6.21	9.69	12.78	15.67	18.45	21.14	23.78
10	6.19	9.68	12.76	15.65	18.43	21.13	23.77
11	6.18	9.66	12.75	15.64	18.42	21.12	23.76
12	6.16	9.65	12.73	15.63	18.41	21.11	23.75
13	6.15	9.64	12.72	15.62	18.40	21.10	23.75
14	6.14	9.63	12.71	15.62	18.40	21.10	23.75
15	6.13	9.62	12.71	15.61	18.39	21.09	23.74
16	6.12	9.61	12.70	15.60	18.39	21.09	23.74
17	6.12	9.60	12.70	15.59	18.38	21.09	23.73
18	6.11	9.60	12.69	15.59	18.38	21.08	23.73
19	6.10	9.59	12.68	15.59	18.37	21.08	23.73
20	6.10	9.59	12.68	15.58	18.37	21.08	23.72
25	6.08	9.57	12.66	15.57	18.36	21.07	23.72
30	6.06	9.56	12.65	15.56	18.35	21.06	23.71

p = 2

M	k = 2	k = 3	k = 4	k = 5	k = 6	k = 7	k = 8
1	15.74	24.25	32.03	39.46	46.70	53.78	60.77
2	14.19	22.26	29.63	36.70	43.56	50.29	56.91
3	13.41	21.27	28.46	35.33	42.02	48.58	55.04
4	12.94	20.67	27.75	34.53	41.12	47.58	53.95
5	12.63	20.28	27.29	34.00	40.52	46.92	53.22
6	12.41	20.00	26.96	33.62	40.10	46.46	52.71
7	12.24	19.78	26.71	33.34	39.79	46.11	52.34
8	12.12	19.62	26.51	33.12	39.54	45.84	52.04
9	12.01	19.49	26.36	32.94	39.35	45.63	51.81
10	11.93	19.38	26.24	32.81	39.19	45.46	51.62
11	11.86	19.29	26.13	32.69	39.06	45.31	51.47
12	11.79	19.22	26.04	32.58	38.95	45.19	51.34
13	11.74	19.15	25.97	32.50	38.86	45.09	51.23
14	11.69	19.09	25.90	32.43	38.77	44.99	51.13
15	11.66	19.04	25.84	32.36	38.70	44.92	51.04
16	11.63	19.00	25.79	32.31	38.64	44.85	50.97
17	11.59	18.96	25.75	32.26	38.59	44.79	50.90
18	11.56	18.93	25.72	32.21	38.54	44.73	50.85
19	11.54	18.90	25.67	32.17	38.49	44.69	50.79
20	11.52	18.87	25.65	32.14	38.45	44.64	50.74
25	11.43	18.76	25.52	32.00	38.29	44.47	50.55
30	11.37	18.69	25.44	31.90	38.19	44.36	50.43

TABLE II (cont'd)

Upper 5% Points of $-2 \log \lambda_1$

			p = 3				
M	k = 2	k = 3	k = 4	k = 5	k = 6	k = 7	k = 8
1	27.27	42.89	57.37	71.35	85.01	98.45	111.75
2	23.95	38.43	51.87	64.83	77.51	89.99	102.31
3	22.26	35.15	49.05	61.51	73.68	85.67	97.51
4	21.22	34.75	47.33	59.47	71.35	83.03	94.58
5	20.53	33.80	46.17	58.10	69.78	81.27	92.62
6	20.03	33.12	45.33	57.11	68.64	79.98	91.20
7	19.66	32.60	44.69	56.37	67.79	79.03	90.13
8	19.35	32.20	44.20	55.79	67.12	78.27	89.31
9	19.12	31.88	43.80	55.32	66.58	77.68	88.64
10	18.92	31.61	43.47	54.93	66.15	77.18	88.09
11	18.76	31.39	43.20	54.61	65.77	76.77	87.64
12	18.62	31.20	42.97	54.33	65.46	76.41	87.25
13	18.50	31.03	42.76	54.10	65.20	76.11	86.91
14	18.40	30.90	42.59	53.90	64.97	75.85	86.62
15	18.31	30.77	42.44	53.72	64.75	75.62	86.36
16	18.22	30.66	42.31	53.56	64.58	75.42	86.15
17	18.15	30.56	42.18	53.42	64.42	75.24	85.94
18	18.09	30.48	42.07	53.29	64.27	75.08	85.76
19	18.03	30.39	41.98	53.18	64.14	74.94	85.60
20	17.98	30.32	41.89	53.08	64.03	74.81	85.45
25	17.78	30.05	41.55	52.69	63.58	74.30	84.90
30	17.64	29.86	41.33	52.41	63.26	73.96	84.52

			p = 4				
M	k = 2	k = 3	k = 4	k = 5	k = 6	k = 7	k = 8
1	41.57	66.34	89.52	111.98	134.01	155.75	177.24
2	36.13	58.85	80.15	100.78	121.03	141.02	160.76
3	33.27	54.89	75.18	94.85	114.15	133.18	152.03
4	31.50	52.42	72.07	91.13	109.84	128.28	146.54
5	30.28	50.72	69.94	88.58	106.86	124.92	142.79
6	29.40	49.48	68.37	86.71	104.70	122.47	140.05
7	26.72	48.53	67.18	85.28	103.04	120.57	137.94
8	28.19	47.78	66.24	84.16	101.74	119.10	136.28
9	27.76	47.17	65.48	83.24	100.68	117.91	134.95
10	27.41	46.67	64.84	82.49	99.80	116.91	133.84
11	27.11	46.25	64.31	81.85	99.08	116.07	132.90
12	26.85	45.89	63.86	81.31	98.44	115.36	132.11
13	26.64	45.58	63.46	80.84	97.90	114.75	131.43
14	26.45	45.31	63.13	80.43	97.44	114.22	130.83
15	26.28	45.06	62.83	80.08	97.02	113.74	130.30
16	26.13	44.86	62.56	79.75	96.65	113.32	129.83
17	26.01	44.66	62.32	79.48	96.32	112.95	129.42
18	25.88	44.50	62.10	79.23	96.03	112.62	129.04
19	25.77	44.34	61.92	78.99	95.76	112.32	128.71
20	25.68	44.20	61.74	78.78	95.52	112.04	128.40
25	25.30	43.67	61.06	77.98	94.58	110.99	127.23
30	25.05	43.30	60.59	77.43	93.94	110.26	126.42

3. LIKELIHOOD RATIO STATISTIC FOR TESTING THE HYPOTHESIS THAT THE MEAN VECTOR AND COVARIANCE MATRIX ARE EQUAL TO SPECIFIED VALUES

In this section, we consider an approximation to the distribution of the likelihood ratio statistic for testing H_2 where

$$H_2: \begin{cases} \Sigma_1 = \Sigma_o \\ \underset{\sim}{\mu}_1 = \underset{\sim}{\mu}_o \end{cases}$$

and Σ_o and $\underset{\sim}{\mu}_o$ are known. The likelihood ratio statistic λ_2 for testing H_2 and the moments of λ_2 are known (see Anderson (1958)) to be

$$\lambda_2 = (e/N_1)^{pN_1/2} \mid S_1\Sigma_o^{-1} \mid^{N_1/2} \text{etr}\left[-\frac{1}{2}\Sigma_o^{-1} \{ S_1 + N_1(\bar{\underset{\sim}{X}}_{1.} - \underset{\sim}{\mu}_o)(\bar{\underset{\sim}{X}}_{1.} - \underset{\sim}{\mu}_o)' \} \right] \tag{3.1}$$

$$E\{\lambda_2^h\} = (2e/N_1)^{pN_1h/2} \frac{1}{(1+h)^{pN_1(1+h)/2}} \prod_{i=1}^{p} \frac{\Gamma[\frac{1}{2}(N_1-i+N_1h)]}{\Gamma[\frac{1}{2}(N_1-i)]} \tag{3.2}$$

Davis and Field (1971) gave tables for the percentage points of $-2\rho \log \lambda_2$ using Cornish-Fisher type inversion of an asymptotic series. Korin and Stevens (1973) made some numerical comparisons of the percentage points of the distribution of $\tilde{\lambda}_2 = -2 \log \lambda_2$ obtained by using an asymptotic expansion with the corresponding values determined by χ^2 and F approximations. The asymptotic expansions used by Davis and Field (1971) and Korin and Stevens (1973) were obtained by using the same method as used by Box (1949). Nagarsenker and Pillai (1974) expressed the distribution of $\tilde{\lambda}_2$ as a series of chi-square terms as well as a series of beta densities.

Using the first four moments, we have approximated the distribution of $\lambda_2^{1/b}$ with Pearson Type I distribution where b is a suitably chosen integer. Table III given below gives a comparison of the percentage points obtained by Pearson type approximation with the exact values. In the table $n_1 = N_1-1$.

TABLE III

Comparison of the Pearson Type Approximation with Exact Values

p = 2				p = 3				p = 6			
n_1	α	Approx.	Exact	n_1	α	Approx.	Exact	n_1	α	Approx.	Exact
6	0.05	13.69	13.69	8	0.05	20.91	20.92	12	0.05	51.70	51.70
7	0.05	13.26	13.27	9	0.05	20.38	20.38	13	0.05	50.48	50.48
10	0.05	12.55	12.55	10	0.05	19.98	19.98	14	0.05	49.50	49.50
11	0.05	12.40	12.40	15	0.05	18.85	18.85	15	0.05	48.69	48.69
13	0.05	12.19	12.19	21	0.05	18.26	18.26	21	0.05	45.79	45.79
21	0.05	11.75	11.75					25	0.05	44.75	44.75

4. LIKELIHOOD RATIO TEST STATISTIC FOR THE HOMOGENEITY OF COMPLEX MULTIVARIATE NORMAL POPULATIONS

Let $\underset{\sim}{Z}_1,\ldots,\underset{\sim}{Z}_k$ be pxl random vectors which are distributed independently as complex multivariate normal with mean vectors $\underset{\sim}{\nu}_1,\ldots,\underset{\sim}{\nu}_k$ and covariance matrices $\Omega_1,\ldots,\Omega_k$ respectively. Also, let the hypothesis H_3 be defined as follows:

$$H_3: \begin{cases} \Omega_1 = \ldots = \Omega_k \\ \underset{\sim}{\nu}_1 = \ldots = \underset{\sim}{\nu}_k \end{cases}$$

The modified likelihood ratio statistics for testing H_3 is given by

$$\lambda_3 = \frac{n^{pn} \prod_{i=1}^{k} |C_i|^{n_i}}{\prod_{i=1}^{k} n_i^{pn_i} \; |C + \sum_{i=1}^{k} N_i(\underset{\sim}{Z}_{i.} - \underset{\sim}{Z}_{..}) \overline{(\underset{\sim}{Z}_{i.} - \underset{\sim}{Z}_{..})}'|^n} \tag{4.1}$$

where

$$N = \sum_{i=1}^{k} n_i, \; n_i = N_i - 1,$$

$$N = \sum_{i=1}^{k} N_i, \; \underset{\sim}{Z}_{..} = \frac{1}{N} \sum_i \sum_j \underset{\sim}{Z}_{ij}, \; \underset{\sim}{Z}_{i.} = \frac{1}{N_i} \sum_{j=1}^{N_i} \underset{\sim}{Z}_{ij}$$

$$C_i = \sum_{j=1}^{N_i} (\underset{\sim}{Z}_{ij} - \underset{\sim}{Z}_{i.}) \overline{(\underset{\sim}{Z}_{ij} - \underset{\sim}{Z}_{i})}', \; C = \sum_{i=1}^{k} C_i$$

and Z_{ij} $(j=1,\ldots,N_i)$ denote j-th independent observation on Z_i. The moments of λ_3 are given by

$$E(\lambda_3^h) = \left(\frac{n^{phn}}{\prod_{i=1}^{k} n_i^{phn_i}} \right) \prod_{i=1}^{p} \left\{ \prod_{j=1}^{k} \frac{\Gamma(n_j+hn_j+1-i)}{\Gamma(n_j+1-i)} \right\} \frac{\Gamma(n+k-i)}{\Gamma(n+hn+k-i)} \tag{4.2}$$

The distribution of $\lambda_3^{1/b}$ is approximated with Pearson Type I distribution where b is a suitably chosen integer. Using this approximation, percentage points of the distribution of $\tilde{\lambda}_3 = -2 \log \lambda_3$ are computed for $\alpha=0.05$, $n_i = n_o$, k=2, 3, 4, 5, p = 1, 2, 3, 4 and $M = n_o - p = 1(1)20, 25, 30$. TABLE IV gives the value of c_3 where c_3 is given by

$$P[\tilde{\lambda}_3 \leq c_3 | H_3] = (1-\alpha). \tag{4.3}$$

TABLE IV

Upper 5% Points of $-2 \log \lambda_3$

	p = 1			
M	k = 2	k = 3	k = 4	k = 5
1	8.12	12.47	16.36	20.06
2	8.03	12.49	16.50	20.30
3	7.97	12.51	16.59	20.45
4	7.94	12.52	16.64	20.55
5	7.92	12.53	16.68	20.62
6	7.91	12.53	16.71	20.67
7	7.89	12.54	16.73	20.71
8	7.89	12.55	16.75	20.74
9	7.88	12.55	16.77	20.77
10	7.87	12.56	16.78	20.79
11	7.87	12.56	16.79	20.81
12	7.86	12.56	16.80	20.82
13	7.86	12.56	16.81	20.84
14	7.86	12.56	16.81	20.85
15	7.85	12.57	16.82	20.86
16	7.85	12.56	16.83	20.87
17	7.85	12.57	16.83	20.88
18	7.85	12.57	16.84	20.89
19	7.85	12.57	16.84	20.89
20	7.85	12.57	16.85	20.90
25	7.84	12.57	16.86	20.92
30	7.84	12.58	16.87	20.94

TABLE IV (cont'd)

Upper 5% Points of $-2 \log \lambda_3$

		p = 2		
M	k = 2	k = 3	k = 4	k = 5
1	19.87	31.66	42.59	53.10
2	18.41	29.83	40.42	50.62
3	17.68	28.92	39.37	49.43
4	17.25	28.39	38.75	48.74
5	16.96	28.03	38.35	48.29
6	16.75	27.78	38.06	47.97
7	16.60	27.60	37.85	47.74
8	16.47	27.45	37.69	47.56
9	16.38	27.33	37.55	47.41
10	16.30	27.24	37.44	47.30
11	16.23	27.16	37.36	47.21
12	16.18	27.09	37.28	47.12
13	16.13	27.04	37.22	47.05
14	16.09	26.99	37.16	47.00
15	16.05	26.94	37.12	46.94
16	16.02	26.90	37.07	46.90
17	15.99	26.87	37.04	46.86
18	15.97	26.84	37.01	46.82
19	15.94	26.81	36.97	46.79
20	15.92	26.79	36.95	46.76
25	15.84	26.69	36.85	46.65
30	15.79	26.63	36.78	46.57

		p = 3		
M	k = 2	k = 3	k = 4	k = 5
1	36.46	59.39	80.89	101.75
2	32.90	54.56	74.91	94.65
3	31.04	52.04	71.80	90.97
4	29.90	50.49	69.88	88.70
5	29.13	49.42	68.57	87.16
6	28.56	48.65	67.62	86.05
7	28.13	48.06	66.90	85.21
8	27.79	47.61	66.34	84.55
9	27.52	47.24	65.89	84.02
10	27.30	46.94	65.52	83.57
11	27.11	46.68	65.19	83.21
12	26.96	46.46	64.94	82.90
13	26.82	46.27	64.71	82.63
14	26.70	46.11	64.50	82.40
15	26.60	45.97	64.33	82.20
16	26.51	45.84	64.17	82.01
17	26.42	45.73	64.04	81.85
18	26.35	45.63	63.92	81.71
19	26.28	45.54	63.80	81.58
20	26.22	45.45	63.70	81.46
25	25.99	45.14	63.31	81.01
30	25.83	44.92	63.05	80.70

TABLE IV (cont'd)

Upper 5% Points of $-2 \log \lambda_3$

	p = 4			
M	k = 2	k = 3	k = 4	k = 5
1	58.06	95.91	131.68	166.49
2	51.79	87.16	120.63	153.23
3	48.40	82.37	114.58	145.98
4	46.26	79.35	110.73	141.37
5	44.77	77.23	108.05	138.13
6	43.68	75.67	106.08	135.78
7	42.83	74.46	104.57	133.95
8	42.17	73.51	103.36	132.51
9	41.63	72.74	102.39	131.33
10	41.19	72.10	101.57	130.37
11	40.81	71.56	100.89	129.54
12	40.50	71.09	100.29	128.84
13	40.22	70.70	99.80	128.22
14	39.97	70.35	99.36	127.70
15	39.76	70.03	98.97	127.23
16	39.57	69.77	98.62	126.83
17	39.41	69.52	98.31	126.45
18	39.25	69.30	98.04	126.13
19	39.12	69.10	97.79	125.82
20	38.99	68.92	97.55	125.54
25	38.51	68.22	96.67	124.50
30	38.18	67.74	96.07	123.76

To check for the accuracy of the entries in TABLE IV, we compared some of these values with the corresponding values obtained by using Box's asymptotic series (1949) of order n^{-13}. These comparisons are given in TABLE V below:

TABLE V

Comparison of the Pearson Type Approximation with Asymptotic Expansion

		p = 2			p = 3		
n_o	k	c_3	α_1	α_2	c_3	α_1	α_2
7	3	28.03	0.05	0.0499	50.49	0.05	0.0497
10	3	27.45	0.05	0.0500	48.06	0.05	0.0498
15	3	27.04	0.05	0.0500	46.46	0.05	0.0499
20	3	26.84	0.05	0.0500	45.73	0.05	0.0500

In the above table, α_1 is the value of α obtained by using Pearson type approximation whereas α_2 is the value of α obtained by using Box's asymptotic expansion of order n^{-13}. This table indicates that the accuracy of Pearson type approximation is sufficient for practical purposes.

5. LIKELIHOOD RATIO STATISTIC FOR TESTING THE HYPOTHESIS THAT THE COVARIANCE MATRIX AND MEAN VECTOR OF COMPLEX MULTIVARIATE NORMAL ARE EQUAL TO SPECIFIED VALUES

In this section, we consider the distribution of the likelihood ratio statistic for testing the hypothesis H_4

where

$$H_4: \begin{cases} \Omega_1 = \Omega_o \\ \underset{\sim}{\nu}_1 = \underset{\sim}{\nu}_o \end{cases}$$

and Ω_o, $\underset{\sim}{\nu}_o$ are known.

The likelihood ratio statistic for testing H_4 is given by

$$\lambda_4 = (e/N_1)^{pN_1} \left| C_1 \Sigma_o^{-1} \right|^{N_1} \operatorname{etr}[-\Sigma_o^{-1} \{C_1 + N_1(\underset{\sim}{Z}_{1.} - \underset{\sim}{\nu}_o)\overline{(\underset{\sim}{Z}_{1.} - \underset{\sim}{\nu}_o)}'\}]$$

The moments of λ_4 are given by

$$E(\lambda_4^h) = \left(\frac{e}{N_1}\right)^{phN_1} \frac{1}{(1+h)^{pN_1(1+h)}} \prod_{i=1}^{p} \frac{\Gamma(N_1 - i + N_1 h)}{\Gamma(N_1 - i)} \qquad (5.1)$$

Using the first four moments, we approximated the distribution of $\lambda_4^{1/b}$ with Pearson Type I distribution where b is a suitably chosen integer. This approximation is used to compute the values of c_4 for $\alpha=0.05$, $M=n_1-p-1 = 1(1)20(2)30$, n_1-N_1 1 and $p=2,3,4,5,6$

where

$$P[\tilde{\lambda}_4 \leq c_4 | H_4] = (1-\alpha) \qquad (5.2)$$

and $\tilde{\lambda}_4 = -2 \log \lambda_4$. The values are given in TABLE VI.

TABLE VI

Upper 5% Points of $-2 \log \lambda_4$

M	p = 2	p = 3	p = 4	p = 5	p = 6
1	20.33	34.19	51.51	72.40	96.91
2	19.20	32.18	48.40	68.00	91.05
3	18.51	30.91	46.39	65.10	87.12
4	18.03	30.02	44.97	63.02	84.25
5	17.69	29.37	43.91	61.45	82.09
6	17.43	28.87	43.09	60.22	80.36
7	17.22	28.47	42.43	59.23	78.96
8	17.06	28.15	41.89	58.41	77.80
9	16.92	27.88	41.44	57.73	76.84
10	16.81	27.65	41.06	57.15	76.00
11	16.71	27.46	40.73	56.64	75.29
12	16.63	27.29	40.45	56.20	74.64
13	16.55	27.15	40.20	55.82	74.09
14	16.49	27.02	39.98	55.48	73.61
15	16.44	26.91	39.79	55.17	73.15
16	16.38	26.81	39.61	54.90	72.76
17	16.34	26.72	39.45	54.65	72.40
18	16.30	26.63	39.31	54.43	72.08
19	16.26	26.56	39.18	54.22	71.78
20	16.23	26.49	39.06	54.04	71.51
22	16.17	26.37	38.85	53.71	71.02
24	16.12	26.27	38.67	53.43	70.60
26	16.07	26.18	38.52	53.18	70.25
28	16.04	26.10	38.39	52.98	69.93
30	16.01	26.03	38.27	52.79	69.65

REFERENCES

[1] ANDERSON, T. W. (1958). *Introduction to Multivariate Analysis*. John Wiley & Sons, Inc., New York.

[2] BOX, G. E. P. (1949). A general distribution theory for a class of likelihood criteria. *Biometrika*, **36**, 317-346.

[3] CHANG, T. C., KRISHNAIAH, P. R. and LEE, J. C. (1975). Approximations to the distributions of the likelihood ratio statistics for testing the hypotheses on covariance matrices and mean vectors simultaneously. ARL TR 75-0176.

[4] DAVIS, A. W. and FIELD, J. B. F. (1971). Tables of some multivariate test criteria. Tech. Rept. #32, Division of Mathematical Statistics, CSIRO, Australia.

[5] KORIN, B. P. and STEVENS, E. H. (1973). Some approximations for the distribution of a multivariate likelihood ratio criterion. *J. Roy. Statist. Soc. Ser. B* , 24-27.

[6] NAGARSENKER, B. N. and PILLAI, K. C. S. (1974). Distribution of the likelihood ratio criterion for testing $\Sigma=\Sigma_o, \mu=\mu_o$. *Journal of Multivariate Analysis*, **4**, 114-122.

[7] WILKS, S. S. (1932). Certain generalizations in the analysis of variance. *Biometrika*, **24**, 471-494.

P.R. Krishnaiah, ed., APPLICATIONS OF STATISTICS

AN ALTERNATING RENEWAL PROCESS WITH AN ABSORBING STATE[1]

C. L. CHIANG AND J. P. HSU
University of California, Berkeley, California

1. INTRODUCTION

Fix and Neyman (1951), in their analysis of survival experience of cancer patients, proposed a stochastic model for the description of recovery from and relapse of the disease [1]. The model consists of two transient states: S_1, healthy state and S_2, illness state; and two absorbing states, R_1 and R_2, corresponding to causes of death. A similar model has been described by Du Pasquier [2]. While the idea was originated for the study of health in a human population, the model has applications in many practical fields. The following are a few examples:

	S_1	S_2
a person	is employed	is unemployed
an automobile	is working	is being serviced
a telephone line	is free	is engaged
a woman	is on the pill	is not on the pill
an IUD	is in place	is not in place
a parking space	is vacant	is occupied

In some of the examples, we may also consider an absorbing state. The model may be schematically depicted as in the following diagram, where the arrows indicate the direction of transitions that may take place.

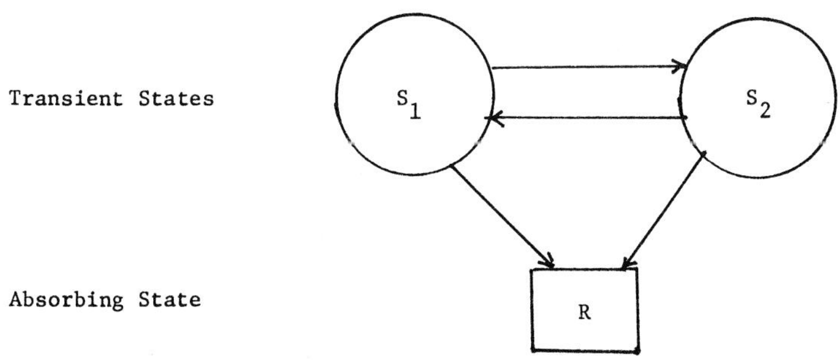

[1]The research reported herein was performed pursuant to a grant (003-7501-P2021) from the Department of Health, Education and Welfare, Washington, D. C. The opinions and conclusions expressed herein are solely those of the authors, and should not be construed as representing the opinion or policy of any agency of the United States government.

The model may be described as an alternating renewal process. Starting from S_1, a sequence of transitions $S_1 \to S_2 \to S_1$ constitutes a renewal. Thus a renewal from S_1 to S_1 can occur only by way of S_2. Similarly, starting from S_2, the sequence $S_2 \to S_1 \to S_2$ is also a renewal. A process terminates as soon as the system enters the absorbing state R. The difference between the present model and the ordinary renewal process lies in the number of states. In the ordinary renewal process, where there is only one state, after the system leaves the state it enters the same state immediately and the process starts all over again. In our alternating renewal process, once the process leaves a state it must enter the other state or the absorbing state and only a change of states is considered a transition. Actually, this model is a special case of the general Markov renewal process where there is a countable number of states with the successive states in a sequence of transitions forming a Markov chain and the duration of stay in each state is a random variable. A comprehensive description of the general model and some theoretical findings are given in Pyke [3] and [4]. Relevant material may be found also in Chung [5] and Takács [6].

Given a system in state S_α at time 0, let $M_{\alpha\alpha}(0,t)$ be the number of renewals of type $S_\alpha \to S_\beta \to S_\alpha$, for $\alpha \neq \beta$; $\alpha,\beta = 1,2$, taking place in the interval $(0,t)$ with ending state S_α at time t, are

$$P_{\alpha\alpha}^{(m)}(0,t) = \Pr\{M_{\alpha\alpha}(0,t) = m\}, \quad \alpha = 1,2;\ m = 0,1,\ldots \tag{1.1}$$

In addition to renewals, we may consider also sequences of transitions where the ending state is different from the initial state, such as $S_1 \to S_2$ or $S_1 \to S_2 \to S_1 \to S_2$, etc. These are called the "first passage," the "second passage," etc. The number of passages occurring in a time interval $(0,t)$ will be denoted by $M_{\alpha\beta}(0,t)$ and the passage probabilities by $P_{\alpha\beta}^{(m)}(0,t)$, so that

$$P_{\alpha\beta}^{(m)}(0,t) = \Pr\{M_{\alpha\beta}(0,t) = m\}, \quad \alpha \neq \beta;\ \alpha,\beta=1,2;\ m=1,\ldots. \tag{1.2}$$

The purpose of this paper is to present explicit formulas for the renewal probabilities in (1.1) and the passage probabilities in (1.2) and to discuss related problems.

A distinction should be made between an exit transition and an entrance transition; they are conceptually different and generally the numbers of the two types of transitions are not equal in a given sequence. In the case of a two state model, however, as soon as a system leaves one state, it enters the other; therefore the number of exit transitions and the number of entrance transitions are equal. In the sequence $S_1 \to S_2$, there is one exit from S_1 and one entrance into S_2; while in the sequence $S_1 \to S_2 \to S_1 \to S_2$, there are two exits from S_1 and two entrances into S_2, or, one exit from S_2 and one entrance into S_1.

Starting from S_α at time 0, a sequence of transitions may lead to the absorbing state R. Let $Q_\alpha^{(m)}(0,t)$ be the probability of leaving S_α m times and entering state R prior to time t. This probability may be derived from the renewal and passage probabilities.

The sums

$$\sum_{m=0}^{\infty} P_{\alpha\alpha}^{(m)}(0,t) = P_{\alpha\alpha}(0,t) \tag{1.3}$$

$$\sum_{m=1}^{\infty} P_{\alpha\beta}^{(m)}(0,t) = P_{\alpha\beta}(0,t) \tag{1.4}$$

and

$$\sum_{m=1}^{\infty} Q_{\alpha}^{(m)}(0,t) = Q_{\alpha}(0,t) \tag{1.5}$$

are the probabilities that at time t the system will be in S_α, S_β, or R, respectively, and satisfy the relationship

$$P_{\alpha\alpha}(0,t) + P_{\alpha\beta}(0,t) + Q_\alpha(0,t) = 1 \quad . \tag{1.6}$$

To derive the formulas for these transition probabilities, we assume the existence of the following derivatives:

$$\frac{\partial}{\partial t} P_{\alpha\beta}(\tau,t)\Big|_{t=\tau} = \nu_{\alpha\beta}, \qquad \alpha,\beta = 1,2, \tag{1.7}$$

$$\frac{\partial}{\partial t} Q_{\alpha}(\tau,t)\Big|_{t=\tau} = \mu_{\alpha}, \qquad \alpha = 1,2, \tag{1.8}$$

so that

$$\nu_{\alpha\alpha} = -[\nu_{\alpha\beta} + \mu_\alpha] , \qquad \alpha\neq\beta \ ; \ \alpha,\beta = 1,2 \quad . \tag{1.9}$$

The assumption that $\nu_{\alpha\beta}$ and μ_α are independent of time is made for the convenience of deriving explicit solutions. This assumption is unnecessary for some theoretical considerations of the process. Formulas for the probabilities on the right hand side of (1.3), (1.4) and (1.5) have been presented in [7]. They are

$$P_{\alpha\alpha}(0,t) = \sum_{i=1}^{2} \frac{\rho_i-\nu_{\beta\beta}}{\rho_i-\rho_j} e^{\rho_i t} \tag{1.10}$$

$$P_{\alpha\beta}(0,t) = \sum_{i=1}^{2} \frac{\nu_{\alpha\beta}}{\rho_i-\rho_j} e^{\rho_i t} \tag{1.11}$$

where

$$\rho_1, \rho_2 = \frac{1}{2}\left[\nu_{11} + \nu_{22} \pm \sqrt{(\nu_{11}-\nu_{22})^2 + 4\nu_{12}\nu_{21}}\right] \tag{1.12}$$

are both negative and

$$Q_\alpha(0,t) = \sum_{i=1}^{2} \frac{\mu_\alpha}{\rho_i(\rho_i-\rho_j)} [e^{\rho_i t} - 1][\nu_{\alpha\beta} + (\rho_i - \nu_{\beta\beta})] ,$$

$$\alpha \neq \beta;\ \alpha, \beta = 1,2;\ j \neq i;\ j = 1,2 . \tag{1.13}$$

2. IDENTITIES AND EXACT FORMULAS

A number of methods are available for deriving formulas for the renewal probabilities and the passage probabilities. The methods of probability generating functions, differential equations, and convolutions are all applicable. We shall, however, derive the solution by way of introducing certain recurrent identities.

For the renewal probability $P_{\alpha\alpha}^{(m)}(0,t)$, let us consider the $(\ell+1)$th exit transition from S_α, for a fixed integer ℓ between 0 and m. The probability that the $(\ell+1)$th transition from S_α to S_β occurs at a particular time τ (or, more precisely, during the infinitesimal time interval $(\tau, \tau+d\tau)$) and the remaining $m-\ell$ transitions occur during (τ,t) is

$$P_{\alpha\alpha}^{(\ell)}(0,\tau)[\nu_{\alpha\beta} d\tau]\, P_{\beta\alpha}^{(m-\ell)}(\tau,t) \quad . \tag{2.1}$$

Since the events corresponding to (2.1) for different τ are mutually exclusive, we integrate (2.1) over the interval (0.t) to obtain the identity

$$P_{\alpha\alpha}^{(m)}(0,t) = \int_0^t P_{\alpha\alpha}^{(\ell)}(0,\tau)\nu_{\alpha\beta}\, P_{\beta\alpha}^{(m-\ell)}(\tau,t)\, d\tau \quad , \quad \ell = 0,\ldots,m-1 . \tag{2.2}$$

Another identity is obtained if we consider the first transition from S_β to S_α after ℓ exits from S_α have been made. This consideration leads to the identity

$$P_{\alpha\alpha}^{(m)}(0,t) = \int_0^t P_{\alpha\beta}^{(\ell)}(0,\tau)\nu_{\beta\alpha}\, P_{\alpha\alpha}^{(m-\ell)}(\tau,t) d\tau \quad , \quad \ell = 1,\ldots,m. \tag{2.3}$$

Using similar reasoning we have two identities for the passage probability:

$$P_{\alpha\beta}^{(m)}(0,t) = \int_0^t P_{\alpha\alpha}^{(\ell)}(0,\tau)\nu_{\alpha\beta}\, P_{\beta\beta}^{(m-\ell-1)}(\tau,t) d\tau \ , \quad \ell = 0,\ldots,m-1, \tag{2.4}$$

and

$$P_{\alpha\beta}^{(m)}(0,t) = \int_0^t P_{\alpha\beta}^{(\ell)}(0,\tau)\nu_{\beta\alpha} \quad P_{\alpha\beta}^{(m-\ell)}(\tau,t)d\tau \ , \quad \ell = 1,\ldots,m-1 \ . \tag{2.5}$$

Identities (2.2) through (2.5) provide a convenient way of deriving the formulas for the renewal probabilities and passage probabilities and have been given in [7]. These identities are also applicable when the process is non-homogeneous.

It is easy to show that the probability of a system remaining in the initial state S_α over an interval (0,t) is given by

$$P_{\alpha\alpha}^{(0)}(0,t) = e^{\nu_{\alpha\alpha}t} \ . \tag{2.6}$$

Introducing (2.6) in (2.4) for m=1 and ℓ=0, we have the first passage probability

$$P_{\alpha\beta}^{(1)}(0,t) = \frac{\nu_{\alpha\beta}}{\nu_{\alpha\alpha}-\nu_{\beta\beta}} \left[e^{\nu_{\alpha\alpha}t} - e^{\nu_{\beta\beta}t}\right] \ . \tag{2.7}$$

Substituting (2.7) in (2.3) and integrating the resulting expression yield the first renewal probability

$$P_{\alpha\alpha}^{(1)}(0,t) = \frac{\nu_{\alpha\beta}\nu_{\beta\alpha}}{\nu_{\alpha\alpha}\ \nu_{\beta\beta}} \left[t\ e^{\nu_{\alpha\alpha}t} - \frac{1}{\nu_{\alpha\alpha}-\nu_{\beta\beta}}(e^{\nu_{\alpha\alpha}t} - e^{\nu_{\beta\beta}t})\right] \ . \tag{2.8}$$

Using identity (2.4) again for m=2 yields the second passage probability

$$P_{\alpha\beta}^{(2)}(0,t) = \nu_{\alpha\beta}^2\nu_{\beta\alpha}\left[\frac{-2}{\left(\nu_{\alpha\alpha}-\nu_{\beta\beta}\right)^3}(e^{\nu_{\alpha\alpha}t} - e^{\nu_{\beta\beta}t}) + \frac{t}{(\nu_{\alpha\alpha}-\nu_{\beta\beta})^2}(e^{\nu_{\alpha\alpha}t} + e^{\nu_{\beta\beta}t})\right] \ . \tag{2.9}$$

and using (2.3) yields the second renewal probability

$$P_{\alpha\alpha}^{(2)}(0,t) = \frac{(\nu_{\alpha\beta}\nu_{\beta\alpha})^2}{\left(\nu_{\alpha\alpha}-\nu_{\beta\beta}\right)^4}[3(e^{\nu_{\alpha\alpha}t} - e^{\nu_{\beta\beta}t}) - (\nu_{\alpha\alpha}-\nu_{\beta\beta})t(2e^{\nu_{\alpha\alpha}t} + e^{\nu_{\beta\beta}t}) + \frac{1}{2}(\nu_{\alpha\alpha}-\nu_{\beta\beta})^2t^2e^{\nu_{\alpha\alpha}t}] \tag{2.10}$$

Repeated applications of formulas (2.3) and (2.4) give the general formulas for the m-th renewal probability

$$P_{\alpha\alpha}^{(m)}(0,t) = \frac{(-\nu_{\alpha\beta}\nu_{\beta\alpha})^m}{(\nu_{\alpha\alpha}-\nu_{\beta\beta})^{2m}} \left[e^{\nu_{\alpha\alpha}t} \sum_{k=0}^{m} \binom{2m-1-k}{m-1} \frac{[-t(\nu_{\alpha\alpha}-\nu_{\beta\beta})]^k}{k!} - e^{\nu_{\beta\beta}t} \sum_{k=0}^{m-1} \binom{2m-1-k}{m} \frac{[t(\nu_{\alpha\alpha}-\nu_{\beta\beta})]^k}{k!} \right]$$

$$\text{for } \alpha\neq\beta;\ \alpha,\beta = 1,2\ ;\quad m = 1,\ldots \qquad (2.11)$$

and the m-th passage probability

$$P_{\alpha\beta}^{(m)}(0,t) = \frac{\nu_{\alpha\beta}(-\nu_{\alpha\beta}\nu_{\beta\alpha})^{m-1}}{(\nu_{\alpha\alpha}-\nu_{\beta\beta})^{2m-1}} \sum_{k=0}^{m-1} \binom{2m-2-k}{m-1} \frac{[t(\nu_{\alpha\alpha}-\nu_{\beta\beta})]^k}{k!} [(-1)^k e^{\nu_{\alpha\alpha}t} - e^{\nu_{\beta\beta}t}]$$

$$\text{for } \alpha\neq\beta\ ;\ m=1,\ldots \qquad (2.12)$$

The formula of the probability $Q_\alpha^{(m)}(0,t)$ consists of two terms depending on the state (S_α or S_β) from which the system enters the absorbing state R. Since m is the number of exits from S_α, we have

$$Q_\alpha^{(m)}(0,t) = \int_0^t [P_{\alpha\alpha}^{(m-1)}(0,\tau)\mu_\alpha + P_{\alpha\beta}^{(m)}(0,\tau)\mu_\beta]d\tau$$

$$= Q_{\alpha\cdot\alpha}^{(m)}(0,t) + Q_{\alpha\cdot\beta}^{(m)}(0,t) \qquad \alpha\neq\beta;\ \alpha,\beta = 1,2;\ m = 1,\ldots \qquad (2.13)$$

Here the first term of the integrand is the probability of (m-1) renewals and an exit from S_α to R, while the second term is the probability of m passages from S_α to S_β followed by an exit from S_β to R. Substituting (2.11) for the renewal probability and (2.12) for the passage probability in (2.13) yields the desired formulas:

$$Q_{\alpha\cdot\alpha}^{(m)}(0,t) = \frac{\mu_\alpha(-\nu_{\alpha\beta}\nu_{\beta\alpha})^{m-1}}{(\nu_{\alpha\alpha}-\nu_{\beta\beta})^{2m-1}}$$

$$\times \left[\sum_{k=0}^{m-1} \binom{2m-3-k}{m-2} \frac{(\nu_{\alpha\alpha}-\nu_{\beta\beta})^{k+1}}{\nu_{\alpha\alpha}^{k+1}} \left(\sum_{i=0}^{k} \frac{(-\nu_{\alpha\alpha}t)^i}{i!} e^{\nu_{\alpha\alpha}t} - 1 \right) \right.$$

$$+ \sum_{k=0}^{m-2} \binom{2m-3-k}{m-1} \frac{[-(\nu_{\alpha\alpha}-\nu_{\beta\beta})]^{k+1}}{\nu_{\beta\beta}^{k+1}} \left(\sum_{i=0}^{k} \frac{(-\nu_{\beta\beta}t)^i}{i!} e^{\nu_{\beta\beta}t} - 1 \right) \Bigg]$$

$$m=1,2,\ldots \tag{2.14}$$

and

$$Q_{\alpha\cdot\beta}^{(m)}(0,t) = \nu_{\alpha\beta}\mu_\beta \frac{(-\nu_{\alpha\beta}\nu_{\beta\alpha})^{m-1}}{(\nu_{\alpha\alpha}-\nu_{\beta\beta})^{2m}} \sum_{k=0}^{m-1} \binom{2m-2-k}{m-1}$$

$$\times \left\{ \frac{(\nu_{\alpha\alpha}-\nu_{\beta\beta})^{k+1}}{\nu_{\alpha\alpha}^{k+1}} \left[\sum_{i=0}^{k} \frac{(-\nu_{\alpha\alpha}t)^i e^{\nu_{\alpha\alpha}t}}{i!} - 1 \right] + \frac{[-(\nu_{\alpha\alpha}-\nu_{\beta\beta})]^{k+1}}{\nu_{\beta\beta}^{k+1}} \left[\sum_{i=0}^{k} \frac{(-\nu_{\beta\beta}t)^i e^{\nu_{\beta\beta}t}}{i!} - 1 \right] \right\},$$

$$m = 1,2,\ \ldots \tag{2.15}$$

3. PROBABILITY GENERATING FUNCTIONS

A usually preferred approach in deriving formulas for a sequence of probabilities is the probability generating function (p.g.f.) since this approach often yields elegant solutions with a minimum amount of computation. For the probabilities discussed in this paper, however, the method of p.g.f. leads to formulas containing an infinite number of terms which are not practical for the purpose of application. Nevertheless, the formulas of the p.g.f.'s will be stated, so that the sum of the probabilities, expectations and variances can be derived with ease. Reference may be made to [7] for detail.

The p.g.f.'s of the renewal probabilities and the passage probabilities,

$$g_{\alpha\beta}(s;t) = \sum_{m=0}^{\infty} s^m P_{\alpha\beta}^{(m)}(0,t) \quad , \quad \alpha,\beta = 1,2 \quad , \tag{3.1}$$

satisfy the following differential equations

$$\frac{\partial}{\partial t} g_{\alpha\alpha}(s;t) = g_{\alpha\alpha}(s;t)\nu_{\alpha\alpha} + g_{\alpha\beta}(s;t)\nu_{\beta\alpha}$$

$$\frac{\partial}{\partial t} g_{\alpha\beta}(s;t) = g_{\alpha\alpha}(s;t)s\nu_{\alpha\beta} + g_{\alpha\beta}(s;t)\nu_{\beta\beta}$$

$$\alpha\neq\beta;\alpha,\beta=1,2, \tag{3.2}$$

with the initial conditions

$$g_{\alpha\alpha}(s;0) = 1, \quad g_{\alpha\beta}(s;0) = 0, \quad \alpha \neq \beta; \alpha, \beta = 1,2 \ . \tag{3.3}$$

The particular solution of (3.2) corresponding to the initial condition (3.3) turns out to be

$$g_{\alpha\beta}(s;t) = \sum_{i=1}^{2} \frac{s\nu_{\alpha\beta}}{\rho_i(s)-\rho_j(s)} e^{\rho_i(s)t}$$

$$g_{\alpha\alpha}(s;t) = \sum_{i=1}^{2} \frac{(\rho_i(s)-\nu_{\beta\beta})}{\rho_i(s)-\rho_j(s)} e^{\rho_i(s)t} \quad , \quad j\neq i; \quad j=1,2, \quad \alpha\neq\beta; \quad \alpha,\beta=1,2 \ , \tag{3.4}$$

where

$$\rho_1(s) = \frac{1}{2}\left[(\nu_{11}+\nu_{22}) + \sqrt{(\nu_{11}-\nu_{12})^2 + 4\nu_{12}\nu_{21}s}\right]$$

$$\rho_2(s) = \frac{1}{2}\left[(\nu_{11}+\nu_{22}) - \sqrt{(\nu_{11}-\nu_{22})^2 + 4\nu_{12}\nu_{21}s}\right] \tag{3.5}$$

Expressing the exponentials in (3.4) in power series, we obtain formulas for the renewal probabilities and the passage probabilities in infinite sums. They can be reduced to formulas in (2.11) and (2.12).

When s=1, the right hand side member in (3.4) becomes the sum of the probabilities. That is

$$\sum_{m=0}^{\infty} P_{\alpha\alpha}^{(m)}(0,t) = P_{\alpha\alpha}(0,t) = \sum_{i=1}^{2} \frac{\rho_i-\nu_{\beta\beta}}{\rho_i-\rho_j} e^{\rho_i t} \quad , \quad j\neq i, \ j=1,2 \tag{3.6}$$

and

$$\sum_{m=0}^{\infty} P_{\alpha\beta}^{(m)}(0,t) = P_{\alpha\beta}(0,t) = \frac{\nu_{\alpha\beta}}{\rho_1-\rho_2}\left(e^{\rho_1 t} - e^{\rho_2 t}\right) \tag{3.7}$$

where ρ_1 and ρ_2 are given in (3.5) with s=1.

4. PROPERIZED PROBABILITY DISTRIBUTIONS

Since $P_{\alpha\alpha}(0,t) < 1$ and $P_{\alpha\beta}(0,t) < 1$, both $M_{\alpha\alpha}(0,t)$ and $M_{\alpha\beta}(0,t)$ are improper random variables. If we consider only those transitions having a specified ending state as well as a specified starting state, then the corresponding numbers of transitions can be proper random variables. Let $\overset{o}{M}_{\alpha\alpha}(0,t)$ be the "properized" number of renewals so that a system is in S_α at 0 and at t, and $\overset{o}{M}_{\alpha\beta}(0,t)$ be the "properized" number of passages that a system is in S_α at 0 and in S_β at t. The corresponding probability distributions are given by

$$\Pr\{\overset{\circ}{M}_{\alpha\alpha}(0,t) = m\} = \overset{\circ}{P}{}^{(m)}_{\alpha\alpha}(0,t)$$

$$= P^{(m)}_{\alpha\alpha}(0,t)/P_{\alpha\alpha}(0,t), \qquad m=0,1,..., \tag{4.1}$$

and

$$\Pr\{\overset{\circ}{M}_{\alpha\beta}(0,t) = m\} = \overset{\circ}{P}{}^{(m)}_{\alpha\beta}(0,t) \quad ,$$

$$= P^{(m)}_{\alpha\beta}(0,t)/P_{\alpha\beta}(0,t) \qquad m=1,2,... \tag{4.2}$$

The expectations of $\overset{\circ}{M}_{\alpha\alpha}(0,t)$ and $\overset{\circ}{M}_{\alpha\beta}(0,t)$ can be found from the corresponding p.g.f.'s in (3.4). We find for the renewal probabilities

$$E[\overset{\circ}{M}_{\alpha\alpha}(0,t)] = \frac{\nu_{\alpha\beta}\nu_{\beta\alpha}}{\rho_1-\rho_2}(t-2) + \frac{2\nu_{\alpha\beta}\nu_{\beta\alpha}(\rho_2-\nu_{\beta\beta})}{(\rho_1-\rho_2)^2 P_{\alpha\alpha}(0,t)} + t\, e^{\rho_2 t} \tag{4.3}$$

and for the passage probability

$$E[\overset{\circ}{M}_{\alpha\beta}(0,t)] = 1 + \frac{\nu_{\alpha\beta}\nu_{\beta\alpha}t}{\rho_1-\rho_2}\;\frac{e^{\rho_1 t} + e^{\rho_2 t}}{e^{\rho_1 t} - e^{\rho_2 t}} - \frac{2\nu_{\alpha\beta}\nu_{\beta\alpha}}{(\rho_1-\rho_2)^2} \tag{4.4}$$

where ρ_1 and ρ_2 are given in (3.5) with s=1, and $P_{\alpha\alpha}(0,t)$ is given in (3.6).

5. ASYMPTOTIC BEHAVIOR

According to the definitions in (1.9), ν_{11} and ν_{22} are negative quantities. Therefore, the exponential $e^{\nu_{\alpha\alpha}t}$ tends to zero as t approaches infinity, for $\alpha = 1,2,$. As a result

$$\lim_{t\to\infty} P^{(m)}_{\alpha\beta}(0,t) = 0 \tag{5.1}$$

and

$$\lim_{t\to\infty} P^{(m)}_{\alpha\alpha}(0,t) = 0 \tag{5.2}$$

and the probabilities in (2.14) and (2.15) become

$$Q^{(m)}_{\alpha\cdot\alpha}(0,\infty) = -\frac{\mu_\alpha(-\nu_{\alpha\beta}\nu_{\beta\alpha})^{m-1}}{(\nu_{\alpha\alpha}-\nu_{\beta\beta})^{2m-1}}$$

$$\times \left[\sum_{k=0}^{m-1} \binom{2m-3-k}{m-2} \frac{(\nu_{\alpha\alpha}-\nu_{\beta\beta})^{k+1}}{\nu_{\alpha\alpha}^{k+1}} + \sum_{k=0}^{m-2} \binom{2m-3-k}{m-1} \frac{[-(\nu_{\alpha\alpha}-\nu_{\beta\beta})]^{k+1}}{\nu_{\beta\beta}^{k+1}} \right] \tag{5.3}$$

$$Q^{(m)}_{\alpha\cdot\beta}(0,\infty) = -\nu_{\alpha\beta}\mu_{\beta} \frac{(-\nu_{\alpha\beta}\nu_{\beta\alpha})^{m-1}}{(\nu_{\alpha\alpha}-\nu_{\beta\beta})^{2m}} \sum_{k=0}^{m-1} \binom{2m-2-k}{m-1} \times \left[\frac{(\nu_{\alpha\alpha}-\nu_{\beta\beta})^{k+1}}{\nu_{\alpha\alpha}^{k+1}} + \frac{[-(\nu_{\alpha\alpha}-\nu_{\beta\beta})]^{k+1}}{\nu_{\beta\beta}^{k+1}} \right]. \tag{5.4}$$

Formulas (5.3) and (5.4) may be simplified to yield

$$Q^{(m)}_{\alpha\cdot\alpha}(0,\infty) = -\frac{\mu_{\alpha}}{\nu_{\alpha\alpha}} \left(\frac{\nu_{\alpha\beta}\nu_{\beta\alpha}}{\nu_{\alpha\alpha}\nu_{\beta\beta}} \right)^{m-1} \tag{5.3a}$$

and

$$Q^{(m)}_{\alpha\cdot\beta}(0,\infty) = \frac{\mu_{\beta}}{\nu_{\beta\alpha}} \left(\frac{\nu_{\alpha\beta}\nu_{\beta\alpha}}{\nu_{\alpha\alpha}\nu_{\beta\beta}} \right)^{m} . \tag{5.4a}$$

Summing over all possible values of m gives

$$\sum_{m=1}^{\infty} Q^{(m)}_{\alpha\cdot\alpha}(0,\infty) = \sum_{m=1}^{\infty} \frac{-\mu_{\alpha}}{\nu_{\alpha\alpha}} \left(\frac{\nu_{\alpha\beta}\nu_{\beta\alpha}}{\nu_{\alpha\alpha}\nu_{\beta\beta}} \right)^{m-1}$$

$$= -\frac{\nu_{\beta\beta}\mu_{\alpha}}{\nu_{\alpha\alpha}\nu_{\beta\beta}-\nu_{\alpha\beta}\nu_{\beta\alpha}}$$

$$= Q_{\alpha\cdot\alpha}(\infty) \tag{5.5}$$

and

$$\sum_{m=1}^{\infty} Q^{(m)}_{\alpha\cdot\beta}(0,\infty) = \sum_{m=1}^{\infty} \frac{\mu_{\beta}}{\nu_{\beta\alpha}} \left(\frac{\nu_{\alpha\beta}\nu_{\beta\alpha}}{\nu_{\alpha\alpha}\nu_{\beta\beta}} \right)^{m}$$

$$= \frac{\nu_{\alpha\beta}\mu_\beta}{\nu_{\alpha\alpha}\nu_{\beta\beta}-\nu_{\alpha\beta}\nu_{\beta\alpha}}$$

$$= Q_{\alpha\cdot\beta}(0,\infty) \tag{5.6}$$

The sum of (5.5a) and (5.6a),

$$Q_{\alpha\cdot\alpha}(0,\infty) + Q_{\alpha\cdot\beta}(0,\infty) = \frac{\nu_{\alpha\beta}\mu_\beta-\nu_{\beta\beta}\mu_\alpha}{\nu_{\alpha\alpha}\nu_{\beta\beta}-\nu_{\alpha\beta}\nu_{\beta\alpha}} , \tag{5.7}$$

is the probability that starting from state S_α the system will terminate eventually. Since (1.9) implies

$$\mu_\alpha = -(\nu_{\alpha\alpha}+\nu_{\alpha\beta}) , \quad \alpha\neq\beta;\ \alpha,\beta=1,2 \tag{5.8}$$

equation (5.7) becomes

$$Q_{\alpha\cdot\alpha}(0,\infty) + Q_{\alpha\cdot\beta}(0,\infty) = 1 , \tag{5.9}$$

as expected.

While the renewal probability $P_{\alpha\alpha}^{(m)}(0,t)$ and the passage probability $P_{\alpha\beta}^{(m)}(0,t)$ tend to zero as $t\to\infty$, the expectations of the properized random variables $\overset{o}{M}_{\alpha\alpha}(0,t)$ in (4.3) and $\overset{o}{M}_{\alpha\beta}(0,t)$ in (4.4) do not vanish. In fact, they both tend to infinity as t increases indefinitely.

Quantities which are of theoretical as well as practical interest are the average number of renewals (or passages) per unit time and the increment of the number of renewals (or passages) in a given time interval. Direct computations show that both the average number of renewals per unit time

$$\lim_{t\to\infty} \frac{E[\overset{o}{M}_{\alpha\alpha}(0,t)]}{t} = \frac{\nu_{\alpha\beta}\nu_{\beta\alpha}}{\rho_1-\rho_2} \quad \alpha\neq\beta;\ \alpha,\beta=1,2 \tag{5.10}$$

and the average number of passages per unit time

$$\lim_{t\to\infty} \frac{E[\overset{o}{M}_{\alpha\beta}(0,t)]}{t} = \frac{\nu_{\alpha\beta}\nu_{\beta\alpha}}{\rho_1-\rho_2} , \quad \alpha\neq\beta;\ \alpha,\beta=1,2 , \tag{5.11}$$

are independent of the initial state S_α. For the increment of the number of renewals in interval (t,t+th), we compute

$$\lim_{t\to\infty} \{E[\overset{o}{M}_{\alpha\alpha}(0,t+th)] - E[\overset{o}{M}_{\alpha\alpha}(0,t)]\} = \frac{\nu_{\alpha\beta}\nu_{\beta\alpha}}{\rho_1-\rho_2} h , \quad \alpha\neq\beta; \ \alpha,\beta=1,2 , \tag{5.12}$$

so that

$$\lim_{t\to\infty}\lim_{h\to 0} \frac{E[\overset{o}{M}_{\alpha\alpha}(0,t+h)] - E[\overset{o}{M}_{\alpha\alpha}(0,t)]}{h} = \frac{\nu_{\alpha\beta}\nu_{\beta\alpha}}{\rho_1-\rho_2} , \quad \alpha\neq\beta; \ \alpha,\beta=1,2. \tag{5.13}$$

Similarly,

$$\lim_{t\to\infty}\lim_{h\to 0} \frac{E[\overset{o}{M}_{\alpha\beta}(0,t+h)] - E[\overset{o}{M}_{\alpha\beta}(0,t)]}{h} = \frac{\nu_{\alpha\beta}\nu_{\beta\alpha}}{\rho_1-\rho_2} . \tag{5.14}$$

Again (5.13) and (5.14) are independent of S_α .

The fact that the quantities on the right hand side of (5.10) and (5.13) are equal is interesting but not unexpected, since both stand for the number of renewals in an infinitesimal. These findings are similar to those in the ordinary renewal process.

6. AN EXAMPLE

The model has been applied to a study of childhood morbidity. Information has been collected on the morbidity, mortality, and medical care of some 20,000 children from birth to early school age in a research project, "Child Health and Development Studies" under the direction of Dr. Barbara van den Berg. These children were born at the Kaiser Hospital, Oakland, California during the period from 1959 to 1967.

For the purpose of illustration, a sample of N = 698 children of exact age of one year was taken from the births occurring in the year 1962. Two health states have been defined according to the need for medical or nursing care; S_1 is the state of no need for medical or nursing care, S_2 is the state in need of care; and there is an absorbing state R, the state of death or loss to follow-up.

The N = 698 children were all in S_1 at the beginning of the second year of life. Their changes in health condition during the second year were recorded.

The two transient-state model has been applied to the data. The maximum likelihood estimates of the intensity functions ν_{ij} and μ_i are displayed as follow follows.

$$[\nu_{ij}] = \begin{pmatrix} -.007765 & .007700 \\ \\ .107335 & -.108788 \end{pmatrix} \tag{6.1}$$

$$[\mu_i] = \begin{pmatrix} .000065 \\ \\ .001453 \end{pmatrix} \tag{6.2}$$

Substituting these estimated values into formula (2.11) we obtain the probabilities for having 0,1,2,... illnesses within the year. They are listed below.

m	$P_{11}^{(m)}(0,t)$
0	.059
1	.171
2	.234
3	.202
4	.124
≥ 5	.093
Total	.883

Here t is written for one year or 365 days.

The expected numbers of transitions during the year are estimated from (4.3) and (4.4). t = 1 year.

$$E\,\overset{o}{M}_{11}(0.t) = 2.85$$

$$E\,\overset{o}{M}_{12}(0,t) = 3.47$$

REFERENCES

[1] FIX, E. and NEYMAN, J. (1951). A simple stochastic model of recovery, relapse, death and loss of patients. *Human Biol.*, **23**, 205-241.

[2] DU PASQUIER, L. G. (1913). Mathematische theorie du invaliditatsversicherung. *Mitt. Verein. Schweiz. Versich-Math.*, **8**, 1-153.

[3] PYKE, R. (1961). Markov renewal processes: Definitions and preliminary properties. *Ann. Math. Stat.*, **32**, 1231-1242.

[4] PYKE, R. (1961). Markov renewal processes with finitely many states. *Ann. Math. Statist.*, **32**, 1243-1259.

[5] CHUNG, K. L. (1960). *Markov Chains with Stationary Transition Probabilities.* Springer-Verlag, Berlin.

[6] TAKÁCS, L. (1954). Some investigations concerning recurrent stochastic processes of a certain type. *Magyar Tud. Aka. Mat. Kutató Int. Közl.*, **3**, 115-128.

[7] CHIANG, C. L. (1968). *Introduction to Stochastic Processes in Biostatistics.* Wiley, New York.

P.R. Krishnaiah, ed., APPLICATIONS OF STATISTICS

STATISTICAL PROBLEMS ARISING IN THE THEORY OF AGGREGATION

JOHN S. CHIPMAN*

Department of Economics
University of Minnesota, Minneapolis, Minnesota

1. INTRODUCTION

The problem of aggregation is as old as quantitative economics itself, but only relatively recently, thanks in large part to the work of Theil [28], Malinvaud [15], and Fisher [8], has it begun to be taken up formally as a subdiscipline in its own right. In some recent papers [5], [3], [4], I have attempted to build a theoretical framework for the analysis of aggregation problems in econometrics. The emphasis so far has been on the modelling and clustering aspects rather than on the statistical aspects. However, the problem of statistical estimation cannot be separated from that of aggregation. In the present paper I take up some of the statistical aspects, and apply the techniques developed to a model of the international transmission of price changes, employing German data for the years 1963-1975.

In section 2 I derive conditions under which blown-up aggregative estimators have lower mean square error than direct least squares estimators, and the corresponding hypothesis test. In section 3, I obtain conditions under which James-Stein estimators in aggregative models have lower mean square error than least squares estimators applied to the same models, as well as the corresponding test. Attention is focussed in section 4 on the impact of international prices (including of course petroleum prices) on the German consumer price index component for electricity, gas, and fuel.

2. AGGREGATION AS A MEANS OF IMPROVING ESTIMATION

Let us suppose that the "true" model has the form of a multivariate multiple regression model[1]

$$Y = XB + E, \quad \mathcal{E}E = 0, \quad \mathcal{E}(\text{row } E)'(\text{row } E) = V \otimes \Sigma, \tag{2.1}$$

*Research supported by Ford Foundation Grant 750-0114. I am much indebted to Messrs. Hans Kullmer and Werner Rostin of the Statistishes Bundesamt, Wiesbaden, for their help in furnishing the data analyzed in section 4, and to Abdessatar Ouanes for his untiring and expert computational assistance. Thanks are also due to P.R. Krishnaiah, H.F. Weinberger, and the referee for their comments, and to Dit Sang Ho for correction of errors and misprints.

[1]Here, "row E" denotes the $1 \times m$ row vector of consecutive rows of the $n \times m$ matrix of errors, E ; the symbol $\otimes$ denotes the Kronecker product $V \otimes \Sigma = [v_{st}\Sigma]$.

where Y is an $n \times m$ matrix of n observations on m dependent (endogenous) variables, X an $n \times k$ matrix of n observations on k independent (exogenous) variables, B is an unknown $k \times m$ matrix of regression coefficients. The expectation $\mathcal{E}$ is conditional on X, which is assumed to have rank r and to be such that its columns are in the column space of the $n \times n$ autocovariance matrix V, i.e., $X = VK$ for some K.[2] Defining the transformations

$$X^* = XG, \qquad Y^* = YH , \tag{2.2}$$

where G and H are $k \times k^*$ and $m \times m^*$ "grouping matrices" with $k^* < k$ and $m^* \leqq m$ respectively, the approximative model is

$$Y^* = X^*B^* + E^*, \quad \mathcal{E}^*E^* = 0, \quad \mathcal{E}^*(\text{row } E^*)'(\text{row } E^*) = V \otimes \Sigma^{**} , \tag{2.3}$$

where $\mathcal{E}^*$ is the assumed, but not "true", expectation operator. It will be assumed in the following development that the $m \times m$ contemporaneous covariance matrix Σ is such that ΣH has rank $m^* \leqq n - r$. The rank of XG will be denoted r^*. It will be assumed that G and H have ranks k^* and m^* respectively; in most practical applications it will be the case that $r = k$ and $r^* = k^*$.

Formulas (2.2) and (2.3) define a mapping GB^* from the k-dimensional space $\mathcal{X}$ of $1 \times k$ vectors x_t of observations on the independent variables, to the m-dimensional space $\mathcal{Y}^*$ of predicted values $\mathcal{E}^*y_t^* = x_tGB^*$, $t = 1, 2, \ldots, n$, of the m dependent variables. Likewise, (2.1) and (2.2) define the mapping $BH: \mathcal{X} \to \mathcal{Y}^*$ furnishing the "true" expected values $\mathcal{E}y_t^* = \mathcal{E}y_tH = x_tBH$. Introducing the matrix indicator[3]

$$D(C_1, C_2) = (C_1 - C_2)'X'V^-X(C_1 - C_2) \tag{2.4}$$

of the distance between any two linear mappings $C_1, C_2: \mathcal{X} \to \mathcal{Y}^*$, the *aggregation bias* is defined as

$$D(BH, GB^*) = (BH - GB^*)'X'V^-X(BH - GB^*) . \tag{2.5}$$

[2]This is a relatively mild requirement, which is automatically fulfilled when V has full rank n. It may be justified on the assumption that X is a realization of a random variable with the same autocovariance structure as $E|X$ (cf. Chipman [5, p. 727]); the condition will then hold with probability 1. The assumption is needed in Theorem 1 (see footnote 3).

[3]Here, V^- denotes a generalized inverse of V in the sense of Rao [19], i.e., a matrix satisfying $VV^-V = V$; note that, from the assumption $X = VK$, we have $VV^-X = X$. It is also assumed that V is symmetric and non-negative definite, as is always possible (cf. [5], [3]). Elsewhere in this paper, for any matrix A, A^- will denote a matrix satisfying $AA^-A = A$. $X^\dagger$ denotes the Moore-Penrose generalized inverse of X; if X is $n \times k$ of rank k, this is $X^\dagger = (X'X)^{-1}X'$.

It was shown in [5, Theorem 2.1] that the minimum of (2.5), in terms of the partial ordering of symmetric non-negative definite matrices, $D_1 \geqslant D_2 \Leftrightarrow x'D_1x \geqq x'D_2x$ for all x, is attained when

$$B^* = G^{\#}BH, \qquad \text{where} \qquad G^{\#} = (G'X'V^{-}XG)^{-}G'X'V^{-}X \ , \tag{2.6}$$

and that

$$\Lambda \equiv \inf_{B^*} D(BH, GB^*) = H'B'(I - GG^{\#})'X'V^{-}X(I - GG^{\#})BH \ . \tag{2.7}$$

Corresponding to (2.4) I adopted in [4] as a scalar measure of the distance between C_1 and C_2 the Mahalanobis distance

$$d(C_1, C_2) = m^{*-1}\ tr(C_1 - C_2)'X'V^{-}X(C_1 - C_2)(H'\Sigma H)^{-1} \ . \tag{2.8}$$

The corresponding measure of minimum aggregation bias is then

$$\lambda \equiv \inf_{B^*} d(BH,GB^*) = m^{*-1}tr\ H'B'(I - GG^{\#})'X'V^{-}X(I - GG^{\#})BH(H'\Sigma H)^{-1}. \tag{2.9}$$

Let the $k \times m^*$ matrix $\hat{C}$ be an estimator of BH. (We shall consider only estimators of the form $\hat{C} = A^*Y^* = A^*YH$.) We adopt the matrix- and scalar-valued loss functions

$$L(\hat{C}) = D(BH, \hat{C}), \qquad \ell(\hat{C}) = d(BH, \hat{C}) \ . \tag{2.10}$$

Defining the generalized least squares estimators

$$\tilde{B} = (X'V^{-}X)^{-}X'V^{-}Y \tag{2.11}$$

and

$$\tilde{B}^* = (X^{*\prime}V^{-}X^*)^{-}X^{*\prime}V^{-}Y^* = G^{\#}\tilde{B}H \ , \tag{2.12}$$

we now pose the following question: when is the "blown-up" aggregative estimator $G\tilde{B}^* = GG^{\#}\tilde{B}H$ a better estimator of BH than the straightforward generalized least-squares estimator $\tilde{B}H$, in terms of the criterion of minimum expected loss?

Theorem 1. In order that $\mathcal{E}L(G\tilde{B}^*) \leqslant \mathcal{E}L(\tilde{B}H)$ (i.e., that $\mathcal{E}L(\tilde{B}H) - \mathcal{E}L(G\tilde{B}^*)$ be non-negative definite), it is necessary and sufficient that

$$\Lambda \leqslant (r - r^*)H'\Sigma H. \tag{2.13}$$

Likewise, a necessary and sufficient condition that $\mathcal{E}\ell(G\tilde{B}^*) \leqq \mathcal{E}\ell(\tilde{B}H)$ is that

$$\lambda \leqq r - r^* \ . \tag{2.14}$$

Proof. Using the fact that for any $n \times n$ matrix A, (2.1) implies that $\mathcal{E}E'AE = (\operatorname{tr} A V)\,\Sigma$, and employing standard properties of generalized inverses (cf. [19], [20], or Lemma 1.4 of [5]), we find (keeping in mind the assumption $X = VK$) that

$$\mathcal{E}L(\tilde{B}H) = rH'\Sigma H, \quad \mathcal{E}L(G\tilde{B}^*) = \Lambda + r^*H'\Sigma H ,$$

from which (2.13) follows immediately. Likewise we find that

$$\mathcal{E}\ell(\tilde{B}H) = r , \quad \mathcal{E}\ell(G\tilde{B}^*) = \lambda + r^* ,$$

yielding (2.14).

Q.E.D.

It is natural to look for a test statistic for testing the hypothesis (2.14), as was done by Toro and Wallace [30] in a related situation. Define the matrices $J = I - X(X'V^-X)^-X'V^-$ and $J^* = I - X^*(X^*'V^-X^*)^-X^*'V^-$, which satisfy $J^2 = JJ^* = J^*J = J$, $J^{*2} = J^*$, hence $(J^* - J)^2 = J^* - J$, and $J'V^-J = V^-J$, $J^*'V^-J^* = V^-J^*$. Define $\tilde{E} = Y - X\tilde{B}$, $\tilde{E}^* = Y^* - X^*\tilde{B}^*$, and the matrices

$$S = Y'V^-JY = \tilde{E}'V^-\tilde{E}, \quad S^* = Y'V^-J^*Y, \quad S^{**} = Y^*'V^-JY^* = \tilde{E}^*'V\,\tilde{E}^* = H'S^*H. \tag{2.15}$$

Replacing B by its estimator $\tilde{B}$ in (2.7) we obtain

$$\tilde{\Lambda} = H'\tilde{B}'X'V^-X(I - GG^{\#})\tilde{B}H = H'(S^* - S)H = S^{**} - H'SH. \tag{2.16}$$

Further, replacing Σ in (2.9) by the unbiased estimator $(n - r)^{-1}S$, we obtain

$$\tilde{\lambda} = \frac{n-r}{m^*}\operatorname{tr}\,[(H'S^*H - H'SH)(H'SH)^{-1}] = (n-r)\left[\frac{\operatorname{tr}H'S^*H(H'SH)^{-1}}{m^*} - 1\right] . \tag{2.17}$$

This suggests employment of the test statistic

$$v = \operatorname{tr}H'(S^* - S)H(H'SH)^{-1}/m^* = [\operatorname{tr}H'S^*H(H'SH)^{-1}]/m^* - 1 . \tag{2.18}$$

Assuming that[4] col $y \sim N_{nm}[(I_m \otimes X)\operatorname{col} B, \Sigma \otimes V]$, it may be verified that $H'(S^* - S)H$ and $H'SH$ are independent. Moreover, if $V = I_n$, they have non-central and central Wishart distributions respectively, given by

$$H'(S^* - S)H \sim W_{m^*}(H'\Sigma H, r-r^*,\Lambda); \; H'SH \sim W_{m^*}(H'\Sigma H, n - r). \tag{2.19}$$

Thus, $m^*(n - r)v = m^*\tilde{\lambda}$ is distributed as Hotelling's [9] generalized T_0^2 statistic;

[4]Here, "col Y" denotes the column vector of consecutive columns of Y ; N_{nm} is the nm-variate normal (possibly singular) distribution.

its percentage points in the non-central case $\Lambda = (r - r^*)H'\Sigma H$ would then furnish an appropriate test of the hypothesis (2.13).[5] The exact distribution of T_0^2 in the central case $\Lambda = 0$ has been obtained by Pillai & Young [17] and Krishnaiah & Chang [11]; expressions for non-central cases are indicated in Krishnaiah & Chattopadhyay [12]. Computation of the appropriate percentage points remains an open problem for future research.

In the case $m = m^* = 1$ and $H = 1$, replacing $\tilde{E}$ and $\tilde{E}^*$ by column vectors e and e*, the statistic (2.18) reduces to

$$v = \frac{e^{*\prime}V^- e^* - e'V^- e}{e'V^- e}, \tag{2.20}$$

a ratio of independent non-central and central chi-square variates. Thus, the statistic $u = (n - r)v/(r - r^*)$ has a non-central F distribution with $p = r - r^*$ and $q = n - r$ degrees of freedom and noncentrality parameter λ. The appropriate test statistic for (2.14) is thus obtained by choosing $\lambda = p$. Defining $\alpha = p/2$, $\beta = q/2$, $\delta = \lambda/2$, the statistic $w = v/(1 + v)$ has the non-central beta distribution

$$I(w;\ \alpha,\beta,\delta) = \sum_{j=0}^{\infty} \frac{e^{-\delta}\delta^j}{j!} \int_0^w \frac{\Gamma(\alpha+j+\beta)}{(\alpha+j)\Gamma(\beta)}\, x^{\alpha+j-1}(1 - x)^{\beta-1}dx, \tag{2.21}$$

with $\delta = \alpha$.[6]

3. LIMITATIONS OF JAMES-STEIN ESTIMATION AS APPLIED TO AGGREGATIVE MODELS

On the face of it, it might seem that the methods of James and Stein [10] could lead to considerable improvement in estimation in large-scale econometric models. Two problems arise, however. The first is that the superiority of James-Stein estimators over least squares estimators is contingent upon the model being correctly specified; when applied to an aggregative model, it turns out that James-Stein estimation no longer has lower mean square error than least squares for all hypothetical values of the regression coefficients, but only for those values yielding a sufficiently small aggregation bias. The precise condition is des-

[5] It may be conjectured that the distribution of v depends on λ.

[6] Compare Toro & Wallace [30], whose statistic differs from this one in that δ (their λ) = 1/2. This would correspond to the special case $p = r - r^* = 1$.

cribed in Theorem 2, for the univariate case.[7] The second problem is that the lower the estimated aggregation bias, and thus the more favorable the conditions for James-Stein estimation, the closer to 1 is the shrinkage factor. In practice, therefore (at least in time series analysis), James-Stein estimators will differ very little from least squares estimators under the very conditions that are most favorable to their use.

Let the detailed and aggregative models both be univariate, so that $m = m^* = 1$ and $H = 1$, and assume as before that $X = VK$ for some K. The detailed, "true", model is given by

$$y \sim N_n(X\beta, \sigma^2 V), \qquad X{:}n \times k \text{ of rank } r, \tag{3.1}$$

and the aggregative, "false", model by

$$y^* \sim N_n(X^*\beta^*, \sigma^{*2} V), \qquad X^* : n \times k^* \text{ of rank } r^*, \tag{3.2}$$

where

$$X^* = XG. \tag{3.3}$$

Defining aggregation bias as in (2.5) by

$$||X^*\beta^* - X\beta||^2 = (\beta - G\beta^*)'X'V^- X(\beta - G\beta^*) \tag{3.4}$$

we see that it reaches a minimum with respect to β^* when

$$\beta^* = G^{\#}\beta, \qquad \text{where } G^{\#} = (G'X'V^- XG)^- G'X'V^- X\ . \tag{3.5}$$

The minimum aggregation bias is then

$$\inf_{\beta^*} ||X^*\beta^* - X\beta||^2 = ||X(I - GG^{\#})\beta||^2 = \beta'(I - GG^{\#})'X'V^- X(I - GG^{\#})\beta\ , \tag{3.6}$$

which may also be expressed as

$$\inf_{\beta^*} ||X^*\beta^* - X\beta||^2 = \beta' X'V^- X(I - GG^{\#})\beta = \beta'X'V^- [I - XG(G'X'V^- XG)^- G'X'V^-]X\beta\ . \tag{3.7}$$

In view of (3.5) we shall adopt as our definition of the loss associated with an estimator $\hat{\beta}^*$ of β^* in the model (3.2) its distance from $G^{\#}\beta$. In order

[7] A generalization to the multivariate case has been obtained by Sclove [23] which, however, is not a full generalization since it reduces to least squares in the case $m \leq 2$. Recent contributions by Stein [26] and Efron & Morris [7] likewise do not appear to provide the required analogue for the multivariate multiple regression model.

to define this distance, we may note that the generalized least squares estimator

$$b^* = (G'X'V^-XG)^-G'X'V^-y \tag{3.8}$$

is distributed (in terms of the "true" model (3.1)) as

$$b^* \sim N_{k^*}(G\#\beta,\ \sigma^2(G'X'V^-XG)^-G'X'V^-XG(G'X'V^-XG)^{-\prime}) . \tag{3.9}$$

Accordingly, we may formulate our estimation problem as that of estimating the mean of the (possibly singular) multivariate normal distribution (3.9).

The problem may in turn be transformed into one of estimating the mean of a spherical r^*-variate normal distribution (see for instance Sclove [22]), as follows. Let $0 = [P, Q]$ be an orthogonal $k^* \times k^*$ matrix such that P is $k^* \times r^*$ and such that $0'G'X'V^-XG0$ is a diagonal matrix whose first r^* diagonal elements are the positive characteristic roots of $G'X'V^-XG$; let Ω be the corresponding diagonal matrix of order r^*. Then $G'X'V^-XG = P\Omega P'$. Defining

$$c = \sigma^{-1}\Omega^{\frac{1}{2}}p'b^*, \qquad \gamma = \mathcal{E}c = \sigma^{-1}\Omega^{\frac{1}{2}}P'G\#\beta, \tag{3.10}$$

we have

$$\mathcal{E}(c - \gamma)(c-\gamma)' = \Omega^{\frac{1}{2}}P'(G'X'V^-XG)^-G'X'V^-XG(G'X'V^-XG)^{-\prime}P\Omega^{\frac{1}{2}} ,$$

which we verify to be idempotent, hence equal to the identity matrix I_{r^*}. Thus, $c \sim N_{r^*}(\gamma, I)$. Stein's loss function for an estimator $\hat{\gamma}$ of γ is (cf. Stein [24])

$$\ell(\hat{\gamma}; \gamma) = ||\hat{\gamma} - \gamma||^2 = (\hat{\gamma} - \gamma)'(\hat{\gamma} - \gamma).$$

Defining $\hat{\beta}^* = \sigma P\Omega^{-\frac{1}{2}}\hat{\gamma}$, and noting from (3.9) that $(\mathrm{Var}\ b^*)^- = \sigma^{-2}G'X'V^-XG$, this goes over into the loss function

$$\ell^*(\hat{\beta}^*; \mathcal{E}b^*, \mathrm{Var}\ b^*) = ||\ \hat{\beta}^* - \mathcal{E}\,b^*||^2_{(\mathrm{Var}\ b^*)^-}$$
$$= (\hat{\beta}^* - G\#\beta)'G'X'V^-XG(\hat{\beta}^* - G\#\beta)/\sigma^2 .$$

Accordingly, our risk function is

$$\mathrm{Risk}\ \hat{\beta}^* = \mathcal{E}(\hat{\beta}^* - G\#\beta)'G'X'V^-XG(\hat{\beta}^* - G\#\beta)/\sigma^2 . \tag{3.11}$$

The James-Stein class of estimators of the mean of (3.9) is defined by (cf. James & Stein [10])

$$\hat{\beta}^{*\nu} = b^* - \frac{\nu}{F^*}(b^* - \bar{\beta}^*) = (1 - \frac{\nu}{F^*})b^* + \frac{\nu}{F^*}\bar{\beta}^*, \tag{3.12}$$

where $\bar{\beta}^*$ is an arbitrary initial guess at the right value of $G\#\beta$ (which may be taken to be $G\#\bar{\beta}$, $\bar{\beta}$ being an arbitrary initial guess at the right value of β), where

$$F^* = S_1^*/S_0^*, \quad S_1^* = (b^* - \bar{\beta}^*)'G'X'V^-XG(b^* - \bar{\beta}^*), \quad S_0^* = e^{*\prime}V^-e^*, \tag{3.13}$$

and

$$e^* = y - X^*b^* = [I - XG(G'X'V^-XG)^-G'X'V^-]y \ , \tag{3.14}$$

and where ν is a constant in the interval

$$0 \leqq \nu \leqq \frac{2(r^* - 2)}{n - r^* + 2} \ . \tag{3.15}$$

In the special case in which

$$X = X\ G\ G\# = X^*(X^{*\prime}V^-X^*)^-X^{*\prime}V^-X \tag{3.16}$$

---so that (3.7) vanishes identically for all β ---it follows from formula (3.25) below that the James-Stein estimators $\hat{\beta}^{*\nu}$ dominate the generalized least squares estimator b^* for all ν in the interior of the interval (3.15). This is a slight generalization of the James-Stein result. Again, when (3.16) holds, the James-Stein estimator with minimum risk (3.11) is the one with

$$\nu = \frac{r^* - 2}{n - r^* + 2} \ , \tag{3.17}$$

and will be denoted

$$\hat{\beta}^* = (1 - \frac{r^* - 2}{n - r^* + 2}\frac{1}{F^*})b^* + \frac{r^* - 2}{n - r^* + 2}\frac{1}{F^*}\bar{\beta}^* \ . \tag{3.18}$$

It follows readily from (3.14) and (3.1) (cf.,e.g., Pringle & Rayner [18, p. 77], Rao & Mitra [20, p. 171]) that $S_0^* \sim \chi^2\ (n - r^*, \lambda)$, where

$$\lambda = \frac{\beta'X'V^-[I - XG(G'X'V^-XG)^-G'X'V^-]X\beta}{} = \frac{\beta'(I - GG\#)'X'V^-X(I - GG\#)\beta}{\sigma^2} \ . \tag{3.19}$$

The numerator of (3.19) is simply the minimum aggregation bias (3.7), which in turn is what the matrix (2.7) reduces to in the special case $m = m^* = H = 1$; (3.19) thus corresponds to the scalar measure (2.9) of aggregation bias in this case.

It follows similarly from (3.9) and (3.13) that $S_1^* \sim \chi^2(r^*, \lambda^*)$, where

$$\lambda^* = (G\#\beta - \bar{\beta}^*)'G'X'V^-XG(G\#\beta - \bar{\beta}^*)/\sigma^2 \ .$$

Moreover, S_1^* and S_0^* are independent, since $\text{Cov}(b^*,e^*) = \mathscr{E}(b^* - G\#\beta)(e^* - X[I - GG\#]\beta)'$

$= 0$. Thus, $(n - r^*)F^*/r^*$ has the doubly non-central F distribution with r^* and $n - r^*$ degrees of freedom and non-centrality parameters λ^* and λ (cf. Tang [27], Malik [14]).

The following theorem actually includes the James-Stein theorem, for the case $r^* = k^* = r$ and $G = I_k$. In fact, to obtain the James-Stein result it is enough to assume that $r = k^* = r^*$ and that G is chosen in such a way that $X = XGG\#$; then it is still the case that $\lambda = 0$ for all β. This is the well known case of aggregating to take advantage of multicollinearity.

Theorem 2. Supposing the model (3.1) to be true, let the James-Stein and least squares estimators of $G\#\beta$ be given by (3.18) and (3.8) respectively, and let the risk function be defined by (3.11). Let $r^* \geqq 3$. Then Risk $\hat{\beta}^* <$ Risk b^* for all β such that

$$\lambda < \sqrt{(n - r^*)(n - r^* + 2)} \quad . \tag{3.20}$$

Proof. From (3.12) and (3.11) we have

$$\text{Risk } \hat{\beta}^{*\nu} = \sigma^{-2}\mathcal{E}[b^* G\#\beta - \frac{\nu}{F^*}(b^* - \bar{\beta}^*)]'G'X'V^-XG[b^* - G\#\beta - \frac{\nu}{F^*}(b^* - \bar{\beta}^*)] \quad .$$

Since $b^* - G\#\beta = (G'X'V^-XG)^-G'X'V^-\varepsilon$, where $\varepsilon = y - X\beta$, this becomes

$$\text{Risk } \hat{\beta}^{*\nu} = r^* - 2\frac{\nu}{\sigma^2}\mathcal{E}\frac{(b^* - G\#\beta)'G'X'V^-XG\ (b^* - \bar{\beta}^*)}{F^*} + \frac{\nu^2}{\sigma^2}\mathcal{E}\frac{(b^* - \bar{\beta}^*)'G'X'V^-XG(b^* - \bar{\beta}^*)}{F^{*2}} \quad .$$

Since b^* and e^* are independent, this simplifies further to

$$\text{Risk } \hat{\beta}^{*\nu} = r^* - 2\ \nu\mathcal{E}S_0^*\mathcal{E}\ \frac{(b^* - G\#\beta)'G'X'V^-XG(b^* - \bar{\beta}^*)}{(b^* - \bar{\beta}^*)'G'X'V^-XG(b^* - \bar{\beta}^*)} + \nu^2\mathcal{E}\ S_0^{*2}\ \mathcal{E}\frac{1}{S_1^*} \quad . \tag{3.21}$$

Analogously to (3.10) let us define

$$z^* = \sigma^{-1}\Omega^{\frac{1}{2}}P'(b^* - \bar{\beta}^*), \quad \zeta^* = \mathcal{E}z^* = \sigma^{-1}\Omega^{\frac{1}{2}}P'(G\#\beta - \bar{\beta}^*) \ ;$$

then $z^* \sim N_{r^*}(\zeta^*, I)$, and by Lemma 2 of the Appendix,

$$\mathcal{E}\frac{(b^* - G\#\beta)'G'X'V^-XG(b^*-\bar{\beta}^*)}{(b^* - \bar{\beta}^*)'G'X'V^-XG(b^*-\bar{\beta}^*)} = \mathcal{E}\frac{(z^* - \zeta^*)'z^*}{z^{*\prime}z^*} = (r^* - 2)\ \mathcal{E}\frac{1}{z^{*\prime}z^*} \ .$$

Since $S_1^* = z^{*\prime}z^*$, (3.21) becomes

$$\text{Risk } \hat{\beta}^{*\nu} = r^* - [2\nu(r^*-2)\,\mathcal{E}\,S_0^* - \nu^2\mathcal{E}\,S_0^{*2}]\,\mathcal{E}\frac{1}{S_1^*}\,. \tag{3.22}$$

Now since $S_0^* \sim \chi^2(n - r^*, \lambda)$, we have

$$\mathcal{E}\,S_0^* = n - r^* + \lambda, \tag{3.23}$$

and from Lemma 3, formula (A6), of the Appendix (with $h(W) = W$) we have[8]

$$\mathcal{E}\,S_0^{*2} = (n - r^*)(n - r^* + 2) + 2(n - r^* + 2)\lambda + \lambda^2. \tag{3.24}$$

From (3.22), (3.23), and (3.24) we therefore have

$$\text{Risk } b^* - \text{Risk } \hat{\beta}^{*\nu} =$$

$$\{2\nu(r^*-2)(n-r^*+\lambda) - \nu^2\,[(n-r^*)(n-r^*+2) + 2(n-r^*+2)\lambda + \lambda^2]\}\,\mathcal{E}\,\frac{1}{S_1^*}\,. \tag{3.25}$$

The expression in braces in (3.25) reaches a maximum with respect to ν when

$$\nu = \frac{(r^* - 2)(n - r^* + \lambda)}{(n - r^*)(n - r^* + 2) + 2(n - r^* + 2)\lambda + \lambda^2}\,; \tag{3.26}$$

this is a decreasing function of λ for $r^* \geqq 3$, and reduces to (3.17) when $\lambda = 0$. In terms of the estimator (3.18), (3.25) then becomes

$$\text{Risk } b^* - \text{Risk } \hat{\beta}^* =$$

$$\left(\frac{r^* - 2}{n - r^* + 2}\right)^2[(n - r^*)(n - r^* + 2) - \lambda^2]\,\mathcal{E}\frac{1}{S_1^*}\,. \tag{3.27}$$

For $r^* \geqq 3$, this is positive if and only if (3.20) holds. ∎

Just as in the preceding section, the hypothesis (3.20) can be tested, since the statistic $u = (n - r)v/(r - r^*)$, where v is given by (2.20), has the non-central F distribution with $p = r - r^*$ and $q = n - r$ degrees of freedom and noncentrality parameter λ. The appropriate test statistic is obtained by setting $\lambda = \sqrt{(n-r^*)(n-r^*+2)} = \sqrt{(p+q)(p+q+2)}$. Defining $\alpha = p/2$, $\beta = q/2$, $\delta = \lambda/2$, the corresponding statistic $w = v/(1 + v)$ has the non-central beta distribution (2.21) with $\delta = \sqrt{(\alpha + \beta)(\alpha + \beta + 1)}$.

Note that, since v and F^* are inversely related, a lower aggregation bias v,

[8] For an alternative method of derivation, see Theil [29, p. 534].

while increasing the acceptability of the James-Stein estimator, causes it to be closer to the Gauss-Markoff estimator (3.8). This will be illustrated in Section 4.

We conclude this section with some brief remarks concerning the generalizations of James-Stein estimation developed by Stein [25], Baranchik [1], Lin & Tsai [13], Efron & Morris [6], and others, leading to a class of estimators of the form

$$\tilde{\beta}^* = \left(1 - \frac{\psi(F^*)}{F^*}\right) b^* + \frac{\psi(F^*)}{F^*} \bar{\beta}^* , \tag{3.28}$$

where ψ is an absolutely continuous non-decreasing function satisfying

$$0 \leqq \psi(F) \leqq \frac{2(r^* - 2)}{n - r^* + 2} \ . \tag{3.29}$$

These include the James-Stein estimators (with $\psi(F^*) \equiv \nu$), as well as the estimators introduced by Khursheed Alam and J. R. Thompson (cf. Baranchik [1]) with $\psi(F) = \nu/(1 + \nu/F^*)$. As a limiting case it includes Stein's "positive part" estimator [25] with $\psi(F) = \min \{F^*, (r^* - 2)/(n - r^* + 2)\}$. The Alam-Thompson and positive-part estimators have the advantage of giving rise to a positive shrinkage factor $1 - \psi(F^*)/F^*$. It is therefore reasonable to inquire whether a result analogous to that of Theorem 2 can be obtained for the more general class of estimators (3.28).

It can be shown by the methods of Efron & Morris [6] and with the aid of Lemma 3 of the Appendix (with $h(W) = \eta(S_1^*/2W)$, where $\eta(F^*) = \psi(F^*)/F^*$ and $W = S_0^*/2$), that

$$\text{Risk } \tilde{\beta}^* = r^* - \mathcal{E}\Big\{\frac{\psi(F^*)}{F^*}[2(r^*-2) - (n - r^* + 2 + \lambda)\psi(F^*)]$$
$$+ 4\psi'(F^*)[1 + \psi(F^*)] - \lambda \sum_{i=1}^{\infty} h^{(i)}(W)\Big\} . \tag{3.30}$$

When (3.16) holds, this immediately yields the Baranchik, Lin-Tsai, Efron-Morris result that for ψ in the interior of the interval (3.29), the estimator (3.28) dominates b^*. However, the nature of the last term in (3.30) makes it apparent that no straightforward generalization of Theorem 2 can be expected.

4. APPLICATION TO A STUDY OF THE RELATION BETWEEN INTERNAL AND EXTERNAL PRICES IN GERMANY, 1963-1975

The Statistisches Bundesamt, Wiesbaden, publishes a series [2] of monthly import and export price indices, divided into about 200 categories each, as well as monthly series of cost-of-living indices (consumer price indices) and wholesale price indices, and a quarterly series of wages and salaries classified by industry group. An appropriate model would allow for a 24-month (or even 36-month) distributed lag of internal prices behind external prices; but because of the limited length of the series, and limited core storage capacity of computers, no more than eight independent variables could be included in such a model, requiring one to deal with highly aggregated series of price indices, the mode of aggregation being determined necessarily on an ad hoc basis. In order to allow the choice of the mode aggregation to depend on the data, it was decided instead to employ a model that ignored lags and treated the dependence of internal and external prices as simultaneous.

The monthly data on import and export prices were reduced from a 36-category series (cf. Rostin [21]) to a 12-category one, on an introspective basis; a 9-group monthly consumer price index (CPI) series, and a 15-group monthly wholesale price index (WPI) series, were used; these monthly series were then averaged to form quarterly series comparable to a 7-group wage series and an 8-group salary series. The data of the model thus consisted of a 52×24 X-matrix of consecutive quarterly observations on the international price indices as independent variables, and a 52×39 Y-matrix of simultaneous consecutive quarterly observations on the 9 CPI's, 15 WPI's, and 15 wages and salaries. We thus have $n=52$, $k=24$, $m=39$. The G matrices were $24 \times k^*$ grouping matrices containing exactly one non-zero entry in each row, this being the relative weight of the index aggregated into a cruder index. Specifically, letting W be the diagonal matrix whose diagonal elements are the weights in TABLE I below, the G matrices were of the form $G_W = WG(G'WG)^{-1}$, where G is a matrix with exactly one unit entry in each row and the remaining entries = 0. In TABLE I are displayed the fourth columns, respectively--corresponding to the consumer price index component for electricity, gas, and fuel--of $\tilde{B} = X^\dagger Y$ and of $G_W B_W^* = G_W(XG_W)^\dagger Y$ for four out of 36 G_W's of order 24×6 that were tested.

The number of partitions of 24 objects into 6 groups is approximately 6.09×10^{15}. An exhaustive search is therefore out of the question. Out of the 36 chosen for comparison, those three with the lowest aggregation bias (as measured by the v-statistic of (2.20)) and the one with the highest aggregation bias are the ones for which the computations are displayed in TABLE I. In TABLE II are displayed upper 5% points $v_{.05}$ for testing the three hypotheses indicated.[9] The first three partitions of TABLE I are accepted by both the James-Stein and the reduced-mean-square-error test, and the fourth rejected. Partition 4 happened to be the best for CPI 1 (food), yielding an aggregation bias of 1.128. Note that the James-Stein estimates, when acceptable, are indistinguishable from the aggregative least squares estimates up to the fourth decimal place.

[9]These are approximate computations based on Pearson's tables [16]. More comprehensive and more accurate tables will be presented in a subsequent paper.

TABLE 1. Least squares and blown-up aggregative estimators of regression coefficients for CPI 4 (Electricity, gas, and fuel, for selected partitions into six groups. Partitions are identified by the identical t-ratios in parentheses

	International price	Weight	Least squares	Partition 1	Partition 2	Partition 3	Partition 4
IMPORTS	1 Agric., forestry & fishery raw produce	.071940	.035 (0.33)	.031 (0.36)	-.304 (-4.48)	.095 (1.48)	.353 (6.53)
	2 Coal, ores, stones & gravel	.025055	-1.076 (-2.79)	.046 (2.61)	.238 (4.60)	.018 (0.25)	.123 (6.53)
	3 Petroleum & petroleum products	.044130	.405 (3.15)	.223 (17.07)	.237 (25.22)	.106 (3.72)	.217 (6.53)
	4 Iron & steel	.043330	.506 (1.03)	.080 (2.61)	-.183 (-4.48)	-.166 (-2.59)	.213 (6.53)
	5 Non-ferrous metals	.039545	.694 (2.07)	.073 (2.61)	.070 (3.63)	.052 (1.48)	.194 (6.53)
	6 Machinery	.040885	-2.083 (-2.10)	.018 (0.36)	-.173 (-4.48)	.054 (1.48)	.201 (6.53)
	7 Road vehicles	.023490	1.926 (1.70)	.044 (2.61)	.223 (4.60)	.056 (3.72)	.115 (6.53)
	8 Electrical & precision goods	.037645	- .639 (-1.24)	-.293 (-5.27)	-.159 (-4.48)	-.144 (-2.59)	.185 (6.53)
	9 Chemicals	.039110	-2.031 (-2.82)	-.305 (-5.27)	-.165 (-4.48)	-.303 (-2.66)	.192 (6.53)
	10 Wood, glass, rubber products & plastics	.041970	.375 (0.54)	.078 (2.61)	.398 (4.60)	.056 (1.48)	.383 (1.76)
	11 Leather & textile products	.049140	1.167 (1.54)	.091 (2.61)	.466 (4.60)	.065 (1.48)	-2.359 (-8.56)
	12 Foodstuffs, beverages, & tobacco	.043760	.318 (1.00)	.081 (2.61)	.077 (3.63)	.058 (1.48)	.400 (1.76)
EXPORTS	1 Agric., forestry & fishery raw produce	.006980	-1.072 (-0.98)	.035 (17.07)	-.029 (-4.48)	.009 (1.48)	.034 (6.53)
	2 Coal, ores, salt, stones & gravel	.014990	2.297 (4.48)	.131 (10.94)	.026 (3.63)	.274 (4.20)	.074 (6.53)
	3 Petroleum products, peat	.005255	- .493 (-0.56)	.027 (17.07)	.009 (3.63)	.096 (4.20)	.026 (6.53)
	4 Iron & steel	.067565	- .546 (-2.71)	-.390 (-4.27)	-.134 (-1.42)	.190 (1.48)	-.763 (-3.02)
	5 Non-ferrous metals	.011370	-1.024 (-1.17)	.058 (17.07)	.020 (3.63)	.008 (0.25)	.056 (6.53)
	6 Machinery	.108820	- .256 (-1.19)	.953 (10.94)	.986 (7.02)	.261 (3.72)	1.273 (5.84)
	7 Road vehicles	.075005	.757 (1.85)	.139 (2.61)	-.316 (-4.48)	.100 (1.48)	.368 (6.53)
	8 Electrical & precision goods	.065185	.304 (0.97)	-.376 (-4.27)	-.129 (-1.42)	.156 (3.72)	-.411 (-1.84)
	9 Chemicals	.071655	.514 (0.90)	.363 (17.07)	-.302 (-4.48)	.095 (1.48)	-.452 (-1.84)
	10 Wood, glass, rubber products & plastics	.035090	-1.045 (-2.65)	-.203 (-4.27)	-.070 (-1.42)	.026 (0.25)	.410 (5.84)
	11 Leather & textile products	.025365	1.902 (1.39)	.047 (2.61)	.241 (4.60)	-.097 (-2.59)	.125 (6.53)
	12 Foodstuffs, beverages, tobacco	.012720	-1.382 (-1.18)	.024 (2.61)	-.054 (-4.48)	.009 (0.25)	.062 (6.53)
	F* [a]		5898.6	2882.9	2822.5	2629.5	479.8
	James-Stein shrinkage factor[b]		.9998757	.9999711	.9999705	.9999683	.9998263
	v (for CPI 4)[c]		0	2.045	2.089	2.242	12.268
	v (for all 9 CPI's)[d]		0	7.097	7.449	5.427	10.802

[a] Formula (3.13) of the text. [b] The factor of b^* in formula (3.16) of the text, when $\bar{\beta}^* = 0$.
[c] Formula (2.20), when e and e^* are the fourth columns of $\tilde{E}$ and $\tilde{E}^*$ respectively. [d] Formula (2.18), when $H' = [I_9, 0, 0]$.

TABLE II

Upper 5% points for the distribution of the v-statistic (2.20)(with V = I) for q = n - k = 28 and selected values of p = k - k* = 24 - k*, for three values of λ .

p	12	13	14	15	16	17	18	19	20
$\lambda = \sqrt{(p+q)(p+q+2)}$	3.53	3.65	3.76	3.88	4.00	4.13	4.26	4.40	4.50
$\lambda =$ p	1.74	1.86	1.98	2.09	2.20	2.33	2.46	2.57	2.70
$\lambda =$ 0	0.91	0.97	1.03	1.09	1.15	1.22	1.28	1.34	1.40

APPENDIX

Some useful and recurring results underlying Stein's theory of estimation are recorded here for convenience. Lemma 1 is due to Stein [26], and provides a simple method of proving Lemma 2 which is the basic result used by James and Stein [10]. Lemma 3 is a generalization to $\delta \geqq 0$ of a result due to Efron and Morris [6] for the case $\delta = 0$.

Lemma 1. Let $h: R \to R$ be an absolutely continuous function with a Lebesgue measurable derivative h' satisfying $h(b) - h(a) = \int_a^b h'(z)dz$ (for all $a < b$) and $\mathcal{E}|h'(Z)| < \infty$, where $Z \sim N(\zeta, 1)$. Then

$$\text{Cov}(Z, h(Z)) = \mathcal{E}(Z - \zeta)h(Z) = \mathcal{E}h'(Z). \tag{A1}$$

Proof. Define $x = z - \zeta$, $X = Z - \zeta$, and $g(x) = h(x + \zeta)$. Then for $x > a > 0$ we have

$$|g(x)-g(a)| = |\int_a^x g'(u)e^{-u^2/2}e^{u^2/2}du| \leqq e^{x^2/2} \int_a^\infty |g'(u)|e^{-u^2/2}du .$$

Given any $\varepsilon > 0$, choose a so large that $\int_a^\infty |g'(u)|e^{-u^2/2}du < \varepsilon/2$ and x so large that $|g(a)e^{-x^2/2}| < \varepsilon/2$; then $|g(x)e^{-x^2/2}| < \varepsilon$. Thus, $\lim_{x\to\infty} g(x)e^{-x^2/2} = 0$. A similar argument shows that $\lim_{x\to-\infty} g(x)e^{-x^2/2} = 0$. Integrating $\mathcal{E}\, g'(X) = (2\pi)^{-1/2}\int_{-\infty}^{\infty} g'(x)e^{-x^2/2}dx$ by parts we then obtain $\mathcal{E}g'(X) = Xg(X)$, which is equivalent to (A1). ∎

Lemma 2. Let $Z_i \sim N(\zeta_i, 1)$ be independently distributed random variables for $i = 1, 2, \ldots, r$, and denote $Z = (Z_1, Z_2, \ldots, Z_r)'$. Then

$$\mathcal{E}\frac{(Z - \zeta)'Z}{Z'Z} = (r - 2)\mathcal{E}\frac{1}{Z'Z} . \tag{A2}$$

Proof. Choosing $h_i(Z) = Z_i/Z'Z$ and denoting $Z^{)i(} = (Z_1, Z_2, \ldots, Z_{i-1}, Z_{i+1}, \ldots, Z_r)'$, we have by Lemma 1

$$\mathcal{E}\frac{(Z_i - \zeta_i)Z_i}{Z'Z} = \mathcal{E}\{\mathcal{E}\{\frac{(Z_i - \zeta_i)Z_i}{Z'Z} \,|Z^{)i(}\}\}$$

$$= \mathcal{E}\{\mathcal{E}\{\frac{\partial}{\partial Z_i}(\frac{Z_i}{Z'Z}) \quad |Z^{)i(}\}\}$$

$$= \mathcal{E}\{\frac{1}{Z'Z} - \frac{2Z_i^2}{(Z'Z)^2}\} \quad .$$

Thus,

$$\mathcal{E}\frac{(Z - \zeta)'Z}{Z'Z} = \sum_{i=1}^{r} \mathcal{E}\frac{(Z_i - \zeta_i)Z}{Z'Z} = (r - 2)\,\mathcal{E}\frac{1}{Z'Z} \; . \quad ■$$

Let $X_j \sim N(\mu_j, \sigma^2)$ be independently distributed random variables for $j = 1, 2, \ldots, q$. Then $S = \Sigma_{j=1}^{q} X_j^2/\sigma^2$ has the non-central chi-square distribution with q degrees of freedom and non-centrality parameter defined by $\lambda = \Sigma_{j=1}^{q}\mu_j^2/\sigma^2$. It will be more convenient instead to deal with the random variable $W = S/2$, whose distribution will be called "non-central gamma" with parameters $a = q/2$ and $\delta = \lambda/2$; its density function is

$$f(w) = \sum_{j=0}^{\infty} \frac{e^{-\delta}\delta^j}{j\,!}\,\frac{1}{\Gamma(a + j\,)}\,w^{a + j - 1}\, e^{-w} \,, \tag{A3}$$

and its moment generating function is

$$m_W(t) = \mathcal{E}e^{tW} = (1 - t)^{-a}\; e^{\delta t/(1 - t)}. \tag{A4}$$

Thus, $\mathcal{E}W = m_W'(0) = a + \delta$ and hence $\mathcal{E}S = q + \lambda$.

Lemma 3. The non-central gamma variable W with density (A3) satisfies

$$\mathrm{Cov}(W, h(W)) = \mathcal{E}(W - \mathcal{E}W)h(W) = \mathcal{E}Wh'(W) + \delta\sum_{i=1}^{\infty}\mathcal{E}h^{(i)}(W) \tag{A5}$$

for any absolutely continuous infinitely differentiable function h such that $\mathcal{E}|Wh'(W)| < \infty$ and $\mathcal{E}|h^{(i)}(W)| < \infty$ for $i=0, 1, \ldots$, where $h^{(i)}$ denotes the ith derivative of h .

Proof. Integrating by parts, we have

$$\mathcal{E}Wh'(W) = \int_0^\infty wf(w)h'(w)\,dw$$

$$= [wf(w)h(w)]_0^\infty - \int_0^\infty [f(w) + wf'(w)]h(w)\,dw.$$

Since wf(w) is unimodal, an argument similar to that used in the proof of Lemma 1 shows that the first term on the right vanishes, since $\mathcal{E}|Wh(W)| < \infty$, and from (A3) we have

$$f(w) + wf'(w) = -wf(w) + \sum_{j=0}^{\infty} (a + j)\frac{e^{-\delta}\delta^j}{j!}\frac{1}{\Gamma(a + j)}w^{a+j-1}e^{-w}.$$

From this we obtain

$$\mathcal{E}Wh'(W) = \mathcal{E}(W - \mathcal{E}W)h(W) + \delta\mathcal{E}h(W) - \psi_0(\delta)$$

where we define, for i = 0, 1, 2,...,

$$\psi_i(\delta) = \sum_{j=0}^{\infty} \frac{e^{-\delta}\delta^j}{j!}\frac{1}{\Gamma(a+j)}\int_0^\infty w^{a+j-1}e^{-w}h^{(i)}(w)\,dw.$$

We verify by integration by parts that these functions satisfy the recursive formulae

$$\psi_i(\delta) = \delta\mathcal{E}h^{(i)}(W) + \psi_{i+1}(\delta)$$

for i = 0, 1, 2, ... Consequently, for any k = 1, 2,..., we have

$$\mathcal{E}(W - \mathcal{E}W)h(W) = \mathcal{E}Wh'(W) - \delta\mathcal{E}h(W) + \delta\sum_{i=0}^{k-1}\mathcal{E}h^{(i)}(W) + \psi_k(\delta).$$

Proceeding to the limit, we obtain (A5).

If we take the case h (W) = W, (A5) yields

$$\mathcal{E}W^2 = (\mathcal{E}W)^2 + \mathcal{E}W + \delta = a(a + 1) + 2(a + 1)\delta + \delta^2.$$

For the chi-square variable S = 2W with q = 2a degrees of freedom and non-centrality parameter $\lambda = 2\delta$ we then have

$$\mathcal{E}S^2 = q(q + 2) + 2(q + 2)\lambda + \lambda^2. \tag{A6}$$

REFERENCES

[1] BARANCHIK, A. J. (1970). A family of minimax estimators of the mean of a multivariate normal distribution. *Ann. Math. Statist.*, **41**, 642-645.

[2] Bundesrepublik Deutschland, Statistisches Bundesamt, Wiesbaden, Fachserie M: Preise, Löhne, Wirtschaftsrechnungen. Reihe 1: Preise und Preisindizes für Aussenhandelsgüter. Reihe 6: Preise und Preisindizes der Lebenshaltung. Reihe 8: Index der Grosshandelsverkaufspreise. Reihe 15, I & II: Arbeitnehmerverdienste in Industrie und Handel. Mainz: Verlag W. Kohlhammer, 1963-1975.

[3] CHIPMAN, JOHN S. (1975). The aggregation problem in econometrics. *Advances in Appl. Probability* (Special Supplement), **7**, 72-83.

[4] CHIPMAN, JOHN S. (1975). Optimal aggregation in large-scale econometric models. *Sankhya, Ser C*, **37**, 4, 121-159.

[5] CHIPMAN, JOHN S. (1976). Estimation and aggregation in econometrics; an application of the theory of generalized inverses. *Generalized Inverses and Applications* (M. Zuhair Nashed, ed.). Academic Press, New York. 549-769.

[6] EFRON, BRADLEY and MORRIS, CARL (1976). Families of minimax estimators of the mean of a multivariate normal distribution. *Annals of Statistics*, **4**, 11-21.

[7] EFRON, BRADLEY and MORRIS, CARL (1976). Multivariate empirical Bayes and estimation of covariance matrices. *Annals of Statistics*, **4**, 22-32.

[8] FISHER, WALTER D. (1962). Optimal aggregation in multi-equation prediction models. *Econometrica*, **30**, 744-769.

[9] HOTELLING, HAROLD (1951). A generalized T test and measure of multivariate dispersion. *Proc. Second Berkeley Symp. Math. Statist. Prob.*, (Jerzy Neyman, ed.). University of California Press, Berkeley and Los Angeles, 23-41.

[10] JAMES, W. and STEIN, CHARLES Estimation with quadratic loss. *Proc. Fourth Berkeley Symp. Math. Statist. Prob.* (Jerzy Neyman, ed.). University of California Press, Berkeley and Los Angeles. I, 361-379.

[11] KRISHNAIAH, P. R. and CHANG, T. C. (1972). On the exact distribution of the traces of $S_1(S_1 + S_2)^{-1}$ and $S_1S_2^{-1}$. *Sankhya, Ser. A* , **34**, 2, 153-160.

[12] KRISHNAIAH, P. R. and CHATTOPADHYAY, A. K. (1975). On some noncentral distributions in multivariate analysis. *South African Statistical Journal*, **9**, 1, 37-46.

[13] LIN, PI-ERH and TSAI, HUI-LIAN (1973). Generalized Bayes minimax estimators of the multivariate normal mean with unknown covariance matrix. *Annals of Statistics*, **1**, 142-145.

[14] MALIK, HENRICK JOHN (1970). An alternative derivation of doubly non-central F distribution. *Metron*, **28**, 2-4, 180-184.

[15] MALINVAUD, E. (1956). L'agrégation dans les modèles économiques. *Cahiers du Séminaire d'Econométrie*, **4**, 69-146.

[16] PEARSON, KARL (1968). *Tables of the Incomplete Beta-Function* (2nd ed.). University Press, Cambridge.

[17] PILLAI, K. C. S. and YOUNG, D. L. (1971). On the exact distribution of Hotelling's T_0^2. *Journal of Multivariate Analysis*, **1**, 90-107.

[18] PRINGLE, R. M. and RAYNER, A. A. (1971). *Generalized Inverse Matrices with Applications to Statistics.* Charles Griffin & Company, Ltd., London.

[19] RAO, C. R. (1966). Generalized inverse for matrices and its application to statistics. *Festschrift for J. Neyman: Research Papers in Statistics*, (F. N. David, ed.). John Wiley & Sons, New York. 263-279.

[20] RAO, C. RADHAKRISHNA and MITRA, SUJIT KUMAR (1971). *Generalized Inverse of Matrices and its Applications.* John Wiley & Sons, Inc., New York.

[21] ROSTIN, WERNER (1974). Die Indices der Aussenhandelspreise auf Basis 1970. *Wirtschaft und Statistik*, 387-394. (Engl. transl.: Indices of foreign trade prices, Studies on Statistics, **30**, W. Kohlhammer, Stuttgart and Mainz.).

[22] SCLOVE, STANLEY L. (1968). Improved estimators for coefficients in linear regression. *J. Amer. Statist. Assoc.*, **63**, 596-606.

[23] SCLOVE, STANLEY L. (1971). Improved estimation of parameters in multivariate regression. *Sankhyā, Ser. A*, **33**, 61-66.

[24] STEIN, CHARLES (1956). Inadmissibility of the usual estimator for the mean of a multivariate normal distribution. *Proceedings of the Third Berkeley Symp. Math. Statist. Prob.*, (Jerzy Neyman, ed.). University of California Press, Berkeley and Los Angeles. **I**, 197-206.

[25] STEIN, CHARLES (1966). An approach to the recovery of inter-block information in balanced incomplete block designs. *Festschrift for J. Neyman: Research Papers in Statistics*, (F. N. David, ed.). John Wiley & Sons, Inc. New York. 351-366.

[26] STEIN, CHARLES (1974). Estimation of the mean of a multivariate normal distribution. *Proceedings of the Prague Symposium on Asymptotic Statistics*, (Jaroslav Hájek, ed.). Charles University, Prague. **II**, 345-381.

[27] TANG, P. C. (1938). The power function of the analysis of variance tests with tables and illustrations of their use. *Statistical Research Memoirs.* (J. Neyman and E. S. Pearson, eds.). Department of Statistics, University College, University of London, London. **II**, 126-149.

[28] THEIL, H. (1954). *Linear Aggregation of Economic Relations.* North-Holland Publishing Co., Amsterdam.

[29] THEIL, H. (1961). *Economic Forecasts and Policy* (2nd ed.). North-Holland Publishing Co., Amsterdam.

[30] TORO-VIZCARRONDO, CARLOS and WALLACE, T. D. (1968). A test of the mean square error criterion for restrictions in linear regression. *J. Amer. Statist. Assoc.*, **63**, 558-572.

METHODOLOGICAL DEVELOPMENTS IN SOME APPLICATIONS OF CLUSTERING

A. COHEN*, R. GNANADESIKAN, J. R. KETTENRING AND J. M. LANDWEHR
Bell Telephone Laboratories, Murray Hill, New Jersey

Four specific applications of cluster analysis are used to motivate a variety of problems that arise in practice. While some of the problems may be unique to the particular application, others appear to be of more general interest. Examples include: (1) finding an appropriate metric for measuring closeness, (2) clustering objects and/or grouping variables especially when one or both of these are large in number, and (3) supplementing outputs of clustering algorithms with graphical and numerical aids for assessing and interpreting the clusters. Solutions to the problems in the context of the four applications are described and some general questions relating to the suggested methods are raised.

1. INTRODUCTION

In the development of multivariate statistical methodology, Fisher's (1936) work leading to the two-group discriminant function and generalizations to the multiple-group case (see, e.g., Rao, 1952, § 9c) have been a major focus both of theoretical and applied interest. As against these situations wherein one's goal is to classify an object with respect to certain prespecified groups, the problems of clustering, where the interest is in letting the data determine the groups, have received relatively little attention until recently. The origins of cluster analysis appear to be in fields such as numerical taxonomy (see, e.g., Sokal & Sneath, 1963) and psychology (see, e.g., Tryon, 1939) rather than in statistics.

In recent years, partly because of the feasibility of complex computations via the modern computer, cluster analysis has received considerable attention. Unfortunately, much of the work seems to have resulted in a spate of different algorithms whose statistical properties are not well understood. Fortunately, some useful survey articles (e.g., Cormack, 1971) and books (e.g., Everitt, 1974, and Hartigan, 1975) are helping to integrate the concerns of and approaches to clustering into the mainstream of statistics. This could result in the statistical aspects of clustering receiving more attention by statisticians.

For convenience, one can distinguish three stages of a cluster analysis with each stage having associated with it several interesting aspects that need new or improved statistical methodology. The first is the *input* stage, where the data

*Now at the Technion, Haifa, Israel.

has to be processed and appropriate representations of it obtained for initiating the cluster analysis. The nature of the data (e.g., metric vs. nonmetric) and the choice of representations (e.g., scaled vs. unscaled variables) can have dramatic impacts on the results. Careful selection of a measure of closeness between objects, as required by some algorithms, is also crucial. The second stage of clustering is the *analytical methods* one, where issues of different approaches and algorithms arise, and in fact this is the phase that has attracted much of the attention. The third stage is associated with *output*, and the need here is for methods that facilitate understanding and interpreting the results obtained from clustering algorithms.

The present authors' involvement in clustering has been with specific data analysis problems in which cluster analytic approaches proved to be useful. In the spirit of the theme for this symposium, this paper will briefly describe four particular applications of cluster analysis in which one or more of the authors was involved. Its main objective is to utilize the contexts of these applications both to motivate the development of specific methodology and to illustrate the uses of the *ad hoc* methods. A minor goal will be to raise some general questions concerning the *ad hoc* techniques as well as other issues for possible future investigation. In terms of the three stages of clustering, this paper will be concerned largely with issues and approaches pertaining to the analytical and the output phases.

More specifically, § 2 outlines the four applications. Section 3 is concerned with methodological aspects associated with the analytic or algorithmic phase while § 4 addresses the problems at the output stage. More general questions are raised in §5 which consists also of concluding remarks.

2. DESCRIPTIONS OF THE FOUR EXAMPLES

The purpose here is to give brief descriptions of each of the examples and to indicate the role cluster analysis played in each of them. Although other statistical techniques were useful in these analyses, the discussion is limited to clustering aspects.

In the first example (Cohen et al., 1975) the goal was to compare the performance of 52 investment portfolios. The comparisons were based on the changes in value of each portfolio across 11 successive quarterly periods. Clustering techniques were used to organize the portfolios into groups according to the similarities of their performances.

In another problem (Chen et al.,1975) the purpose was to compare 48 subsidiary companies according to several measures of costs and revenues. Since the subsidiaries operate in different environments, a straightforward comparison of these measures could be very misleading. Thus, the subsidiaries were first clustered into homogeneous groups using values on 25 environmental-type variables. Then cost and revenue comparisons were made within each cluster rather than across the whole population.

The third example (Kettenring et al., 1976) involved the training needs of workers engaged in a very complex job assignment. A preliminary study identified 169 different components of this job. To study training needs relative to these components, 452 workers completed a questionnaire in which they answered whether or not training would be helpful to them in each of the items. This yielded a 452×169 matrix with each entry a 0 (no) or 1 (yes). Clustering was used both to reduce the number of columns, by looking for clusters of variables representing similar training needs, and to reduce the number of rows, by looking for clusters of people with similar training needs profiles. This process led to a manageable and meaningful summarization of the data.

The overall goal in the last example (Chen et al., 1974) was to use data on several economic measures of industrial and utility companies for forming cohesive groups of corporations. Superficially, this example resembles the subsidiary example, but in fact some of the problems involved are quite different. For instance, there is much a priori information on what the groupings might be - chemical companies, steel companies, etc. Also there is extreme heterogeneity among the clusters, e.g., the industrials as a group are much more diverse than the utilities. Moreover, there are a very large number of firms to be clustered - about 500 industrials and 100 utilities - but some groups are of more interest than others.

3. PROBLEMS AND METHODOLOGICAL DEVELOPMENTS AT THE ANALYTICAL METHODS STAGE

3.1. Large Number of Variables

The presence of a large number, p, of variables can be an advantage for purposes of clustering a collection, n, of objects. Indeed, p may need to be fairly large if one is to obtain an interesting separation of the objects.

Nevertheless, a large p can also be troublesome. First there is the need to interpret the clusters in terms of the variables (§4.3). Which are the important ones? Are any of them redundant? Second, the presence of many variables may be a burden on the clustering algorithm. This is true, for example, in a modified version (Warner, 1968) of the ISODATA algorithm developed by Ball and Hall (1967).

The algorithm, which is iterative, allows for the repeated consideration of whether or not any cluster is too large. In the modified version of the algorithm, this decision is based on the maximum eigenvalue of the (p×p) dispersion matrix of the candidate cluster. If it is "too large", the cluster is split into two. The locations of the new clusters depend upon the eigenvector associated with this eigenvalue. Consequently, the overall effort mounts rapidly as p increases.

Thus, it may be necessary to reduce the number of variables while being careful to preserve those which would be useful for clustering. Although classical principal components analysis can be useful for this purpose, it does not guarantee that the best variables (linear combinations) for clustering will be chosen. Another approach would be to develop procedures akin to stepwise regression but designed around a clustering objective. Methods of this type do not appear to be currently available. The tactics followed in the examples below are based on (a) clustering the variables and then (b) selecting simple linear combinations of the clusters as proxies for the original variables.

In the training needs example, for clustering the workers by the modified ISODATA algorithm, the number of variables had to be reduced from $p=169$. Specifically, hierarchical cluster analysis was applied to the variables and resulted in a consolidation of them into 24 groups. An average variable was constructed to represent each cluster and then scaled by a measure of its "importance" in the overall sample, the percentage of yes responses in the cluster. The resulting (452×24) matrix formed the input to ISODATA for clustering the 452 workers.

In the subsidiaries example, an initial reduction of the $p=25$ variables was also made but mainly for purposes of removing redundancies since the dimensionality was already manageable. First, using a robust correlation coefficient to measure similarity, the variables were clustered as in the training needs example. Four clusters of variables resulted. Then, rather than simply choosing an average variable, principal components analysis within each cluster was used to suggest one or two simple and interpretable linear combinations (e.g., ones with ± 1 weights for the variables) to represent each cluster. In all, seven linear combinations were formed from the four clusters.

An interesting question regarding these linear combinations is how they compare against the more familiar principal components in terms of the variability explained by them. One way of making this comparison is in terms of the canonical correlations between the set of linear combinations and the set of principal components. The first four canonical correlations all exceeded 0.9, thus indicating substantial closeness between the spaces spanned by the simple linear combinations and by the principal components. This idea as well as others for comparing an arbitrary set of linear combinations against principal components are described by Landwehr (1976).

3.2. Large Number of Objects

As with a large p, a large n can be both a hindrance and a help. If too large, it can discourage the application of many of the hierarchical clustering algorithms which typically require $\binom{n}{2}$ inter-object dissimilarity measures as input. The main advantage which can be reaped from a large n is a more definitive determination of major clusters.

In the training needs example, the large number of workers (n=452) presented a problem since it was intended to group them by hierarchical analysis. One approach to such a situation would be to cluster a stratified sample of the workers, but the most appropriate strata would be the unknown clusters themselves! The initial reduction was accomplished instead by the ISODATA procedure, mentioned in §3.1, since large n is less trouble for such non-hierarchical algorithms. The parameters governing the iterations were selected to produce a large number - about 100 - of small clusters, and these in turn were further processed using a hierarchical algorithm.

In one facet of the corporation grouping problem, the presence of a reasonably large number of firms meant that the metric defining inter-company distances could be calculated so as to reflect external information on appropriate groupings of these companies. Since the firms were drawn from standard lists of chemical, drug, domestic oil, and steel companies, it was felt that these *a priori* classifications should influence the analysis. To accomplish this, within-group dispersions were calculated for each of these four groups (after deleting outliers), and generalized distances among the companies were computed as

$$d_{ij}^2 = (\underset{\sim}{x}_i - \underset{\sim}{x}_j)'\underset{\sim}{M}(\underset{\sim}{x}_i - \underset{\sim}{x}_j), \tag{1}$$

where $\underset{\sim}{x}_i$ and $\underset{\sim}{x}_j$ denote the observations for companies i and j and $\underset{\sim}{M}^{-1}$ is either the individual or a pooled-within-group dispersion matrix. The clustering was then based on these distances. More ambitiously, one could refine the metric in light of the initial clustering and repeat the process.

In another facet of the corporation grouping problem, the goal was to identify the nearest neighbors of AT&T among roughly 500 industrial companies. The associated, but slightly different, clustering problem would be to find the cluster of companies containing AT&T. The nearest neighbor calculation was based on p=5 variables, and it was known from other analyses that the companies really do not separate into very distinct groups. In such a case, a clustering algorithm can do little more than effect a sensible partition of the data and the distinction between "nearest neighbors" and "clusters" becomes obscure.

One can exploit the large n in this situation by identifying neighbors of AT&T through an iterative scheme which leads to a local metric for measuring distances of companies to AT&T. The main advantage of this strategy as compared, for example, to an ISODATA approach is that one is forming only groups that are of primary interest.

Specifically, the technique is to (i) find some number, n_0 say, of companies closest to the object of interest (i.e., AT&T) using Eq. (1) with $\underset{\sim}{M}^{-1}$ the overall dispersion based on all n companies; (ii) replace $\underset{\sim}{M}^{-1}$ with an estimate of the "local dispersion" based on the n_0 companies; (iii) find a new list of n_0 neighbors in this local metric; and (iv) iterate. Figure 1 shows a hypothetical example of how the procedure might work when real clusters are present. The procedure could be repeated to find other clusters, but computation might become prohibitive if many clusters need to be found.

In the case of AT&T and with n_0=100, convergence to a stable list of neighbors took only five iterations. The neighbors included many chemical, paper, and food-related companies plus representatives of several other basic industries.

In a methodologically similar vein, Rohlf (1970), Chernoff (1972), and Maronna and Jacovkis (1974) also deal with the issue of allowing each cluster to possess its own metric.

3.3. *Graphical Methods*

Perhaps the most perplexing aspect of multivariate analysis is the difficulty in visualizing patterns in the data. This has prompted many proposals for novel graphical representations such as star diagrams (Monkhouse and Wilkinson, 1973,

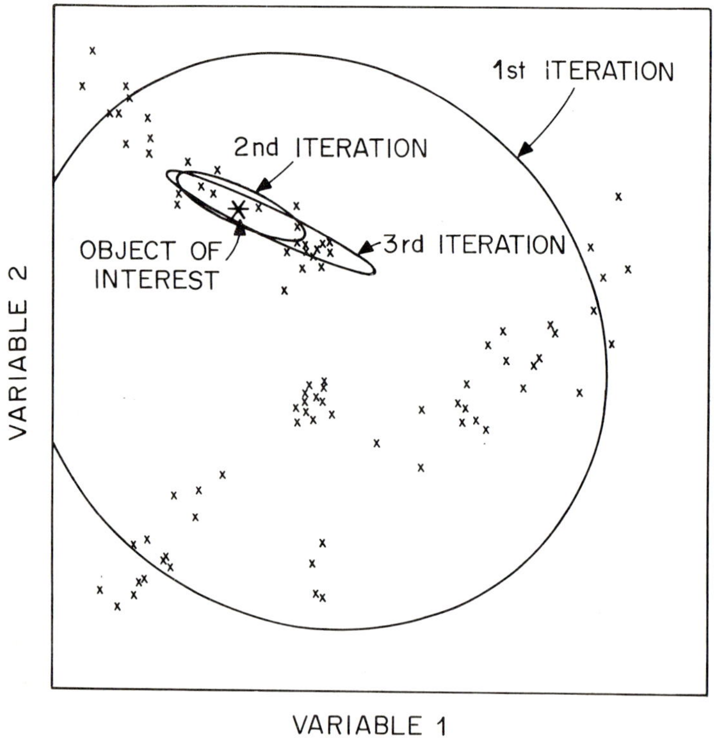

Fig. 1. Artificial example of iterating towards a local metric.

p. 36), metroglyphs (Anderson, 1957), function plots (Andrews, 1972), and faces (Chernoff, 1973). These and other graphical displays are relevant to the clustering problem because they can often suggest natural groupings of the data.

Various forms of Andrews' function plots were produced in the portfolio and corporations examples. The procedure proposed by Andrews is to plot, for each observation vector $\underset{\sim}{x}_i$, $i=1,\ldots,n$,

$$f_i(t) = [\underset{\sim}{a}(t)]'\underset{\sim}{x}_i, \quad t_1 \leq t < t_2,$$

as a function of t. One attractive choice for the coefficient vector is

$$[\underset{\sim}{a}(t)]' = (1/\sqrt{2}, \sin t, \cos t, \sin 2t, \cos 2t, \ldots), \; t_1 = -\pi, \; t_2 = \pi.$$

For a specific value of t, the plot shows projections of the data onto a particular one-dimensional subspace. The plot as a whole exhibits projections across a smoothly varying spectrum of linear combinations. To handle the situation when n is large, an adaptation of the function plot called a quantile contour plot is described by Gnanadesikan (1973); see §4.3 below for an example.

In the portfolio example, plots were made of $f_i(t)-f(t)$ versus t, where $f(t) = \underset{i=1,\ldots,52}{\text{median}} f_i(t)$ to emphasize differences among the portfolios. Nine of the portfolios formed strikingly similar curves indicative of a cluster. A few (e.g. #s 47 and 52) behaved as outliers deviating sharply from the median behavior line. Figure 2 shows some of the 52 plots, including the nine similar ones, ordered for convenience of display and laid out in strips one on top of the other.

Another plot used in the portfolio example is constructed from the inter-object distances, d_{ij}. The plot is of

$$(\underset{i}{\text{median}}\; d_{i(j)},\; d_{i(j)}),\; i=1,\ldots,n;\; j=1,\ldots,n-1,$$

where $d_{i(j)}$ is the jth largest value among $d_{i1},\ldots,d_{i,i-1},d_{i,i+1},\ldots,d_{in}$. Thus, the first column of the plot displays the empirical distribution of the n nearest neighbor distances plotted against their median. The second column shows the n second nearest neighbor distances, and so on. A graph of this type, together with output identifying i and j for each point, can be helpful in detecting certain types of clusters and outliers.

The plot for the portfolio data is presented in Figure 3. The largest two distances in all but the last four columns involve either portfolio #47 or #52, i.e., $\max_i d_{i(j)} = d_{47(j)}$ or $d_{52(j)}$, $j=1,\ldots,47$. These two along with two others are nearly always associated with the four largest distances in each column and hence may be outliers.

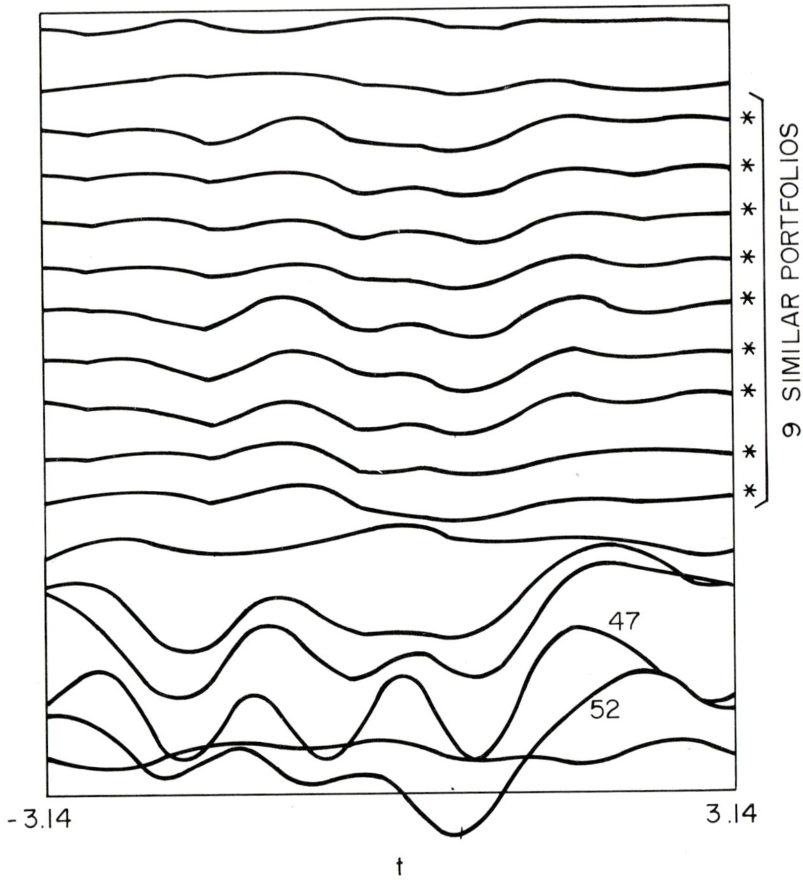

Fig. 2. High-dimensional plot in strip form for portfolios data.

Moreover, there is a noticeable blob in the lower left section of the figure. In particular, the smallest eight distances in the first eight columns involve only the nine portfolios mentioned previously. This suggests, once again, that these nine cluster together.

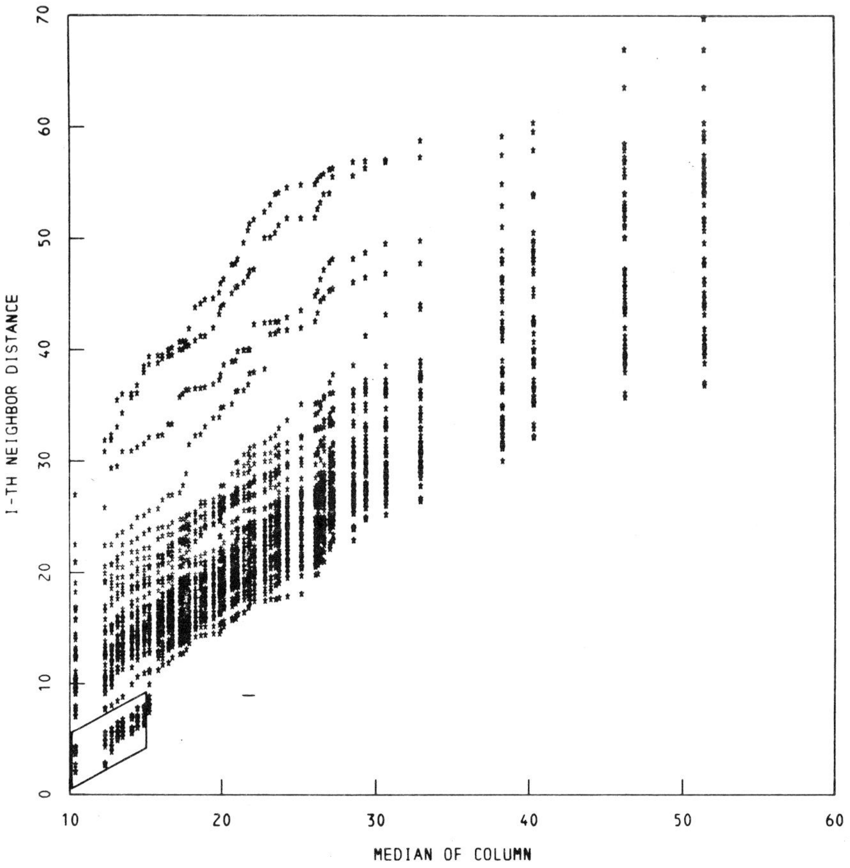

Fig. 3. Ordered distances for portfolios data.

4. PROBLEMS AND METHODOLOGICAL DEVELOPMENTS AT THE OUTPUT STAGE

4.1. Assessment of Statistical Significance

Statistical models for clustering which are both realistic and tractable are needed to help in the assessment of cluster significance. The models used in two of the present examples are reasonably tractable in that they can be studied by permutation methods. However, the models are less realistic than one might like.

An informal graphical procedure making use of permutations was used in the portfolio example. First, the observed values of each variable (p=11) were replaced by their ranks within that variable, thus changing the original data matrix to a new n×p matrix each of whose columns contained integers from 1 to n. The transformation to ranks enables the use of a non-parametric approach. With no cluster structure in the data and if the p variables are independent (a plausible assumption for these data), each permutation of the integers 1,2,...,n in a column is equally likely, and the columns are independent; this is the null model. Under non-null conditions, all permutations would not be equally likely. Therefore, to assess the significance of cluster structure in the data, the clusters from the ranked data can be compared to clusters obtained from simulations of the null model. MacQueen (1965) has considered a related permutation idea.

A graphical method was developed in the portfolio application for comparing the actual clusters to these null simulations. A hierarchical clustering was done of the n=52 portfolios resulting in 51 cluster strengths. Then 250 randomly permuted data ranks matrices were used as input to the same cluster method, leading to 51 new strengths each time. Figure 4 uses box plots (cf. Tukey, 1970) to dis-

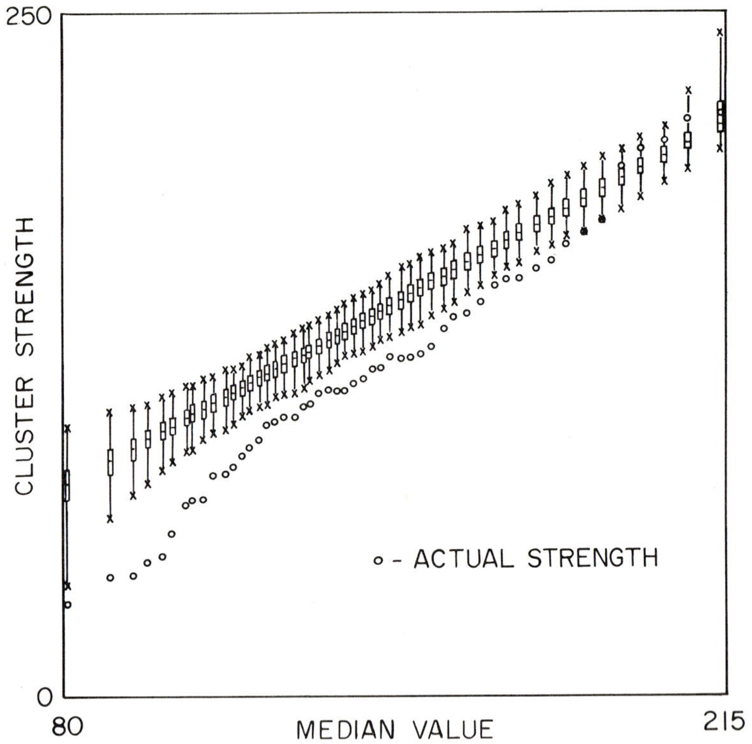

Fig. 4. Comparison of strengths for portfolios data. 52 portfolios.

play the distributions of the 250 values of each of the 51 strengths, with the ith box corresponding to the ith strength. The boxes are placed along the x-axis so that the medians fall on the 45° line. Superimposed on the box plots are the strengths derived from the actual data ranks matrix for the portfolios. These strengths are lower than the "null regions", suggesting smaller distances between certain portfolios than would be expected by random chance.

Recalling the discussion of the portfolio example in § 3.3, there was evidence for a cluster of nine portfolios and four "outliers". Removing these portfolios (plus two additional possible "outliers" revealed by other methods), and then repeating the above steps on the remaining 37 portfolios gave Figure 5. These portfolios look more like a random sample from one parent distribution.

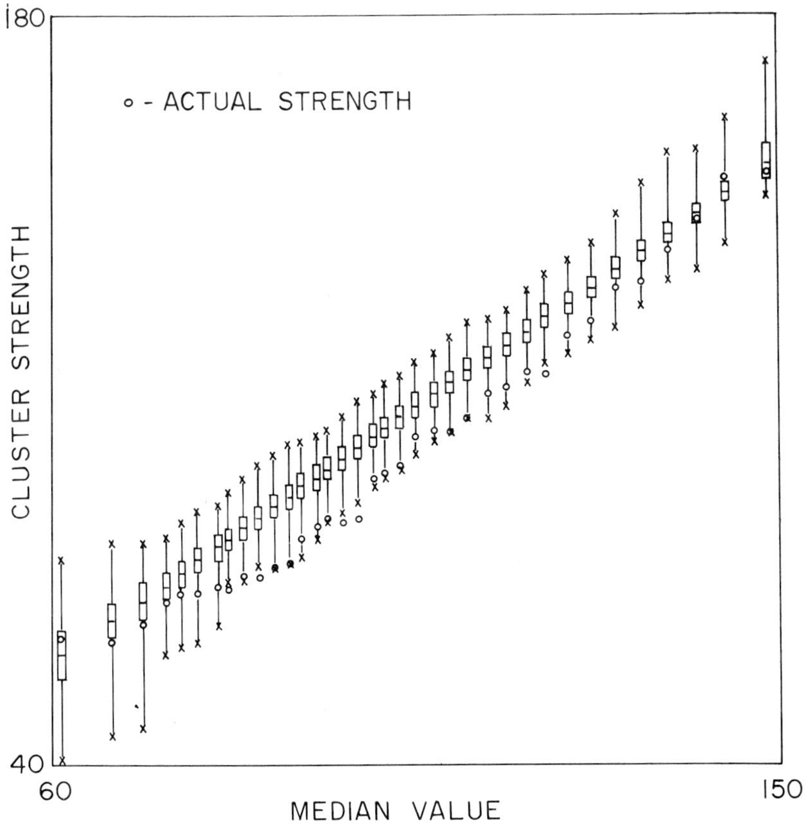

Fig. 5. Comparison of strengths for portfolios data. 37 portfolios.

A somewhat different permutation idea was used in the subsidiaries example. The set of similarities used as input for the hierarchical clustering algorithm was randomly permuted several times, and each time the resulting similarity matrix was clustered. Then informal comparisons were made between the sizes of the clusters obtained from the actual similarities and the sizes of clusters from the permuted similarities. Several authors (e.g., Ling, 1973; Hubert, 1974; and Baker and Hubert, 1975) have considered related procedures which are more formal. The graphical method described above for comparing actual strengths with simulated strengths could also be used here.

The above significance assessment methods may be useful crutches, but they have limitations even for detection of cluster structure (see § 5). Moreover, insight into and interpretation of the structure may be of greater interest than the mere statistical significance of the existence of clusters.

4.2. Examination of Clusters Using Internal Distances

Methods for studying the internal cohesiveness of clusters can be useful in interpreting cluster analyses. First, consider the relative tightness of a k-point group (a potential cluster) compared to other k-point neighborhoods in the data. For each data point, find its k-1 closest neighbors; then form the average interpoint distance d_i among these k objects over all k(k-1)/2 pairs. Data points contained in "real" k-point clusters should give d_i values substantially smaller than data points not in "real" clusters. A reasonable starting point for comparing the d_i's, for a given k, is by means of a normal probability plot. A tight cluster of size j should produce j points with nearly equal d_i's which are well separated from the others at the bottom of the plot for k=j. Figure 6 shows normal probability plots of the d_i's for the portfolio data when k=2,5,9, and 14. The behavior at the lower ends of these plots suggests the existence of a group of size approximately k=9 in these data.

The preceding method can be modified to study the isolation, rather than the internal cohesiveness, of a given group of k points relative to the rest of the data. For a specific k-point group, let $\bar{e}$ denote the average intra-group distance. For each point in this group, replace it by its closest neighbor outside the specified group. Next calculate the average intra-group distance e_i for this new, slightly modified, k-point group. If the initial group is in fact a well-isolated cluster, then $\bar{e}$ should be much smaller than $e_1,\ldots,e_k$. A normal probability plot of $\bar{e}$, $e_1,\ldots,e_k$ may again help in this evaluation. Figure 7 gives such a plot for the potential cluster of size k=9 in the portfolio data. The lowest point corresponds to $\bar{e}$ and is quite apart from the rest. The curvature in the plot suggests nonnormality of $e_1,\ldots e_9$; this is possibly due to strong dependencies among these values.

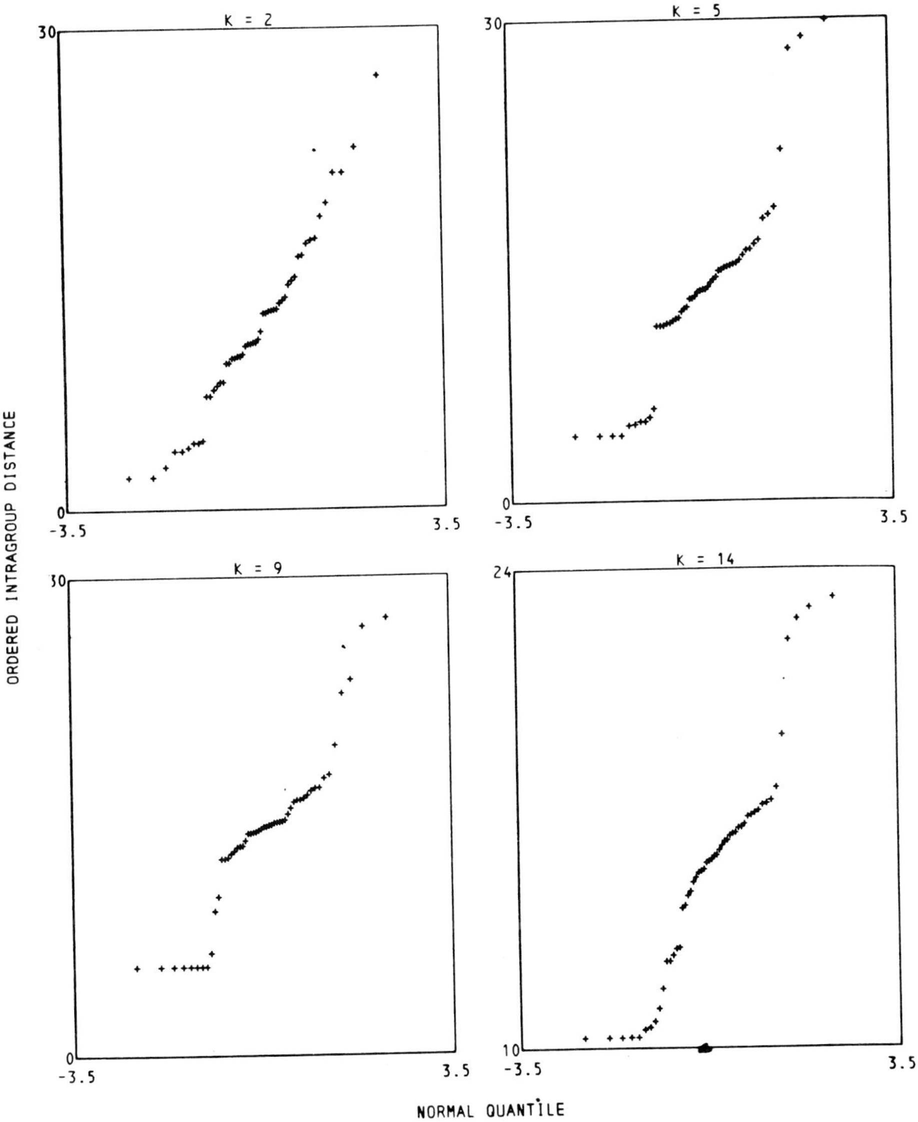

Fig. 6. Normal probability plots of intragroup average distances.

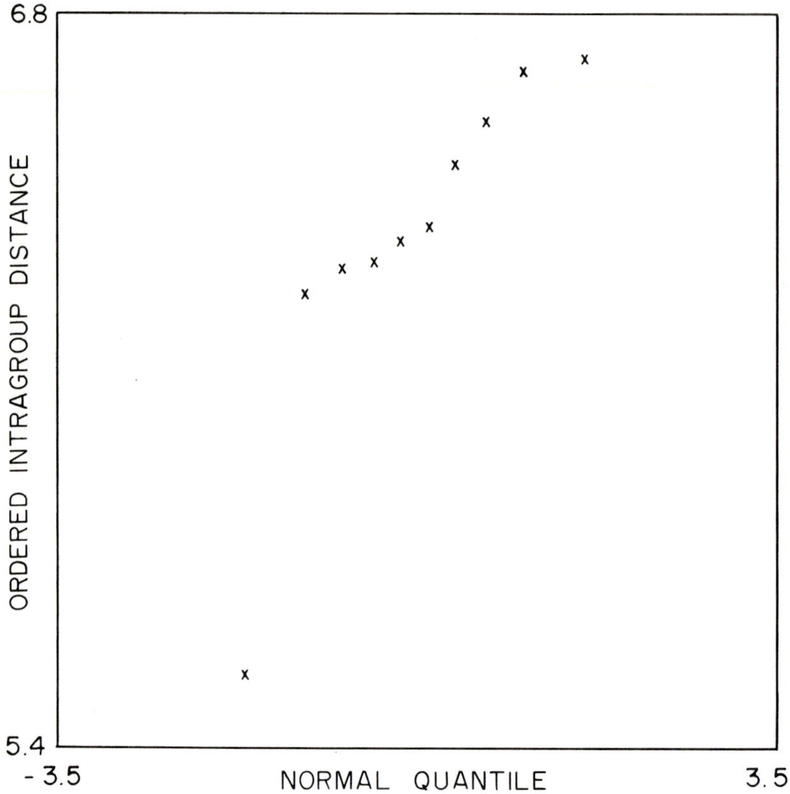

Fig. 7. Normal probability plot of intragroup average distances after replacements.

Another way of examining the internal cohesiveness of a cluster is illustrated in Figure 8 with data from the subsidiaries example. It shows the squared distances from certain cluster centroids to entities that are near that centroid. For each cluster centroid, the distances of every entity from that centroid are plotted above the cluster identification shown on the x-axis. (The y-axis was truncated so as to permit the display of every entity in the column corresponding to its own cluster and a few other nearby entities.) The symbol plotted is the cluster to which the entity was assigned. Information in Figure 8 includes, for example, the following: clusters F and H are extremely well separated from neighboring entities; an entity not assigned to cluster E is still reasonably close to cluster E; and there is a large distance between the members of cluster A farthest from their centroid and the next closest entities not assigned to A.

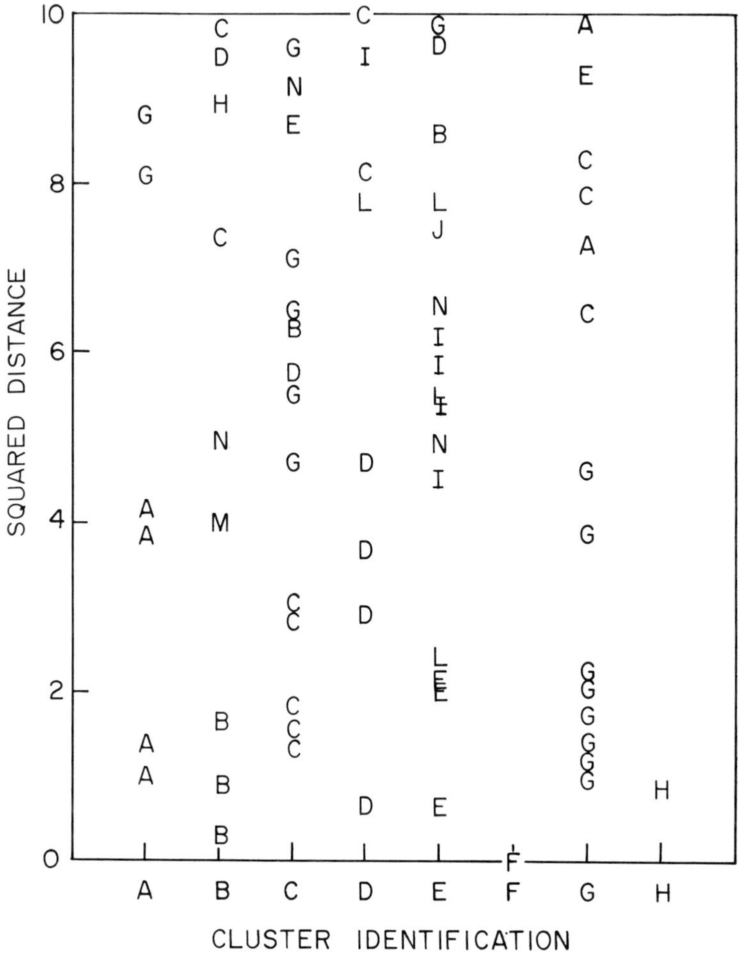

Fig. 8. Distances of entities to nearby cluster centroids.

4.3. Examination of Clusters Using Individual Variables and Linear Combinations

This section discusses numerical and graphical methods for examining the clusters in terms of either variables used to form clusters or other variables of interest.

The t-statistic is often helpful in interpreting a hierarchical clustering tree. At those stages in the tree where two clusters are merged, a two-sample t can be computed for each of the variables. Then log p, where p is the attained significance level calculated in the usual way, is a useful measure of the separation of the clusters, especially when the t-statistics are based on different "degrees of freedom". The t-statistic cannot, of course, be given its usual formal significance level interpretation since the necessary assumptions are not satisfied; rather, it is intended as a descriptive measure. In the training-needs problem this method was used on variables both internal and exogenous to the cluster analysis to focus on groups of workers which were well separated.

A graphical method used in the subsidiaries example to compare an exogenous variable across several clusters is illustrated in Figure 9. The clusters are

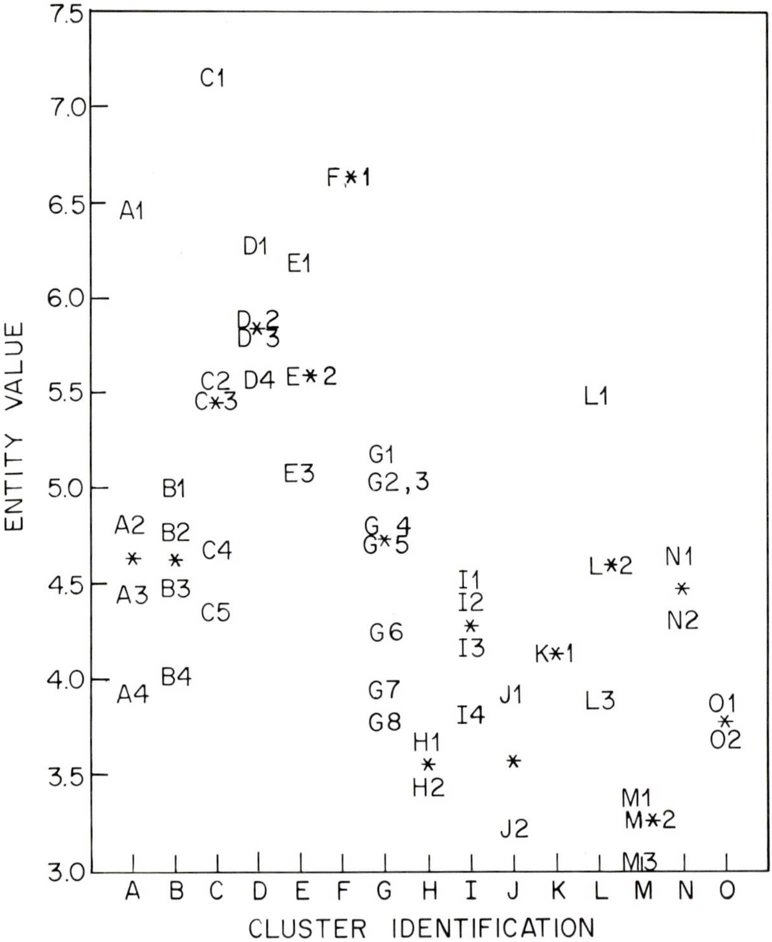

Fig. 9. Entity values by cluster identification.

identified by labels along the x-axis, and above each label the values on a certain variable for each entity in that cluster are plotted. The median for the cluster is plotted as a star. This enables one to use the variable in question for comparing several entities grouped in the same cluster and to make multiple comparisons across clusters. Figure 9 shows, for example, that clusters A and B are quite similar except for one entity (A1) in cluster A; cluster E tends to have large entities with moderate spread, while cluster M has much smaller values with small spread.

While the preceding plot was for one variable and several clusters, a plot for studying one cluster across variables can also be useful. Figure 10 shows, for one cluster, a plot of the deviation of the cluster median from the median of all the cluster medians for each of 15 exogenous variables. To aid comparison across variables, each variable was scaled by the median absolute deviation of the cluster medians for that variable. Using this scaling, the distances measured across the 15 variables are directly comparable to one another. Such plots were used in the subsidiaries application to aid in the characterization of the clusters of entities across certain interesting variables. In the present example this cluster lies below the median value on eight of the 15 variables; on variable 10 it is particularly low, but on variable two it is relatively large. This multivariable plot was used on exogenous variables, but - as with Figure 9 - such a plot could also be used with internal variables.

The quantile contour plot, referred to in §3.3, is another graphical vehicle for revealing information related to clustering. In the corporation grouping example, a version of this plot was used to check whether AT&T fit in better with the industrials or the utilities.

Figure 11 shows a quantile contour plot for 495 industrials based on 1969 data on 13 standardized variables. Instead of plotting a curve, $[\underset{\sim}{a}(t)]' \underset{\sim}{x}_i$ vs. t, for i=1,..., 495, only selected quantiles are indicated for each of 100 equally spaced values of t. Moreover, the column medians have been shifted to be zero, and the curve for AT&T has been superimposed. A similar plot for the utilities is given in Figure 12. The two figures together suggest that AT&T resembles the industrials more than the utilities. For instance, AT&T falls within the lower and upper quartiles about 50% of the time for the industrials but only about 10% of the time for the utilities. For additional discussion of this example, see Gnanadesikan (1977, §6.2).

5. CONCLUDING REMARKS

The primary emphasis in this paper has been on the description and illustration of methodology developed for the clustering aspects of four data analysis problems. Clearly, the general applicability and statistical scope of these

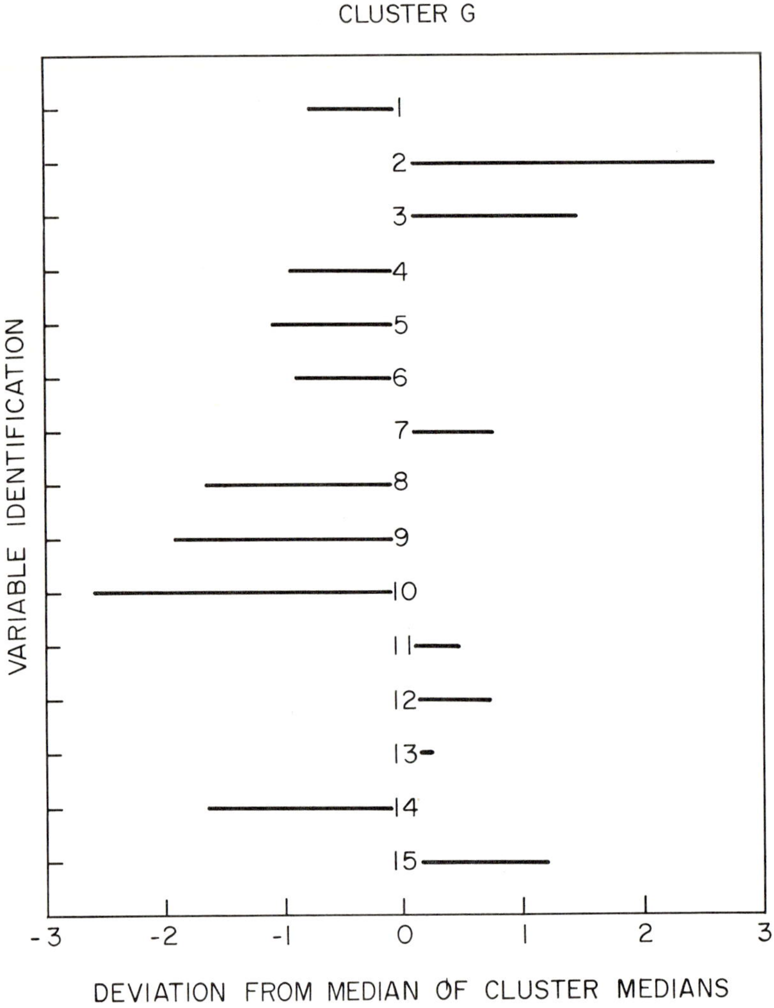

Fig. 10. Summary of exogenous variables for individual cluster.

largely *ad hoc* procedures need to be investigated further.

For instance, the method of selecting nearest neighbors based on a local metric, which was applied in the example on grouping corporations, requires modification and refinement for the more difficult problem of cluster formation. The procedure must be made effective not only when the object of focus is near the middle of the cluster but also in the trickier situation when it is near the "edge".

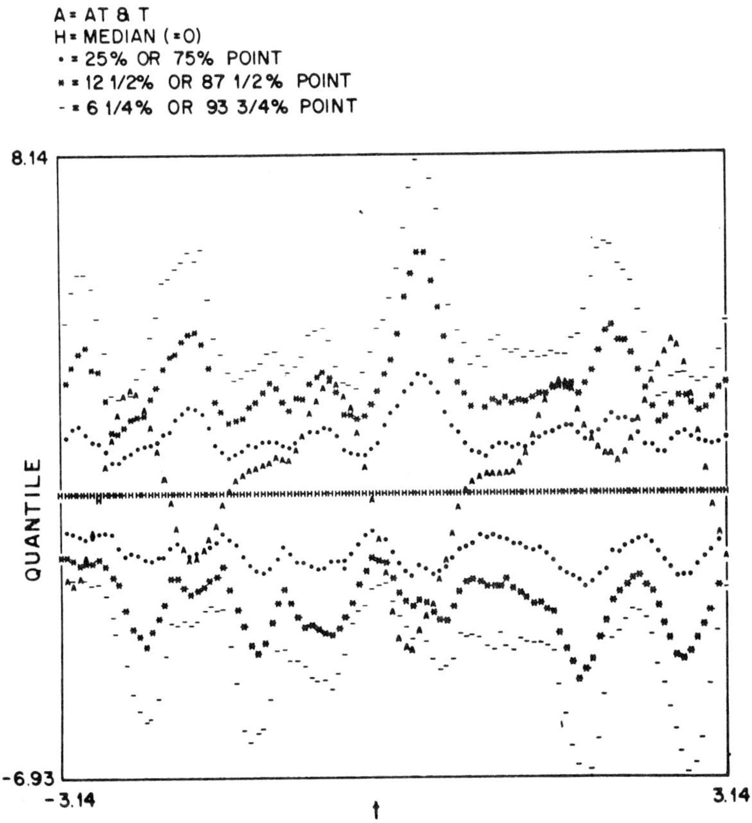

Fig. 11. Quantile contour plot of industrials.

A second instance involves the graphical box plot display for assessing the significance of clusters, which was conceived especially for the portfolios example. Departures from a null configuration on such a plot could be caused by such things as dependencies among the variables, as well as by the presence of clusters. Methods ought to be explored for removing, or mitigating the influence of, causes for departure other than cluster structure. For example, dependencies among the variables might be removed by an initial transformation.

Other questions of a less specific nature also need to be addressed. What models are appropriate for the mathematical constraints (e.g., on the admissible permutations of an inter-object distance matrix) and statistical dependencies (e.g., among cluster strengths) that are used in or result from a cluster analysis?

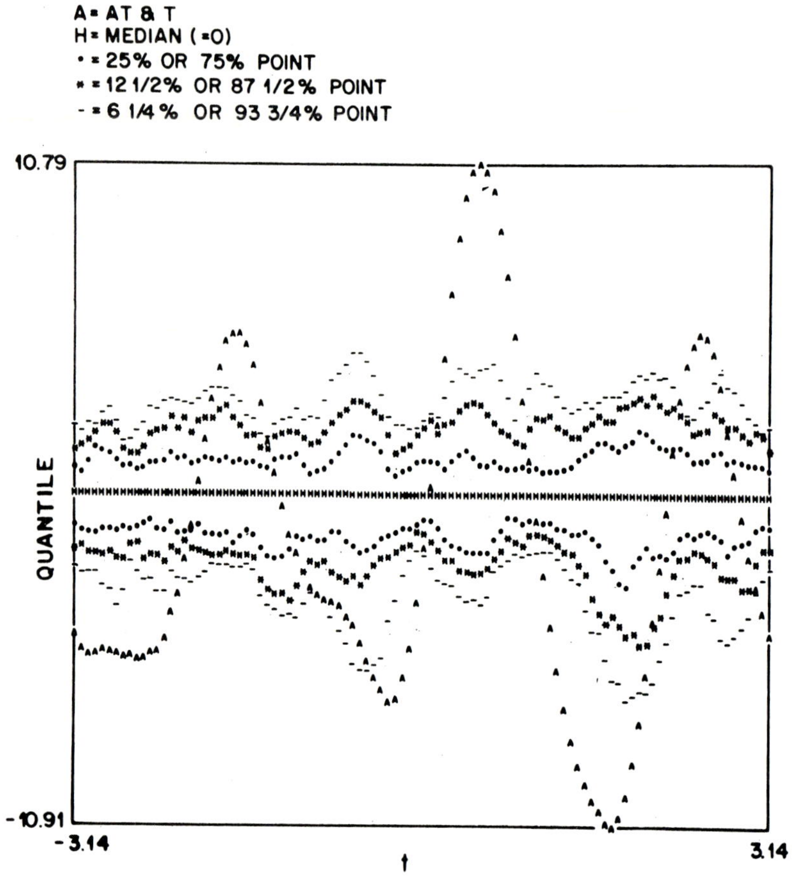

Fig. 12. Quantile contour plot of utilities.

What effect do these constraints and dependencies have on existing procedures?

What should be the role of robustness in clustering? This could involve - as in two of the present examples - simply the use of robust measures as inputs. A deeper consideration would be how to balance the main goal of robustness (minimizing the impact of a small fraction of outliers) against the main goal of clustering (identifying group structure including that of outliers).

Clustering techniques are rapidly becoming routine tools of data analysis. Answers to questions such as those raised here should help to raise the level of sophistication and understanding of these analyses.

REFERENCES

[1] ANDERSON, E. (1957). A semi-graphical method for the analysis of complex problems. *Proc. Nat. Acad. Sci., USA*, **43**, 923-927.

[2] ANDREWS, D. G. (1972). Plots of high-dimensional data. *Biometrics*, **28**, 125-136.

[3] BALL, G. H. and HALL, D. J. (1967). A clustering technique for summarizing multivariate data. *Behav. Sci.*, **12**, 153-155.

[4] BAKER, F. B. and HUBERT, L. J. (1975). Measuring the power of hierarchical cluster analysis. *J. Amer. Statist. Assoc.*, **70**, 31-38.

[5] CHEN, H., DUNN, D. M. and LANDWEHR, J. M. (1975). Grouping companies based on their operating environment. *Proc. of Bus. and Econ. Statist. Sect. Amer. Statist. Assoc.*, 278-283.

[6] CHEN, H., GNANADESIKAN, R. and KETTENRING, J. R. (1974). Statistical methods for grouping corporations. *Sankhyā, Ser. B*, **36**, 1-28.

[7] CHERNOFF, H. (1972). Metric considerations in cluster analysis. *Proc. 6th Berkeley Symp. Math. Statist. Prob.*, **1**, 621-629.

[8] CHERNOFF, H. (1973). Using faces to represent points in k-dimensional space graphically. *J. Amer. Statist. Assoc.*, **68**, 361-368.

[9] COHEN, A., GROSS, A. M. and TERPENNING, I. J. (1975). Personal communication.

[10] CORMACK, R. M. (1971). A review of classification. *J. Roy. Statist. Soc. Ser. A*, **134**, 321-367.

[11] EVERITT, B. (1974). *Cluster Analysis*. John Wiley & Sons, Inc., New York.

[12] FISHER, R. A. (1936). The use of multiple measurements in taxonomic problems. *Ann. Eugen.*, **7**, 179-188.

[13] GNANADESIKAN, R. (1973). Graphical methods for informal inference in multivariate data analysis. *Bull. Int. Statist. Inst.*, **45**, 4, 195-206.

[14] GNANADESIKAN, R. (1977). *Methods for Statistical Data Analysis of Multivariate Observations*. John Wiley & Sons, Inc., New York.

[15] HARTIGAN, J. A. (1975). *Clustering Algorithms*. John Wiley & Sons, Inc., New York.

[16] HUBERT, L. (1974). Approximate evaluation techniques for the single-link and complete-link hierarchical clustering procedures. *J. Amer. Statist. Assoc.*, **69**, 698-704.

[17] KETTENRING, J. R., ROGERS, W. H., SMITH, M. E. and WARNER, J. L. (1976). Cluster analysis applied to the validation of course objectives. *J. Educat. Statist.*, **1**, 39-57.

[18] LANDWEHR, J. M. (1976). Measuring variability for purposes of reducing dimensionality. (Unpublished manuscript.)

[19] LING, R. F. (1973). A probability theory of cluster analysis. *J. Amer. Statist. Assoc.*, **68**, 159-164.

[20] MACQUEEN, J. (1965). Some methods for classification and analysis of multivariate observations. *Proc. 5th Berkeley Symp. Math. Statist. Prob.*, **1**, 281-297.

[21] MARONNA, R. and JACOVKIS, P. M. (1974). Multivariate clustering procedures with variable metrics. *Biometrics*, **30**, 499-505.

[22] MONKHOUSE, F. J. and WILKINSON, H. R. (1973). *Maps and Diagrams* (3rd ed.). Methuen, London.

[23] RAO, C. R. (1952). *Advanced Statistical Methods in Biometric Research.* John Wiley & Sons, Inc., New York.

[24] ROHLF, F. J. (1970). Adaptive hierarchical clustering schemes. *Syst. Zool.*, **19**, 58-81.

[25] SOKAL, R. R. and SNEATH, P. H. A. (1963). *Principles of Numerical Taxonomy.* Freeman, San Francisco.

[26] TRYON, R. C. (1939). *Cluster Analysis.* Edward Brothers, Ann Arbor, Michigan.

[27] TUKEY, J. W. (1970). *Exploratory Data Analysis.* (Limited preliminary edition). Addison Wesley, Reading, Massachusetts.

[28] WARNER, J. L. (1968). An adaptation of ISODATA-POINTS, an *I*terative *S*elf-*O*rganizing *D*ata *A*nalysis *T*echnique *A*. (Unpublished Bell Laboratories technical memorandum).

P.R. Krishnaiah, ed., APPLICATIONS OF STATISTICS

SOURCE CORRELATION EFFECTS ON STRUCTURAL RESPONSE

STEPHEN H. CRANDALL

AND

ALEXANDER P. KULVETS

Acoustics and Vibration Laboratory
Massachusetts Institute of Technology
Cambridge, Massachusetts

When a mechanical structure is driven by stationary wide-band random point forces the resulting stationary random vibration depends on the number and location of exciting forces and on the joint statistical properties of the random force processes. When all forces have identical spectra, the mean square response distribution depends on the cross-correlations between processes. This problem is analyzed in detail for one-and two-dimensional structures (e.g., beams and plates) driven by two forces with band-limited white spectra. The mean square response is substantially uniform over most of the structure except for zones or lanes of enhanced or diminished response. The number and strength of these zones depends on the correlation of the input force processes. Experimental results for rectangular flat plates confirm the analytical predictions.

1. INTRODUCTION

When a uniform rectangular membrane or plate is acted on by a single point force whose time history is a sample function of a stationary wide-band random process the mean-square velocity response is surprisingly uniform over most of the structure with the exception of a tic-tac-toe pattern of lanes of enhanced response. The location and strength of these lanes can be predicted analytically [1-5]. The presence of these lanes can be verified quantitatively, by traversing a non-contact displacement transducer across the surface [1,3], or qualitatively, by sprinkling salt on the surface and observing the migration of salt away from the lanes of enhanced response [1-4].

In the present paper we consider similar structures excited by two or more point forces whose time histories are sample functions of stationary wide-band random processes with known cross-correlations. The mean-square velocity response is again quite uniform except for certain patterns of lanes and zones of enhanced or diminished response. These patterns are more complicated than for a single source of excitation and depend on the location of, and the cross correlations between, the several sources.

In the following sections the analysis is developed in some detail for the case of a one-dimensional structure, such as a taut string or a flexible beam, excited by n sources. An approximate procedure for analyzing two-dimensional structures, such as membranes or plates, is briefly outlined and analytical predictions for a square plate are compared with the corresponding salt patterns obtained in the laboratory.

2. ONE-DIMENSIONAL STRUCTURES

Consider either a taut string with fixed ends or a simply supported beam, with uniform properties and length L, which is acted upon by n transverse forces $f_\nu(t)$ at locations $x = a_\nu$. The equation of motion for the transverse displacement $w(x,t)$ in either case has the form

$$\mathcal{L}[w] + c_1 \frac{\partial w}{\partial t} + \rho A \frac{\partial^2 w}{\partial t^2} = \sum_{\nu=1}^{n} \delta(x-a_\nu)\, f_\nu(t) \tag{1}$$

where ρA is the mass per unit length and c_1 is the viscous damping force per unit length, per unit velocity. The operator $\mathcal{L}[\ ]$ is $-T_0\, \partial^2/\partial x^2$ for a string with tension T_0 and $EI\, \partial^4/\partial x^4$ for a beam with flexural modulus EI. By superposition the solution to (1) for the transverse velocity is

$$v(x,t) = \frac{\partial w}{\partial t} = \sum_\nu \int_{-\infty}^{\infty} g_\nu(x,\theta)\, f_\nu(t-\theta)\, d\theta \tag{2}$$

where the Green's function $g_\nu(x,t-t_0)$ is the velocity response at x, at time t, due to a unit impulse applied at a_ν, at time t_0.

We take the forces $f_\nu(t)$ to be stationary wide-band random processes with known cross correlations

$$E\left[f_\mu(t)\, f_\nu(t+\tau)\right] = R_{\mu\nu}(\tau) \tag{3}$$

The mean square of the stationary velocity response at x is then given by squaring (2) and taking the expectation

$$E\left[v^2(x)\right] = \sum_\mu \sum_\nu \int_{-\infty}^{\infty} \int_{-\infty}^{\infty} g_\mu(x,\theta_1)\, g_\nu(x,\theta_2)\, R_{\mu\nu}(\theta_1-\theta_2)\, d\theta_1\, d\theta_2 \tag{4}$$

The formal solution (4) may be cast in an alternative form by introducing the Fourier transform of the Green's function

$$G_\nu(x,\omega) = \int_{-\infty}^{\infty} g_\nu(x,t)\, e^{-i\omega t}\, dt \tag{5}$$

and the Wiener-Khintchine relations

$$S_{\mu\nu}(\omega) = \frac{1}{2\pi} \int_{-\infty}^{\infty} R_{\mu\nu}(\tau)\, e^{-i\omega\tau}\, d\tau$$

$$R_{\mu\nu}(\tau) = \int_{-\infty}^{\infty} S_{\mu\nu}(\omega)\, e^{i\omega\tau}\, d\omega \qquad (6)$$

between the cross correlations and the cross spectral densities. Inserting (5) and (6) into (4) we find

$$E\left[v^2(x)\right] = \sum_{\mu}\sum_{\nu} \int_{-\infty}^{\infty} G_{\mu}(x,-\omega)\, S_{\mu\nu}(\omega)\, G_{\nu}(x,\omega)\, d\omega \qquad (7)$$

These formal solutions can be developed further by introducing explicit representations for the $g_{\nu}(x,t)$ and the $G_{\nu}(x,\omega)$. In terms of the natural modes

$$\psi_j(x) = \sqrt{2}\, \sin j\pi x/L \qquad (8)$$

the Green's function is

$$g_{\nu}(x,t) = \sum_{j=1}^{\infty} \psi_j(x)\, \psi_j(a_{\nu})\, \frac{d}{dt}\, h_j(t) \qquad (9)$$

where the modal displacement impulse response function $h_j(t)$ is

$$h_j(t) = \exp\left\{-c_1 t/2\rho A\right\} \sin\left[(\omega_j^2 - c_1^2/4\rho^2A^2)^{\frac{1}{2}}\, t\right] \qquad (10)$$

and the natural frequency ω_j corresponding to the natural mode (8) is either $(T_0/\rho A)^{\frac{1}{2}}\, j\pi/L$ for the string or $(EI/\rho A)^{\frac{1}{2}}\, j^2\pi^2/L^2$ for the beam. The Fourier transform (5) of the Green's function is

$$G_{\nu}(x,\omega) = \sum_{j=1}^{\infty} \psi_j(x)\, \psi_j(a_{\nu})\, i\omega\, H_j(\omega) \qquad (11)$$

where the modal displacement frequency response $H_j(\omega)$ is

$$H_j(\omega) = \left[\rho AL(\omega_j^2 + i\omega c\ /\rho A - \omega^2)\right]^{-1} \qquad (12)$$

When (9) and (11) are inserted in (4) and (7) respectively we obtain explicit formulas for the mean-square velocity response. For example, substitution of (11) into (7) yields

$$E\left[v^2(x)\right] = \sum_{\mu}\sum_{\nu}\sum_{j}\sum_{k} \psi_j(x)\, \psi_k(x)\, \psi_j(a_\mu)\, \psi_k(a_\nu)\, I_{\mu\nu jk} \tag{13}$$

where $I_{\mu\nu jk}$ is the integral

$$I_{\mu\nu jk} = \int_{-\infty}^{\infty} \omega^2\, H_j(-\omega)\, S_{\mu\nu}(\omega)\, H_k(\omega)\, d\omega \tag{14}$$

As an example, we consider the particular case of two band-limited white noise forces for which

$$\begin{aligned} S_{11} = S_{22} &= \begin{cases} S_0, & |\omega| < \omega_c \\ 0, & \omega_c < |\omega| \end{cases} \\ S_{12} = S_{21} &= \begin{cases} rS_0, & |\omega| < \omega_c \\ 0, & \omega_c < |\omega| \end{cases} \end{aligned} \tag{15}$$

where r is the excitation correlation coefficient $(-1 \leq r \leq 1)$ and ω_c is the cut-off frequency which is taken to include N structural modes; i.e., $\omega_N < \omega_c < \omega_{N+1}$. If we assume the damping to be light enough the only values of $I_{\mu\nu jk}$ which cannot be neglected are those for which ω_j and ω_k are both less than ω_c and for which $\omega_j \approx \omega_k$. In uniform strings and beams there are no repeated roots and the natural frequencies are well separated. Thus the only non-zero values of $I_{\mu\nu jk}$ are those for which $j = k \leq N$. In these cases the integration (14) yields

$$\left.\begin{aligned} I_{11_{jj}} = I_{22_{jj}} &= \frac{\pi S_0}{c_1 \rho A L^2} \\ I_{12_{jj}} = I_{21_{jj}} &= \frac{r\pi S_0}{c_1 \rho A L^2} \end{aligned}\right\} \; j = 1, \ldots, N \tag{16}$$

if in the integrand the approximation is made that $S_{\mu\nu}(\omega)$ is ideally white. The error involved in this approximation is small for light damping [6]. Inserting (16) into (13) we find

$$E\left[v^2(x)\right] = V_0^2\, F_N(x;\, a_1,\, a_2;\, r) \tag{17}$$

where

$$V_0^2 = 2S_0\omega_c \cdot \frac{1}{c_1 L} \cdot \frac{\pi N}{2\rho A L \omega_c}$$

$$= E\left[f_\nu^2\right] G_0 \, \bar{G}_\infty \qquad (18)$$

is the space-averaged mean-square velocity due to a single force f_ν [1,5], G_0 is the rigid-body conductance of the structure, $\bar{G}_\infty$ is the frequency-averaged radiation conductance of an infinite structure with the same inertial and elastic parameters [5], and the dimensionless spatial distribution function is

$$F_N(x;\, a_1,\, a_2;\, r) = \frac{1}{N} \sum_{j=1}^{N} \psi_j^2(x) \left[\psi_j^2(a_1) + 2r\psi_j(a_1)\, \psi_j(a_2) + \psi_j^2(a_2)\right] \qquad (19)$$

Once a_1, a_2, and N have been fixed, F_N becomes a function only of x and r, which for brevity we may write as

$$F_N(x;\, a_1,\, a_2;\, r) = F_{11}(x) + 2rF_{12}(x) + F_{22}(x) \qquad (20)$$

where the sums $F_{ij}(x)$ are defined by (19). For the natural modes (8) these sums can be evaluated in closed form [1] or summed on a computer. The resulting sums for the case where $a_1 = L/5$ and $a_2 = 2L/5$ are displayed in Figure 1 for two values of N. When the driving forces are uncorrelated (r=0) the response spatial distribution function reduces to $F_{11} + F_{22}$. The spatial effect of source correlation is indicated by the function $2F_{12}$. Depending on the value of $r(-1 \leq r \leq 1)$ some fraction of $2F_{12}$ is added to or subtracted from $F_{11} + F_{12}$ to give the complete distribution $F_N(x;\, a_1,\, a_2;\, r)$. The wider the excitation band width (i.e., the larger the value of N) the clearer is the response pattern.

For a pair of uncorrelated driving forces (r=0) the mean-square velocity distribution is essentially the superposition of the distributions which result when each force acts alone. The mean-square velocity is substantially uniform over the structure except at the ends, where it falls to zero, and at the drive points a_1 and a_2 and their symmetric image points $L-a_1$ and $L-a_2$, where it is locally enhanced. The relative enhancement is not as strong as is the case for excitation by a single force. For a single force the local enhancement is 50% (1.76 dB) while for two forces of equal strength the local enhancement is 25% (0.97 dB). When the driving forces are correlated ($r \neq 0$) there are additional local enhancements, or decreases, in response at the points $(a_2 - a_1)/2$ and $(a_2 + a_1)/2$ and their symmetric image points. In the extreme cases ($r = \pm 1$) these enhancements can be as large as 50% (1.76dB) which is greater than the enhancement at the drive points. Examples of the resulting spatial distributions $F_N(x;\, a_1,\, a_2;\, r)$ are shown in Figure 2.

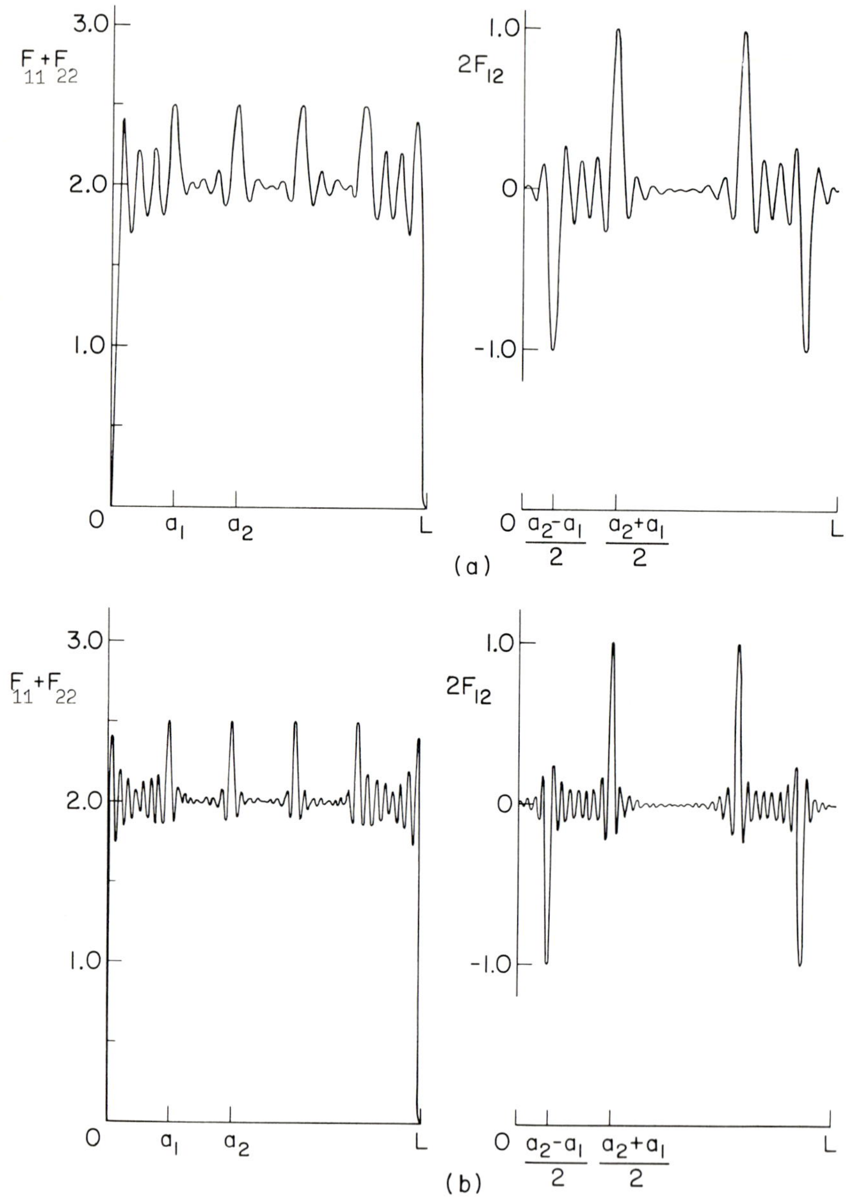

Fig. 1. Spatial distribution functions for response of a one-dimensional structure driven by a pair of random forces which excite N modes; (a) N = 20, (b) N = 40.

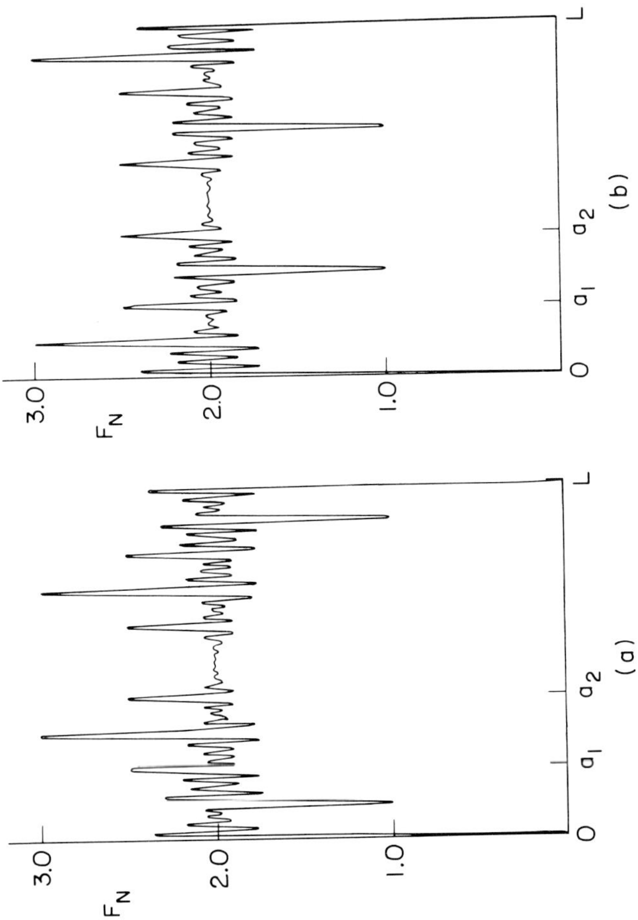

Fig. 2. Mean-square velocity response distribution for a pair of correlated driving forces (N = 40); (a) r = +1, (b) r = -1.

When N is large the sums in (19) can be quickly estimated by the following asymptotic approximation method. Trigonometric identities are used to transform each sum into a number of sums of the form

$$\frac{1}{N}\sum_{j=1}^{N} \cos j2\pi(x-b)/L \tag{21}$$

which are assumed to vanish for all x within $0 < x < L$ except in the neighborhood of $x = b$ where there is a narrow peak of unit height. As an illustration of this method consider the sum $F_{12}(x)$ in (20) where the $\psi_j(x)$ are defined by (8).

$$\begin{aligned} F_{12}(x) &= \frac{1}{N}\sum_{j=1}^{N} 4 \sin^2 j\pi x/L \sin j\pi a_1/L \sin j\pi a_2/L \\ &= \frac{1}{N}\sum_{j=1}^{N}\Big[\cos j\pi(a_2 - a_1)/L - \cos j\pi(a_2 + a_1)/L \\ &\quad + \frac{1}{2}\cos j\pi(2x + a_2 + a_1)/L + \frac{1}{2}\cos j\pi(2x - a_2 - a_1)/L \\ &\quad - \frac{1}{2}\cos j\pi(2x - a_2 + a_1)/L - \frac{1}{2}\cos j\pi(2x + a_2 - a_1)/L\Big] \end{aligned} \tag{22}$$

Using the asymptotic approximation the first two terms are neglected, the third and fourth terms give narrow positive peaks of amplitude one-half at $x = L - (a_2 + a_1)/2$ and $x = (a_2 + a_1)/2$ respectively, and the last two terms give narrow negative peaks of amplitude one-half at $x = (a_2 - a_1)/2$ and $x = L - (a_2 - a_1)/2$ respectively. The resulting distribution is sketched in Figure 3. This should be compared with the exact sum for finite N displayed in Figure 1. The form of $F_{12}(x)$ is as shown in Figure 3 for all combinations of a_1 and a_2 as long as a_2 and a_1 are not symmetric images of one another; i.e., as long as $a_2 + a_1 \neq L$. If $a_1 + a_2 = L$ the peaks corresponding to the third and fourth terms of (22) superimpose to become a single positive peak of unit amplitude at $x = L/2$ as indicated in Figure 4. This implies that in the extreme case of complete correlation ($r = 1$) the relative enhancement at $x = L/2$ is 100% (3dB).

The foregoing results show that when uncorrelated wide-band driving forces excite a uniform one-dimensional structure the mean-square velocity response tends to be uniformly distributed except for local intensification of response in the neighborhood of the drive points and their symmetric image points. When the sources are correlated there may be additional local intensifications at other points (e.g., midway between the drive points). In the extreme case of symmetrically located forces of equal strength, which are perfectly correlated, the mean-square response at the middle of the structure is double the background level.

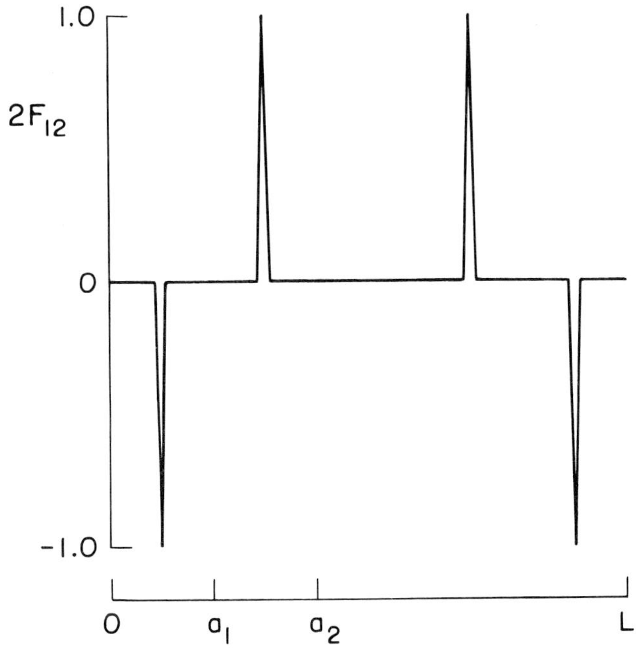

Fig. 3. Asymptotic approximation of spatial distribution function for general location of driving points.

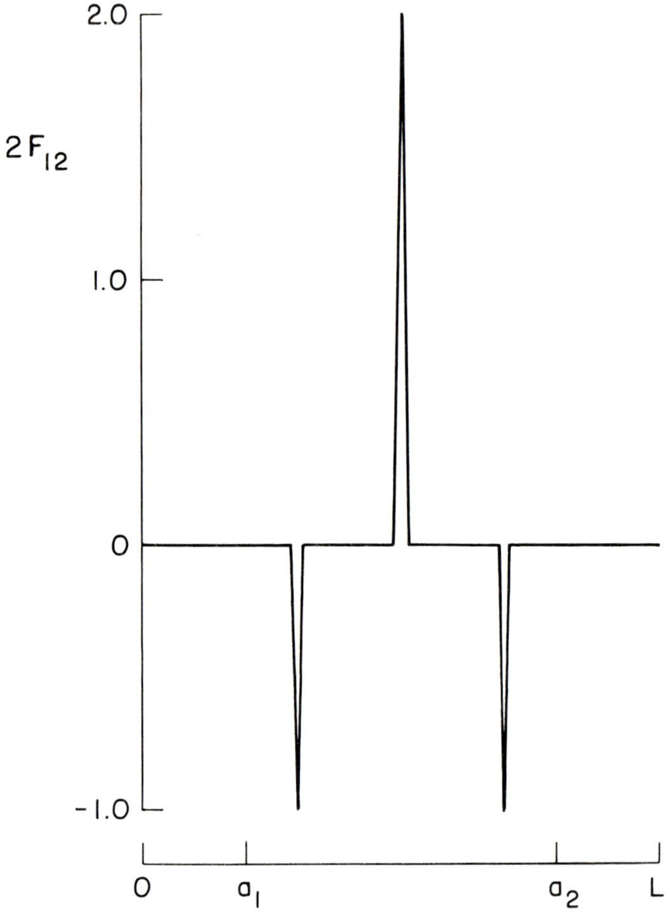

Fig. 4. Asymptotic approximation of spatial distribution function for case when $a_1 + a_2 = L$.

3. TWO-DIMENSIONAL STRUCTURES

The formal analysis for one-dimensional structures is readily extended to two-dimensional structures. The mean-square velocity response of a rectangular membrane with fixed edges, or plate with simply supported edges, when driven by n transverse stationary random forces can be represented by the sum (13) if the meaning of certain symbols is suitably extended. For two dimensions it is necessary to consider the loading points a_μ and a_ν, and the response point x, to be two-dimensional vectors with x and y components. The natural modes $\psi_j(x)$ with a single subscript must be taken to mean an appropriate ordering of the two-parameter family

$$\psi_{jp}(x,y) = 2 \sin j\pi x/L_x \sin p\pi y/L_y \tag{23}$$

which applies to rectangular structures with sides of length L_x and L_y. If the plate or membrane has mass per unit area ρh and viscous damping force per unit area c_2, per unit velocity, the modal displacement frequency response is

$$H_{jp}(\omega) = \left[\rho h L_x L_y\left(\omega_{jp}^2 + i\omega c_2/\rho h - \omega^2\right)\right]^{-1} \tag{24}$$

and the natural frequency ω_{jp} corresponding to the natural mode (23) is

$$\omega_{jp} = \left[\left(\frac{j\pi}{L_x}\right)^2 + \left(\frac{p\pi}{L_y}\right)^2\right]^{\frac{1}{2}} \left(\frac{T_1}{\rho h}\right)^{\frac{1}{2}} \tag{25}$$

for a membrane with uniform surface tension T_1 per unit length, or

$$\omega_{jp} = \left[\left(\frac{j\pi}{L_x}\right)^2 + \left(\frac{p\pi}{L_y}\right)^2\right] \left(\frac{D}{\rho h}\right)^{\frac{1}{2}} \tag{26}$$

for a plate with uniform flexural modulus D. In the case of two exciting forces with the statistical properties (15) the only values of $I_{\mu\nu jk}$ which cannot be neglected are those for which ω_{jp} and ω_{kq} are both less than the cut-off frequency ω_c and for which $\omega_{jp} \simeq \omega_{kq}$.

For two dimensional structures, the natural frequencies are not necessarily all well separated and repeated roots are possible. If out of a very large number of modes there are only a few accidental modal overlaps they may be neglected without serious error in evaluating the mean-square response (13). On the other hand, if the aspect ratio of the rectangular structure takes certain special values there is a systematic occurrence of repeated roots and it is necessary to consider their contribution to (13). To illustrate a case of this sort we shall consider a square structure ($L_x = L_y = L$) for which $\omega_{jp} = \omega_{kq}$ not only for the identical mode when $j = k$ and $p = q$ but also for the transposed mode when $j = q$ and $p = k$.

The requirement that the summation (13) extend over all modes ψ_{jp}, ψ_{kq} for which the corresponding natural frequencies ω_{jp}, ω_{kq} are smaller than the cut-off frequency ω_c is conveniently visualized in a wave number space, Figure 5, where the mode ψ_{jp} is represented by the point $k_x = j\pi/L_x$, $k_y = p\pi/L_y$. All modes with

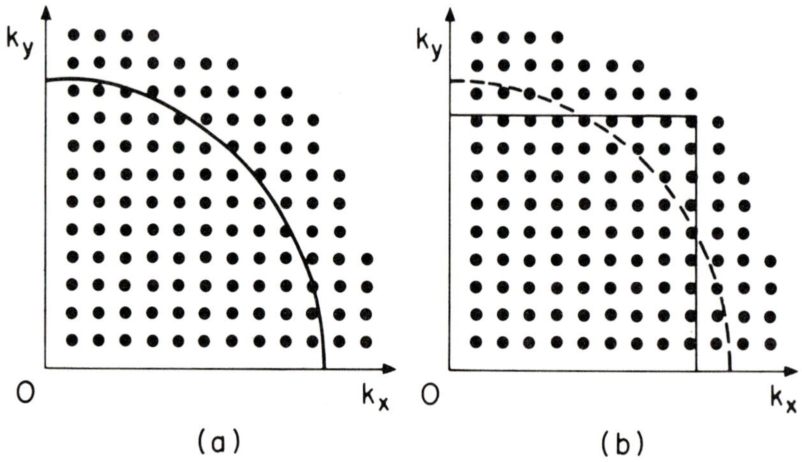

Fig. 5. Array of plate modes in wave-number space: (a) modes excited by band-limited excitation lie within quarter circle of radius k_c, (b) in separable sum approximation, sum is taken over modes in square of equal area.

frequency less than ω_c are represented by the points which lie within the quarter circle of radius k_c where $k_c = (\rho h/T_1)^{\frac{1}{2}}\ \omega_c$ for the membrane (25) and $k_c^2 = (\rho h/D)^{\frac{1}{2}}\ \omega_c$ for the plate (26). For the present, we do not attempt to sum over all these modes. Instead we adopt the approximation, which appears to give qualitatively correct results [1], of summing over the modes included within the square of side $(\pi/4)^{\frac{1}{2}}\ k_c$ which contains the same area as the quarter circle. The composition of the population of modes included in the summation is altered by this approximation although the total number of modes is essentially unchanged. If N is the largest integer for which $N\pi/L < (\pi/4)^{\frac{1}{2}}\ k_c$, the mean-square velocity at (x,y) in a square plate of side L due to forces $f_1(t)$ and $f_2(t)$ with statistical properties (15), acting at (a_1,b_2) and (a_2,b_2) is

$$E\left[v^2(x,y)\right] = V_0^2\left[F_{N^2}(x,y;\ a_1,b_1,a_2,b_2;r) + F'_{N^2}(x,y;\ a_1,b_1,a_2,b_2;r)\right] \quad (27)$$

where

$$
\begin{aligned}
V_0^2 &= 2S_0\omega_c \cdot \frac{1}{c_2L^2} \frac{\pi N}{2\rho h L^2 \omega_c} \\
&= E\left[f_\nu^2\right] G_0 \, \bar{G}_\infty
\end{aligned}
\tag{28}
$$

is the space-averaged mean-square velocity due to a single force f_ν, G_0 is the rigid body conductance of the structure, $\bar{G}_\infty$ is the frequency-averaged radiation conductance of an infinite structure with the same inertial and elastic parameters [5] and the dimensionless spatial distribution functions are

$$
\begin{aligned}
F_{N^2} &= \frac{1}{N^2} \sum_j^N \sum_p^N \psi_{jp}^2(x,y) \left[\psi_{jp}^2(a_1,b_1) + 2r\psi_{jp}(a_1,b_1)\,\psi_{jp}(a_2,b_2) + \psi_{jp}^2(a_2,b_2)\right] \\
F'_{N^2} &= \frac{1}{N^2} \sum_j^N \sum_p^N{}' \psi_{jp}(x,y)\quad \psi_{pj}(x,y) \left[\psi_{jp}(a_1,b_1)\,\psi_{pj}(a_1,b_1) + \right. \\
&\quad + r\psi_{jp}(a_1,b_1)\,\psi_{pj}(a_2,b_2) + r\psi_{jp}(a_2,b_2)\,\psi_{pj}(a_1,b_1) + \\
&\quad \left. + \psi_{jp}(a_2,b_2)\,\psi_{pj}(a_2,b_2)\right]
\end{aligned}
\tag{29}
$$

where the prime on the second summation indicates that terms with $j = p$ are omitted. The distribution F_{N^2} results from the terms where $\omega_{jp} = \omega_{kq}$ because $j = k$ and $p = q$. A similar distribution is involved for every rectangular structure. The distribution F'_{N^2} results from the repeated roots peculiar to a square structure; i.e., the terms where $\omega_{jp} = \omega_{kq}$ because $j = q$ and $p = k$ (and $j \neq p$).

The sums (29) can be evaluated by introducing (23) and separating each term into a product of a j-summation multiplied by a p-summation. The summations can be expressed in closed form [1], although we shall not write them out here. The asymptotic approximation described in connection with (21) can be used to predict the main features of the response distribution. Once (a_1,b_1), (a_2,b_2) and N have been fixed the spatial distribution functions (29) become functions of x,y, and r, which for brevity we may write as

$$
\begin{aligned}
F_{N^2} &= F_{11}(x,y) + 2r\,F_{12}(x,y) + F_{22}(x,y) \\
F'_{N^2} &= F'_{11}(x,y) + 2r\,F'_{12}\,(x,y) + F'_{22}(x,y)
\end{aligned}
\tag{30}
$$

where the $F_{ij}(x,y)$ and $F'_{ij}(x,y)$ are defined by (29). The distributions (30) predicted by the asymptotic approximation for general drive points P_1 and P_2 are shown in Figure 6. If the driving forces are uncorrelated (r = 0) the total distribution $F_{N^2} + F'_{N^2}$ is just the superposition of (a) and (c) in Figure 6. In (a) the eight lines shown (which either pass through P_1 and P_2 or through the symmetric image points) represent lanes of intensified response with magnitude 2 1/2 standing out from a uniform background with magnitude 2. In (c) the solid lines represent lanes with positive contributions of magnitude 1/4 while the dotted lines represent lanes with negative contributions of the same magnitude. The background in (c) has a null contribution. When the driving forces are correlated ($r \neq 0$) r-multiples of (b) and (d) are superposed on (a) and (c) in Figure 6. In (b) and (d) the backgrounds have null contributions and in the asymptotic approximation there are only isolated points with non-zero contributions. In each case there are eight points with + 1/2 contribution and eight points with - 1/2 contribution. This is the common case when the driving points have arbitrary location. For special orientations of the driving points, however, the local intensifications due to load correlation can extend to lines.

For example, Figure 7 shows the same spatial distribution functions for the case where the driving points are symmetrically located with respect to the diagonal DD. Now in (d) there is a +2 contribution along the entire length of the diagonal DD. This implies that, as the load correlation varies from r = -1 to r = +1, the resultant mean-square velocity spatial distribution function will vary from $F_{N^2} + F'_{N^2} = 0$ to $F_{N^2} + F'_{N^2} = 4$.

This prediction is born out by an experiment in which an aluminum plate 1.59 mm (1/16 in) thick and 0.91 m (3 ft) square was excited by loud speaker voice coils driven by wide-band random noise generators. Salt sprinkled on the vibrating plate moves randomly but tends to migrate away from zones of large mean-square response and accumulate in zones of lower mean-square response. Figure 8 shows the resulting salt pattern when the driving forces are uncorrelated (r = 0). There is no indication of intensified response along the diagonal. Figure 9 shows the salt pattern when r = +1. The salt has clearly migrated away from the diagonal indicating a lane of intensified response. Finally, Figure 10 shows the salt pattern when r = -1. The salt has now accumulated along the diagonal indicating a zone of very low response.

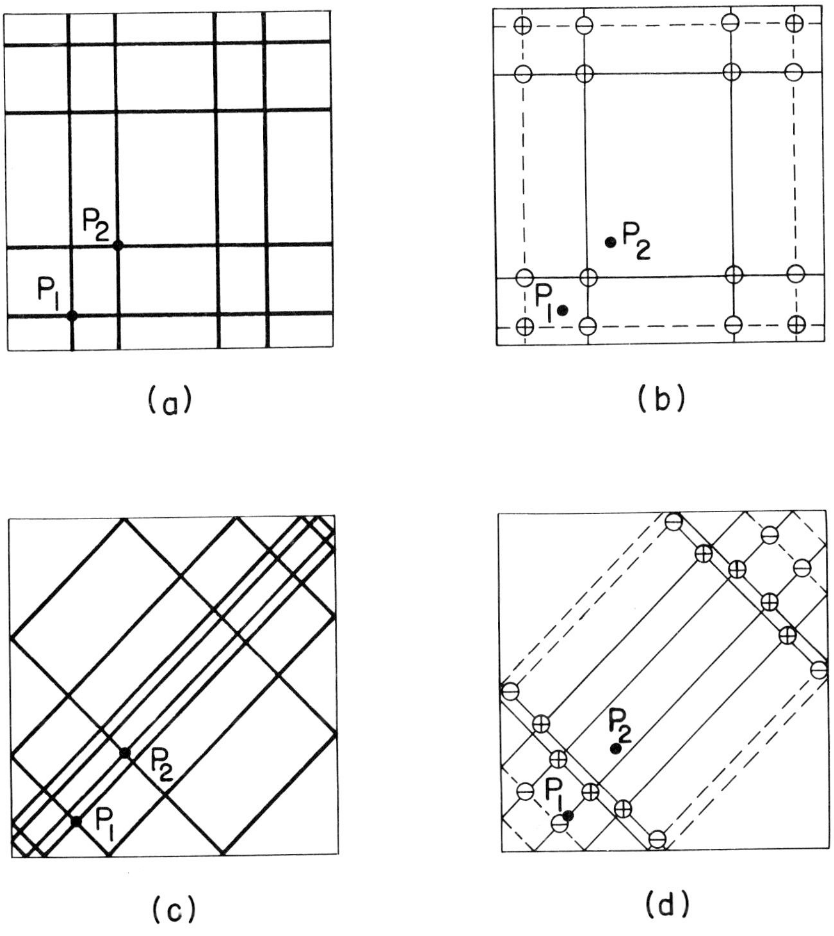

Fig. 6. Spatial distribution functions for general location of driving points: (a) $F_{11} + F_{22}$, lines have magnitude $2\frac{1}{2}$, background has magnitude 2; (b) $2F_{12}$, isolated points with contribution $\pm \frac{1}{2}$; (c) $F_{11}' + F_{22}'$, lines contribute $+\frac{1}{4}$; (d) $2F_{12}'$, isolated points with contribution $\pm \frac{1}{2}$.

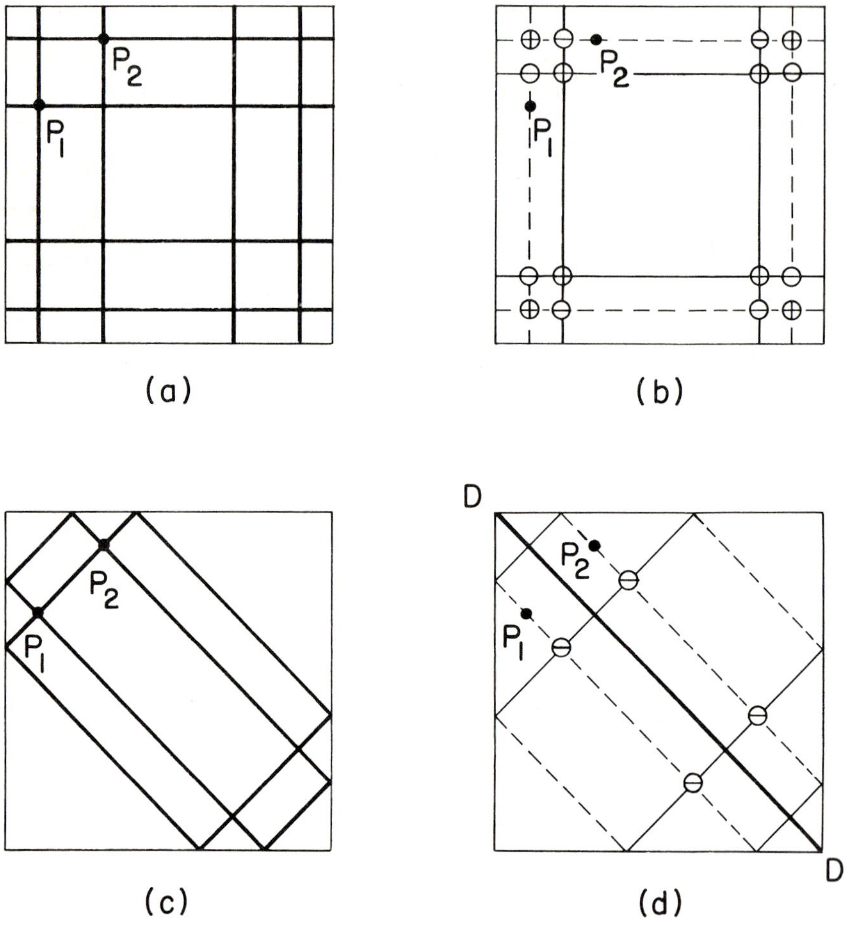

Fig. 7. Spatial distribution functions for driving points symmetrically located with respect to diagonal DD; (a) $F_{11} + F_{22}$, lines have magnitude $2\frac{1}{2}$, background has magnitude 2; (b) $2F_{12}$, isolated points with contribution $\pm\frac{1}{2}$; (c) $F_{11}' + F_{22}'$ lines contribute $+\frac{1}{2}$; $2\,F_{12}'$, diagonal DD contributes 2, isolated points contribute $-\frac{1}{2}$.

Fig. 8. Salt pattern corresponding to Fig. 7 for r = 0.

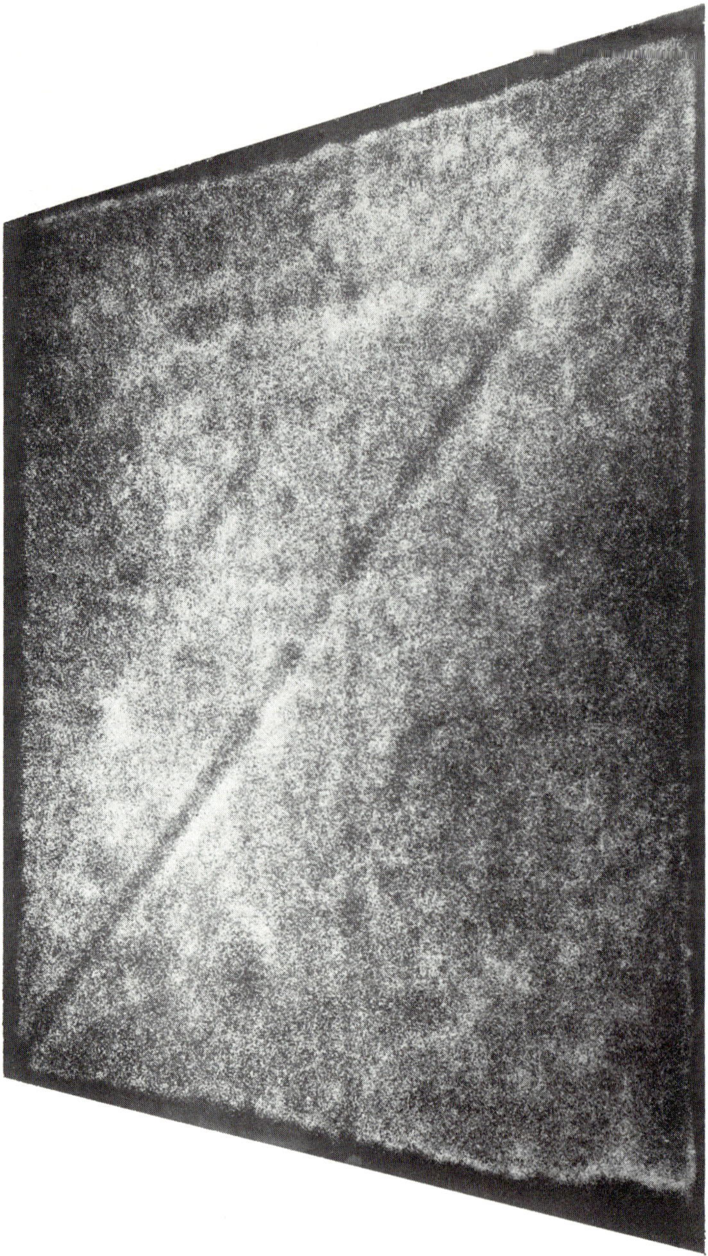

Fig. 9. Salt pattern corresponding to Fig. 7 for r = +1.

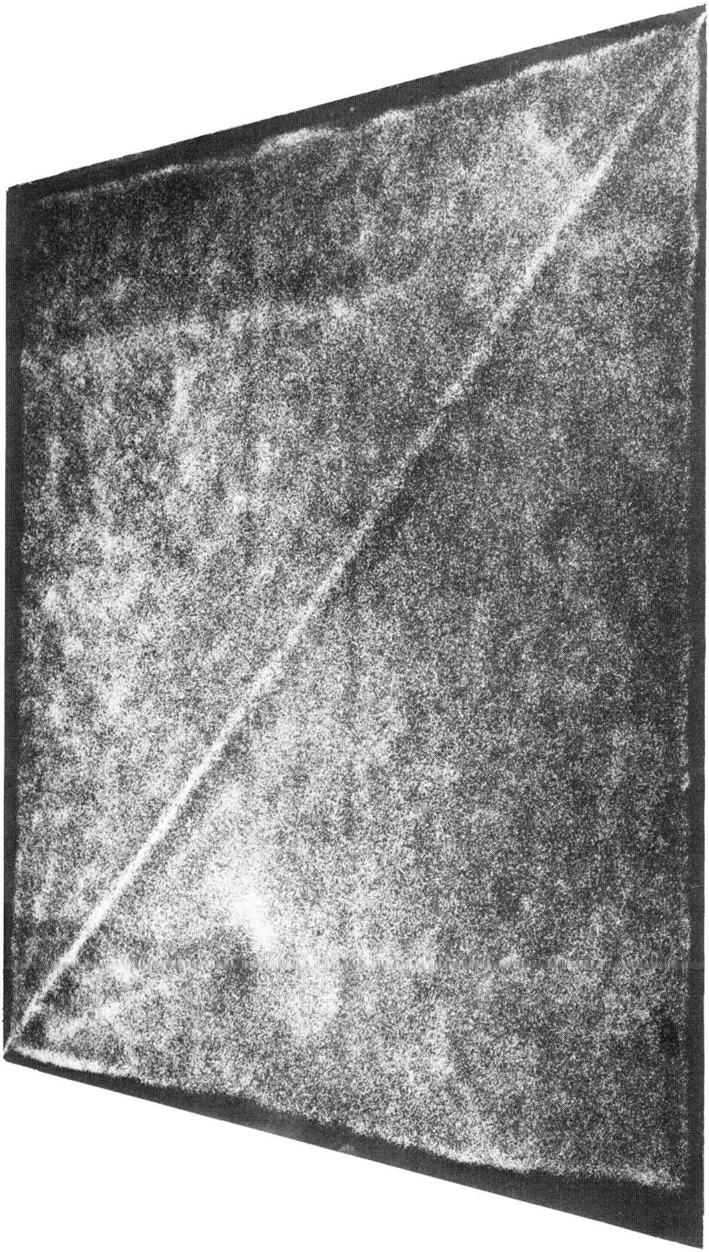

Fig. 10. Salt pattern corresponding to Fig. 7 for r = -1.

ACKNOWLEDGEMENT

This research was supported by a Grant from the National Science Foundation and by the International Research and Exchanges Board.

REFERENCES

[1] CRANDALL, S. H. and WITTIG, L. E. (1972). Chladni's patterns for random vibrations of a plate. *Dynamic Response of Structures* (G. Herrman and N. Perrone, eds.), 55-71. Pergamon Press, New York.

[2] CRANDALL, S. H. (1974). Wide-band random vibration of structures. *Proc. 7th U. S. National Congress of Applied Mechanics*, 131-138. ASME, New York.

[3] YOUNG, L. E. (1975). On the dynamic response of a damped membrane to a point stochastic load. M.I.T. Acoustics and Vibration Laboratory Rept. No. 81678-2.

[4] LEE, S. S. (1976). Lanes of intensified response in structures excited by wide-band random excitation. *Ph.D. Thesis*. Department of Mechanical Engineering, M.I.T., Cambridge, Massachusetts.

[5] CRANDALL, S. H. (1976). Structured response patterns due to wide-band random excitation. *Proc. IUTAM Symposium on Stochastic Problems in Dynamics, July 19-23, 1976*. To appear.

[6] BOLOTIN, V. V. (1969). *Statistical Methods in Structural Mechanics*. Holden Day, San Francisco.

P.R. Krishnaiah, ed., APPLICATIONS OF STATISTICS

SOME APPLICATIONS OF STOCHASTIC LANGUAGES*

K. S. FU
School of Electrical Engineering
Purdue University
West Lafayette, Indiana

1. INTRODUCTION

Formal languages and their corresponding automata and parsing procedures have been used in the modeling and analysis of natural and computer languages [2, 6], and the description and recognition of patterns [11]. In some applications, a certain amount of uncertainty exists in the process under study. For example, noise or distortion occurs in the communication, storage or retrieval of information, or in the measurement of processing of physical patterns. Under these situations, in order to model the process more realistically, the approach of using stochastic languages has been suggested [8, 11, 15]. For every sentence or string x in a language L, a probability p(x) can be assigned such that $0 < p(x) \leq 1$ and $\sum_{x \in L} p(x) = 1$. Thus, the function p(x) is a probability measure defined over L, and it can be used to characterize the uncertainty and randomness of L. In this paper, after a brief introduction of stochastic languages, three applications are described. They are (1) communication and coding, (2) syntactic pattern recognition, and (3) error-correcting parsing. Other applications of stochastic languages include language learning [12, 27] and digital system design [4].

2. REVIEW OF STOCHASTIC LANGUAGES

Two natural ways of extending the concept of formal languages to stochastic languages are to randomize the productions of grammars and the state transitions of recognition devices (acceptors) respectively. In this section, some major results in stochastic grammars and stochastic syntax analysis are reviewed.

Definition 1: **A stochastic phrase structure grammar** (or simply stochastic grammar) is a four-tuple $G_s = (V_N, V_T, P_s, S)$ where V_N and V_T are finite sets of nonterminals and terminals; $S \in V_N$ is the start symbol†; P_s is a finite set of stochastic productions each of which is of the form

$$\alpha_i \xrightarrow{p_{ij}} \beta_{ij}, \qquad j=1,\ldots,n_i, \qquad i=1,\ldots,k \tag{1}$$

*This work was supported by the AFOSR Grant 74-2661.

†In a more general formulation, instead of a single start symbol, a start symbol (probability) distribution can be used.

where

$$\alpha_i \in (V_N \cup V_T)^* V_N (V_N \cup V_T)^*$$

$$\beta_{ij} \in (V_N \cup V_T)^*$$

and p_{ij} is the probability associated with the application of this stochastic production,

$$0 < p_{ij} \leq 1 \text{ and } \sum_{j=1}^{n_i} p_{ij} = 1^{‡} \tag{2}$$

V* denotes the set of all strings composed of symbols of V, including the empty string λ.

Suppose that $\alpha_i \xrightarrow{p_{ij}} \beta_{ij}$ is in P_s. Then the string $\xi = \gamma_1\alpha_i\gamma_2$ may be replaced by $\eta = \gamma_1\beta_{ij}\gamma_2$ with the probability p_{ij}. We shall denote this derivation by

$$\xi \xRightarrow{p_{ij}} \eta$$

and we say that ξ directly generates η with probability p_{ij}. If there exists a sequence of strings $\omega_1,\ldots,\omega_{n+1}$ such that

$$\xi = \omega_1, \quad \eta = \omega_{n+1}, \quad \omega_i \xRightarrow{p_i} \omega_{i+1}, \quad i=1,\ldots,n,$$

then we say that ξ generates η with probability $p = \prod_{i=1}^{n} p_i$ and denote this derivation by

$$\xi \overset{p}{\underset{*}{\Longrightarrow}} \eta$$

The probability associated with this derivation is equal to the product of the probabilities associated with the sequence of stochastic productions used in the derivation. It is clear that $\overset{p}{\underset{*}{\Longrightarrow}}$ is the reflexive and transitive closure of the relation $\xRightarrow{p}$.

The stochastic language generated by G_s is

‡If these two conditions are not satisfied, p_{ij} will be denoted as the weight w_{ij} and P_s the set of weighted productions. Consequently, the grammar and the language generated are called weighted grammar and weighted language, respectively [9].

$$L(G_s) = \{(x, p(x)) \mid x \in V_T^*, \; S \overset{p_j}{\underset{*}{\Longrightarrow}} x, \quad j=1,\ldots, k$$

$$\text{and } p(x) = \sum_{j=1}^{k} p_j\} \tag{3}$$

where k is the number of all distinctively different derivations of x from S and p_j is the probability associated with the jth distinctive derivation of x.

In general, a stochastic language L (G_s) is characterized by (L,p) where L is a language and p is a probability distribution defined over L. The language L of a stochastic language $L(G_s)$ = (L,p) is called the characteristic language of $L(G_s)$.

Since the productions of a stochastic grammar are exactly the same as those of the non-randomized grammar except for the assignment of the probability distribution, the language L generated by a stochastic grammar is the same as that generated by the non-randomized version.

Example 1: Consider the stochastic finite-state grammar

$$G_s = (V_N, V_T, P_s, S)$$

where $V_N = \{S, A, B\}$, $V_T = \{0, 1\}$ and

$$P_s: \; S \xrightarrow{1} 1A$$

$$A \xrightarrow{0.8} 0B$$

$$A \xrightarrow{0.2} 1$$

$$B \xrightarrow{0.3} 0$$

$$B \xrightarrow{0.7} 1S$$

A typical derivation, for example, would be

$$S \Longrightarrow 1A \longrightarrow 10B \longrightarrow 100$$

$$p(100) = 1 \times 0.8 \times 0.3 = 0.24$$

The stochastic language generated by G_s, $L(G_s)$ is as follows:

String generated x	*p(x)*
11	0.2
100	0.24
$(101)^n 11$	$0.2 \times (0.56)^n$
$(101)^n 100$	$0.24 \times (0.56)^n$

It is noted that

$$\sum_{x \in L(\bar{G}_s)} p(x) = 0.2 + 0.24 + \sum_{n=1}^{\infty} (0.2 + 0.24)(0.56)^n = 1$$

Definition 2: If, for a stochastic grammar G_s,

$$\sum_{x \in L(G_s)} p(x) = 1 \tag{4}$$

then G_s is said to be a consistent stochastic grammar.

The condition for a stochastic context-free grammar to be consistent is briefly described [5, 11]. Whether or not a stochastic context-sensitive grammar is consistent under certain conditions has not yet been known. The consistency condition for stochastic finite-state grammars is trivial since every stochastic finite-state grammar is consistent.

For a context-free grammar, all the productions are of the form

$$A \to \alpha, \quad A \in V_N, \quad \alpha \in V^+ = V^* - \{\lambda\} .$$

In this case, a nonterminal at the left side of a production may directly generate zero or a finite number of nonterminals. The theory of multitype Galton-Watson branching processes [16] can be applied to study the language generation process by stochastic context-free grammars. The zero-th level of a generation process corresponds to the start symbol S. The first level will be taken as β_1 where β_1 is the string generated by the production $S \to \beta_1$. The second level will correspong to the string β_2 which is obtained from β_1 by applying appropriate productions to every nonterminal in β_1. If β_1 does not contain any nonterminal, the process is terminated. Following this procedure, the jth level string β_j is defined to be the string obtained from the string β_{j-1} by applying appropriate productions to every nonterminal of β_{j-1}. Since all nonterminals are considered simultaneously in going from the (j-1)th level to the jth level, only the probabilities associated with each productions $A_i \to \alpha_r$, p_{ir}, need to be considered. Define the equivalence class C_{A_i} = {all productions of P with left side A_i} Thus,

$$P = \bigcup_{i=1}^{k} C_{A_i}$$

For each equivalence class C_{A_i},

$$\sum_{C_{A_i}} p_{ir} = 1$$

Definition 3: For each C_{A_i}, i=1,...,k, define the k-argument generating function $f_i(s_1,s_2,\ldots,s_k)$ as

$$f_i(s_1,\ldots,s_k) = \sum_{C_{A_i}} p_{ir}\, s_1^{\mu_{i1}(\alpha_r)}\, s_2^{\mu_{i2}(\alpha_r)} \cdots s_k^{\mu_{ik}(\alpha_r)} \tag{5}$$

where $\mu_{i\ell}(\alpha_r)$ denotes the number of times the nonterminal A_ℓ appears in the string α_r of the production $A_i \to \alpha_r$, and $S = A_1$.

Definition 4: The jth-level generating function $F_j(s_1,\ldots,s_k)$ is defined recursively as

$$F_0(s_1,\ldots,s_k) = s_1$$

$$F_1(s_1,\ldots,s_k) = f_1(s_1,\ldots,s_k)$$

$$F_j(s_1,\ldots,s_k) = F_{j-1}(f_1(s_1,\ldots,s_k),\ldots,f_k(s_1,\ldots,s_k)) \tag{6}$$

Example 2:

$$G_s = (V_N,\ V_T,\ P_s,\ S)$$

where $V_N = \{A_1,A_2\}$

$V_T = \{a,b\}$

$S = A_1$

and P_s: $A_1 \xrightarrow{p_{11}} aA_1A_2$

$A_1 \xrightarrow{p_{12}} b$

$A_2 \xrightarrow{p_{21}} aA_2A_2$

$A_2 \xrightarrow{p_{22}} a\ a$

The generating functions for C_{A_1} and C_{A_2} are respectively:

$$f_1(s_1,s_2) = p_{11}s_1s_2 + p_{12}$$

$$f_2(s,s_2) = p_{21}s_2^2 + p_{22}$$

The jth-level generating functions for j=0,1,2 are:

$$F_0(s_1,s_2) = s_1$$

$$F_1(s_1,s_2) = f_1(s_1,s_2) = p_{11}s_1s_2 + p_{12}$$

$$F_2(s_1,s_2) = F_1(f_1(s_1,s_2),f_2(s_1,s_2))$$

$$= p_{11}f_1(s_1,s_2)f_2(s_1,s_2) + p_{12}$$

$$= p_{11}^2p_{21}s_1s_2^3 + p_{11}^2p_{22}s_1s_2 + p_{11}p_{12}p_{21}s_2^2 + p_{11}p_{12}p_{22} + p_{12}$$

After examining the previous example, we can express $F_j(s_1,\ldots,s_k)$ as

$$F_j(s_1,\ldots,s_k) = G_j(s_1,\ldots,s_k) + K_j$$

where $G_j(s_1,\ldots,s_k)$ denotes the polynominal of $s_1,\ldots,s_k$ without including the constant term. The constant term, K_j, corresponds to the probability of all the strings $x \in L(G_s)$ that can be derived in j or fewer levels. This leads to the following theorem.

Theorem 1: A stochastic context-free grammar G_s is consistent if and only if

$$\lim_{j\to\infty} K_j = 1 \tag{7}$$

Note that if the above limit does not equal to 1 there is a finite probability that a generation process may never terminate. Thus, the probability measure defined over L will be less than 1 and, consequently, G_s will not be consistent. On the other hand, if the limit is equal to 1 then there exists no such infinite (nonterminating) generation process since the limit represents the probability of all the strings which are generated by applications of a finite number of productions. Consequently, G_s is consistent. The problem of testing the consistency of a given stochastic context-free grammar can be solved by using the following testing procedure developed for branching processes.

Definition 5: The expected number of occurrences of the nonterminal A_j in the production set C_{A_i} is

$$e_{ij} = \left.\frac{\partial f_i(s_1,\ldots,s_k)}{\partial s_j}\right|_{s_1,\ldots,s_k = 1} \tag{8}$$

Definition 6: The first moment matrix E of the generation process corresponding to a stochastic context-free grammar G_s is defined as

$$E = [e_{ij}] , \quad 1 \leq i,j \leq k \tag{9}$$

where k is the number of nonterminals in G_s.

Theorem 2: For a given stochastic context-free grammar G_s, order the eigen-values or the characteristic roots $\rho_1,\ldots,\rho_k$ of the first moment matrix according to the descending order of their magnitudes; that is,

$$|\rho_i| \geq |\rho_j| \quad \text{if } i < j \tag{10}$$

Then G_s is consistent if $\rho_1 < 1$ and it is not consistent if $\rho_1 > 1$.

For the stochastic context-free grammars G_s given in Example 2,

$$e_{11} = \left.\frac{\partial f_1(s_1,s_2)}{\partial s_1}\right|_{s_1,s_2 = 1} = p_{11}$$

$$e_{12} = \left.\frac{\partial f_1(s_1,s_2)}{\partial s_2}\right|_{s_1,s_2 = 1} = p_{11}$$

$$e_{21} = \left.\frac{\partial f_2(s_1,s_2)}{\partial s_1}\right|_{s_1,s_2 = 1} = 0$$

$$e_{22} = \left.\frac{\partial f_2(s_1,s_2)}{\partial s_2}\right|_{s_1,s_2 = 1} = 2p_{21}$$

Thus,

$$E = \begin{bmatrix} p_{11} & p_{11} \\ 0 & 2p_{21} \end{bmatrix}$$

The characteristic equation associated with E is

$$\Phi(x) = (x - p_{11})(x - 2p_{21})$$

Hence, G_s will be consistent as long as $p_{21} < \frac{1}{2}$.

3. APPLICATION TO COMMUNICATION AND CODING

An information source of a communication system is defined as a generative device which produces, at a uniform rate, sequences of symbols from an alphabet. In channel coding, these symbols are usually assumed to be randomly generated and therefore they do not represent any specific meanings other than their own identifications. We shall call sources of this kind lexicographical, to emphasize the total lack of regularity or "structure" in the formation of source messages. The realm of algebraic coding has hitherto been concerned almost exclusively with these sources and we shall also call the encoder, decoder as well as the coding technique lexicographical, in order to emphasize the contrast between these and the new type of syntactic decoder introduced in this section. Other types of information sources, of course, do exist. In particular, Markov information source models are widely accepted in source coding.

Several attempts have been made to treat the problem of human communication from the narrow, strictly probabilistic point of view, based on the measure of information introduced by Shannon in 1948. Recently, it was pointed out that information theory fell quite short of an acceptable description for the process of transference of ideas that occurs when intelligent beings communicate [26]. It has been demonstrated that syntactic information about the source is useful for designing a "syntactic decoder" which supplements an ordinary decoder [13].

On the other hand, Hutchins [18] showed that we can do source coding for some formal languages by removing a certain amount of "redundancy" in the languages. Smith [25] suggested that the existence of "built-in redundancy" in formal languages can be utilized in achieving error-detection without the obligatory addition of redundancy. Problems of a similar nature were recently studied by Hellman [17] who developed a method of combining source coding (i.e., data compression) and channel coding.

A brief discussion about the maximum-likelihood syntactic decoder is given in this section. Let the linguistic source be characterized by a context-free grammar $G = (V_N, V_T, P, S)$ which generates a language $L(G) \subset V_T^*$. The source generates a sequence of sentences of $L(G)$. A special kind of synchronization mechanism is provided so that there is no error in locating the beginning and the end of every sentence issued by the source. With this assumption, we may proceed to consider syntactic decoding for individual sentences.

Refer to Fig. 1, the output of the source is encoded into binary form which is transmitted over a binary symmetric memoryless channel N with bit error rate p. Output symbols V_T of the source are encoded by C_n. Suppose V_T has ℓ symbols, i.e., $V_T = \{\sigma_1, \sigma_2, \ldots, \sigma_\ell\}$, then $n \geq [\log_2 \ell]$. Let $\bar{u}_i = C_n(\sigma_i)$ be the codeword for the ith symbol σ_i of the alphabet, which can be expressed as a binary sequence $\bar{u}_i = (u_1^i, u_2^i, \ldots, u_n^i)$. Then the conditional probability of a binary sequence $\bar{v} = (v_1, v_2, \ldots, v_n)$ from the output of the channel given that the input sequence is $\bar{u} = (u_1, u_2, \ldots u_n)$ is

$$P(\bar{v}|\bar{u}) = P(s_0 = v_1 | s_1 = u_1) \ldots P(s_0 = v_n | s_1 = u_n)$$

$$= (1-p)^{\sum_{i=1}^{n} \delta(u_i, v_i)} \cdot p^{\, n - \sum_{i=1}^{n} \delta(u_i, v_i)} \tag{11}$$

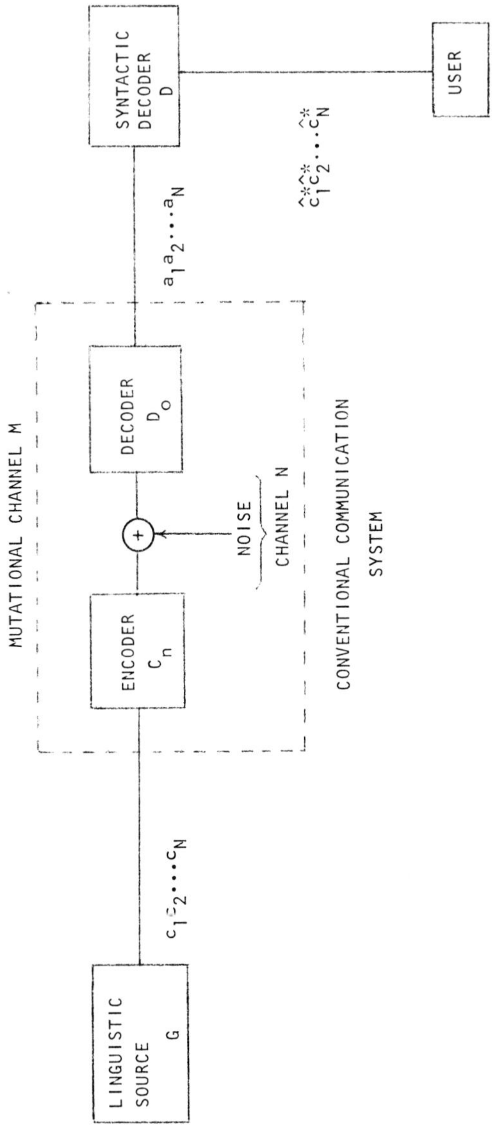

Fig. 1. A communication model for linguistic sources.

where $P(s_0 = v_i \mid s_1 = u_i) = \begin{cases} p & \text{if } v_i \neq u_i \quad \text{for } i=1,\ldots,n, \\ 1-p & \text{otherwise} \end{cases}$

and $\delta(\ ,\ )$ is the Kronecker delta function.

At the receiver a decoder assigns a symbol σ to every binary sequence $\bar{v} \in \{0,1\}^n$. More precisely, let B be a clock code defined by

$$B = \{ (\bar{u}_1, U_1), (\bar{u}_2, U_2), \ldots, (\bar{u}_\ell, U_\ell) \}$$

where the $\bar{u}_i$'s are codewords for the alphabet V_T, and U's are disjoint subsets of $\{0,1\}^n$ such that $U_1 \cup \cdots\cdots \cup U_\ell = \{0,1\}^n$. A lexicographical decoder D_0 is a mapping from $\{0,1\}^n$ into V_T, defined by

$$D_0(\bar{v}) = \sigma_j \quad \text{if and only if } \bar{v} \in U_j. \tag{12}$$

Designs of the code as well as its encoders and decoders are studied extensively in information theory (e.g., [19]). We shall present in the following a decoding scheme which in a certain sense is superior to the lexicographical decoding scheme.

The encoder, binary symmetric channel N and the decoder D_0 can be combined and represented by a mutational channel M characterized by the following transition probabilities:

$$\begin{aligned} Q_B(\beta = \sigma \mid \alpha = \sigma_i) &= P_B(\bar{v} \in U_j \mid \bar{u}_i) \\ &= \sum_{\bar{v} \in U_j} P_B(\bar{v} = (v_1, v_2, \ldots, v_n) \mid \bar{u}_i) \\ &= \sum_{\bar{v} \in U_j} (1-p)^{\sum_{k=1}^{n} \delta(u_k^i, v_k)} \cdot p^{[n - \sum_{k=1}^{n} \delta(u_k^i, v_k)]} \end{aligned} \tag{13}$$

for every $i,j = 1,2,\ldots,\ell$. Here α and β are respectively the input and output of the mutational channel M, and $Q_B(\beta = \sigma_j \mid \alpha = \sigma_i)$ is the probability of observing a symbol σ_j at the output of M when symbol σ_i is the true symbol at the input of M. Note that $\sum_j Q_B(\beta = \sigma_j \mid \alpha = \sigma_i) = 1$ for all i.

As illustrated in Figure 1, a syntactic decoder D is introduced to the output of the mutational channel, i.e., to the output of the lexicographical decoder D_0, before the messages are conveyed to the user. Suppose the source grammar G generates a sentence $w = c_1 c_2 \cdots c_N \in L(G)$. The symbols $c_1, \ldots, c_N$ are fed into the encoder C_n sequentially such that a sequence of binary codewords $C_n(c_1), \ldots, C_n(c_N)$ come out of the encoder and are modulated and transmitted over the noisy channel N. Let $\bar{u}_i = C_n(c_i)$, $i=1,\ldots,N$, and $\bar{v}_i$ be the binary sequence of length n coming out of the channel N corresponding to the transmitted codeword $\bar{u}_i$.

These binary sequences $\bar{v}_i,\ldots,\bar{v}_N$ are decoded according to (12) so that the lexicographical decoder outputs a string of symbols in V_T, $z = a_1a_2\ldots a_N$ where $a_i = D_0(v_i)$, $i=1,2,\ldots,N$. We can symbolically describe this coding scheme by $z = a_1a_2\ldots a_N = M(c_1)\,M(c_2)\ldots M(c_N)$. The block code B is designed in such a way that B is optimal in some sense. For example, B results in minimal average probability of error for each individual symbol, among all possible block codes of codeword length n. That is, for every symbol σ_i transmitted, the average probability of error per symbol,

$$\overline{P(e|B)} = \frac{1}{\ell}\sum_{i=1}^{\ell} Q_B\,(\beta \neq \sigma_i | \alpha = \sigma_i) \qquad (14)$$

is minimized.

However, it is immediately obvious that the above optimal block code B is based on the performance of the code for individual symbols. For linguistic messages, we may take advantage of our knowledge about the syntactic structure among the transmitted symbols to improve the efficiency of the communication system. Since the utility of the communicated linguistic messages depends mostly on the correctness of individual sentences, the average probability of obtaining a correct sentance for the source grammar will be considered to be a significant factor in evaluating the efficiency of the communication system.

Let $z = a_1a_2\ldots a_N$ be the decoded sequence of symbols from the lexicographical decoder D_0. The syntactic decoder D is a mapping from V_T^* into L(G), which is optimal according to the maximum likelihood criterion:

$$P(z=a_1a_2\ldots a_N \mid D(z) = \hat{c}_1^*\hat{c}_2^*\ldots\hat{c}_N^*) \geq P(z=a_1a_2\ldots a_N|\hat{w}=\hat{c}_1\hat{c}_2\ldots\hat{c}_N) \qquad (15)$$

for all $\hat{w} \in L(G) \cap V_T^N$. Here, $D(z) = \hat{c}_1^*\hat{c}_2^*\ldots\hat{c}_N^*$ is the output sequence of the syntactic decoder with input sequence z. Since the channel N is memoryless, and since the symbols of a transmitted sentence are encoded and decoded sequentially, (15) can be rewritten as

$$\prod_{i=1}^{N} Q(\beta-a_i \mid \alpha-\hat{c}_i^*) \geq \prod_{i=1}^{N} Q(\beta-a_i \mid \alpha-\hat{c}_i) \qquad (16)$$

Define $d(\sigma_i, \sigma_j) = -\log Q(\beta=\sigma_j \mid \alpha = \sigma_i)$. Then (16) is equivalent to

$$\sum_{i=1}^{N} d(a_i,\hat{c}_i^*) \leq \sum_{i=1}^{N} d(a_i,\hat{c}_i) \qquad (17)$$

Extend the definition of $d(\sigma_j,\sigma_i)$ to the case of two strings $z = a_1a_2\ldots a_N$ and $w = c_1c_2\ldots c_N$ which are both of length N, $N = 2,3,\ldots$. Then

$$d\,(z = a_1a_2\ldots a_N;\; w = c_1c_2\ldots c_N) = \sum_{i=1}^{N} d(a_i, c_i) \tag{18}$$

Similarly, we can also define

$$d(z;L(G)) = \inf_{w\,\varepsilon\,L(G)} d(z;w) \tag{19}$$

Hence, using (17) and (18) we can rewrite (15) as

$$D(z) = \{\, \hat{w}^* \mid \hat{w}^* = \hat{c}_1^*\hat{c}_2^*\ldots\hat{c}_N^* \;\varepsilon\; L(G) \text{ such that } d(z;\hat{w}^*) \leq d(z;\hat{w}) \text{ for all sentences } \hat{w} \;\varepsilon\; L(G)\} \tag{20}$$

It follows from (19) that $\hat{w}^* \;\varepsilon\; D(z)$ implies that

$$d(z;\hat{w}^*) = d(z;L(G)) \tag{21}$$

We say that $\hat{w}^*$ of the source language L(G) is the "nearest" to the string z.

An algorithm based on the Cocke-Younger-Kasami parsing scheme has been proposed for the syntactic decoder [17]. The syntactic decoder is basically an error-correcting parser which attempts to produce a collection of strings $D(z) \subset L(G)$ for any string z not in the language L(G) so that every string $\hat{w}^*$ in D(z) is "nearest" to z as compared with other strings in L(G). An example of applying such a syntactic decoder to computer networks is given in [14].

4. APPLICATION TO SYNTACTIC PATTERN RECOGNITION

In syntactic or linguistic pattern recognition [11], patterns are described by sentences of a language characterized by a grammar. Typically, each class of patterns is characterized by a grammar which generates sentences describing each of the patterns in the class. The recognition of a pattern then becomes a syntax analysis or parsing of the sentence describing the pattern with respect to the grammar characterizing each pattern class.

In some practical applications, however, noise and distortion occur in the measurement or processing of physical patterns. Languages used to describe the noisy and distorted patterns under consideration are often ambiguous in the sense that a sentence (a pattern) can be generated by more than one grammar which characterizes the patterns (sentences) generated from a particular pattern class. In terms of statistical pattern recognition [7], this is the case of overlapping pattern classes; patterns belonging to different classes may have the same descriptions or measurement values (but with different probabilities of occurrence). In these situations, in order to model the process more realistically, the approach of using stochastic languages has been suggested [8, 11, 15].

For example, three types of chromosome patterns and their corresponding string representations are shown in Fig. 2. It is easy to notice that with noise and distortion one type of chromosomes (e.g., submedian) could become another type (e.g., median or acrocentric). In other words, the grammar characterizing submedian chromosomes should sometimes (with small probabilities) also generate median and acrocentric chromosomes. Similar conclusions can also be drawn for grammars characterizing median and acrocentric chromosomes. One approach to model the situation is to use one grammar which generates all the three types of chromosome patterns but to use three different sets of production probabilities. The following is a grammar generating median, submedian, and acrocentric chromosome patterns. However, three different production probability assignments results in three different stochastic grammars which characterize median, submedian, and acrocentric chromosome patterns, respectively. Presumably, the stochastic grammar for median chromosome patterns will have high probability of generating median strings than submedian and acrocentric strings; the stochastic grammar for submedian chromosome patterns will have high probability of generating submedian strings, etc. In practice, the production probabilities will have to be inferred from string probabilities or assigned subjectively by the designer [10, 11].

Chromosome Grammar $G_s = (V_N, V_T, P_s, S)$

where V_N = {S, A, B, D, H, J, E, F, W, G, R, L}

V_T = { 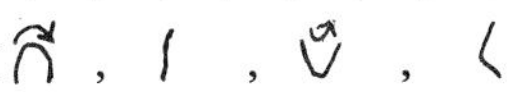 }

a b c d

and P_s:

$S \xrightarrow{1} AA$	$R \xrightarrow{p_3} RE$	$W \xrightarrow{p_3} WE$
$A \xrightarrow{1} cb$	$R \xrightarrow{p_4} HDJ$	$W \xrightarrow{p_4} d$
$B \xrightarrow{p_1} FBE$	$D \xrightarrow{p_1} FDE$	$F \xrightarrow{1} b$
$B \xrightarrow{p_2} HDJ$	$D \xrightarrow{p_2} d$	$E \xrightarrow{1} b$
$B \xrightarrow{p_3} RE$	$D \xrightarrow{p_3} FG$	$H \xrightarrow{1} a$
$B \xrightarrow{p_3} FL$	$D \xrightarrow{p_3} WE$	$J \xrightarrow{1} a$
$L \xrightarrow{p_3} FL$	$G \xrightarrow{p_3} FG$	
$L \xrightarrow{p_4} HDJ$	$G \xrightarrow{p_4} d$	

$p_1 + p_2 + 2p_3 = 1$

$p_3 + p_4 = 1$

By associating probabilities with the strings, we can impose a probabilistic structure on the language to describe noisy patterns. The probability distribution characterizing the patterns in a class can be interpreted as the probability

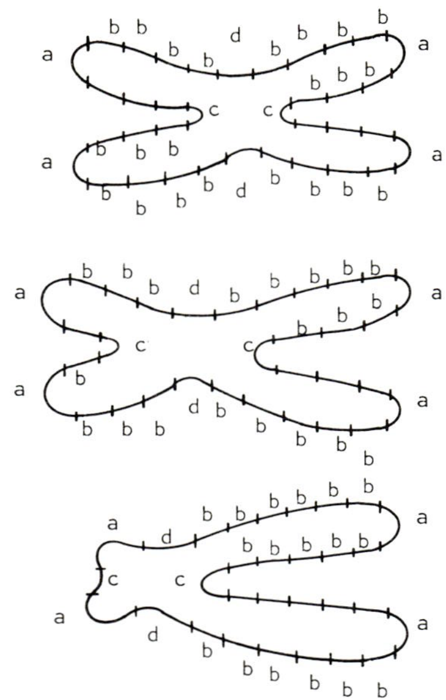

(a) Median String Representation
cbbbabbbbdbbbbabbbcbbbabbbbdbbbbabbb

(b) Submedian String Representation
cbabbbdbbbbbabbbcbbbabbbbbbdbbbab

(c) Acrocentric String Representation
cadbbbbbbabbbbbcbbbbbabbbbbbda

Fig. 2. Three sample chromosomes.

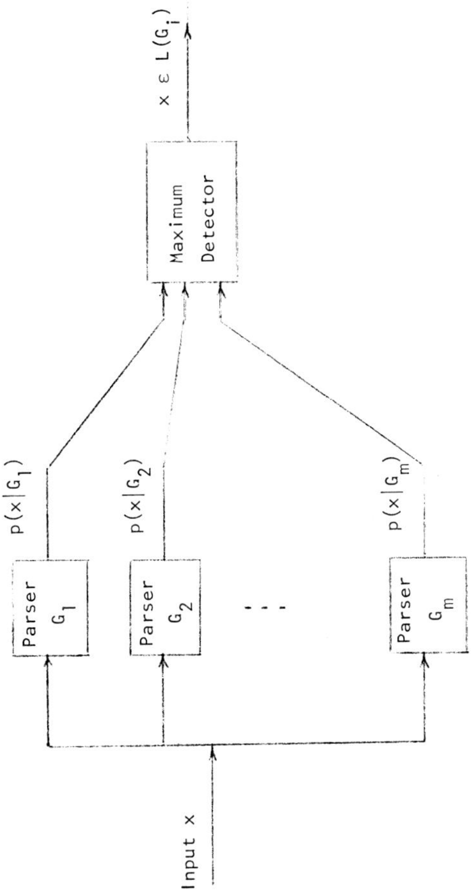

Fig. 3. Maximum-likelihood syntactic pattern recognition system.

distribution associated with the strings in a language. Thus, statistical decision rules can be applied to the classification of a pattern under ambiguous situations (for example, use the maximum-likelihood or Bayes decision rule). A block diagram of such a recognition system using maximum-likelihood decision rule is shown in Fig. 3. Furthermore, because of the availability of the information about production probabilities, the speed of syntactic analysis can be improved through the use of this information [21, 24].

5. APPLICATION TO ERROR-CORRECTING PARSING

The syntactic decoder discussed in Section 3 is actually implemented as an error-correcting parser with substitution error. In language parsing, in addition to substitution error; that is, the substitution of a symbol by another symbol, two other types of syntactic error also occur. They are the insertion error (the insertion of an extra symbol) and the deletion error (the deletion of a symbol). Application examples of this kind include error-correcting compiler [20] and the recognition of continuous speech [3, 22]. In this section, error-correcting parsing for the three types of error is briefly described.

Consider that the three types of error are represented by three error transformations; namely, T_S for substitution error transformation, T_I for insertion error transformation, and T_D for deletion error transformation. Then, T_S, T_I, and T_D are defined as mappings from a given language L to the subset of V^*_T. Specifically,

(1) xby is in T_S (xay) for all $b \neq a$ in V_T,

(2) xay is in T_I (xy) for all $a \in V_T$,

(3) xy is in T_D (xay) for all $a \in V_T$,

where x and y are strings in V_T^*. A sentence not in L can be transformed from some sentence in L by a sequence of error transformations. When the language L is non-stochastic, the minimum number of error transformations required to transform a sentence in L to a given erroneous sentence (not in L) is often used as a decision criterion (minimum distance criterion [1]).

When the error statistics are known or can be estimated, the error transformations become probabilistic. For instance, we may know that the probability of substituting terminal a by b or the probability associated with T_S is $q_S(b|a)$, the probability of inserting terminal b in front of a is $q_I(b|a)$, and the probability of deleting terminal a is $q_D(\lambda|a)$ where λ is the empty sentence or sentence with length zero. With the information of these error transformation probabilities, we can apply the maximum-likelihood decision criterion to obtain an error-corrected sentence in L. If the language L is also stochastic, Bayes decision rule can be used [23] The implementation of such error-correcting parsers is a modified Earley's parsing algorithm for context-free languages [23] or a modified Cocke-Younger-Kasami algorithm for context-sensitive languages [28], based on an expanded grammar by taking the error transformations and their associated probabilities into consideration.

6. CONCLUSIONS AND REMARKS

Stochastic languages are introduced and some of their applications to coding, pattern recognition, and language analysis are briefly described in this paper. With probabilistic information about the process under study, maximum-likelihood and Bayes decision rules can be directly applied to the coding/decoding and analysis of linguistic source and the classification of noisy and distorted linguistic patterns. Stochastic finite-state and context-free languages are easier to analyze compared with stochastic context-sensitive languages; however, their descriptive power of complex processes is less. The consistency problem of stochastic context-sensitive languages is still a problem of investigation. This, of course, directly limits the practical applications of stochastic context-sensitive languages except in a few special cases (e.g., stochastic context-sensitive languages generated by stochastic context-free programmed grammars [11]). Only very limited results in grammatical inference are practically useful [12]. Efficient inference algorithms are definitely needed before we can design a system to automatically infer a grammar from sample sentences of a language.

REFERENCES

[1] AHO, A. V. and PETERSON, T. G. (1972). A minimum distance error-correcting parser for context-free languages. *SIAM J. Computing*, 4.

[2] AHO, A. V. and ULLMAN, J. D. (1972). *Theory of Parsing, Translation and Compiling*, Vol. 1. (Vol. 2 - 1973). Prentice-Hall.

[3] BAHL, L. R. and JELINEK, F. (1975). Decoding for channels with insertion, deletion and substitutions with applications to speech recognition. *IEEE Trans. Information Theory*, IT-21, 4.

[4] BOOTH, T. L. (1974). Design of minimal expected processing time finite-state transducers. *Proc. IFIP Congress 74.* North Holland Press.

[5] BOOTH, T. L. (1969). Probability representation of formal languages. *IEEE 10th Annual Symposium on Switching and Automata Theory.*

[6] CHOMSKY, N. (1956). Three models for the description of language. *IEEE Trans. Information Theory*, IT-2, 113-124.

[7] FU, K. S. (1968). *Sequential Methods in Pattern Recognition and Machine Learning.* Academic Press.

[8] FU, K. S. (1972). On syntactic pattern recognition and stochastic languages. *Frontiers of Pattern Recognition*, (S. Watanabe, ed.). Academic Press.

[9] FU, K. S. and HUANG, T. (1972). Stochastic grammars and languages. *International Journal on Computer and Information Sciences*, 1, No. 2.

[10] FU, K. S. (1973). Stochastic languages for picture analysis. *Computer Graphics and Image Processing*, 2, 4, 433-453.

[11] FU, K. S. (1974). *Syntactic Methods in Pattern Recognition*. Academic Press.

[12] FU, K. S. and BOOTH, T. L. (1975). Grammatical inference: Introduction and survey. I - II, *IEEE Trans. Systems, Man and Cybernetics*, SMC-5, 95-111 and 409-423.

[13] FUNG, L. W. and FU, K. S. (1975). Maximum-likelihood syntactic decoding. *IEEE Trans. Information Theory*, IT-21.

[14] FUNG, L. W. and FU, K. S. (1976). An error-correcting syntactic decoder for computer networks. *International Journal of Computer and Information Sciences*, 5, No. 1.

[15] GRENANDER, U. (1969). Foundation of pattern analysis. *Quart. Appl. Math.*, 27, 1-55.

[16] HARRIS, T. E. (1963). *The Theory of Branching Processes*. Springer-Verlag, Berlin.

[17] HELLMAN, M. E. (1973). Joint source and channel encoding. Stanford Electronics Lab., Stanford University.

[18] HUTCHINS, S. E. (1970). Stochastic sources for context-free languages. Ph.D. Dissertation, University of California, San Diego.

[19] JELINEK, F. (1968). *Probabilistic Information Theory*. McGraw-Hill.

[20] LAFRANCE, J. E. (1971). Syntax-directed error-recovery for compilers. Rept. #459, Dept. of Comp. Science, University of Illinois.

[21] LEE, H. C. and FU, K. S. (1972). A stochastic syntax analysis procedure and its application to pattern recognition. *IEEE Trans. Computers*, C-21.

[22] LIPTON, R. J. and SNYDER, L. (1974). On the optimal parsing of speech. Res. Rept. #37, Dept. of Comp. Science, Yale University.

[23] LU, S. Y. and FU, K. S. (1976). Efficient error-correcting syntax analysis for recognition of noisy patterns. Tech. Rept. EE-76-9, Purdue University.

[24] PERSOON, E. and FU, K. S. (1975). Sequential classification of strings generated by SCFG's. *International Journal of Computer and Information Sciences*, 4, 3.

[25] SMITH, W. B. (1970). Error detection in formal languages. *J. Compt. Syst. Sci.*, (October).

[26] SOUZA, C. R. and SCHOLTZ, R. A. (1969). Syntactical decoders and backtracking S-grammars. The ALOHA System Rept. A69-9, University of Hawaii.

[27] SUPPES, P. (1970). Probabilistic grammars for natural languages. *Synthese*, 22, 95-116.

[28] TANAKA, E. and FU, K. S. (1976). Error-correcting parsers for formal languages. Tech. Rept. EE-76-7, Purdue University.

P.R. Krishnaiah, ed., APPLICATIONS OF STATISTICS

STATISTICAL PROBLEMS IN THE SIZE EFFECT ON MATERIAL STRENGTH

H. LEON HARTER

Air Force Flight Dynamics Laboratory
Wright-Patterson Air Force Base, Ohio

Nearly 500 years ago, Leonardo da Vinci observed that long wires are weaker than short wires of the same diameter. Over 100 years ago, Karmarsch found that the tensile strength of wires of a given length can be represented by an expression of the form $F = A + B/d$, where d is the diameter and A and B are positive constants. In the present century, studies of the size effect have proliferated. Weibull (1939 a,b) was the first to give a reasonably satisfactory explanation of the volume effect on material strength. Daniels (1945) studied the statistical theory of bundles of threads. The statistical theory of extreme values ("weakest-link" theory) plays a very important role in studies of the size effect; competing theories include the energy theory and the technological theory. High lights are given of a literature survey on the size effect. Since this survey was motivated by concern about the reliability of large composite aircraft structures, which are now coming into use, special attention is given to the size effect on composite materials and structures. An attempt is made to summarize the present state of knowledge and to identify unsolved problems requiring further research.

1. INTRODUCTION

The statistical theory of extreme values ("weakest-link" theory) plays an important role in studies of the size effect on material strength. Chaplin (1880, 1882) applied this theory to the effect of length on the tensile strength of metal bars of constant cross-sectional area. Weibull (1939 a,b) also used extreme-value theory to give the first reasonably satisfactory explanation of the volume effect on material strength. In so doing, he introduced the statistical distribution which now bears his name. This distribution is also known as the third asymptotic distribution of smallest values, and presumably would be known exclusively by that name except for the influence of Weibull. Nevertheless, engineering statisticians owe him a great debt of gratitude for pointing out its broad applicability to material strength and many other phenomena. Tucker (1941) questioned the application of the "weakest-link" theory to elements in parallel, and proposed instead the so-called "strength-summation" theory, which he found to work much better for beams differing in width (presumably the same is also true for those differing in depth). Daniels (1945) reached a similar conclusion concerning the strength of bundles of threads or fibers, where failure of one thread does not necessarily result in failure of the bundle, but only in redistribution of the load to the remaining threads. Still further complications arise in the case of composites with fiber reinforcement, since a portion of the load is carried by the matrix in conjunction with the broken fibers. Though the "weakest-link" theory is directly applicable only on condition that the failure of the first element results in failure of the entire specimen or structure, statistical considerations enter into determination of the static strength, whether or not that condition is satisfied, as well as the fatigue life.

One should not make the mistake of assuming that statistical theory alone can explain all the size-effect phenomena that have been observed. Davidenkov (1960) enumerated four competing theories: 1. *The statistical theory*, which explains the influence of dimensions by the presence of chance inhomogeneities in a solid, where the largest of these inhomogeneities determines the strength. 2. *The energy theory*, which states that the brittle failure depends on the elastic energy which is stored in the specimen-machine system and contributes to the specimen failure. 3. *The technological theory*, which ascribes strength differences to unequal treatment conditions, for instance, to the effect of treatment by cutting. 4. *The theory proposed by Lavrov (1958)*, according to which -- at least in bending -- failure is caused by the initial microcrack on the stretched side of a beam, the dimensions and deformation of which do not depend on the beam dimensions. Lavrov's theory has never received wide acceptance, but the energy theory and the technological theory must be synthesized with the statistical theory in order to attain a comprehensive theory which will explain all the observed phenomena, as proposed, for example, by Volkov (1960).

A survey has been made of the literature on the size effect on material strength. The results have been published in an Air Force technical report [Harter (1977)] divided into three parts, each arranged in chronological order of publication -- a summary of each relevant document, a list of the references, and a supplementary bibliography on extreme-value theory. Some highlights of the literature survey are given in this paper, with sections on the effects of length, cross-sectional area, and volume (or surface area), on the strength of fiber bundles and of composites, and on the size effect in fatigue. Some unsolved problems concerning the strength of composite materials and the reliability of composite structures are identified.

2. EFFECT OF LENGTH

Timoshenko (1953) has traced the history of the strength of materials from the times of the ancient Egyptians, Greeks and Romans. Knowledge of the size effect goes back at least as far as Leonardo da Vinci (ca. 1500), who made experimental studies of the dependence of the strength of iron wires on their length and observed that long wires are weaker than short wires of the same diameter. Trautwine (1872) noted the effect of the length of a specimen of constant cross section on its strength, and Chaplin (1880, 1882) applied the "weakest-link" theory to the effect of length on the tensile strength of metal bars of constant cross-sectional area. He stated that, according to the laws of probability, if the probability is P that the tensile strength of a bar of given length and cross section exceeds a certain value, the probability that the strength of all of n bars of the same dimensions exceeds that value is P^n. Equivalently, the probability that the strength of the weakest of the bars (or the strength of a single bar of the same cross section but n times as long) exceeds the specified value is P^n. He worked out a quantitative relation between the average strength of a bar of given length and one n times as long, assuming normality. If the former is S, the latter is S-ap, where $p = .6745\sigma \doteq .6745s$ and $a = \Phi^{-1}[\sqrt[n]{.5}]/.6745$, p being the probable error, σ the population standard deviation, and s the usual estimate of σ from the sample, with $\Phi(x) = \int_{-\infty}^{x} \phi(x)\,dx$ and $\phi(x) = e^{-x^2/2}/\sqrt{2\pi}$. Chaplin (1882) tabulated values of a for n = 1(1)30(5)60(10)100(50)200(100)1000. These results were summarized in a textbook by Slocum and Hancock (1906), but then in the course of time they were apparently forgotten, only to be rediscovered by Peirce (1926), who observed that if specimens of length ℓ break under normally distributed loads, then specimens of length $n\ell$, where n is an integer greater than one, will break under loads whose distribution is negatively skewed, the more so the larger the value of n. Peirce determined the decrease in the mode and the median of the tensile strength corresponding to an n-fold increase in length, and estimated the corresponding decrease in the mean by making use of the fact that the difference between the mean and the mode is approximately three times that between the mean and the median. He also pointed out the applicability to this problem of the theoretical results of Tippett (1925) on extreme values of samples from normal distributions. Epstein (1948 a) summarized the work of Peirce, and stated that, as far as he knew, Peirce was the first worker to realize the close connection between the strength of a specimen and the distribution of smallest values; apparently neither Peirce nor Epstein was aware of the work of Chaplin. Lieblein (1954) called attention to the two papers by Chaplin, summarized the first of them, and interpreted it in modern symbols and terminology. In a footnote, he acknowledged indebtedness to Churchill Eisenhart for calling attention to an old engineering text [Slocum & Hancock (1906)] in which these articles were cited. Lévi (1949) concluded [see Peirce (1926)] , since the distribution of the smallest of n values ($n \geq 2$) drawn at random from a normal distribution is not normally distributed, that if the strength of the links of a chain (or of short segments of a long bar) is normally distributed, that of the chain (or bar) is not, and proposed use of the lognormal distribution as an approximation to the distribution of the strength of bars or chains. Weibull (1949, 1951), in his treatment of the effect of length, assumed a Weibull distribution of strength, which is better than assuming a normal or a lognormal distribution. This is true because, if the strength of a specimen of length L_1 has a Weibull cdf (cumulative distribution function) $F_1 = 1 - e^{-k(S-U)^m}$, where U is the load for which there is no probability of failure, the corresponding cdf for a specimen of length L_2 is given by $F_2 = 1 - e^{-(L_2/L_1)\, k\, (S-U)^m}$, which is the cdf of a Weibull distribution with the same location and shape parameters, U and m, respectively; only the scale parameter has changed. During the past quarter century, many authors have given expository treatments of the effect of the length of a specimen on its strength. Among the few who have made significant new contributions to the theory are Bartenev & Sidorov (1966), who proposed a statistical theory for strength of glass fibers which takes account of different types of surface defects and their distribution along the length of the fiber. They assumed that the probability of not encountering a defect in the length ℓ is given by $\exp[-(\lambda\ell)^n]$, where $n = 1$ for an exponential distribution of the defects along the length, $n > 1$ for any other distribution, and $n = \infty$ for defects occurring at a fixed interval $1/\lambda$ along the length. Defects of two kinds -- microcracks and microbreaks of the surface layer -- are encountered. A fiber has strength σ_A if neither kind of defect is present, σ_B if a defect of the first type is encountered in it but not one of the second type, and σ_C if a defect of the second type is encountered. Bartenev & Sidorov showed that the mean strength of a fiber of length ℓ is given by $\overline{\sigma}_\ell = (\sigma_A - \sigma_B) \exp[-(\lambda_1\ell)^{n_1} - (\lambda_2\ell)^{n_2}] + (\sigma_B - \sigma_C) \exp[-(\lambda_2\ell)^{n_2}] + \sigma_C$, which can easily be generalized to the case of m different strengths (m - 1 kinds of defects).

3. EFFECT OF CROSS-SECTION

Timoshenko (1953) has pointed out that Leonardo da Vinci (ca. 1500) made experimental studies of the dependence of the strength of beams on their width and of columns on their cross-section, also that Galileo Galilei (1638) enunciated and elaborated upon the similarity principle, which states that geometrically similar structures become weaker and weaker as their dimensions increase. Karmarsch (1859) represented the tensile strength (per unit cross-sectional area) of metal wires by an expression of the form $F = A + B/d$, where d is the diameter and A and B are positive constants. Griffith (1920, 1924) reported theoretical and experimental results on the strength of glass and other solids.

His theory is based partly on the "weakest-link" or "largest-flaw" concept, but he found this concept inadequate to explain the extremely high strength of freshly-drawn fine glass threads which he observed. To explain this phenomenon, he postulated that fracture results from extension of pre-existing microscopic or submicroscopic cracks (which have come to be called "Griffith cracks"), with stability controlled by a balance between released elastic-strain energy and absorbed surface energy (solid-state surface tension). His experimental values of the strength of glass fibers are approximated by $F = 22,400 \; [(4.4 + d)/(0.06 + d)]$, where F is in lbs/in^2 and d is in thousandths of an inch. He noted that, within the range of diameters available to Karmarsch (1859), this expression differs little from $F = 22,400 + 98,600/d$, which is of the form given by Karmarsch for metal wires. Similar results have been given by Reinkober (1931) and by Alexandrov and Žurkov (1933), who attributed the effect of cross section on strength to the fact that the greater the cross section the greater the probability of a dangerous surface crack. Slightly different results were obtained by Kontorova (1940) and by Kontorova & Frenkel (1941), who developed a statistical theory based on the length of the most dangerous crack -- the larger the specimen the greater the probability that it contains a dangerous inhomogeneity. Tucker (1941) questioned the application of the "weakest-link" theory to elements in parallel, and proposed instead the "strength-summation" theory, which he used in analyzing the strength of beams differing in width, with excellent results. Gerald Pickett, in his contribution to the discussion of Tucker's paper, stated that he saw no reason why the "strength-summation" theory cannot also be used for beams of varying depth. Tucker (1945) asserted that the "weakest-link" theory predicts the effects of an increase in the depth of specimens subjected to flexure, while the "strength-summation" theory appears to predict reasonably well the effects of an increase in the cross-sectional area of specimens subjected to tension or compression and of an increase in width of a flexural specimen. Weibull (1949) asserted that the "weakest-link" theory applies to two identical metal bars coupled side by side, thus leaving the length unchanged but doubling the cross section, because if one of the bars fails, the load on the other will be doubled and it will almost certainly fail also. However, if the two bars are allowed to melt together to form a single specimen with a doubled cross section, the validity of the "weakest-link" theory is not at all self-evident; it depends on the manner in which the fracture is spread over the area and also on the homogeneity of the material, so it is not a trivial problem to determine the effect of changes in the cross-sectional area. Meyersberg (1952) and Weibull (1961) have pointed out that the effect of the cross section of metal bars or wires on their strength is not entirely statistical in nature, but depends also on differences in the cooling rates and in the effects of heat treatments for large and for small specimens, and have made attempts to separate these other effects from the statistical size effect. Gücer & Gurland (1962) have proposed a "dispersed fracture" model which is based on the following assumptions: (1) Cylindrical bodies, loaded in simple tension in a direction parallel to the axis of the cylinders, consist of identical volume elements subject to a homogeneous, uniaxial stress. (2) The fracture strengths (fracture loads per unit area) of the elements are independent of each other and all elements fracture independently, i.e. the fracture does not propagate from one element to the next. (3) A body has failed if all volume elements intersecting a cross-sectional plane are broken. Under these assumptions, their analysis shows that, for a Weibull distribution of the strengths of the individual volume elements, the strength of a specimen of constant length increases with cross-sectional area, the increase being most pronounced for small values of the Weibull shape parameter. Experimental results, however, regularly show that the fracture strength decreases as the cross-sectional area increases, as predicted by the "weakest-link" theory, which leads one to believe that very few, if any, real materials satisfy the assumptions underlying the "dispersed fracture" model, especially as regards independence (assumption 2).

4. EFFECT OF VOLUME

Weibull (1939a) presented a statistical ("weakest-link") theory of the effect of volume on the strength of materials. According to his theory, the probability of rupture (S) at any given distribution of stress (σ) over a volume (V) is determined by the equation $\log (1-S) = -\int_V n(\sigma)\, dV$, where $n(\sigma)$ is a function characteristic of each particular material. For statistically homogeneous materials, the material function may be expressed by the formula $n(\sigma) = [(\sigma - \sigma_u)/\sigma_0]^m$, where σ_u, σ_0 and m are constants characteristic of the material -- location, scale and shape parameter, respectively, of the Weibull distribution, whose cdf is given by $F(\sigma) = 1 - \exp\{-[(\sigma - \sigma_u)/\sigma_0]^m\}$. The scale (threshold) parameter, σ_u, below which there is no probability of failure, is often set equal to 0. Weibull (1939b) reported the results of experimental investigations intended to verify his theory. Somewhat different results were obtained by the Russian writers Afanas'ev (1940, 1941), Kontorova (1940), and Kontorova & Frenkel (1941). Afanas'ev assumed a distribution of strength having cdf $F(z) = z^k/(A^k + z^k)$, where A is a positive constant and $z \geq 0$, and Kontorova & Frenkel assumed a normal distribution. The latter authors also took account of the number of inhomogeneities and the size of the largest one. Fowler (1945) gave a statistical explanation of the size (volume) effect based on the Weibull theory. Davidenkov, Shevandin & Wittmann (1947) reviewed the work of Weibull and that of Kontorova & Frenkel and pointed out that both are based on the assumption that brittle fracture is determined by the local stress at the most dangerous defect; the larger the piece, the lower the strength of its weakest element and therefore the lower the strength of the piece itself. They gave the theory of Kontorova & Frenkel the advantage over that of Weibull in the sense of stricter physical foundation, but found the latter more convenient for practical purposes, and gave qualitative experimental verification of it for static tension and bending of cylindrical specimens of brittle phosphorous steel in liquid air. Epstein (1948 b) summarized the work of the above authors and related work on the statistical theory of extreme values, which is the basis of the "weakest-link" theory. Chechulin (1954) criticized the theory of Kontorova & Frenkel (1941) for its use of a normal distribution of local strength, which admits negative values for sufficiently large volumes and infinitely large values for vanishingly small volumes, and that of Weibull (1939 a) for its lack of a rigorous physical basis, while admitting that it does not give bad agreement with experimental results. He proposed a new theory, based on the Pearson Type III distribution, for which he claimed greater rigor than the previous theories, and showed that Weibull's theory is a special case of his own when the number of defects in the body is large. Bartenev & Tsepkov (1958, 1960) pointed out that the bending strength of glass and other brittle materials depends primarily on the surface area rather than the volume, since failure starts at the surface, and suggested that, when this is true, the working volume V should be replaced by the working surface S in Weibull's theory. Robinson (1965) summarized the Weibull theory, and pointed out that the strength in uniform tension of specimens differing in size from the test specimen may be predicted if m and σ_0 remain unchanged (normally a valid assumption, since they are constants of the material), in which case the survival probabilities at the mean values are the same, so that (if σ_u is taken to be zero) we get $\exp[-(\sigma_1/\sigma_0)^m V_1] = \exp[-(\sigma_2/\sigma_0)^m V_2]$, and therefore $\sigma_1/\sigma_2 = (V_2/V_1)^m$. Thus the larger specimen is weaker and this size effect increases as m increases. Batdorf & Crose (1974) [see also Batdorf (1974, 1975)] introduced a new physically based statistical theory of fracture for isotropic brittle materials, which (they claim) eliminates in large measure the long-standing rift between "weakest-link" theories and the main body of fracture mechanics. Their theory allows extension of the theory of strength for a stressed volume under uniaxial tension to the cases of biaxial and triaxial tension, and gives simple expressions for the probabilities of failure under equibiaxial and equitriaxial tension, but is not applicable when the stress is predominantly compressive. Jones (1975) summarized the volume effect on strength in relation to the distribution of the smallest value as a function of the sample size, whose general features he summarized for rectanglar, Cauchy, Laplace, Gauss (normal) and Weibull distributions. Vardar & Finnie (1975) pointed out that the Weibull multiaxial treatment of brittle strength contains limitations which are not present in the more familiar uniaxial formulation, and asserted that if these limitations are satisfied, it is possible to use tension or bending data to predict multiaxial behavior when at least one principal stress is tensile.

5. STRENGTH OF FIBER BUNDLES

Peirce (1926) considered the strength of cotton yarns consisting of parallel threads and Daniels (1945) studied the statistical theory of the strength of bundles of threads. If a load S is applied to a bundle of n threads or fibers, it is assumed that each fiber carries an equal part of the load, S/n for each fiber. If one fiber, having strength less than S/n, breaks, the load is redistributed to the remaining fibers, S/(n-1) for each. This may cause more fibers to break. After r fibers have broken, the load on each of the remaining fibers is S/(n-r). The process of breaking and load redistribution continues either until all n fibers have broken or until k fibers, each having strength greater than S/k, remain. Daniels developed the probability distribution of the strength of bundles of n threads, and found that asymptotically (as $n \to \infty$), this distribution tends to normality. Winkler (1954) approximated the strength of bundles of fibers by taking the total length as the product of the number of fibers in the bundle and the the test length, and applying "weakest-link" theory. He reported close agreement of the resulting theoretical strength with experimental results. Coleman (1958) applied the theory of Daniels to infinite bundles composed of fibers which obey the Weibull distribution. He found that, in general, the tensile strength (per unit initial cross-sectional area) of a large bundle has the same order of magnitude as, but is less than, the mean strength of the component fibers, and decreases monotonically with increasing dispersion of the fiber strength. Sedrakyan (1958) used elementary combinational probability methods to determine the probability, if a cable is braided of n wires of equal length and cross-section whose strength distribution function is given, that $c \leq n$ wires will break under the influence of a given external load. Corten (1967) plotted normalized bundle strength as a function of the Weibull shape parameter, m (a material constant), and Rosen (1967) plotted normalized bundle strength as a function of normalized bundle length. Sen, Bhattacharyya & Suh (1969, 1973) pointed out that, if a bundle of fibers breaks under a load L, then the inequalities $nX_{n,1} \leq L, (n-1)X_{n,2} \leq L, \ldots, X_{n,n} \leq L$, where $X_{n,1} \leq X_{n,2} \leq \ldots \leq X_{n,n}$ are the ordered values of the strengths of the individual fibers, are simultaneously satisfied and that, consequently, the bundle strength can be represented as $B_n = \max\{nX_{n,1}, (n-1)X_{n,2}, \ldots, X_{n,n}\}$. They observed that $n^{-1} B_n$ can be written as $\sup_{x \geq 0} x[1-S_n(x)]$ where $S_n(x)$ is the empirical cdf for the strength of individual fibers, and proceeded to generalize the results of Daniels on the limiting distribution of bundle strength. Suh, Bhattacharyya & Grandage (1970) developed, by probabilistic argument, small-sample and large-sample properties, as well as certain moment properties, of the bundle strength B_n. Hoshiya (1972) approximated, by Monte Carlo methods, the probability that exactly i of m parallel cables with a single load S will fail, where the component strengths are mutually independent and identically distributed with lognormal distribution and share equally the external load S, also lognormally distributed. He considered both the case of brittle fracture, where the failed components become inactive, and the case of ductile failure, where they still carry loads equal to their yield strength. He tabulated the results of 2,000 trials each for ductile and brittle systems with m = 2(1)14. Phoenix & Taylor (1973) determined the asymptotic distribution of tensile strength of a bundle of parallel fibers in terms of the statistical characteristics of the individual fibers, extending the results of Daniels (1945), Sen et al. (1969, 1973) and Suh et al. (1970) to cover the case of bundles composed of a mixture, in fixed proportions, of several fiber types. They showed that such bundles can be weaker than similar bundles of any single component. Sen (1973 a,b) developed the theory of sequential confidence procedures and sequential tests for the strength of bundles of fibers. Kreider (1974) converted the plot of normalized bundle strength as a function of the Weibull shape parameter, m, given by Corten (1967) to a plot of normalized bundle strength as a function of the coefficient of variation of the underlying Weibull distribution (itself a function of m). Phoenix (1975) gave asymptotic results for the distribution of tensile strength of a special class of classic fiber bundles composed of several sub-bundles wherein the random variables associated with the loading and failure of any two fibers in the same sub-bundle are allowed to be probabilistically dependent. He modified Daniels' model to include both random fiber and random sub-bundle slack, and found that random slack results in a loss in asymptotic bundle mean strength and a change in asymptotic variance, both of which are approximately proportional to the random fiber slack variance.

6. STRENGTH OF COMPOSITES

Peirce (1926) mentioned composite materials in connection with his study of the strength of fiber bundles, but serious study of composites did not begin until about 20 years ago. Parratt (1960) noted that plastics reinforced with glass and other fibers had gained acceptance as structural materials. He attempted to connect the mechanical behavior of individual glass fibers and the strength of glass fiber moldings, hoping to gain a more detailed mechanical picture on which to base work on glass fibers in resin and other fiber reinforced materials. He concluded that the reinforcing strength obtained from a fiber of a given diameter is the strength corresponding to a certain "critical length" of fiber at which the strength of the fiber exceeds the maximum stress it can transmit. The thicker the fiber, the greater is the "critical length", which means that thicker fibers make a weaker reinforcement. Bell (1961) reported the results of a study to determine the effect of glass fiber geometry (defined to include the length and cross-sectional area of the fiber as well as its orientation in the matrix) on composite material strength. A preliminary analysis indicated that, for current fiber diameters (0.0004 in.) and matrix shear strengths (1000 psi), a minimum fiber length of approximately one-quarter inch is required to develop full fiber strength. Rosen (1964) presented a theoretical and experimental treatment of the failure of a composite, consisting of a matrix stiffened by uniaxially oriented fibers, when subjected to a uniaxial tensile load parallel to the fiber direction. He treated fibers as having a statistical distribution of flaws or imperfections, which result in fiber failure at various stress levels. Composite failure occurs when the remaining unbroken fibers, at the weakest cross section, are unable to resist the applied load. Thus, composite failure results from tensile fracture of the fibers. This model contrasts with that of Parratt (1960), according to which failure occurs when the accumulation of fiber fractures, resulting from increasing load, shortens the fiber lengths to the point that further increases in load cannot be transmitted to the fibers because the maximum matrix shear stress is exceeded, so that composite failure results when a shear failure of the matrix or the interface occurs. Rosen evaluated the composite strength as a function of the statistical strength characteristics of the fiber population and of the significant parameters defining composite geometry. Among these is the "ineffective length", δ , which is the fiber length, measured from an internal fiber break, over which the stress σ is less than some given fraction ϕ (= 0.9, say) of the undisturbed fiber stress σ_{f0}. Rosen's model consists of a chain of bundles of fiber links, each of length δ. After the numerical value of δ has been calculated from an equation given by Rosen and the distribution of fiber stress has been determined experimentally, the distribution of bundle strength can be found from the theory of Daniels; then the composite can be treated as a chain of bundles, and "weakest-link" theory can be applied to obtain the composite strength. Schuerch (1964) studied the effect of diameter upon elastic properties in thin fibers, motivated by the remarkable mechanical properties of filament-wound structures made from endless, thin glass fibers bonded with various organic resins. He pointed out that these properties result, in part, from the size effect (increase in strength observed in thin glass fibers as compared with the bulk strength of glass), combined with an effective crack barrier function of the bonding material. Riley & Reddaway (1968) derived a theory for predicting the tensile strengths and failure mechanisms of fiber composites where the fibers are aligned in the direction of tensile loads and are flawed to some extent. They pointed out that several modes of failure can occur, depending on the length of the fibers and the degree to which they are flawed; the presence of defects leads to size effects, with the strength of the fibers (and hence of the composite) decreasing as either the length or the diameter of the fibers increases. Wadsworth & Spilling (1968) presented the results of a theoretical and experimental study of the transfer of loads from broken fibers in composite materials (high-modulus carbon fibers placed in a resin matrix on an aluminum backing). Zweben (1968) investigated two modes of composite tensile failure, one involving composites containing a planar array of parallel fibers which exhibit a large number of isolated fiber breaks before failure, and the other involving monolayer and multilayer unidirectional composites that do not display many isolated breaks before failure. He reported that the failure loads predicted by these analyses were significantly closer to experimental data than the predictions of other theories. He established failure criteria for each mode, and reported that the analyses, based on the "weakest-link" theory and making use of the Weibull distribution, predict, for both cumulative and non-cumulative failure modes, that composite strength decreases with the size of the body, whereas the theory of Rosen (1964) predicts that composite strength is essentially independent of size. McKee & Sines (1969) presented a statistical model for the tensile fracture of parallel-fiber composites based on a stress criterion for crack propagation. The stresses in fibers surrounding a crack nucleus and the number and size of such nuclei are evaluated, and failure is predicted when, statistically, a new break is expected at any crack nucleus. The predictions of this model are in good agreement with measured tensile strength, the variation of strength with size, the variability of tensile fracture strength, and the observed mechanism of fracture of fiberglass composites. The model predicts that reinforced fiber structures will exhibit a regular decrease in fracture stress as volume is increased and that the variability of tensile strength of large specimens will be only slightly less than that of the individual fibers, in agreement with experimental results. Zweben & Rosen (1970) compared the "weakest-link" model of Weibull (1939 a), according to which failure occurs by the propagation of a crack through the material, and the "dispersed fracture" model of Gücer & Gurland (1962), according to which failure occurs by the inability of the overall cross-section to resist the applied load. They pointed out that these two criteria provide lower and upper bounds on expected material strength, but that the bounds may be too far apart for any practical use, especially for large specimens, since the former predicts a strong size effect and the latter is relatively insensitive to size. Zweben & Rosen suggested a new model, based on the multiple break criterion, according to which fracture propagation results from the first multiple break for continuous fiber composites or the first triple break for whisker-reinforced composites, and applied the results to composite materials, obtaining good agreement with available experimental data. They plotted a graph of the size effect predicted by the multiple break criterion. Anthony (1972) pointed out that, even though statistical models that essentially assume series behavior have given good fits to experimental data for composite materials, one cannot help being somewhat apprehensive when small specimen data are used in designing large assemblies. Fortunately, he wrote, scaling based on a series statistical model tends to be conservative if the real situation involves parallel behavior; however, unduly large conservatisms may result in serious penalties. Cruse, Konish & Waszczak (1972) attempted to adapt metals technology (in particular, linear elastic fracture mechanics) for use with composite materials. They pointed out that recent work has shown that scaling of failure loads for geometrically similar coupons, which can be done for metals, is not always possible for composite materials. They stated that the development of sufficient experimental data to correct for size effects should improve prediction capabilities, though conservative strength estimates can be and have been made without such corrections. They identified three characteristic material dimensions that may be important in fracture testing of composites: Ply thickness, fiber diameter, and size of "initial damage" zone. Eisenman, Kaminski, Reed & Wilkins (1972, 1973) advocated development of a new set of design rules for composite structures, taking account of scaling and complexity, which are not addressed by the conventional approach developed over the past forty years for metallic structures. Lifshitz & Rotem (1972) pointed out that, while some carbon/epoxy composites fail by fracture at a single cross section as assumed by Rosen (1964), experimental results with glass reinforced plastics (e.g., epoxy or polyester) show very complicated fracture surfaces. They proposed a new failure model that agrees with the observed failure geometry. The results of their analysis, based on this model, show that a small decrease in strength is expected when the size is increased by a few orders of magnitude. Parratt (1972) summarized the accumulated knowledge concerning various aspects, including the size effect, of the strength of fibers, fiber bundles, and fiber-reinforced composites. Rosen & Dow (1972) presented a similar summation, and proposed a new model for the strength of composites reinforced with brittle fibers. This model differs from that of Rosen (1964) mainly because it uses a different definition of "ineffective length" proposed by Friedman (1967) and includes the defined ineffective region on both sides of the break to establish the link length. Waddoups (1972) discussed the implications of various models for the strength of composites and their correlation with experimental data. Armenàkas & Sciammerella (1973) discussed the two best-known statistical models employed in predicting the strength of fiber-reinforced composite materials, the "cumulative weakening model" [Rosen (1964)] and the "crack-propagation model" [Zweben (1968)]. They compared the results given by the two theories with available experimental results, and concluded that there is a great need for an extensive, systematic experimental program to evaluate those theories. Owen & Bishop (1973) called attention to the possibility that glass reinforced plastics structural components may fail in a manner similar to brittle fracture in metals; if the structural element is large enough the loaded element may store sufficient elastic energy to propagate a fracture from a defect, accidental local damage or a design detail. As evidence, they cited two failures of large structures where the failures took the form of narrow cracks, with the material almost entirely damage free on either side of the fracture surface, in contrast with the progressive damage (debonding and resin cracking), extending through the stressed region, observed in testing small laboratory specimens. Wright & Iannuzzi (1973) measured the strength of individual carbon fibers and used the results to calculate the lower bound of strength expected from carbon-fiber reinforced epoxy matrix materials. The tensile strength of the composite was found to be greater than one would expect by assuming that the fibers functioned as a simple load carrying bundle; in effect, a synergistic strengthening effect was observed when the matrix surrounded the fiber. The following experimental results were also observed: (1) The shorter fibers exhibited the larger mean strengths, as predicted by theory; (2) The strengths of bundles of fibers also increased as the fiber length decreased; and (3) The bundle strengths were always less than the mean strength of the individual fibers. Argon & Bailey (1974) showed that if the surface flaw distribution in the brittle reinforcement layers of a laminate is a Weibull distribution with shape parameter, m, less than 10, the average strength of the laminate is greater than that of the isolated element (for a given laboratory test length), while for $m > 10$, the inequality is reversed; in nearly all cases of reinforcement using glass, boron, or carbon fibers, $m < 10$ and the laminate is stronger than the elements. Hahn & Tsai (1974) gave a set of generalized graphs, for all practical laminates of a given composite, for determining changes in stiffness and strength resulting from varying the ply orientation and total laminate thickness. This knowledge is required in designing composite laminates.

7. SIZE EFFECT IN FATIGUE

Among the earliest authors to discuss the size effect in fatigue were Peterson (1930), Peterson & Wahl (1936), Weibull (1939 a,b), and Afanas'ev (1940, 1941). Freudenthal (1946) reported on a study of the statistical aspect of the fatigue of materials. He pointed out that the size effect in the crack initiation stage (the probability of rupture under a definite load amplitude increases with increasing number of bonds, which is proportional to specimen size) is opposed by a size effect in the crack propagation stage (at equal rate of propagation the small cross section will be destroyed more rapidly); the resulting size effect depends on the relative magnitudes of the two opposing effects. Aphanasiev [Afanas'ev] (1948) noted that an increase in the dimensions (diameter) of test specimens lowers their fatigue limit, and that this phenomenon becomes more apparent when stress concentrations are present, as in notched specimens. He developed a theory to explain the greater size effect on fatigue strength for notched specimens as compared with unnotched ones. Epstein (1948 a) pointed out a fallacy in applying "weakest-link" theory to the study of fatigue of materials [see Freudenthal (1946)] -- the specimen is changing in time and its distribution of strengths due to flaws is also changing, so that any essentially static approach which uses the "weakest-link" concept without modification leaves out certain basic features of the process. Stulen (1951) summarized the elementary theory of extreme values as related to the size effect on the endurance limit. He pointed out that, for specimens in which the stress is not uniform, allowance must be made for the fact that different volumes are stressed at different levels. He estimated the effective volume of a specimen as that portion of the volume where the oscillating stress is within 15 per cent of the maximum stress on the specimen. Afanas'ev (1953) explained the statistical theory of fatigue stability of metals previously published by him (not the "weakest-link" theory, but one based on stresses in the crystalline structure of the metal) and related it to the size effect. Freudenthal & Gumbel (1953) [see also Derman, Kwo & Gumbel (1954)] presented a statistical interpretation of fatigue tests, including the size effect, based on the distribution of largest cracks (extreme-value theory). Weibull (1961) discussed, in some detail, the effect of size on fatigue strength -- a complex problem which depends upon both structural (metallurgical) changes in the material and statistical considerations. The metallurgical size-effect can result from differences in cooling rate and/or differences in the effects of heat treatment and fabrication processes for large and for small specimens. The "weakest-link" concept, initially intended as an explanation of size effects in brittle materials [see Weibull (1939 a,b)], is directly applicable to the fatigue strength at an arbitrarily preassigned life N (cycles), but not to the fatigue life. When crack initiation occurs mainly on the surface, the surface area, and not the volume, is the appropriate "size" to be taken into consideration. Weibull stated that it is safe to conclude that the size effect in fatigue exists, but is not easy to establish its magnitude, because many irrelevant factors mask the result and, even more significantly, the laws of size effect are quite different for the pre-crack and the post-crack stages of the fatigue process. Grover (1966) claimed that, while statistical considerations are important in fatigue studies, the statistical approach cannot provide a wholly satisfactory theory of fatigue. Probability considerations imply lower strengths for larger specimens, but they do not directly imply the magnitude of the size effect or the factors which govern the magnitude -- fatigue strength might be low at a thickness of a few hundred atomic distances, with no further decrease of engineering significance as thickness increases over practical ranges. This information is not likely to come, in any unique sense, from the statistical approach. Moreover, material factors, such as differences in surface finish and in residual stress, play an important part in the size effect. It was shown by de Kazinczy (1969) that the ratio between the endurance limits of two flawed specimens of cast steel with different stressed volumes (w) can be expressed by: $\sigma_1/\sigma_2 = (w_2/w_1)^{1/m + n/p}$ where 1/m is the exponent (reciprocal of the Weibull shape parameter for the material) in the absence of defects, n is obtained from the size distribution of defects, and p is related to their notch effect. Buch (1972) surveyed the literature (32 references) on test results and analyses of the effect of the size of specimen (unnotched) on the fatigue strength in the case of fatigue tests under reversed bending, reversed twisting, and reversed tension-compression stresses; the maximum value of the plain geometrical effect of size, effect of the surface hardening, the surface roughness and the statistical material defects; and estimation of the fatigue strength of structural components. Harris & Lee (1974) examined existing data and data from controlled experiments to determine the effect of fiber diameter on the fatigue strength of metal matrix composites. Several Russian authors have studied the size effect on fatigue strength of steel in sea water and other corrosive media; corrosion affects small-diameter specimens more severely than large-diameter ones, so the size effect may be opposite in direction to that observed in air.

8. UNSOLVED PROBLEMS

The statistical theory of the size effect on both static and fatigue strength has been well developed, but other theories (especially the energy theory and the technological theory) must be synthesized with the statistical theory in order to attain a comprehensive theory which will explain more of the existing phenomena, even in the case of static strength. In the case of fatigue strength, account must also be taken of the theory of fracture mechanics. Experimental results on the size effect on both static and fatigue strength of the older structural materials (such as metals, wood, and concrete) abound in the literature, but such results for the newer fiber-reinforced composite materials are inadequate. In connection with the current trend toward greater use of composite materials in large aircraft structural members, the following efforts are required: (1) Synthesis of the statistical theory of extreme values with engineering theories of fracture and fatigue to predict the size effect on the static strength and fatigue life of composites and other structural materials, and the reliability of large composite structural members; (2) Experimental testing of small and large composite specimens to verify the results given by the unified theory.

REFERENCES

[1] DA VINCI, LEONARDO (ca. 1500). Notebooks (unpublished).

[2] GALILEI, GALILEO (1638). *Dialogues Concerning Two New Sciences* (Italian and Latin). Elzevirs, Leyden. English translation by Henry Crew and Alfonzo de Salvio, The Macmillan Company, London-New York, 1914. Reprinted by Dover Publications, Inc., New York, 1952.

[3] KARMARSCH, KARL (1859). Ueber die absolute Festigkeit der Metalldrähte. *Mittheilungen des Gewerbe Vereins für Hannover* (N.F.), **1859** (3) 137-156.

[4] TRAUTWINE, JOHN C. (1872). *The Civil Engineer's Pocket-book*. Claxton, Remsen and Haffelfinger, Philadelphia. Other editions: 1876, 1881, 1895, 1907, 1909, 1919.

[5] CHAPLIN, W. S. (1880). The relation between the tensile strengths of long and short bars. *Van Nostrand's Engineering Magazine* , **23**, 441-444.

[6] CHAPLIN, W. S. (1882). On the relative tensile strengths of long and short bars. *Proceedings of the Engineer's Club, Philadelphia* , **3**, 15-28.

[7] SLOCUM, C. E. and HANCOCK, E. L. (1906). *Text-Book on the Strength of Materials*. Ginn & Co., Boston. Second edition, 1911.

[8] GRIFFITH, A. A. (1920). The phenomena of rupture and flow in solids. *Philos. Trans. Roy. Soc. London Ser. A* , **221**, 163-198.

[9] GRIFFITH, A. A. (1924). The theory of rupture. *Proceedings of the First International Congress for Applied Mechanics* (Delft), pp. 55-63.

[10] TIPPETT, L. H. C. (1925). On the extreme individuals and the range of samples taken from a normal population. *Biometrika* , **17**, 364-387.

[11] PEIRCE, FREDERICK THOMAS (1926). Tensile strength for cotton yarns. v. "The weakest link" -- theorems on the strength of long and composite specimens. *Journal of the Textile Institute* , **17**, T355-T368.

[12] PETERSON, R. E. (1930). Fatigue tests of small specimens with particular reference to size effect. *Transactions of the American Society for Steel Treating* , **18**, 1041-1053; discussion, 1053-1056.

[13] REINKOBER, O. (1931). Die Zerreissfestigkeit dünner Quarzfäden. *Physikalische Zeitschrift* , **32**, 243-250.

[14] ALEXANDROV, A. P. and ŽURKOV [JURKOV], S. N. (1953). *The Phenomenon of the Brittle Rupture*. (Russian). Moscow-Leningrad.

[15] PETERSON, R. E. and WAHL, A. M. (1936). Two- and three-dimensional cases of stress concentration, and comparison with fatigue tests. *Journal of Applied Mechanics* , **3**, A-15 - A-22; discussion, A-146 - A-150.

[16] WEIBULL, W. (1939 a). A statistical theory of the strength of materials. *Ingeniörs Vetenskaps Akademien Handlingar*, Number 151. Generalstabens Litografiska Anstalts Förlag, Stockholm.

[17] WEIBULL, W. (1939 b). The phenomenon of rupture in solids. *Ingeniörs Vetenskaps Akademien Handlingar*, Number 153. Generalstabens Litografiska Anstalts Förlag, Stockholm.

[18] AFANAS'EV, N. N. (1940). Statistical theory of the fatigue strength of metals. (Russian). *Zhurnal Tekhnicheskoi Fiziki* [*Journal of Technical Physics*. (USSR)]., 10, 1553-1568.

[19] KONTOROVA, T. A. (1940). Statistical theory of strength. (Russian). *Zhurnal Tekhnicheskoi Fiziki* [*Journal of Technical Physics*. (USSR)]., 10, 886-890.

[20] AFANAS'EV, N. N. (1941). Stress-strain curves of metals and their sensitivity to notching under variable load. (Russian). *Zhurnal Tekhnicheskoi Fiziki* [*Journal of Technical Physics*. (USSR)]., 11 (4), 349-358.

[21] KONTOROVA, T. A. and FRENKEL, J. I. (1941). Statistical theory of brittle strength of real crystals. (Russian). *Zhurnal Tekhnicheskoi Fiziki* [*Journal of Technical Physics*. (USSR)]., 11 (3), 173-183; English translation, *Journal of Physics*. (USSR)., 7, (1943), (3), 108-114.

[22] TUCKER, JOHN, JR. (1941). Statistical theory of the effect of dimensions and of method of loading upon the modulus of rupture of beams. *Proceedings of the American Society for Testing Materials*, 41, 1072-1088; discussion, 1089-1094.

[23] DANIELS, H. E. (1945). The statistical theory of the strength of bundles of threads. I. *Proc. Roy. Soc. London Ser. A*, 183, 405-435.

[24] FOWLER, F. H., JR. (1945). On fatigue failure under triaxial static and fluctuating stresses and a statistical explanation of size effect. *Transactions of the American Society of Mechanical Engineers*, 67, 213-215; discussion, 215-216.

[25] TUCKER, JOHN, JR. (1945). Effect of dimensions of specimens upon the precision of strength data. *Proceedings of the American Society for Testing Materials*, 45, 952-959; discussion, 960.

[26] FREUDENTHAL, A. M. (1946). The statistical aspect of fatigue of materials. *Proc. Roy. Soc. London Ser. A*, 187, 416-429.

[27] DAVIDENKOV, N., SHEVANDIN, E. and WITTMANN, F. (1947). The influence of size on the brittle strength of steel. *Journal of Applied Mechanics*, 14, A-63-A-67.

[28] APHANASIEV [AFANAS'EV], N. N. (1948). The effect of shape and size factors on the fatigue strength. *Engineer's Digest*, 5, 132-136.

[29] EPSTEIN, BENJAMIN (1948 a). Statistical aspects of fracture problems. *Journal of Applied Physics*, 19, 140-147.

[30] EPSTEIN, BENJAMIN (1948 b). Application of the theory of extreme values in fracture problems. *J. Amer. Statist. Assoc.*, 43, 403-412.

[31] LÉVI, ROBERT (1949). Calculs probabilistes de la sécurité des constructions. *Annales des Ponts et Chaussées*, 119, 493-539.

[32] WEIBULL, WALODDI (1949). Statistical representation of fatigue failures in solids. *Transactions of the Royal Institute of Technology*. (Stockholm), Separate No. 27.

[33] STULEN, FOSTER B. (1951). On the statistical nature of fatigue. *Symposium on Statistical Aspects of Fatigue Tests*. (Special Technical Publication No. 121), pp. 23-40; discussion, 41-44; general discussion, 59-64. American Society for Testing Materials, Philadelphia.

[34] WEIBULL, WALODDI (1951). A statistical distribution function of wide applicability. *Journal of Applied Mechanics*, 18, 293-297; discussion, 19 (1952), 233-234.

[35] MEYERSBERG, G. (1952). Grösseneinfluss und Randeinfluss auf die Festigkeit der Werkstoffe. *Tekniska Hogskolan Handlingar.* (Stockholm), No. 58.

[36] AFANAS'EV, N. N. (1953). *The Statistical Theory of Fatigue Resistance in Metals.* (Russian). Izd-vo Akademii Nauk USSR, Kiev.

[37] FREUDENTHAL, A. M. and GUMBEL, E.J. (1953). On the statistical interpretation of fatigue tests. *Proc. Roy. Soc. London Ser. A* , **216**, 319-322.

[38] TIMOSHENKO, STEPHEN P. (1953). *History of Strength of Materials.* McGraw-Hill Book Co., Inc., New York-Toronto-London.

[39] CHECHULIN, B. B. (1954). On the statistical theory of brittle strength. (Russian). *Zhurnal Tekhnicheskoi Fiziki* , **24** (2), 292-298.

[40] DERMAN, C., KWO, T. T. and GUMBEL, E. J. (1954). Standard Errors of Estimate of Parameters of Fatigue Failure Survivorship Functions. Technical Report T-2A, Department of Industrial Engineering, Columbia University, New York.

[41] LIEBLEIN, JULIUS (1954). Two early papers on the relation between extreme values and tensile strength. *Biometrika* , **41**, 559-560.

[42] WINKLER, F. (1954). Relation between bundle strength and strength of individual fibers or yarns. (German). *Faserforschung und Textiltechnik*, **5** (9), 398-403.

[43] BARTENEV, G. M. and TSEPKOV, L. P. (1958). Scale factor and strength of glass. (Russian). *Dokl. Akad. Nauk. SSSR* , **121** (2), 260-263; English translation, *Soviet Physics -- Doklady* , **4**, 822-825.

[44] COLEMAN, B. D. (1958). On the strength of classical fibres and fibre bundles. *J. Mech. Phys. Solids* , **7**, 60-70.

[45] LAVROV, V. V. (1958). The nature of the scale effect in ice and strength of the ice sheet. (Russian). *Dokl. Akad. Nauk. SSSR* (N.S.)., 122 (4), 570-574; English translation, *Soviet Physics -- Doklady* , **3** (1959), (2), 934-937.

[46] SEDRAKYAN, L. G. (1958). Two problems of the statistical strength theory. (Russian-Armenian summary). *Doklady Akademii Nauk Armianskoi SSR* , 26 (3), 135-140. English translation, *Selected Translations in Mathematical Statistics and Probability* , **4** (1963), pp. 289-294. American Mathematical Society, Providence, Rhode Island.

[47] BARTENEV, G. M. and TSEPKOV, L. P. (1960). On the nature of the scale effect in ice. (Russian). *Zavodskaia Laboratoriia* , **26** (3), 330-331; English translation, *Industrial Laboratory* , **26** (3), 352-354.

[48] DAVIDENKOV, N. N. (1960). Effect of the dimension factor on the mechanical properties of specimens. (Russian). *Zavodskaia Laboratoriia* , **26** (3), 319-320. English translation, *Industrial Laboratory* , **26** (3), 338-339.

[49] PARRATT, N. J. (1960). Defects in glass fibres and their effect on the strength of plastic mouldings. *Rubber and Plastics Age* , **41**, 263-266.

[50] VOLKOV, S. D. (1960). Failure kinetics and the scale effect. (Russian). *Zavodskaia Laboratoriia* , **26** (3), 323-329; English translation, *Industrial Laboratory* , **26** (3), 342-351.

[51] BELL, JAMES E. (1961). Effect of glass fiber geometry on composite material strength. *ARS Journal* , **31** (9), 1260-1265.

[52] WEIBULL, W. (1961). *Fatigue Testing and Analysis of Results.* Pergamon Press, New York-Oxford-London-Paris.

[53] GÜCER, D. E. and GURLAND, J. (1962). Comparison of the statistics of two fracture modes. *J. Mech. Phys. Solids* , **10**, 365-373.

[54] ROSEN, B. WALTER (1964). Tensile failure of fibrous composites. *AIAA. J.*, **2** (11), 1985-1991.

[55] SCHUERCH, H. (1964). Effect of diameter upon elastic properties in thin fibers. *AIAA. J.*, 2 (3), 569-571.

[56] ROBINSON, E. Y. (1965). Engineering Application of Weibull Statistics. UCRL-12405, AEC Contract No. W-7405-eng-48, University of California Lawrence Radiation Laboratory, Livermore, California.

[57] BARTENEV, G.M. and SIDOROV, A. B. (1966). Statistical theory of the strength of glass fibers. (Russian-English summary). *Mekhanika Polimerov*, 2 (1), 74-81. English translation, *Polymer Mechanics*, 2 (1), 52-56.

[58] GROVER, HORACE J. (1966). *Fatigue of Aircraft Structures*.(NAVAIR 01-1A-13). U. S. Government Printing Office, Washington.

[59] CORTEN, HERBERT T. (1967). Micromechanics and fracture behavior of composites. *Modern Composite Materials*,(edited by Lawrence J. Broutman and Richard H. Krock), pp. 27-105. Addison-Wesley Publishing Co., Reading, Massachusetts.

[60] FRIEDMAN, E. (1967). A tensile failure mechanism for whisker reinforced composites. *Proceedings of the SPI 22nd Annual Conference*, Sect. 4-A, pp. 1-10. Reinforced Plastics Division, Society of Plastics Industry, New York.

[61] ROSEN, B. WALTER (1967). The strength and stiffness of fibrous composites. *Modern Composite Materials*,(edited by Lawrence J. Broutman and Richard H. Krock), pp. 106-119. Addison-Wesley Publishing Co., Reading, Massachusetts.

[62] RILEY, V. R. and REDDAWAY, J. L. (1968). Tensile strength and failure mechanics of fibre composites. *Journal of Materials Science*, 3 (1), 41-46.

[63] WADSWORTH, N. J. and SPILLING, I. (1968). Load transfer from broken fibres in composite materials. *British Journal of Applied Physics* [*Journal of Physics*, (D)] (2) 1 (7), 1049-1058.

[64] ZWEBEN, CARL (1968). Tensile failure of fiber composites. *AIAA. J.*, 6 (12), 2325-2331.

[65] DE KAZINCZY, F. (1969). Size effect in fatigue of cast steel. *Journal of the Iron and Steel Institute*, 207 (1), 40-43.

[66] MCKEE, R. B., JR. and SINES, GEORGE (1969). A statistical model for the tensile fracture of parallel fiber composites. *Journal of Electroplastics*, 1, 185-189.

[67] SEN, P. K., BHATTACHARYYA, B. B. and SUH, M. W. (1969). Limiting Behavior of the Extremum of Certain Sample Functions. Mimeo Series No. 628, Institute of Statistics, University of North Carolina, Chapel Hill, North Carolina.

[68] SUH, M. W., BHATTACHARYYA, B. B. and GRANDAGE, A. (1970). On the distribution and moments of the strength of a bundle of filaments. *J. Appl. Probability*, 7, 712-720; correction: 8 (1971), 830.

[69] ZWEBEN, C. and ROSEN, B.W. (1970). A statistical theory of material strength with application to composite materials. *J. Mech. Phys. Solids*, 18, 189-206.

[70] ANTHONY, FRANK M. (1972). Discussion of the Colloquium on Structural Reliability. *Proceedings of the Colloquium on Structural Reliability: The Impact of Advanced Materials on Engineering Design.* (Carnegie-Mellon University, Pittsburgh, Pennsylvania, October 1972) (edited by J. L. Swedlow, T. A. Cruse and J. C. Halpin), pp. 527-534. Department of Mechanical Engineering, Carnegie-Mellon University, Pittsburgh, Pennsylvania.

[71] BUCH, ALFRED (1972). Auswertung und Beurteilung des Grösseneinflusses bei Dauerschwingversuchen mit ungerkerbten Proben and Bauteilen. *Archiv für das Eisenhüttenwesen*, 43 (12), 895-900.

[72] CRUSE, T. A., KONISH, H. J., JR. and WASZCZAK, J. P. (1972). Strength analyses for design with composite materials using metals technology. *Proceedings of the Colloquium on Structural Reliability: The Impact of Advanced Materials on Engineering Design*. [see Ref. 70], pp. 222-236.

[73] EISENMAN, J. R., KAMINSKI, B. E., REED, D. L. and WILKINS, D. J. (1972). Toward reliable composites: an examination of design methodology. *ibid* , pp. 359-381.

[74] HOSHIYA, MASARU (1972). Monte Carlo evaluation of safety of indeterminate structure. *Proceedings of the Japan Society of Civil Engineers*. No. 205, pp. 147-154.

[75] LIFSHITZ, JACOB M. and ROTEM, ASSA (1972). Longitudinal tensile failure of unidirectional fibrous composites. *Journal of Materials Science* , 7 (8), 861-869.

[76] PARRATT, N. J. (1972). *Fibre-Reinforced Materials Technology*. Van Nostrand Reinhold Company, London-New York-Cincinnati-Toronto-Melbourne.

[77] ROSEN, B. WALTER and DOW, NORRIS F. (1972). Mechanics of failure of fibrous composites. *Fracture: An Advanced Treatise, Volume VII: Fracture of Nonmetals and Composites*. (edited by H. Liebowitz), pp. 611-674. Academic Press, New York-London.

[78] WADDOUPS, M. E. (1972). Characteristics of advanced composite materials. *Proceedings of the Colloquium on Structural Reliability: The Impact of Advanced Materials on Engineering Design*. [see Ref. 70], pp. 535-588.

[79] ARMENÀKAS, A. E. and SCIAMMERELLA, C. A. (1973). Experimental investigation of the failure mechanism of fiber-reinforced composites subjected to uniaxial tension. *Experimental Mechanics* , 13 (2), 49-58.

[80] OWEN, M. J.and BISHOP, P. T. (1973). Critical stress intensity factors applied to glass reinforced polyester resin. *Journal of Composite Materials* , 7 (2), 146-159.

[81] PHOENIX, S. LEIGH and TAYLOR, HOWARD M. (1973). The asymptotic strength distribution of a general fiber bundle. *Advances in Appl. Probability* , 5, 200-216.

[82] SEN, PRANAB KUMAR (1973a). On fixed size confidence bands for the bundle strength of filaments. *Annals of Statistics* , 1, 526-537.

[83] SEN, PRANAB KUMAR (1973 b). An asymptotically efficient test for the bundle strength of filaments. *J. Appl Probability* , 10, 586-596.

[84] SEN, P. K., BHATTACHARYYA, B. B. and SUH, M. W. (1973). Limiting behavior of the extremum of certain sample functions. *Annals of Statistics* , 1, 297-311.

[85] WRIGHT, M. A. and IANNUZZI, F. A. (1973). The application of the principles of linear elastic fracture mechanics to unidirectional fiber reinforced composite materials. *Journal of Composite Materials* , 7, 430-447.

[86] ARGON, A. S. and BAILEY, D. G. (1974). Failure in laminates in tension under increasing stress, constant stress, and cyclic stress. *Journal of Materials Science* , 9 (2), 201-211.

[87] BATDORF, S. B. (1974). Weibull statistics for polyaxial stress states. *Journal of the American Ceramic Society* , 57 (1), 44-45.

[88] BATDORF, S. B. and CROSE, J. G. (1974). A statistical theory for the fracture of brittle structures subjected to nonuniform polyaxial stresses. *Trans. ASME Ser. E. J. Appl. Mech.*, 41, 459-464.

[89] HAHN, H. T. and TSAI, S. W. (1974). Graphical determination of stiffness and strength of composite laminates. *Journal of Composite Materials* , 8 (2), 160-177.

[90] HARRIS, S. J. and LEE, R. E. (1974). Effect of fibre spacing on the fatigue behaviour of metal matrix composites. *Composites* , 5 (3), 101-106.

[91] KREIDER, KENNETH G. (1974). Introduction to metal-matrix composites. *Composite Materials, Volume 4: Metallic Matrix Composites*. (edited by Kenneth G. Kreider), pp. 1-35. Academic Press, New York-London.

[92] BATDORF, S. G. (1975). Observations on Approximate and Exact Treatments of Fracture Statistics for Polyaxial Stress States. SAMSO-TR-75-289, Space and Missile Systems Organization, Los Angeles Air Force Station, California.

[93] JONES, BRIAN H. (1975). Probabilistic design and reliability. *Composite Materials, Volume 8: Structural Design and Analysis, Part II*. (edited by C. C. Chamis), pp. 33-72. Academic Press, New York-San Francisco-London.

[94] PHOENIX, S. L. (1975). Probabilistic inter-fiber dependence and the asymptotic strength distribution of classic fiber bundles. (French, Italian, German and Russian summaries). *Internat. J. Engrg. Sci.*, 13, 287-304.

[95] VARDAR, Ö. and FINNIE, I. (1975). An analysis of the Brazilian disk fracture test using the Weibull probabilistic treatment of brittle strength. (French and German summaries). *International Journal of Fracture* , 11 (3), 495-508.

[96] HARTER, H. LEON (1977). A Survey of the Literature on the Size Effect on Material Strength. AFFDL-TR-77-11. Air Force Flight Dynamics Laboratory, Wright-Patterson Air Force Base, Ohio.

P.R. Krishnaiah, ed., APPLICATIONS OF STATISTICS

BISPECTRA OF ATMOSPHERIC AND WIND TUNNEL TURBULENCE

K. N. HELLAND, K. S. LII AND M. ROSENBLATT

University of California, San Diego
La Jolla, California

and

Northwestern University
Evanston, Illinois

A bispectral analysis of turbulence velocity-derivative data gathered in the atmosphere (high Reynolds number) and in a wind tunnel (low Reynolds Number) is carried out over the range of 25 to 1000 Hz. The two estimated bispectra are compared and can be readily distinguished in terms of their shape. Approximate relationships between the bispectra and second order spectra (power spectra) are suggested. Integrations of the bispectra are in satisfactory agreement with directly measured skewness values. The computations suggest that contributions of wave-number triplets are non-local in wave-number space in both cases studied. Evidence is found which suggests that the contribution to the total skewness from the dissipation scales of the atmospheric measurements are comparable to the total skewness obtained in low Reynolds number grid turbulence measurements.

1. INTRODUCTION

The bispectrum has been used as a tool in a number of areas. One of the earliest papers is that of Hasselman, Munk and MacDonald (1963) on ocean waves. There have also been large sample theoretical expositions, for example, in Brillinger and Rosenblatt (1967), Rosenblatt (1971) and discussion of related questions in Lumley and Takeuchi (1976). In the recent paper of Lii, Rosenblatt and Van Atta (1976) hot-wire data obtained from an experiment carried out in an atmospheric measurement program described by Haugen, Kaimal and Bradley (1971) was used to obtain a bispectral estimate over a limited range with effective statistical resolution. The data is analyzed over a much broader range in this paper and compared with a corresponding analysis of wind tunnel data. Sharp differences in the character of the bispectra estimated are noted.

2. DEFINITION OF BISPECTRUM

This paper presents the results of bispectral analyses (a harmonic analysis of third order moments) of atmospheric and wind tunnel turbulence and comments on the similarities and differences. One hopes that a continuing investigation in this direction will give further understanding of the nonlinear mechanism of turbulence. The atmospheric measurements used were obtained during the 1968 Kansas expedition of the Air Force Cambridge Research Laboratories. The wind tunnel measurements were carried out at UCSD in 1975.

We first give a brief discussion of the basic concepts used in the analysis of the data. A more detailed consideration of the data analysis and the instrumentation will be given later on. In both the atmospheric and laboratory contexts the velocity derivative is analyzed.

In the usual model of homogeneous turbulence one assumes that one has a random velocity field $\underset{\sim}{u}(\underset{\sim}{x},t)$ stationary with respect to spatial displacement with $\underset{\sim}{u}(\underset{\sim}{x},t)$ satisfying the Navier-Stokes equation

$$\partial \underset{\sim}{u}/\partial t + \underset{\sim}{u} \cdot \nabla \underset{\sim}{u} = - \rho^{-1} \nabla p + \nu \nabla^2 \underset{\sim}{u}.$$

Here p is pressure, ρ density and ν kinematic viscocity. Given incompressibility, the continuity equation is

$$\nabla \cdot \underset{\sim}{u} = 0$$

Assume appropriate moments exist. The homogeneity (spatial stationarity) of $\underset{\sim}{u}(\underset{\sim}{x},t)$ implies that one has a spatial harmonic analysis

$$\underset{\sim}{u}(\underset{\sim}{x},t) = \int \exp(i\underset{\sim}{k} \cdot \underset{\sim}{x})\, dz(\underset{\sim}{k},t)$$

with $\underset{\sim}{z}(\underset{\sim}{k}) = \underset{\sim}{z}(\underset{\sim}{k},t)$ a process of orthogonal increments in $\underset{\sim}{k}$. The equation for $f_{ii}(\underset{\sim}{k})$, the spectral energy density associated with $\underset{\sim}{k}$,

$$\frac{\partial}{\partial t}\left[\frac{1}{2} f_{ii}(\underset{\sim}{k})\right] = \int Q(\underset{\sim}{k},\underset{\sim}{k}')\, d\underset{\sim}{k}' - \nu \underset{\sim}{k}^2 f_{ii}(\underset{\sim}{k})$$

where

$$Q_{\alpha,\beta}(\underset{\sim}{k},\underset{\sim}{k}')\, dkdk' = \operatorname{Im} E\, \{k_\alpha\, dz_\alpha(\underset{\sim}{k}-\underset{\sim}{k}')\, dz_\beta(-\underset{\sim}{k})\, dz_\beta(\underset{\sim}{k}')\}$$

and

$$Q(\underset{\sim}{k},\underset{\sim}{k}') = \sum_{\alpha,\beta} Q_{\alpha,\beta}(\underset{\sim}{k},\underset{\sim}{k}').$$

Except for the two preceding equations the standard summation convention is used, i.e. sum over all values of an index if it appears twice. The equation for $f_{ii}(\underset{\sim}{k})$ is derivable from the Navier-Stokes system after Fourier analysis. A related equation for the time rate of change of mean square vorticity $E\{w^2\}$ derivable from the Navier-Stokes system is

$$\frac{\partial}{\partial t} E\,\{\underset{\sim}{w}^2\} = E\,\{w_i\, w_j \frac{\partial u_i}{\partial x_j}\} - \nu E\,(\partial w_i/\partial x_j)^2.$$

If the turbulence is homogeneous and isotropic

$$E\left\{w_i w_j \frac{\partial u_i}{\partial x_j}\right\} = \frac{35}{8} E(\partial u/\partial x)^3 = \frac{35}{8} S \frac{\{E u^2\}^{3/2}}{\lambda^3}$$

where

$$S = E(\partial u/\partial x)^3 / \{E(\partial u/\partial x)^2\}^{3/2}$$

is the skewness factor of $\partial u/\partial x$ and

$$\lambda = \{Eu^2/E(\partial u/\partial x)^2\}^{1/2}$$

is the Taylor microscale.

The rate of change of spectral energy and the production of vorticity are related to the transfer of spectral energy from small wavenumbers to large wavenumbers by the nonlinear mechanism in turbulence. The transfer is determined by interactions of triplets of wave number vectors summing to the null vector. One would like to determine which triplets interact most strongly to produce spectral transfer of energy. The bispectrum of $\partial u/\partial x$, $B'(k_1,k_2)$ is a natural tool to use in such an analysis. Since the velocity field is sampled one-dimensionally, our wavenumber k is the longitudinal wave number corresponding to the mean velocity direction. Thus

$$B'(k_1,k_2) = \frac{1}{(2\pi)}2 \iint E\left\{\frac{\partial u}{\partial x}(x) \frac{\partial u}{\partial x}(x+r_1) \frac{\partial u}{\partial x}(x+r_2)\right\}$$
$$\exp\{-ik_1r_1 - ik_2 r_2\}\, dr_1\, dr_2 .$$

$B'(k_1,k_2)$ is related to the bispectrum $B(k_1,k_2)$ of $u(x)$ by

$$B'(k_1,k_2) = -ik_1k_2(k_1+k_2)B(k_1,k_2) .$$

Notice that one of the terms in $Q(k,k')$ is

$$\mathrm{Im}\{k_1 E\{dz_1(k-k')dz_1(-k)dz_1(k')\}\} .$$

If this is integrated over k_2, k_3, k_2', k_3' an expression proportional to the imaginary part of the one-dimensional bispectrum is obtained. The bispectrum B has the symmetries

$$B(k_1,k_1') = B(k_1',k_1) = B(k_1,-k_1-k_1') = \overline{B(-k_1,-k_1')}.$$

3. DATA ACQUISITION

The data were obtained from two Reynolds number experiments. Low Reynolds number laboratory grid-generated turbulence experiments were performed to obtain velocity and velocity derivative data at conditions nearly identical to those used by Van Atta and Chen (1969). High Reynolds number velocity derivative data were obtained from the atmospheric-surface-layer measurement program described by Haugen, et al. (1971).

The low Reynolds number laboratory measurements were made in the UCSD closed circuit wind tunnel which has a 76 cm × 76 cm × 10 m test section. A biplane grid with square mesh (mesh size M = 5.08 cm) and cylindrical rods (0.953 cm in diameter) was located two tunnel diameters downstream from the entrance to the test section. The tunnel was operated at a mean speed of 7.7 m/s giving a grid Reynolds number of 27,000. The mean velocity U and the velocity and velocity derivative fluctuations, u and $\frac{\partial u}{\partial t}$, respectively, in the longitudinal direction were measured at a downstream location of X/M = 48. The fluctuating velocity was measured with a Thermo-Systems, Inc. cylindrical film sensor (-10A) which has a short, 0.25 mm, sensing length appropriate for velocity derivative measurements in the laboratory. The Kolmogorov microscale η was 0.047 cm. $\eta = (\nu^3/\varepsilon)^{\frac{1}{4}}$ where ν is the kinematic viscocity and ε is the mean dissipation rate of energy. The sensor was operated in the constant-temperature mode using a DISA Model 55M01 hot-wire anemometer system and was linearized with a DISA Model 55D10 linearizer. An analog circuit was used to electrically generate the velocity derivative signal from the linearized velocity signal. The velocity derivative was digitally recorded in real time with the AMES digitizer at a sample rate of 4172 samples per second with 12-bit resolution. The derivative signal was low-pass filtered at 2 kHz (24 dB/octave) before digital recording.

The high Reynolds number atmospheric data used in the present calculations were part of the data used in previous studies by Lii, et al. (1976) for the bispectral calculations at low-low frequency interactions (upper limit of 150 Hz). The data also formed the basis for studies by Van Atta and Yeh (1973, 1975) and by Wyngaard and Pao (1972). A single vertically oriented hot wire 5 μm in diameter and 1.2 mm long was operated in the linearized constant-temperature mode using a DISA 55D05-15 system at a height of z = 5.66 m. The mean velocity was 3.76 m/s and the Kolmogorov microscale was 0.08 cm. The streamwise velocity signal was differentiated electrically and filtered with a low-pass filter (24 dB/octave) before recording on analog magnetic tape. The analog data were played back at UCSD and digitally recorded with the AMES digitizer at a sample rate of 4172 samples per second and 12-bit resolution.

4. COMPUTATION AND STATISTICAL ANALYSIS

The computation was carried out by using a CDC 7600 digital computer at Lawrence Berkeley Laboratory at Berkeley through a terminal in the Norpax Building at La Jolla.

The figures in Tables 1 and 2 of Lii, et al. (1976) were obtained as indicated in the section on computation in that paper. These figures had to be multiplied by 64 (due to initial averaging of blocks of eight) and then divided by the factors in Table 6 of Lii, et al, before being used in the computations for Figure 3.

The digitized tape of the atmospheric data contains 700 records, each record containing 8192 readings in approximately two seconds. There is a small gap between each record. In computing the whole range bispectrum from 25 Hz to 975 Hz records 301 to 400 were used. In each record the first 4096 readings were used and the remaining readings were discarded. The computational procedure ran as follows:

(1) Let the first 4096 readings of each record be written as

$$x_1,\ x_2,\ldots,x_{4096}\ .$$

Each record used had its first half fast Fourier transformed with the mean removed, that is one computed

$$F(\lambda) = \sum_{j=1}^{n} e^{ij\lambda 2\pi/n}\, x_j, \qquad \lambda = 1,\ldots,n-1\ ,$$

$n = 4096$ and set $F(0) = 0$. By symmetry, computation of $F(\lambda)$ is only required for $\lambda = 0,1,\ldots,2048$.

(2) The bispectral estimate

$$b_k(\lambda_1,\lambda_2) = \frac{1}{(2\pi)^2 n} F(\lambda_1)F(\lambda_2)F(-\lambda_1-\lambda_2)$$

$$\lambda_1,\lambda_2 = 0,1,\ldots,2048$$

was computed for each record $k = 301,\ldots,400$.

(3) To reduce the variance we averaged at each point (λ_1,λ_2) of interest within a square neighborhood of 51×51, that is, for the kth record

$$\bar{b}_k(\lambda_1,\lambda_2) = \frac{1}{2601} \sum_{j_1=\lambda_1-25}^{25} \sum_{j_2=\lambda_2-25}^{25} b_k(j_1,j_2)\ .$$

(4) To reduce the variance further, 100 records were averaged to get the estimates

$$\bar{\bar{b}}(\lambda_1, \lambda_2) = \left(\frac{1}{100}\right)^2 \sum_{k=301}^{400} \bar{b}_k(\lambda_1, \lambda_2) .$$

The variance of the real and imaginary parts of this statistic were then estimated. The formula is just noted in the real case

$$\mathrm{var}(\lambda_1, \lambda_2) = \frac{1}{100} \sum_{k=301}^{400} [\mathrm{Re}\ \bar{b}_k(\lambda_1, \lambda_2) - \mathrm{Re}\ \bar{\bar{b}}(\lambda_1, \lambda_2)]^2,$$

with an estimate of the standard deviation given by the square root of this expression.

The digital tape for the wind tunnel data contains 800 records, each record containing 8192 in approximately two seconds with a small gap between each record. The computation for the whole range bispectrum from 25 Hz to 975 Hz was carried out just as it was for the atmospheric data except that records 101 to 200 were used.

The smoothing and interpolation procedures used in the perspective plots are described in Akima (1974). Scalings of the x, y and z axes as well as the rotation angle φ about the z-axis and the elevation angle ψ are indicated.

The feature leading to greater difficulty in estimation of the bispectrum as contrasted to an ordinary spectral estimate can be seen in the following asymptotic results. Under appropriate conditions the covariance of the spectral periodogram

$$\mathrm{cov}\left(\frac{1}{2\pi n} \left| F\left(\frac{2\pi s}{n}\right)\right|^2 , \frac{1}{2\pi n}\left| F\left(\frac{2\pi \ell}{n}\right)\right|^2\right)$$

$$= \delta_{s,\ell} \ f^2\left(\frac{2\pi s}{n}\right) + 0\left(\frac{\log n}{n}\right)$$

$s, \ell = 1, \ldots , [n/2]$, for large n. Here $[x]$ is the greatest integer less than or equal to x. Also f is the spectral density of the process. By contrast the covariance of a bispectral periodogram

$$\mathrm{cov}\left(\mathrm{Re}\ \{\frac{1}{(2\pi)^2 n} F\left(\frac{2\pi s_1}{n}\right) F\left(\frac{2\pi s_2}{n}\right) F\left(\frac{-2\pi(s_1+s_2)}{n}\right)\},\right.$$

$$\left.\mathrm{Re}\ \{\frac{1}{(2\pi)^2 n} F\left(\frac{2\pi \ell_1}{n}\right) F\left(\frac{2\pi \ell_2}{n}\right) F\left(\frac{-2\pi(\ell_1+\ell_2)}{n}\right)\}\right)$$

$$= \frac{n}{4\pi} f\left(\frac{2\pi s_1}{n}\right) f\left(\frac{2\pi s_2}{n}\right) f\left(\frac{2\pi(s_1+s_2)}{n}\right) \delta_{s_1,\ell_1} \delta_{s_2,\ell_2} + 0\ (\log n)^2,$$

$s_1, s_2, \ell_1, \ell_2 = 1,\ldots,[n/2]$, $s_1 < s_2$, $\ell_1 < \ell_2$, for n large. A completely analogous result holds for the imaginary part of the bispectral periodogram. The proportionality of the variance of the bispectral periodogram to the sample size n means that much more time and spatial averaging (in the frequency or wave number variable)is required to get statistical resolution, but yet not so much that great effects of bias appear.

5. RESULTS OF THE COMPUTATION

Log-log plots of the atmospheric and wind tunnel spectral densities are given in Figures 1 and 2. Notice that the atmospheric velocity derivative spectrum starts at a high level already (10^3) at virtually 1 Hz, increases to a maximum of about 7×10^3 at 60 Hz and starts decreasing from this magnitude at 100 Hz to 10 at 1000 Hz. This amounts to a very large concentration of spectral mass at low wave number for the atmospheric velocity spectrum. On the other hand, the velocity derivative spectrum for the wind tunnel data starts at a low level, is about 10^2 at 10 Hz, increases to about 10^4 at 200 Hz and decreases slowly from there on to 3×10 at 1000 Hz and 3×10^2 at 2000 Hz. The velocity derivative spectrum for the wind tunnel data is decidedly unimodal with the mode shifted to decidedly higher wave numbers than in the case of the atmospheric velocity derivative spectrum.

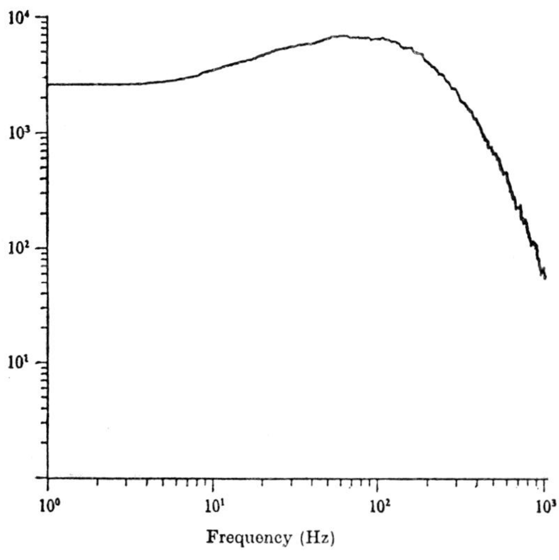

Fig. 1. Estimated spectral density of the atmospheric velocity derivative (record 301-400): log-log plot.

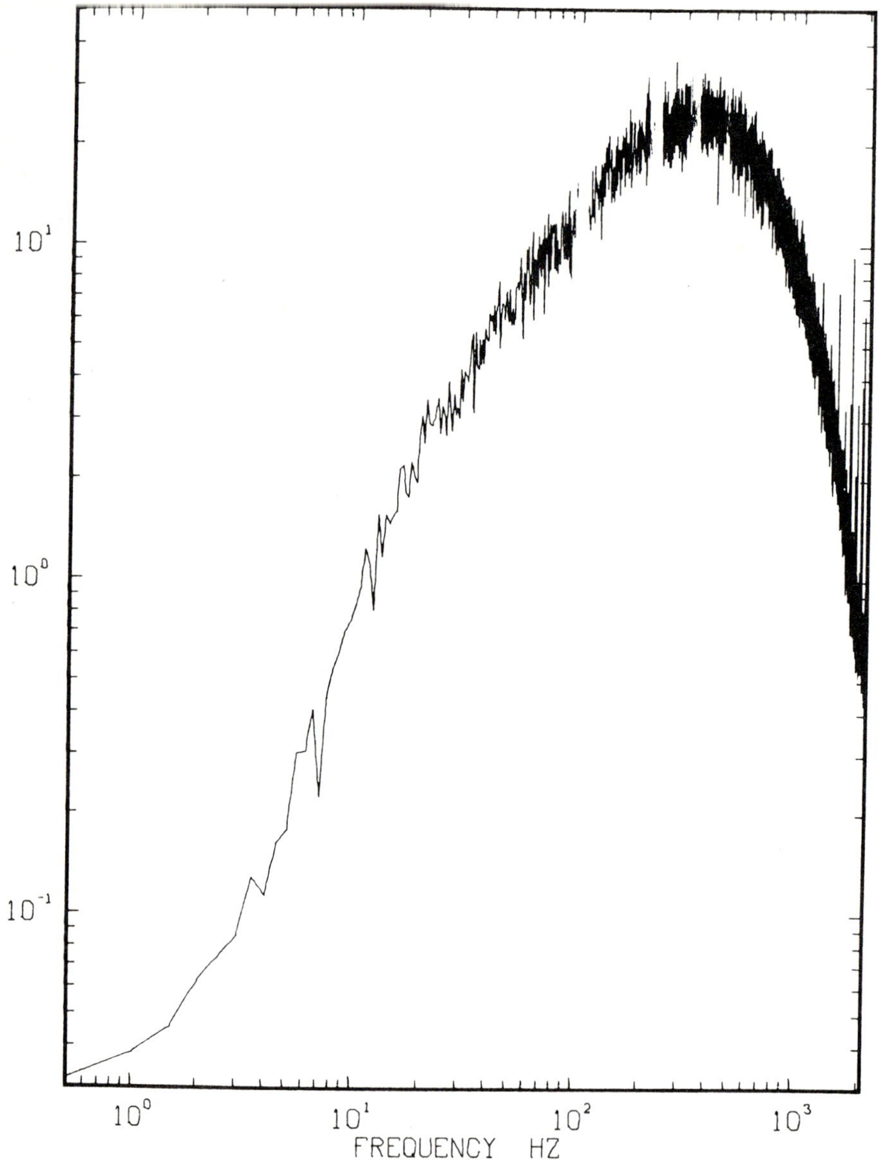

Fig. 2. Estimated spectral density of the wind tunnel velocity derivative.

Also the spectral mass is relatively much lower at low wave numbers than for the atmospheric data and decreases much more gradually for high wave numbers.

The bispectra for both the atmospheric and wind tunnel derivatives were computed in a low wave number range and then for a global wave number range. Tables 1 and 2 of Lii, et al. (1976) give the real and imaginary part of the average bispectral mass in approximately 15 Hz $\times$ 15 Hz squares for the atmospheric wind velocity derivative in a range 0 to 9 in each wave number variable where a unit represents 15.276 Hertz. A similar computation was made of the real and imaginary part of the average bispectral mass in 10 Hz $\times$ 10 Hz squares of the wind tunnel velocity derivative readings for a range of 1 to 10 in each wave number variable where a unit represents 15 Hz. It should be noted that generally the real parts of the bispectra are well resolved in both the atmospheric and wind tunnel context as one can see by comparing the bispectral estimates with estimates of their standard deviations. However, this is typically not the case with the imaginary parts of the bispectra which are either unresolved or too close to zero. This is consistent with isotropy since isotropy would imply that the imaginary part of a velocity derivative bispectrum would be zero. Three dimensional perspective plots of the real parts of the bispectra of the atmospheric and wind tunnel velocity derivatives respectively are given in Figures 3 and 4 with the scaling indicated above. The graphical representations are obtained by using the computations and applying a bicubic interpolation procedure. The result of the interpolation is then plotted using the AMESPLOT program on the UCSD CDC 3600. The real part of the bispectra is negative. To get a better view the numerical value of the bispectra has been plotted.

The real part of the bispectral mass in 50 Hz $\times$ 50 Hz squares of the atmospheric velocity derivative and the wind tunnel velocity derivative over a range of 25 Hz to 975 Hz in increments of 50 Hz has been given in Tables 1 and 2 respectively. Each figure represents an average bispectral mass (real) in a 50 Hz $\times$ 50 Hz square centered at the wavenumber doublet appropriate. As is usual, an estimate of the standard deviation of the estimate is noted below in parenthesis. The imaginary part of the bispectrum is not given since it is as noted above either unmeasurable or too close to zero. Two perspective plots of the real part of the bispectrum for the atmospheric data are given in Figures 5 and 6 for the global range of 25 Hz to 975 Hz. Figures 7 and 8 are three dimensional plots of the real part of the bispectrum for the wind tunnel data over the same range.

As we can see by looking at the 3-dimensional plot (Figure 3) of the real part of the atmospheric bispectrum in the low wave number range, there is a representation of a rather flat slowly varying surface. There are a number of oscillations up and down superimposed on the surface but this would appear to be due to the local error (or variability) in the estimation procedure. The maximal readings of the real part of the bispectrum are well within this low wave number range

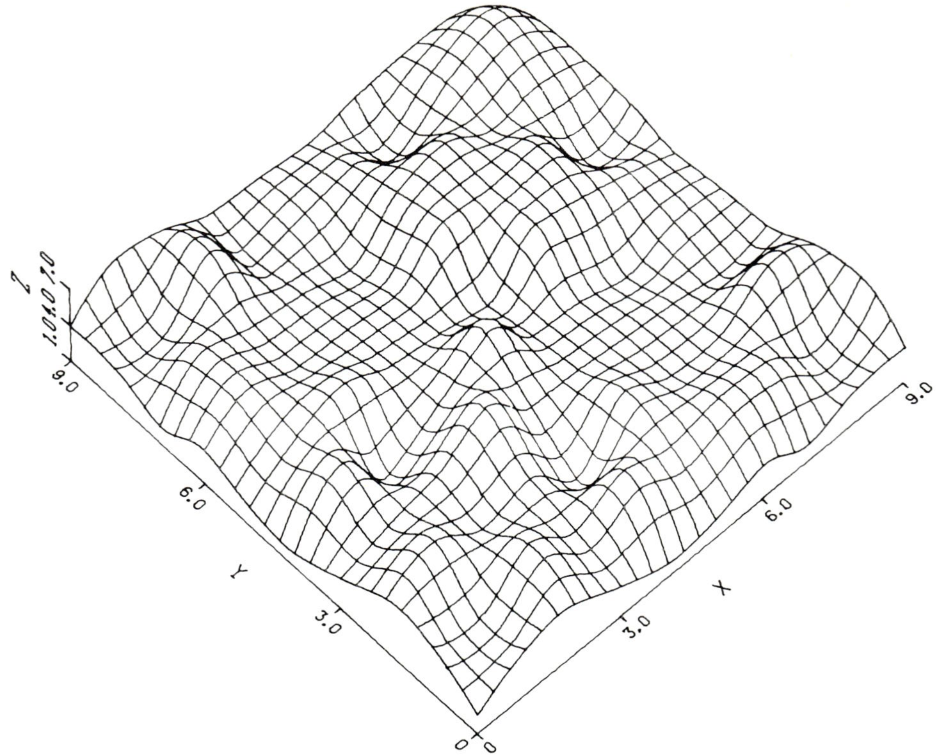

Fig. 3. Surface plot of magnitude of real part of bispectral mass, atmospheric data, low versus low. Estimates of mass in 15.3 Hz × 15.3 Hz squares. Range 1 to 9 with each unit 15.3 Hz. x, y, z axis scaling (6, 6, 1.5). Rotation angle 45°, elevation angle 60°.

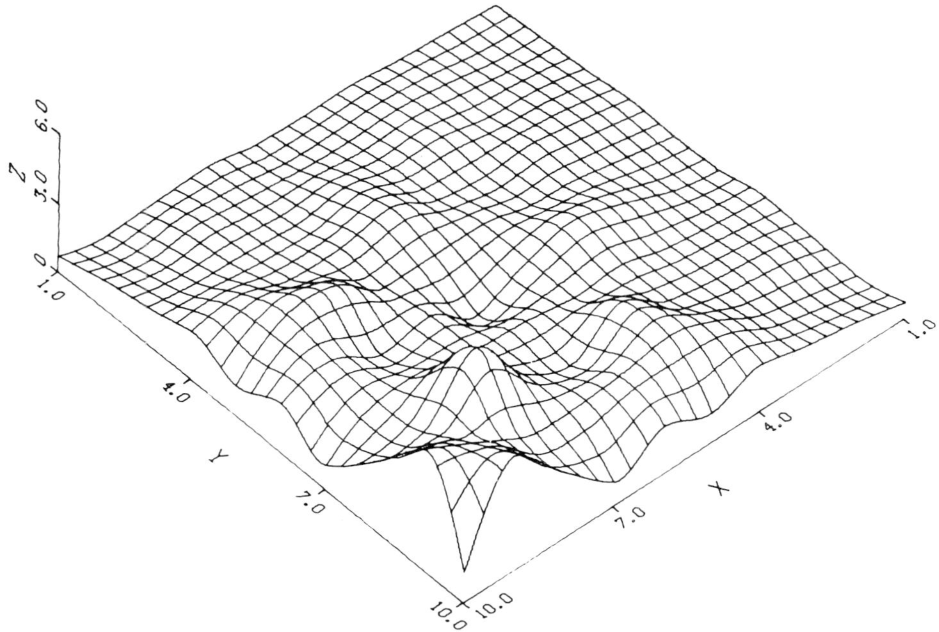

Fig. 4. Surface plot of magnitude of real part of bispectral mass, wind tunnel data, low versus low. Estimates of mass in 15 Hz by 15 Hz squares. Range 1 to 10 with each unit 15 Hz. x, y, z scaling (6, 6, 2). Rotation angle 135°, elevation angle 45°.

TABLE I

Real part of the bispectrum velocity derivative. Wind tunnel data. From 25 Hz to 975 Hz in blocks of 50 Hz. Estimated standard deviation in parenthesis.

	25	75	125	175	225	275	325	375	425	475
25	-2.70+04 (3.49+03)	-6.50+04 (6.51+03)	-6.44+04 (8.73+03)	-7.14+04 (9.78+03)	-7.18+04 (1.12+04)	-7.80+04 (1.27+04)	-4.86+04 (1.07+04)	-6.07+04 (1.35+04)	-3.92+04 (1.08+04)	-3.42+04 (9.34+03)
75		-1.27+05 (1.54+04)	-1.33+05 (1.50+04)	-1.60+05 (1.74+04)	-1.52+05 (1.75+04)	1.22+05 (2.29+04)	-1.41+05 (2.06+04)	-1.51+05 (1.96+04)	-1.27+05 (2.14+04)	-9.27+04 (1.70+04)
125			-1.40+05 (2.44+04)	-1.61+05 (2.34+04)	-1.48+05 (2.25+04)	-1.64+05 (2.55+04)	-1.73+05 (2.34+04)	-1.55+05 (2.13+04)	-1.44+05 (1.97+04)	-1.18+05 (1.82+04)
175				-1.59+05 (3.07+04)	-1.45+05 (2.48+04)	-1.31+05 (2.68+04)	-1.58+05 (2.62+04)	-1.13+05 (3.06+04)	-1.70+05 (2.34+04)	-1.39+05 (2.18+04)
225					-1.62+04 (3.66+04)	-1.46+05 (2.30+04)	-1.69+05 (2.50+04)	-1.79+05 (2.35+04)	-1.00+05 (2.13+04)	-1.18+05 (2.09+04)
275						-1.51+05 (3.92+04)	-1.54+05 (2.74+04)	-1.11+05 (2.61+04)	-1.44+05 (2.18+04)	-1.33+05 (2.12+04)
325							-1.13+05 (3.03+04)	-1.52+05 (2.63+04)	-1.20+05 (2.19+04)	-1.01+05 (1.81+04)
375								-1.13+05 (2.97+04)	-8.56+04 (2.34+04)	-9.80+04 (1.74+04)
425									-1.00+05 (2.54+04)	-7.05+04 (1.85+04)
475										-1.19+05 (2.41+04)

Real part of the bispectrum velocity derivative. Wind tunnel data. From 25 Hz to 975 Hz in blocks of 50 Hz. Estimated standard deviation in parenthesis. (continued)

	525	575	625	675	725	775	825	875	925	975
25	-4.75+04 (9.95+03)	2.66+04 (8.74+03)	-4.70+04 (8.10+03)	-2.67+04 (6.80+03)	-2.32+04 (6.12+03)	2.24+04 (5.31+03)	-2.67+04 (5.35+03)	-1.58+04 (4.63+03)	-1.59+04 (3.61+03)	-1.74+04 (3.97+03)
75	-5.91+04 (1.73+04)	-7.87+04 (1.34+04)	-9.91+04 (1.23+04)	-9.34+04 (1.22+04)	-7.67+04 (1.11+04)	-5.90+04 (2.65+03)	-6.48+04 (9.11+03)	-5.22+04 (7.03+03)	-3.55+04 (7.35+03)	-4.86+04 (5.53+03)
125	-1.03+05 (1.82+04)	-1.06+05 (1.45+04)	-1.02+05 (1.48+04)	-9.75+04 (1.16+04)	-1.04+05 (1.20+04)	-9.59+04 (1.09+04)	-6.96+04 (1.10+04)	-7.71+04 (8.49+03)	-7.26+04 (9.36+03)	-5.52+04 (6.75+03)
175	-9.38+04 (1.97+04)	-1.08+05 (1.88+04)	-9.96+04 (1.54+04)	-1.27+05 (1.55+04)	-8.22+04 (1.45+04)	-7.73+04 (1.31+04)	-9.85+04 (1.06+04)	-6.97+04 (9.75+03)	-7.26+04 (8.49+03)	-4.58+04 (7.38+03)
225	-9.09+04 (1.99+04)	-1.15-04 (1.70-04)	-9.53+04 (1.75+04)	-1.23+05 (1.55+04)	-8.72+04 (1.39+04)	-8.28+04 (1.18+04)	-7.65+04 (9.86+03)	-8.17+04 (9.77+03)	-7.61+04 (7.36+03)	-6.47+04 (6.74+03)
275	-1.16+05 (2.07+04)	-1.16-05 (1.80-04)	-7.03+04 (1.71+04)	-9.27+04 (1.47+04)	-9.62+04 (1.44+04)	-7.95+04 (1.28+04)	-6.18+04 (9.21+03)	-6.84+04 (9.01+03)	-5.77+04 (8.99+03)	-6.14+04 (7.27+03)
325	-8.84+04 (1.93+04)	-8.97+04 (1.66+04)	-8.67+04 (1.37+04)	-6.23+04 (1.31+04)	8.23+04 (1.30+04)	-5.94+04 (9.61+03)	-5.30+04 (1.08+04)	-4.96+04 (8.21+03)	-5.85+04 (8.52+03)	-4.74+04 (7.87+03)
375	-9.18+04 (1.84+04)	-6.75+04 (1.05+04)	-9.56+04 (1.67+04)	-7.79+04 (1.23+04)	-7.96+04 (1.22+04)	-6.84+04 (1.01+04)	-5.52+04 (1.09+04)	-5.73+04 (7.57+03)	-5.39+04 (7.83+03)	-4.94+04 (6.46+03)
425	-8.15+04 (1.72+04)	-5.58+04 (1.54+04)	-5.20+04 (1.36+04)	-7.00+04 (1.12+04)	-6.79+04 (1.04+04)	-4.35+04 (8.72+03)	-5.83+04 (8.48+03)	-4.64+04 (7.98+03)	-4.78+04 (7.31+03)	-4.87+04 (6.07+03)
475	-8.27+04 (1.66+04)	-1.20+05 (1.49+04)	-8.00+04 (1.24+04)	-6.34+04 (1.11+04)	-5.22+04 (8.87+03)	4.56+04 (1.01+04)	-3.90+04 (8.25+03)	-4.93+04 (6.79+03)	-4.10+04 (5.92+03)	-3.49+04 (4.92+03)
525	-7.26+04 (1.75+04)	-5.51+04 (1.30+04)	-6.89+04 (1.11+04)	-3.59+04 (9.98+03)	-4.76+04 (9.21+03)	-2.77+04 (9.04+03)	-3.03+04 (7.56+03)	-2.87+04 (6.53+03)	-2.60+04 (6.01+03)	-3.36+04 (4.99+03)

Real part of the bispectrum velocity derivative. Wind tunnel data. From 25 Hz to 975 Hz in blocks of Hz. Estimated standard deviation in parenthesis. (continued)

	525	575	625	675	725	775	825	875	925	975
525	-7.26+04 (1.75+04)	-5.51+04 (1.30+04)	-6.89+04 (1.11+04)	-3.59+04 (9.98+03)	-4.76+04 (9.21+03)	-2.77+04 (9.04+03)	-3.03+04 (7.56+03)	-2.87+04 (6.53+03)	-2.60+04 (6.01+03)	-3.36+C4 (4.99+C3)
575		-5.88+04 (1.52+04)	-5.94+04 (9.62+03)	-3.70+04 (8.64+03)	-3.05+04 (8.44+03)	-2.79+04 (6.41+03)	-3.10+04 (6.51+03)	-3.83+04 (6.41+03)	-2.74+04 (4.46+03)	-2.24+C4 (4.76+C3)
625			-6.89+04 (1.44+04)	-3.63+04 (7.75+03)	-2.75+04 (7.86+03)	-3.24+04 (6.44+03)	-1.68+04 (5.18+03)	-3.79+04 (4.77+03)	-2.08+04 (4.37+03)	-1.89+C4 (3.87+C3)
675				-3.08+04 (9.63+03)	-2.75+04 (6.19+03)	-2.75+04 (5.83+03)	-2.02+04 (4.94+03)	-2.34+04 (4.37+03)	-1.89+04 (3.82+03)	-1.56+C4 (3.56+C3)
725					-2.78+04 (8.54+03)	-2.09+04 (5.17+03)	-1.83+04 (4.58+03)	-2.18+04 (3.70+03)	-8.19+03 (3.51+03)	-1.31+C4 (3.12+C3)
775						-1.79+04 (6.22+03)	-1.23+94 (4.13+03)	-1.70+04 (3.20+03)	-1.11+04 (2.90+03)	-8.70+C3 (2.43+C3)
825							-8.30+03 (5.07+03)	-1.51+04 (3.30+03)	-1.06+04 (2.85+03)	-8.72+C3 (2.40+C3)
875								-1.32+04 (4.23+03)	-8.13+03 (2.54+03)	-1.18+C4 (2.32+C3)
925									-4.21+03 (3.23+03)	-7.05+C3 (2.02+C3)
975										-5.42+C3 (2.52+C3)

TABLE II

Real part of the bispectrum of the velocity derivative. Atmospheric data. From 25 Hz to 975 Hz in blocks of 50 Hz. Estimated standard deviation in parenthesis.

	25	75	125	175	225	275	325	375	425	475
25	-3.01+05 (5.24+04)	-2.76+05 (4.58+04)	-2.39+05 (3.43+04)	-2.10+05 (4.20+04)	-1.34+05 (3.43+04)	-1.11+05 (2.49+04)	-8.76+04 (1.70+04)	-7.26+04 (1.51+04)	-4.21+04 (1.16+04)	-4.81+04 (1.26+04)
75		-3.22+05 (7.21+04)	-1.03+05 (4.94+04)	-1.33+05 (4.56+04)	-1.22+05 (2.96+04)	-8.72+04 (2.34+04)	-8.13+04 (1.92+04)	-5.03+04 (1.36+04)	-5.81+04 (1.31+04)	-4.45+04 (9.99+03)
125			-8.24+04 (7.29+04)	-1.54+05 (4.05+04)	-1.33+05 (4.11+04)	-6.83+04 (2.14+04)	-5.74+04 (1.69+04)	-6.08+04 (2.09+04)	-4.68+04 (1.29+04)	-4.73+04 (1.13+04)
175				-1.44+05 (4.80+04)	-1.06+05 (3.44+04)	-4.75+04 (1.88+04)	-5.95+04 (1.74+04)	-5.24+04 (1.61+04)	-3.47+04 (1.47+04)	-2.67+04 (9.44+03)
225					-2.63+04 (2.08+04)	-4.11+04 (2.42+04)	-6.43+04 (2.12+04)	-2.93+04 (1.43+04)	-6.66+03 (1.65+04)	-3.92+03 (1.40+04)
275						-2.03+04 (1.79+04)	-1.31+04 (1.68+04)	-1.84+04 (1.18+04)	-8.40+03 (9.35+03)	-9.30+03 (5.83+03)
325							-1.58+02 (1.04+04)	-2.51+04 (7.80+03)	-1.23+04 (9.03+03)	-3.77+03 (6.02+03)
375								-8.49+03 (6.95+03)	-5.45+03 (7.38+03)	-5.71+03 (2.89+03)
425									3.36+03 (7.36+03)	2.13+03 (4.45+03)
475										-1.47+03 (3.45+03)

Real part of the bispectrum of the velocity derivative. Atmospheric data. From 25 Hz to 975 Hz in blocks of 50 Hz. Estimated standard deviation in parenthesis. (continued)

	525	575	625	675	725	775	825	875	925	975
25	-4.21+04 (1.05+04)	-2.25+04 (5.27+03)	-1.74+04 (4.87+03)	-1.71+04 (3.86+03)	-1.24+04 (2.72+03)	-1.31+04 (3.96+03)	-1.07+04 (2.92+03)	-6.46+03 (1.96+03)	-4.89+03 (2.67+03)	-4.96+03 (1.21+03)
75	-3.46+04 (7.77+03)	-2.60+04 (9.56+03)	-2.36+04 (9.28+03)	-2.14+04 (7.00+03)	-2.09+04 (5.99+03)	-1.20+04 (5.38+03)	-7.05+03 (4.41+03)	-7.68+03 (2.24+03)	-7.36+03 (2.23+03)	-3.54+03 (1.63+03)
125	-3.68+04 (7.89+03)	-1.95+04 (6.99+03)	-2.08+04 (5.31+03)	-1.73+04 (3.61+03)	-1.83+04 (4.66+03)	-1.11+05 (2.66+03)	-7.06+03 (3.45+03)	-6.41+03 (2.62+03)	-7.61+03 (1.87+03)	-4.21+03 (1.50+03)
175	-8.83+03 (8.63+03)	-1.55+04 (7.18+03)	-2.36+04 (7.86+03)	-2.39+04 (7.40+03)	-1.75+04 (5.16+03)	-1.31+04 (5.32+03)	-8.87+03 (2.88+03)	-7.99+03 (2.47+03)	-9.24+03 (2.57+03)	-4.51+03 (1.78+03)
225	-2.18+04 (1.31+04)	-1.55+04 (6.34+03)	-8.90+03 (5.12+03)	-1.44+04 (5.57+03)	-9.78+03 (4.37+03)	-1.10+04 (3.58+03)	-9.60+02 (5.56+03)	-6.97+03 (2.19+03)	-4.00+03 (1.92+03)	-8.73+02 (8.15+02)
275	-9.59+03 (6.21+03)	-8.27+03 (5.83+03)	-6.44+03 (6.15+03)	-1.06+04 (3.50+03)	-5.13+03 (3.11+03)	-5.00+03 (2.07+03)	-7.53+02 (5.26+03)	-4.28+03 (2.21+03)	-5.30+03 (1.91+03)	-3.24+03 (1.34+03)
325	-5.35+03 (4.07+03)	-8.53+03 (5.14+03)	-9.64+03 (5.61+03)	-8.08+03 (3.37+03)	-7.73+03 (2.44+03)	-6.25+03 (2.88+03)	-3.93+03 (1.37+03)	-2.46+03 (7.53+02)	-1.90+03 (1.02+03)	-1.63+03 (6.86+03)
375	-8.43+03 (5.05+03)	-2.06+03 (1.40+03)	-2.17+03 (1.74+03)	-2.96+03 (1.04+03)	-1.91+03 (6.42+02)	-7.68+02 (9.52+02)	-7.84+01 (1.07+03)	2.80+02 (1.10+03)	2.47+02 (1.37+03)	-7.07+02 (4.09+02)
425	-7.45+03 (4.76+03)	-6.10+02 (2.47+03)	4.41+02 (3.25+03)	-1.56+03 (1.65+03)	-2.22+03 (1.47+03)	-1.17+03 (8.74+02)	2.13+03 (2.69+03)	1.07+03 (1.49+03)	-7.86+02 (6.14+02)	-6.76+02 (3.04+02)
475	-1.83+03 (2.37+03)	-1.14+03 (2.04+03)	9.96+02 (2.92+03)	-1.88+03 (8.98+02)	-1.04+02 (1.40+03)	6.05+02 (2.77+03)	-4.10+01 (1.23+03)	-6.72+02 (5.19+02)	5.75+02 (1.59+03)	-1.53+02 (5.43+02)
525	-4.77+03 (2.55+03)	-1.40+03 (1.70+03)	-1.03+03 (1.95+03)	-1.71+03 (1.44+03)	-9.03+02 (1.06+03)	-3.03+03 (1.94+03)	-6.82+02 (3.98+02)	-1.29+03 (7.39+02)	-1.52+03 (1.08+03)	-5.72+02 (2.63+02)

Real part of the bispectrum of the velocity derivative. Atmospheric data. From 25 Hz to 975 Hz in blocks of 50 Hz. Estimated standard deviation in parenthesis. (continued)

	525	575	625	675	725	775	825	875	925	975
575		8.86+01 (1.14+03)	2.17+02 (1.10+03)	-3.84+02 (6.99+02)	1.74+02 (5.95+02)	2.10+02 (7.18+02)	-3.51+02 (5.64+02)	-3.02+02 (3.78+02)	2.37+02 (3.83+02)	-2.41+02 (2.86+02)
625			2.37+03 (3.71+03)	-1.12+02 (6.07+02)	1.26+03 (7.64+02)	1.67+03 (2.52+03)	8.41+02 (1.14+03)	3.20+02 (3.82+02)	4.76+02 (9.08+02)	-6.41+01 (3.43+02)
675				2.05+02 (4.20+02)	1.39+02 (2.59+02)	-1.40+02 (2.64+02)	2.89+02 (2.96+02)	8.44+01 (3.27+02)	-1.06+02 (4.07+02)	-1.22+02 (1.57+02)
725					2.67+02 (4.74+02)	6.25+02 (3.67+02)	6.12+02 (4.19+02)	-6.37+01 (2.44+02)	2.65+02 (2.81+02)	5.52+01 (1.74+02)
775						5.86+02 (8.12+02)	8.27+02 (1.11+03)	-2.32+02 (2.24+02)	-1.46+02 (2.06+02)	-6.15+01 (1.80+02)
825							-6.35+01 (2.52+02)	-1.74+02 (2.50+02)	3.60+02 (5.87+02)	-1.01+01 (1.73+02)
875								-7.02+01 (2.73+02)	-1.21+02 (1.51+02)	-1.69+02 (1.41+02)
925									-4.21+02 (2.18+02)	-2.71+02 (1.47+02)
975										-1.09+02 (8.79+01)

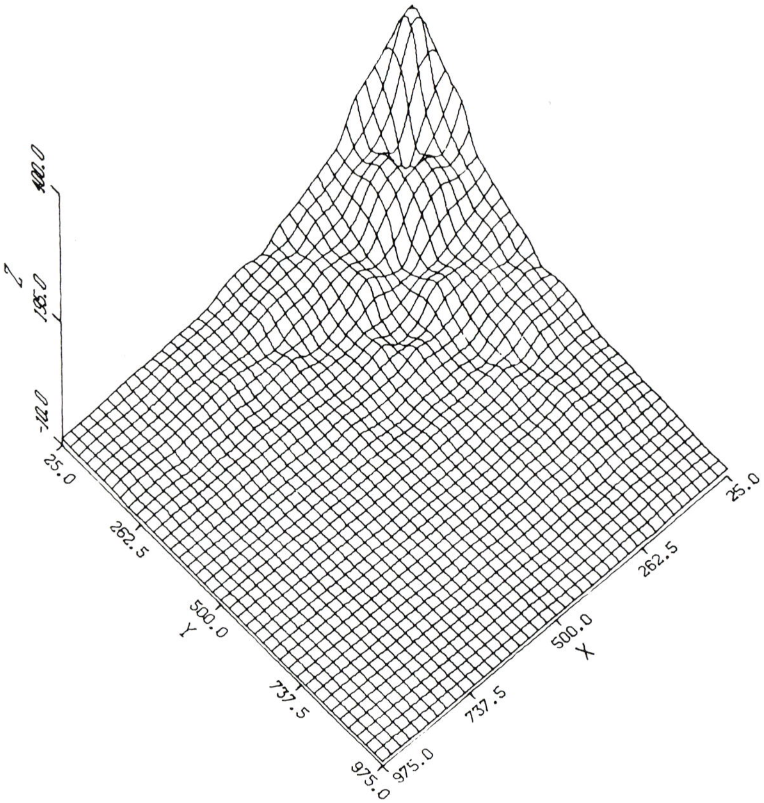

Fig. 5. Global plot of magnitude of real part of bispectral mass, atmospheric data. Estimates of mass in 50 Hz × 50 Hz squares. Linear range of of 25 Hz to 975 Hz. x, y, z axis scaling (6, 6, 6). Rotation angle 135°, elevation angle 60°.

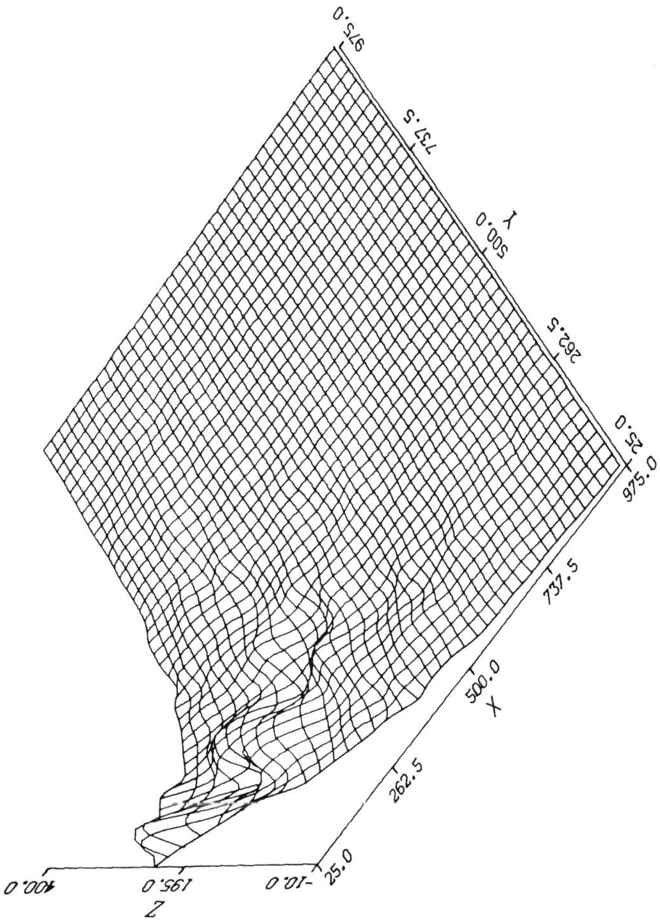

Fig. 6. Different plot of data of Fig. 5. x, y, z scaling (6, 6, 4). Rotation angle 45°, elevation angle 45°.

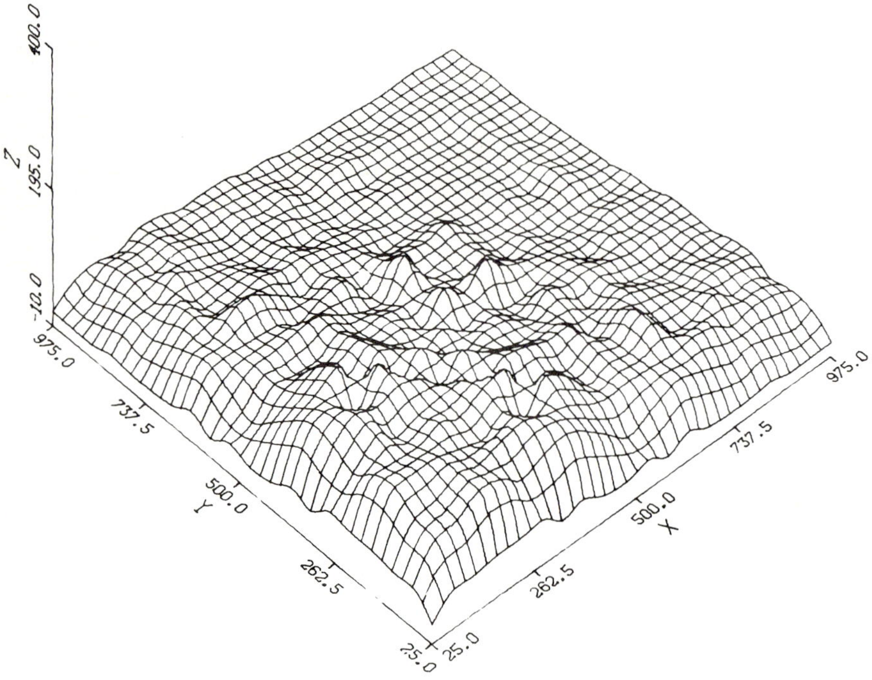

Fig. 7. Global plot of magnitude of real part of bispectral mass, wind tunnel data. Estimates of mass in 50 Hz × 50 Hz squares. Linear range of 25 Hz to 975 Hz. x, y, z axis scaling (6, 6, 4). Rotation angle +45°, elevation angle 45°.

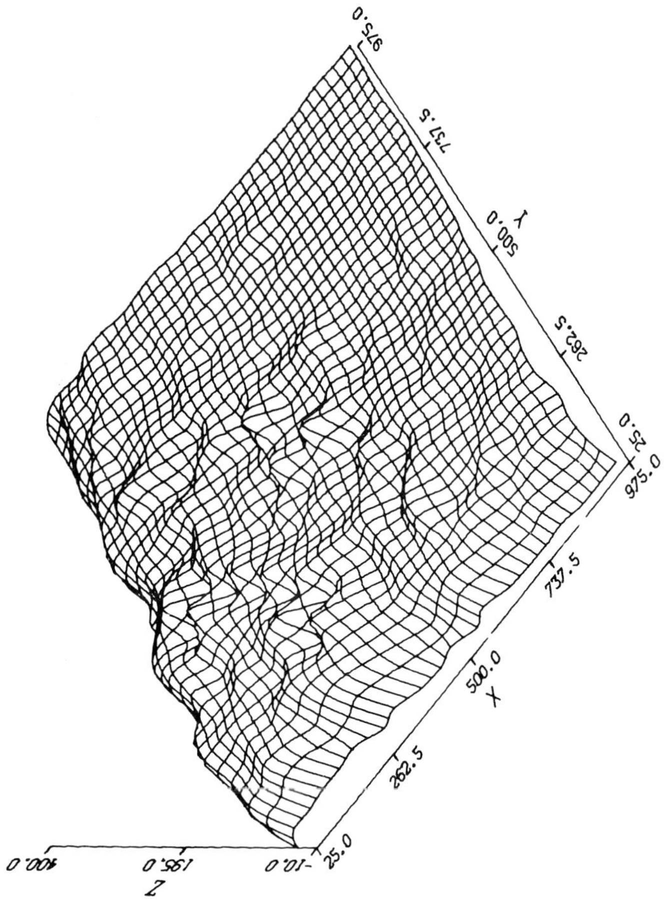

Fig. 8. Different plot of data of Fig. 7. x, y, z scaling (6, 6, 4). Rotation angle -45°, elevation 45°.

(not too far from 60 Hz × 60 Hz). The global graphical representation of the real part of the atmospheric bispectrum as given in Figures 5 and 6 indicates that the surface is high in the low wave number range and decreases rapidly as the wave number variables increase. The decrease is greatest along the diagonal. The decrease is somewhat slower along the edge of the square where one wave number is small while the other is large.

In the case of the wind tunnel data, the bispectral density starts effectively at or near zero when both wave numbers are zero (see Figure 4) and slowly increases up to a broad ridge around $\lambda + \mu = 450$ Hz and then slowly decreases as $\lambda + \mu$ increases (as given in Figures 7 and 8). The bispectral surface for the wind tunnel contrasts with that for the atmospheric data since it is small (rather than large) for low wave number, the maximal values are attained at much higher wave numbers than in the atmospheric case and it decreases much more slowly than in the atmospheric case.

The real bispectral estimates obtained for blocks of 50 Hz by 50 Hz were normalized by the square root of the product of the average spectral values over the 50 Hz intervals centered at λ, μ and $\lambda + \mu$ respectively. It is quite interesting to note that these ratios are relatively stable over the range of roughly 1000 Hz by 1000 Hz in the case of wind tunnel data as well as that of atmospheric data except for the occasional entries in which there is effectively no resolution of the real part of the bispectrum. In the case of the wind tunnel data the ratio ranges between 1.5 and 4.5. There is wider variation of the ratio for the atmospheric data, for the most part between 1 and 10. It would be interesting to see whether a similar crude approximation would be valid in the case of readings on the temperature derivative, that is, whether one has approximately

$$b(\lambda,\mu) \cong c\{f(\lambda)f(\mu)f(\lambda+\mu)\}^{1/2} ,$$

where b is the real part of the bispectrum, f is the spectrum and c is some constant.

An example of a linear but nonGaussian process satisfying an exact relationship for the bispectrum of the form above is given by

$$x(t) = \int a(t - \tau)d\xi(\tau)$$

with ξ a nonGaussian (e.g. Poisson) process of independent and stationary increments and $a(\cdot)$ a symmetric function in L^2.

Estimates of the skewness have been obtained directly from the data in the atmospheric case (-.86) and for the wind tunnel (-.41). One can alternatively get estimates of the third moment by integrating (or summing) our estimates of the bispectrum and then normalizing by the cube of an estimated standard deviation.

This was done by summing over the estimated bispectra in the range 1000 Hz by 1000 Hz given in TABLES I and II. The estimate thereby derived for the skewness in the atmospheric case was -.714 and for the wind tunnel was -.309 (the sum has to be multiplied by a factor of 6 to get the contribution from the whole plane of wave number doublets). A more accurate estimate for the wind tunnel of -.394 is gotten by using additional estimates corresponding to the larger triangle $0 \leq \lambda,\mu$ with $\lambda + \mu \leq 2000$ Hz.

It is interesting to decompose the bispectra into subregions in order to quantify the relative contributions to the skewness from the various range of frequency interactions. We have chosen to subdivide the bispectrum into three regions in the first quadrant: 1. Low-low interactions, Type I, bounded by $\lambda + \mu \leq \lambda^*$, 2. Low-high interactions, Type II, $\lambda + \mu > \lambda^*$ with λ or $\mu \leq \lambda^*/2$, and high-high interactions, type III, with $\lambda > \lambda^*/2$, $\mu > \lambda^*/2$. A diagram of the decomposition is given in Figure 9. This figure shows that extent of the bispectral calculations relative to the Nyquist frequency, $\lambda + \mu = 2000$ Hz. The dividing frequency, λ^*, defined in the diagram is sufficient to characterize the decomposition. The contribution to the skewness was determined for each region by summing over the appropriate frequency doublets using the bispectral estimates of TABLES I and II. Various values for λ^* were selected and the results for the atmospheric and laboratory data are presented in TABLE III. Note that the bispectral regions marked II' and III' in Figure 9 are not currently available but will be in a subsequent report. Comparisons between the atmospheric and wind tunnel data should be made using Kolmogorov nondimensionalization. This will be done in a future study after the outer triangle bispectra are available.

A column of frequencies nondimensionalized by the Kolmogroov frequency $\lambda_k = U/2\pi\eta$ is included in TABLE III. It is tempting to interpret the skewness of the velocity derivative as a good measure of frequency interactions at moderate to high frequencies because the differentiation operation emphasizes the high frequencies in the ordinary second-order spectrum. TABLE III shows that such a simple interpretation is not possible. Both the atmospheric and wind tunnel bispectra have large contributions to the skewness arising from low frequency interactions of Type I. For example at $\lambda^*/\lambda_k \approx 0.3$ the Type I fractional contributions are 29% and 35% for the atmospheric and wind tunnel data respectively. These preliminary results suggest that great caution must be exercised before interpreting skewness of derivative statistics as reliable indicators of purely high frequency phenomena. Bispectra should provide a powerful statistical tool for evaluating third-order conditions for local isotropy in a turbulent flow.

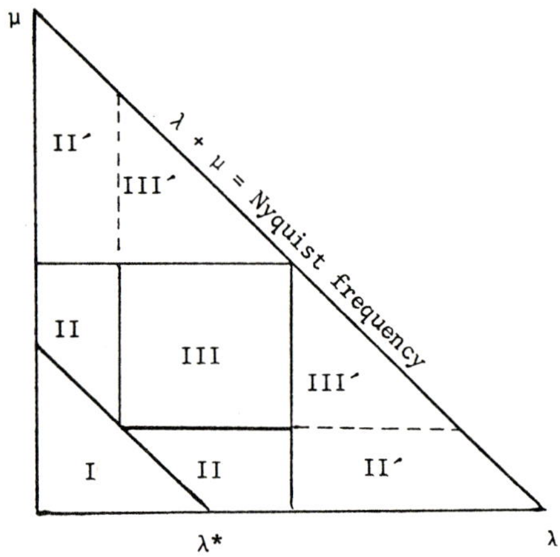

Fig. 9. Diagram of bispectral subdivisions.

Region I. Low-low
Region II. Low-high
Region III. High-high

TABLE III

Fractional Contributions to Skewness from 3 Subregions

$\lambda^* = \lambda + \mu$ Hz	λ^*/λ_k	Low-low	Low-high	High-high
Atmospheric[1]				
50	0.066	0.028	0.258	0.544
150	0.199	0.155	0.335	0.340
250	0.332	0.287	0.345	0.198
350	0.465	0.431	0.301	0.098
450	0.598	0.533	0.247	0.050
Wind Tunnel[2]				
50	0.019	0.001	0.042	0.711
150	0.058	0.011	0.128	0.615
250	0.096	0.038	0.207	0.509
350	0.134	0.079	0.261	0.413
450	0.173	0.137	0.293	0.324
550	0.211	0.203	0.302	0.248
650	0.249	0.274	0.290	0.190
750	0.288	0.351	0.262	0.140

[1]Total bispectral mass 17% low, λ_k = 752 Hz.
[2]Total bispectral mass 25% low, λ_k = 2607 Hz.

6. CONCLUDING REMARKS

It has been demonstrated that bispectral estimates of turbulence signals can provide useful and interesting insight into the frequency interactions occurring within the turbulent flow field. The bispectral estimates of grid turbulence shows dramatically different features when compared to similar estimates from atmospheric measurements. In particular the bispectral level for the laboratory data is relatively large for all frequency interactions considered herein while the low-low frequency interactions for the atmospheric measurements are significantly larger than the high-high interaction region.

The satisfactory agreement between the skewnesses determined directly from the velocity derivative signals and those found by integration of the bispectra provide confidence in the feasibility of these techniques and should encourage additional efforts along these lines. Some very interesting results must surely exist for other calculations, especially the derivative of temperature which has direct implications for the existence of local isotropy.

ACKNOWLEDGEMENT

This research was partially supported by the Office of Naval Research Contract N00014-75-C-0428 and National Science Foundation Grant MCS 76-03942. We thank Stewart Strait who carried out some of the computations and John Wyngaard for the loan of tape W-9.

REFERENCES

[1] AKIMA, H. (1974). Bivariate interpolation and smooth surface fitting based on local procedures. Communications of the ACM, **17**, 1, 26.

[2] BRILLINGER, D. R. and ROSENBLATT, M. (1967). Computation and interpretation of kth order spectra. In *Spectral Analysis of Time Series* (B. Harris, ed.), 189-232. Wiley.

[3] HASSELMAN, K., MUNK, W. and MACDONALD, G. (1963). Bispectra of ocean waves. In *Time Series Analysis* (M. Rosenblatt, ed.), 125-139. Wiley & Sons, New York.

[4] HAUGEN, D. A., KAIMAL, J. C. and BRADLEY, E. F. (1971). An experimental study of Reynolds stress and heat flux in the atmospheric boundary layer. *Quart. J. Roy. Met. Soc.*, **97**, 1968.

[5] LII, K. S., ROSENBLATT, M. and VAN ATTA, C. (1976). Bispectral measurements in turbulence. *J. Fluid Mech.*, **77**, 45-62.

[6] LUMLEY, J. L. and TAKEUCHI, K. (1976). Application of central-limit theorems to turbulence and higher order spectra. *J. Fluid Mech.*, **74**, 433-468.

[7] ROSENBLATT, M. (1971). Curve estimates. *Ann. Math. Statist.*, **42**, 1815-1842.

[8] VAN ATTA, C. W. and CHEN, W. Y. (1969). Measurements of spectral energy transfer in grid turbulence. *J. Fluid Mech.*, **38**, 743.

[9] VAN ATTA, C. W. and YEH, T. T. (1973). The structure of internal intermittency in turbulent flows at large Reynolds number: experimental scale similarity. *J. Fluid Mech.*, **59**, 537-559.

[10] VAN ATTA, C. W. and YEH, T. T. (1975). Evidence for scale similarity of internal intermittency in turbulent flows at large Reynolds numbers. *J. Fluid Mech.*, **71**, 417-440.

[11] WYNGAARD, J. C. and PAO, Y. H. (1972). Some measurements of the fine structure of large Reynolds number turbulence. In *Statistical Models and Turbulence, Lecture Notes in Physics* (M. Rosenblatt and C. Van Atta, eds.), **12**, 384-401. Springer.

P.R. Krishnaiah, ed., APPLICATIONS OF STATISTICS

URN MODELS: A USEFUL TOOL IN APPLIED STATISTICS

N. L. JOHNSON *
University of North Carolina, Chapel Hill, North Carolina 27514

SAMUEL KOTZ †
Temple University, Philadelphia, Pennsylvania 19122

We show by means of a series of examples how urn models may be useful in constructing easily understood models for certain problems in applied statistical work.

Among the topics considered are genetics, capture-recapture models, learning models, sampling surveys and military applications.

1. INTRODUCTION

The use of devices such as cards, coins, dice, etc. to make concrete the ideas in otherwise abstract formulations of probability models is well established. In the present paper we try to present a case for use of urn models for this purpose. Although such models are already commonly used, along with others, the point we wish to make is that urn systems have an inherent flexibility which makes them applicable over a very wide range of situations.

Urn models are included among the basic and most general models used in probability theory. In fact, a practically adequate model of a finite probability space can always be constructed in the form of an urn model. Models of this kind are continuously used in courses on applied and theoretical probability as didactic tools, and in statistical analysis as a means of formulating usable approximations to real situations in mathematical terms.

As an example, consider the problem of "inverse probability." In our context, we have to estimate the contents of an urn on the "empirical basis" of a random sample therefrom. The classical formulation of this problem by T. Bayes supposes we have to choose among a number of urns of known compositions - corresponding to various hypotheses. A random sample has been drawn from one urn, and the problem is to guess from which urn the sample has been taken.

In this survey paper we will give examples of application of urn models in a number of different fields, demonstrating the flexibility and value of these kinds of models.

*Research supported by the Army Research Office under contract DAHC04-74-C-0030.

†Research supported by the U. S. AFOSR, Grant 75-2837.

2. BIOLOGICAL MODELS

We first consider a discrete analogue of Volterra's model for "struggle for existence," proposed by Itoh (1973).

We suppose that an urn contains balls of 3 colors, white, red and black. When balls of the same color encounter one another they are unchanged, but when balls of dissimilar colors meet, the color of one changes to that of the other according to the following scheme:

Colors before encounter	Colors after encounter
Red, black	Both red
White, red	Both white
Black, white	Both black

This model is intended to represent a situation in which there are 3 species (colors) with dominance relations in individual encounters of the kind described in the above table. Note that "dominance" is not transitive in this scheme.

It is assumed that "encounters" occur at a uniform rate and each possible pair of balls in the urn is equally likely to be involved in each encounter. (Itoh proceeds to study the variation in time of the number of balls of each color (i.e. the abundance of the species) for the above system of 3 species, and for systems with more species, with various kinds of dominance relationships.)

Another biological application has been described by Joel Cohen (1966). A *biotope*, a place inhabited by living organisms such as an island or a desert oasis, is studied in hopes of being able to make predictions regarding relative frequencies of species. For the purpose of constructing the model, the biotope is partitioned into subniches. A subniche could be a certain category of types of places within the biotope such as wooded areas or swamp lands. It could be a certain way in which a group of animals utilize the environment such as an amount of vegetation serving as food. Cohen (1966) and others (e.g. Bailey (1972)) have discussed the difficulty of precisely defining subniche and other unavoidable basic inaccuracies of the model which follow. Despite its faults, it has been found useful in some applications of model-building in biology.

Let the biotope be represented by a collection of m urns. Each urn corresponds to a disjoint subniche. The life process is represented by a game involving $s(\leq m)$ players who take turns throwing balls at the collection of urns. Each "player" represents a different species and each "ball" represents a member of that species. Every time a player throws a ball, it lands at random in one of the m urns. Each urn is capable of holding an unlimited number of balls, and each player has an unlimited number of balls available to him. The play proceeds according to the following rules:

(1) Each player must throw at least one ball.

(2) Each urn must (eventually) contain at least one ball.

(3) No two players may "inhabit" the same number of urns (although they may share certain urns).

(4) Each player throws the minimum number of balls necessary to satisfy (1),(2) and (3).

Note that the rules (3) and (4) imply that the j-th player must eventually "inhabit" just j urns.

The relative abundance of the various species in the different subniches correspond to the number of balls from the various players in the different urns at the conclusion of play. The expected distribution of relative abundance can be evaluated for the stated conditions.

We now describe a few genetic models and formulate them as urn models. (We acknowledge gratefully the advice and assistance of Dr. W. J. Ewens in this matter.)

Consider a population comprising a fixed number m of genes in any generation. Each gene is of one out of two genetic types A_1, A_2; each generation will consist of some number i of genes of type A_1 and m-i genes of type A_2. Suppose the genetic composition of a daughter generation is derived by binomial sampling from the genes of the parent generation. Then if, in the parent generation, there are i genes of type A_1, the probability p_{ij} that in the daughter generation there will be j such genes is

$$p_{ij} = \binom{m}{j} \left(\frac{i}{m}\right)^j \left(1 - \frac{i}{m}\right)^{m-j} . \qquad (1)$$

While many models have been put forward to describe the genetic evolution of a population, here we will consider only the above binomial model.

In terms of urn models, we imagine an urn containing m balls, each ball being either (say) red or green. A second urn is now filled with m balls, as follows: balls are drawn one by one, with replacement, from the first urn. As each ball is drawn, a new ball of the same color as that drawn is placed in the second urn; this process continues until there are m balls in the second urn. Equation (1) yields the probability that the second urn will contain j red balls, given that the first urn contained i red balls. In accordance with the evolutionary processes considered in genetics, the first urn is now discarded, a third urn filled by sampling from the second urn, and so on.

Evolutionary studies concern the long-term behavior of such a system. Normally the mathematical apparatus used for such studies is the theory of Markov chains. However, a number of important questions can be discussed without resort to this, using only elementary arguments.

We first consider the conditional probability, given the initial urn contains i red balls, that the urns eventually contain only red balls. (In evolutionary studies, this is the probability that a certain type of gene eventually takes over in the entire population.) If at any stage all the balls in the urn are of one color (either red or green) they will so remain permanently. Whatever the initial proportion, there is a non-zero probability that this will happen even after one "generation." The probability that it will eventually happen is therefore one.

Equation (1) shows that the expected number of red balls ($m \cdot im^{-1} = i$) does not change from one urn to the next. The limit of expected number of red balls in urn t (as $t \to \infty$) is

$$0(1-p) + mp = mp ,$$

where p is the required probability. Equating this to the initial number of red balls gives

$$p = im^{-1}. \tag{2}$$

The probability that the urns eventually contain only green balls is $1 - im^{-1}$.

We next consider the rate at which color uniformity will be approached. Perhaps the most straighforward approach to this question is to find the mean number (E_i) of urn replacements needed until uniformity is achieved. No simple explicit answer to this has been found. For realistically large values of m, the expression for E_i is completely unwieldy, even using the symmetry relation $E_i = E_{m-i}$.

We consider rather an approach common in the evolutionary study of organisms where each individual carries two genes, namely (in the urn context) to find the probability F_t that two balls drawn at random from urn t should be of the same color. Clearly $F_t \to 1$ as $t \to \infty$, and the value of F_t gives a measure of color uniformity of the balls in urn t.

To calculate F_t, notice that two balls drawn from urn t will be of the same color if both are copies of the same ball in urn t-1 (probability m^{-1}) or are copies of different balls in urn t-1 which were of the same color. Evidently

$$F_t = m^{-1} + (1-m^{-1})F_{t-1} . \tag{3}$$

It follows easily from (3) that

$$F_t = 1 - (1-m^{-1})^{t-1}(1-F_1). \tag{4}$$

The value of F_1 is $\{\binom{i}{2} + \binom{m-i}{2}\}/\binom{m}{2} = \{i(i-1) + (m-i)(m-i-1)\}/\{m(m-1)\}$. Equation (4) shows that loss of color variation is rather slow: in the genetic context, this fact is of great importance in reconciling the Mendelian mechanism of inheritance with the Darwinian theory of natural selection, since it shows that genetic variation tends to be preserved for long time intervals and thus allows ample time for selective forces to act.

A second generalization allows for the possibility of gene mutation or, in the urn context, of an error in copying the color of any ball from that of its "parent." Suppose that two colors exist, red and green, and that a green ball will be mis-copied as red with probability u, and a red ball miscopied as green with probability v. (In genetics, u and v are typically of the order 10^{-5}.) In this case we are usually interested in the stationary distribution of the number of red balls (i.e. in urn t as $t \to \infty$), and the rate of approach to stationarity. Let μ_t be the mean number of red balls in urn t. Then

$$\mu_t = \mu_{t-1}(1-v) + (m-\mu_{t-1})u \ . \tag{5}$$

Solving the difference equation (5) we obtain

$$\mu_t = A(1-v-u)^t + \frac{mu}{u+v} \ . \tag{6}$$

(A is an arbitrary constant, defined by μ_1.)

Setting $t \to \infty$ we find

$$\lim_{t\to\infty} \mu_t = \mu_\infty = mu/(u+v) \ . \tag{7}$$

Equation (7) gives the mean of the stationary distribution of the number of red balls, and equation (6) shows that the rate of approach to stationarity is geometrical with rate $1-u-v$: for values of u and v arising in genetics, this rate is extremely slow. Similar arguments show that the variance σ_∞^2 of the stationary distribution of the number of red balls is given, to a sufficient approximation, by

$$\sigma_\infty^2 = m^2uv[(u+v)^2(2mu + 2mv+1)]^{-1} \ . \tag{8}$$

The complete stationary distribution of the number of red balls is not known exactly, although excellent approximations are available (Ewens (1969, Ch. 6).

Experimental verification of these results is not easily obtained. Indeed, it may well be that observed *deviations* from predicted values are of most value in indicating new lines of research.

As in most applied statistical work urn models usually represent an over-simplified version of reality: their function is often to act as standards against which experience can be assessed. It is a virtue of urn models that the nature of the simplification is clearly apparent.

3. LEARNING MODELS

This is especially so in the many models for learning theory which have been proposed. In the earlier development of the subject, models of learning curves played a leading role. Restle and Greeno (1970) describe many of them, and we consider just a few here.

The two elementary mathematical models of learning, involving learning curves, which are most commonly used in practice are: 1) the simple *replacement* model and 2) the simple model of *accumulative* learning. Both can be expressed as urn models (or equivalently as a Markov chain model). In our opinion the urn model has the significant advantages of simplicity and visual clarity in the present context. Furthermore, it allows a number of simpler mathematical derivations.

Moreover, as in other cases, the urn scheme helps to make concrete one's intuitive ideas about learning processes.

a) Replacement learning (linear operator model)

In replacement learning some "habits" (or "stimulus elements") are replaced by others. In the simplest case the total number of entities is constant and they are substituted one by one. The process of learning is quantitatively summarized by the number of elements of various types and by the numbers and kinds of elements transferred from one type (or state) to another.

These transfers are conveniently visualized by a model consisting of urns and balls and calculations from such a model represent - to a high degree of approximation - the flow of habits provided the assumption of random sampling is satisfied.

A subject in a learning experiment is regarded as being associated with an urn S containing black and white balls. On each trial balls are taken from the urn. A black ball is supposed to correspond to a correct response, a white ball to an error. Let m be the number of balls in the urn. According to the "replacement" idea m is fixed and *learning is regarded as a process of replacing white balls by black ones.*

In addition to the subject's urn S, we also assume the existence of an experimenter's urn E with an "infinite" (very large) number of balls corresponding to an inexhaustible source of new "habits." A proportion a (≤ 1) of the balls in E is black. This proportion is fixed and if $a < 1$, then the training process is imperfect. The simple replacement learning process consists in selecting some fixed number, k, of balls from S at each trial and replacing them with an equal number of balls from E so that the number of balls in S remains at m. The sampling from both S and E is supposed to be *random*.

Let B_n and W_n be the number of black and white balls in S on the n-th trial. Clearly

$$E[B_{n+1}|B_n] = B_n(1-km^{-1}) + ka \tag{9}$$

and so

$$E[B_{n+1}] = E[B_n](1-km^{-1}) + ka \tag{10}$$

(since the balls taken from S on the average include kB_n/m black balls and kW_n/m white balls and the k balls added to S consist on the average of ka black balls and k(1-a) white balls).

The probability of a correct response on trial n is $P_n = B_n/m$ (which is the proportion of black balls in the urn S).

Let $\theta = k/m$ be the proportion sampled from S. Then

$$E[P_{n+1}|P_n] = E[m^{-1}B_{n+1}|P_n] \tag{11}$$

and from (9)

$$E[P_{n+1}|P_n] = (1-\theta)P_n + \theta a .$$

This explains the alternative name of *linear model*.

We also evidently have (taking expected values with respect to P_n)

$$E[P_{n+1}] = E[P_n](1-\theta) + \theta a. \tag{12}$$

Whence

$$E[P_n] = a - (a-b)\ (1-\theta)^{n-1} \tag{13}$$

where $b = P_1 = B_1/m$.

(c.f. eq. (4) in Section 2 dealing with genetic applications.) Note that as $n \to \infty$, P_n tends to the asymptotic value a. (Bower (1961) gives some numerical results.)

b) Accumulation model

This is essentially the model suggested by L. L. Thurstone in 1919 and 1930. Here we have as before two urns S and E. However in the present model *no* balls are removed from S. A constant number k of balls is transferred from E to S at each trial. An accumulation process becomes "stiffer" and more resistant to change as more elements accumulate in S. (A replacement process on the other hand does not "stiffen" with training.) Note that m, the total number of balls in S is equal to $B_1 + W_1 + (n-1)k$ at the n-th trial. Under these assumptions

$$E[B_{n+1}] + F[B_n] + ka \tag{14}$$

and

$$E[W_{n+1}] = E[W_n] + k(1-a) . \tag{14)'$$

Thus if $P_n \equiv B_n/(B_n+W_n)$ is the proportion of black balls in S after n trials, then

$$E[P_n] = \frac{B_1 + (n-1)ka}{B_1+W_1 + (n-1)k} . \tag{15}$$

Dividing the numerator and denominator by B_1+W_1 and substituting as before, $b = P_1 = B_1/(B_1+W_1)$ and $\theta = k/m = k/(B_1+W_1)$ we have

$$E[P_n] = \frac{b + \theta a(n-1)}{1 + \theta(n-1)} = \frac{b(n-1)^{-1} + \theta a}{(n-1)^{-1} + \theta} . \tag{16}$$

As $n \to \infty$ we have, as in the "linear" case,

$$\lim_{n\to\infty} E[P_n] = P = \frac{\theta a}{\theta} = a. \tag{17}$$

Hence both learning curves start from $P_1 = b$ and end up at $P_\infty = a$ but they follow different routes. (Numerical comparisons of replacement and accumulation curves are given in Restle and Greeno (1970).)

c) Other models

Audley and Jonckheere (1956) have suggested that the process determining the sequence of responses in a learning experiment is analogous to the process involved in the following urn scheme. An urn initially contains B black balls and W white balls. A ball is chosen randomly. The color of the ball determines how many balls of each color are added according to the table:

TABLE I

Color of ball drawn	Number of black balls added	Number of white balls added
black	a_1	$c_1 - a_1$
white	a_2	$c_2 - a_2$

The drawn ball is replaced along with the added balls, and the procedure is repeated n times. This will be recognized as a generalized Polyá-Eggenberger scheme.

A more extensive general system has been constructed by Millward and Wickens (1974). This, also, is intended to model the performance of a subject in a series of experiments in each of which a stimulus is supplied and a response recorded. The principles underlying the performance are expressed in terms of "hypotheses" regarding a correct (or adequate) response to a given stimulus.

The "learning process" consists of the selection of a hypothesis (s) from a set S of hypotheses currently of interest. S is itself chosen from a set A of available hypotheses; there is also a set U of "unavailable" hypotheses. This includes hypotheses previously in A or in S but "rejected" as a result of earlier experience on the consequences of responses based on them. As elements of S are used up and rejected, it is supposed that new hypotheses are brought into A (and eventually S) from U to replenish the "stock." Various possible rules are suggested for deciding when replenishment is necessary (e.g. when S is empty, when A falls below a certain number, when there are fewer than α hypotheses in A, etc.)

Construction of an urn model of this situation commences by assigning urns to the sets S, A and U with balls in the urns corresponding to individual hypotheses. Models are then differentiated by different rules for transferring balls from one urn to another. It may be necessary to introduce rules modifying the randomness of choice from a given urn. For example, although a hypothesis rejected from S and returned to U may eventually be once more chosen to be included in A (and ultimately in S), it appears reasonable to impose some kind of time lag for unfavorable memories of the rejected hypothesis to fade. This might be allowed for by requiring rejected hypotheses not to be available for selection from U for the first k selections after being returned there. A more elaborate arrangement, allowing for a more gradual fade would be to divide the contents of U among several urns $U_1, U_2, \ldots, U_h$, say according to length of sojourn, and choose first one of the U_i's with appropriately weighted probabilities, finally selecting a ball (hypothesis) from the chosen urn at random.

Many other learning models are discussed in a survey paper by Estes (1972).

4. SAMPLING SCHEMES

This latter type of model is also useful in connection with the construction of systems of sampling in sample surveys and similar enquiries. When sampling human, animal or plant populations it is essential to take account, and if possible advantage, of the non-homogeneous nature of these populations. For example human populations can be partitioned according to geographical locations (state, county, city, block, dwelling unit, etc.) within which the variation might be expected to be less than it is overall.

For practical purposes it is often convenient to select first, not the ultimate individuals who will be measured but some larger unit within which individual sampling will be applied. This is called *multistage sampling*. Often there is more than one stage of selection of units, subunits, etc. before the ultimate selection of individuals occurs. For simplicity, however, we will discuss only *one* intermediate stage.

We can imagine one large urn (the population) containing μ smaller urns (the units) marked with numbers 1 to μ. The first stage of sampling selects randomly m of these smaller urns, and the second (and final, in our case) stage selects, at random, n_i individuals from the i-th urn if it is chosen. It is possible, and indeed usual, for the smaller urns to contain different numbers $\nu_1, \nu_2, \ldots, \nu_\mu$, say of balls. While simple random sampling may be used to select the m urns from among the μ available, it is also possible (and often desirable) to select them at random, but with probability of selection proportional to the numbers of balls in the urns. This is called *"probability proportional to size"* ("PPS") *sampling* . Sampling within each urn is supposed random; the ratio n_i/ν_i is the *sampling fraction* for the i-th urn.

We suppose the value of the observed character for an individual to be inscribed on the corresponding ball. Denote by $x_{i1}, x_{i2}, \ldots, x_{i\nu_i}$ the numbers on the ν_i balls in the i-th of the smaller urns.

The sample is of total size $N = n_{A_1} + n_{A_2} + \ldots + n_{A_m}$, where $A_1, A_2, \ldots, A_m$ are the numbers marked on the chosen urns. Note that if the n_i's are fixed in advance then N will be a random variable.

If *all* the smaller urns are always selected we have a situation corresponding to *stratified sampling*, in which the population is divided into μ strata and n_g individuals chosen from the g-th stratum ($g = 1, 2, \ldots, \mu$).

If we do not select all the small urns, but for each of those that are chosen we observe the numbers on every ball contained in them, we have a form of *cluster sampling*, in which the sampling is applied to naturally occurring groups (households, for example) and all members of the chosen groups are observed.

5. CAPTURE-RECAPTURE PROBLEMS

In these investigations we are usually interested in estimating the sum (or average) of the numbers inscribed on the balls either in some of the small urns, or in the large urn as a whole. Knowing how many balls there are in the urns we can use theoretical results to suggest suitable sampling schemes.

There are circumstances, however, in which we do not know the number of balls in an urn, and wish to estimate this number. Such is the case in models of "capture-recapture" problems, in which we wish to estimate the sizes of natural populations. An early example of such a problem was given by Schnabel (1938) who discussed estimation of the number of fish (f) in a lake. The method proposed was, essentially:

(i) catch n_0 fish, tag them, and return them to the lake,

(ii) catch a new sample of n_1 fish and observe the number T_1, say of tagged fish,

(iii) equate the proportions of tagged fish in the lake before taking the second sample, and in the second sample, giving

$$n_0/f = T_1/n_1 \tag{18}$$

and leading to the estimator

$$\hat{f} = n_0 n_1/T_1 \tag{19}$$

for f.

The slightly modified version

$$\hat{f}^* = (n_0+1)(n_1+1)(T_1+1)^{-1} - 1 \tag{20}$$

is a very nearly unbiased estimator of f, and avoids infinite values.

There is a natural, and even self-evident analogy to the above problems in terms of urns and balls. We start with an urn containing f indistinguishable white balls. We take n_0 balls from the urn (without replacement) and dye each one black, then return the black balls to the urn. We then take a random sample (again without replacement) of n_1 balls from the urn (which contained $(f-n_0)$ white and n_0 black balls), and observe the number, T_1, of black balls. It is required to estimate f, the number of balls originally in the urn.

There is clearly much oversimplification in this model. It does not allow for natural increase (by birth, and perhaps immigration) or decrease (by death, and perhaps emigration). It supposes that a genuine random sample of fish is attainable and also that tagging does not affect the catchability of a fish, relative to the rest of the fish in the lake, on a future occasion.

Elaboration of the model - e.g. by supposing a constant rate of augmentation of the number of balls in the urn, or by assigning different probabilities of choice to a ball according to its history of previous catches - can allow for some of these effects, though there is a point beyond which greater complexity is too great a price to pay for greater fidelity. Also, as we have already remarked, *deviations* from a relatively simple model are in themselves often important.

6. OCCUPANCY DISTRIBUTIONS

Urn models play a natural role in the development of occupancy distributions. We have a set of m urns. Balls are thrown, one at a time, in such a way that each has (independently of the others) an equal probability (m^{-1}) of falling into any specified one of the m urns. We are interested in the distribution of the number (X) of occupied urns after n balls have been thrown. It can be shown that

$$\Pr[X=x] = \binom{m}{x} m^{-n} \Delta^{x} 0^{n} \qquad (x = 0,1,\ldots,m) . \tag{21}$$

This is called the *classical occupancy distribution*. There are many variants of this distribution. In particular there is the *randomized occupancy distribution* which results when there is a probability θ say that a ball thrown into an urn will not be counted (will "fall through" or "bounce out" or be ineffective). The corresponding distribution of the number of occupied urns after n balls have been thrown is

$$\Pr[X=x] = \binom{m}{x} m^{-n} \Delta^{x} \{\theta m + (1-\theta)\cdot 0\}^{n} \qquad (x = 0,1,\ldots,m) . \tag{22}$$

An interesting application of occupancy distributions in computer theory is implicit in the work of Denning and Schwartz (1972). They use the concept of a current *working set* $W(t,\tau)$ at time t when using a sequence of *references*. If each reference is defined as a "page number" then it is supposed that $W(t,\tau)$ only includes those page numbers referenced on at least one of the last τ *occasions*. The distribution of the number of distinct pages in the working set, $W(t,\tau)$, is then an occupancy distribution, with τ balls, each possible "page" being an "urn." If it is supposed that the probability of referencing a specific urn is constant (and independent of other references) then the appropriate distribution is that of the classical occupancy problem.

Occupancy distributions are used in many military applications, especially in regard to (a) distributions of hits on targets, or sets of targets and (b) to assess distribution of forces assigned to specific tasks. In regard to (b), this is often an example of the kind, already described, where distributions based on simplified assumptions are used not because they are believed accurate representations, but because they serve as a standard with respect to which deviations may be measured and studied.

We now give a few examples of (a). If there are m targets altogether, each equally likely to be hit (probability m^{-1} in each case) then the number of targets hit at least once in n shots is clearly given by the classical occupancy distribution (21), provided the assumptions are satisfied.

If g targets have already been hit (by the first n shots say) the distribution of the number of targets (Y), among the so far unhit (m-g) targets, hit by a further s shots is

$$\Pr[Y=y] = m^{-s}\binom{m-g}{y}\Delta^{y}\{g + (m-g)\cdot 0\}^{s} . \qquad (23)$$

This is relevant to consideration of continued firing when a number of targets have already been hit.

Generally speaking we can identify targets with urns and missiles with balls. If some missiles may miss the targets altogether - say with probability θ - then the randomized occupancy distribution applies. Harkness (1970) describes another situation in which this distribution might be expected to arise: "In 'air-battle' theory (or predator-prey models) one identifies the m urns with 'bombers', 'missiles' (or more generally, with 'offensive forces') and the n balls with 'fighters' or ' anti-missiles' (or more generally, with 'defensive forces') ... p is a 'kill' parameter corresponding to the probability that a fighter is able to intercept and destroy a bomber."

The number of occupied urns then corresponds to the number of bombers destroyed. If the randomized occupancy distribution is used, it implies, among other things, an assumption that each fighter will engage only one bomber. It is, of course, possible to allow for multiple engagements by means of appropriate modification of the model. This will in general lead to other distributions, though one might hope, under some circumstances, to be able to use the randomized occupancy distribution, with an appropriately modified (increased) value of n, as a useful approximation.

Situations in which the urns (targets) are classified according to the probability that a ball (missile) will hit a given target correspond to situations in which targets are, for example, of different sizes and so cannot be assumed to have equal chances of being hit.

Sprott (1957) gives a useful account of pioneering work on some of these problems.

Urn models can be constructed to represent more complicated situations. For example, if it is supposed that a hit on one target can produce serious damage on another target, one might suppose each urn to contain a number of cards, each card stating which other (neighboring) urns are to be regarded as also having received a ball (or possibly, having received a "fraction of a ball.") When a ball is assigned to an urn, a card is chosen at random from the urn, and the instructions on the card followed for assigning additional balls (or fractions of balls) to the specified urns.

7. SOME OTHER APPLICATIONS

We now mention briefly two other applications in which urn models have been found useful. A computer application has already been described in Section 6. Another kind of application has recently been described by Fagin (1975).

Suppose we have n balls, labeled 1,...,n and we have a method of choosing these numbers (and the corresponding balls) with probabilities $p_1^{(n)}, p_2^{(n)}, \ldots, p_n^{(n)}$ $(\sum_{j=1}^{n} p_j^{(n)} = 1)$. (This might itself be done using a suitable urn system.) There is an urn which can contain a proportion β_n of the balls (i.e. $n\beta_n$ balls *in all*). We commence by choosing $n\beta_n$ of the balls at random with the appropriate probabilities and putting them in the urn. Then numbers (1 through n) continue to be chosen at random. If the corresponding ball is not in the urn, it is placed in the urn and the ball which has been longest in the urn (without its number being selected) is removed from the urn. Note that if a ball is already in the urn when its number is selected there is no transfer of balls into or out of the urn, but the ball now becomes the most recently selected and so cannot be removed until at least $n\beta_n$ further selections have been made. The urn thus always contains the $n\beta_n$ balls which have been most recently selected.

We consider the sum of the selection probabilities for the numbers on the balls in the urn - called the "weight" of the urn - after m selections.

The weight is equal to the probability that a chosen number will be on a ball already in the urn, and that no transfer will be necessary. Regarding the urn as a computer's fast memory and the balls as pages of "either data or instructions" the weight is the probability that a chosen page is already available in the fast memory. The expected value of the weight is called, in computer sciences, the "least-recently-used hit ratio."

The organization of filing systems is another topic in which urn models have been found useful. If there are items in a file which are in greater demand than others it appears desirable to keep them nearer the top of the file. One way of doing this is simply to place each item taken from the file on top of the file when returning it. This can be modelled by supposing we have a number of urns placed in line, corresponding to positions in the file, with the extreme left urn corresponding to the topmost position. Each urn contains one ball. When a particular ball is requested we look in the urns sequentially from left to right until we find the specified ball. It is then removed; all balls to the left are moved one urn to the right, and the specified ball is put into the extreme left urn. The effectiveness of the procedure is measured by the average number of urns needed to be examined before finding the specified ball, (the "expected delay").

Burville and Kingman (1973) have shown that the expected delay is reduced (as compared with random replacement) by the system described.

There are many other possibilities.* The use of urn models in the formal development of Markov chain theory, for instance, was suggested by Feller (1957). Generally the value of urn model representation lies in the concreteness of the representation, and its flexibility, which enables relatively simple modifications to cover many cases of practical and theoretical interest.

REFERENCES

[1] AUDLEY, R. J. and JONCKHEERE, A. R. (1956). The statistical analysis of the learning process, II. *British Journal of Statistical Psychology*, 9, 87-94.

[2] BAILEY, R. C. (1972). A Montage of Diversity. Ph.D. thesis, Emory University, Atlanta, Georgia.

[3] BOWER, G. H. (1961). An application of a model to paired-associate learning. *Psychometrika.*, 26, 255-280.

[4] BURVILLE, P. J. and KINGMAN, J. F. C. (1973). On a model for storage and search. *J. Appl. Probability*, 10, 697-701.

[5] COHEN, J. E. (1966). *A Model of Simple Competition.* Harvard University Press, Cambridge, Massachusetts.

[6] DENNING, P. J. and SCHWARTZ, S. C. (1972). Properties of the working-set model. *Commun. of the ACM*, 15, 191-98.

[7] ESTES, W. K. (1972). Research and theory on the learning of probabilities. *J. Amer. Statist. Assoc.*, 67, 81-102.

[8] EWENS, W. J. (1969). *Population Genetics.* Methuen, London.

[9] FAGIN, R. (1975). Asymptotic miss ratios over independent references. *IBM Research Report RC* 5415, Yorktown Heights, New York. (See also *Notices Amer. Math. Soc.*, Nov. 1975., 22, 7, A-715).

[10] FELLER, W. (1957). *An Introduction to Probability Theory and its Applications.* 2nd ed. Vol. I, Wiley, New York.

[11] HARKNESS, W. L. (1970). The classical occupancy problem revisited. *Random Counts in Physical Sciences.* (G. P. Patil, ed.). Penn State University Press, 107-126.

[12] ITOH, Y. (1973). On a ruin problem with interaction. *Ann. Inst. Statist. Math.*, 25, 635-41.

[13] MILLWARD, R. B. and WICKENS, T. D. (1974). Concept-identification models. *Contemporary Developments in Mathematical Psychology.* (D. H. Krantz et al., ed.). 1, W. H. Freeman, San Francisco.

[14] RESTLE, F. and GREENO, J. E. (1970). *Introduction to Mathematical Psychology.* Addison Wesley, Reading, Massachusetts.

[15] SCHNABEL, Z. (1938). The estimation of the total fish population of a lake. *Amer. Math. Monthly*, 45, 348-52.

*Very recently Wei (1976) used Friedman's urn model (cf. Feller (1957)) to develop a class of designs for sequential clinical trials with optimal properties in the sense that they force small-sized experiments to be more balanced but tend to yield a complete randomization for large scale experiments.

[16] SPROTT, D. A. (1957). Probability distributions associated with direct hits on targets. *Bull. Math. Biophys.*, 19, 163-70.

[17] TUKEY, J. W. (1949). Moments of random group size distributions. *Ann. Math. Statist.*, 20, 523-539.

[18] THURSTONE, L. L. (1919). The learning curve equation. *Psych. Review, Monograph 26*, No. 3.

[19] THURSTONE, L. L. (1930). The learning function. *J. Gen. Psychol.*, 3, 469-91.

[20] WEI, LEE-JEN (1976). A class of designs for sequential clinical trials (manuscript) University of South Carolina, Columbia.

[21] WOLTER, D. G. and EARL, R. W. (1972). The asymptotic distributions of response probabilities in the Audley-Jonckheere learning model. *Psychometrika*, 37, 167-77.

(A more comprehensive bibliography may be obtained by writing the authors).

P.R. Krishnaiah, ed., APPLICATIONS OF STATISTICS

STRUCTURAL EQUATION MODELS IN THE SOCIAL SCIENCES: SPECIFICATION, ESTIMATION AND TESTING *

KARL G. JÖRESKOG

Department of Statistics
University of Uppsala, Uppsala, Sweden

1. INTRODUCTION

One of the most difficult problems for a social scientist, when it comes to the formulation of a causal model, arises because many of the concepts and constructs that he/she wants to work with are not directly measurable (see e.g. Torgerson, [22], Chapter 1 and Goldberger, [11]). A sociologist works with concepts like social (or socioeconomic) status, occupational aspiration, motivations and attitudes. A psychologist uses concepts like intelligence, verbal ability, spatial ability, etc. An educational researcher studying the effect of home factors on school achievements may be using a construct such as "preschool verbal stimulation in the home." Also in economics one uses hypothetical constructs like price level, permanent income, economic expectations, etc. Although such hypothetical concepts and constructs, or latent variables, as we shall call them, cannot be directly measured, a number of variables can be used to measure various aspects of these latent variables more or less accurately. Thus, while the latent variables cannot be directly observed, they have operational implications for relationships among observed variables. We may regard the observed variables as indicators of the latent variables. Each indicator has a relationship with the latent variable, but if we take one indicator alone to measure the latent variable we would obtain a poor measurement. By using several indicators of each latent variable we get a better measurement of these latent variables.

Another reason for using latent variables in socioeconomic studies is that most of the measurements employed contain sizeable errors of measurement (observational errors). Errors of measurement arise because of imperfection in the various measurement instruments (questionnaires, tests, etc), that are used to measure such abstractions as people's behavior, attitudes, feelings and motivations. Even if we could construct valid measurement of these it is usually impossible to obtain perfectly reliable variables.

In recent years there has been a strong interest in more complex models in-

* The research reported in this paper has been supported by the Bank of Sweden Tercentenary Foundation under project *Structural Equation Models in the Social Sciences*, Karl G. Jöreskog, project director. I thank Dag Sörbom for helpful assistance.

volving specified causal structures among the latent variables. These have usually been formulated in terms of path analysis, a term and a technique introduced by Wright [27] and further developed by him in a series of papers (see e.g. Wright, [28] and references therein). For sociological applications of path analysis, see Duncan [4] and for psychological applications, see Werts and Linn [23]. More recent studies are Goldberger [10, 11], Hauser and Goldberger [13], Jöreskog and Goldberger [18], and Werts, Jöreskog and Linn [24].

In this paper we shall describe a general method for estimating the unknown coefficients in a set of linear structural equations. The variables in the equation system may be latent variables and there may be multiple indicators or causes of each latent variable. Also, the method allows for both errors in equations (residuals, disturbances) and errors in the observed variables (errors of measurement, observational errors) and yields estimates of the residual covariance matrix and the measurement error covariance matrix as well as estimates of the unknown coefficients in the structural equations, provided that all these parameters are identified. The method covers a wide range of models useful in the social and behavioral sciences. It is particularly designed to handle models with latent variables, measurement errors and reciprocal causation (simultaneity, interdependence). The general model on which the method is based is a generalization of that considered in a previous paper [17].

2. GENERAL THEORY

2.1. *The General Model*

Consider random vectors $\eta' = (\eta_1, \eta_2, \ldots, \eta_m)$ and $\xi' = (\xi_1, \xi_2, \ldots, \xi_n)$ of latent dependent and independent variables, respectively, and the following system of linear structural relations

$$B\eta = \Gamma\xi + \zeta \tag{1}$$

where $B(m \times m)$ and $\Gamma(m \times n)$ are coefficient matrices and $\zeta' = (\zeta_1, \zeta_2, \ldots, \zeta_m)$ is a random vector of residuals (errors in equations, random disturbance terms). Without loss of generality it may be assumed that $E(\eta) = E(\zeta) = 0$ and $E(\xi) = 0$. It is furthermore assumed that ζ is uncorrelated with ξ and that B is nonsingular.

The vectors η and ξ are not observed but instead vectors $y' = (y_1, y_2, \ldots, y_p)$ and $x' = (x_1, x_2, \ldots, x_q)$ are observed, such that

$$y = \Lambda_y \eta + \varepsilon , \tag{2}$$

$$x = \Lambda_x \xi + \delta , \tag{3}$$

where ε and δ are vectors of errors of measurement in y and x, respectively. We take y and x to be measured as deviations from their means. The matrices $\Lambda_y(p \times m)$ and $\Lambda_x(q \times n)$ are regression matrices of y on η and of x on ξ, respectively. It is convenient to refer to y and x as the observed variables and η and ξ as the latent variables. The errors of measurement ε and δ are assumed to be uncorrelated and uncorrelated with the latent variables.

Let Φ $(n \times n)$ and Ψ $(m \times m)$ be the covariance matrices of ξ and ζ, respectively, and let Θ_ε and Θ_δ be the covariance matrices of ε and δ, respectively. Then it follows, from the above assumptions that the covariance matrix $\Sigma[(p + q) \times (p + q)]$ of $z = (y',x')'$ is

$$\Sigma = \begin{bmatrix} \Lambda_y(B^{-1}\Gamma\Phi\Gamma'B'^{-1} + B^{-1}\Psi B'^{-1})\Lambda_y' + \Theta_\varepsilon & \Lambda_y B^{-1}\Gamma\Phi\Lambda_x' \\ \Lambda_x\Phi\Gamma'B'^{-1}\Lambda_y' & \Lambda_x\Phi\Lambda_x' + \Theta_\delta \end{bmatrix} \tag{4}$$

The elements of Σ are functions of the elements of Λ_y, Λ_x, B, Γ, Φ, Ψ, Θ_δ and Θ_ε. In applications some of these elements are fixed and equal to assigned values. In particular, this is so for elements Λ_y, Λ_x, B and Γ, but we shall allow for fixed values in the other matrices also. For the remaining non-fixed elements of the six parameter matrices one or more subsets may have identical but unknown values. Thus elements in Λ_y, Λ_x, B, Γ, Φ, Ψ, Θ_δ and Θ_ε are of three kinds:

(i) *fixed parameters* that have been assigned given values,

(ii) *constrained parameters* that are unknown but equal to one or more other parameters and

(iii) *free parameters* that are unknown and not constrained to be equal to any other parameter.

Equations (1), (2) and (3), with the accompanying assumptions, define the general model. Equations (2) and (3) constitute the measurement model and equation (1) constitutes the structural equation model. The measurement model can be written more compactly as

$$z = \Lambda f + e \, , \tag{5}$$

where

$$z = (y',x')', \qquad f = (\eta',\xi')', \qquad e = (\varepsilon',\delta')'$$

and

$$\Lambda = \begin{bmatrix} \Lambda_y & 0 \\ 0 & \Lambda_x \end{bmatrix}$$

From (5) it can be seen that the model is a restricted factor analysis model [16] in which the factors η and ξ satisfy a linear structural equation system of the form (1). It should be noted, however, that there is no requirement that $m < p$, $n < q$ and that Θ_ε and Θ_δ be diagonal as in traditional factor analysis. The only requirement is that Σ in (4) is nonsingular and that the model is identified (see next section).

2.2. *Identification of Parameters*

Before an attempt is made to estimate a model of this kind the identification problem must be examined. Identifiability depends on the choice of model and on the specification of fixed, constrained, and free parameters. Under a given specification, a given structure Λ_y, Λ_x, B, Γ, Φ, Ψ, Θ_ε, Θ_δ generates one and only one Σ but there may be several structures generating the same Σ. If two or more structures generate the same Σ, the structures are said to be equivalent. If a parameter has the same value in all equivalent structures, the parameter is said to be identified. If all parameters of the model are identified, the whole model is said to be identified.

We assume that the distribution of the observed variables is sufficiently well described by the moments of first and second order, so that information contained in moments of higher order may be ignored. In particular, this will hold if the distribution is multivariate normal. Since the mean vector is unconstrained, the distribution of $z = (y', x')'$ is described by the independent parameters in Λ_y, Λ_x, B, Γ, Φ, Ψ, Θ_ε, Θ_δ, Let θ be a vector of all the independent, free and constrained parameters (i.e. counting each distinct constrained parameter once only) and let s be the order of θ. The identification problem then is the problem of whether or not θ is determined by Σ.

If a parameter is not identified it will not be possible to find a consistent estimator of it.

To examine the identification problem for a particular model consider the equations in (4) of the form

$$\sigma_{ij} = f_{ij}(\theta) \;, \; i \leq j \; . \tag{6}$$

There are (1/2) (p + q) (p + q + 1) equations and s unknown elements in θ. Hence a necessary condition for identification of all parameters is that

$$s < (1/2)\,(p + q)\,(p + q + 1) \tag{7}$$

If a parameter θ can be determined from Σ, this parameter is identified, otherwise it is not. Often some parameters can be determined from Σ in different ways. This gives rise to overidentifying conditions on Σ which must hold if the model is true. The solution of (6) is often complicated and tedious and explicit solutions for all θ's seldom exist. Examples of how the identification problem can be resolved are given in section 3.

2.3. Estimation of the Model

Let $\underset{\sim}{z}_1, \underset{\sim}{z}_2, \ldots, \underset{\sim}{z}_M$ be independent observations of $\underset{\sim}{z} = (\underset{\sim}{y}', \underset{\sim}{x}')'$ and let $\bar{\underset{\sim}{z}} = (\bar{\underset{\sim}{y}}', \bar{\underset{\sim}{x}}')$be the sample mean vector, and

$$\underset{\sim}{S} = \frac{1}{N} \sum_{\alpha=1}^{N} (\underset{\sim}{z}_\alpha - \bar{\underset{\sim}{z}})\ (\bar{\underset{\sim}{z}}_\alpha - \bar{\underset{\sim}{z}})', \tag{8}$$

with $N = M - 1$, the sample covariance matrix.

We assume that the distribution of the observed variables can be described by the mean vector and the covariance matrix. The mean vector is unconstrained, so the estimation problem is essentially that of fitting the $\underset{\sim}{\Sigma}$ imposed by the model to the sample covariance matrix $\underset{\sim}{S}$. As a fitting function we use

$$F = \log|\underset{\sim}{\Sigma}| + \mathrm{tr}(\underset{\sim}{S}\ \underset{\sim}{\Sigma}^{-1}) - \log|\underset{\sim}{S}| - (p + q) \ , \tag{9}$$

which is to be minimized with respect to $\underset{\sim}{\theta}$. If the distribution of $\underset{\sim}{z} = (\underset{\sim}{y}, \underset{\sim}{x}')'$ is multimormal this yields maximum likelihood estimates which are efficient in large samples. If the distribution deviates far from the multimormal it is probably wise to "robustify" the variances and covariances in $\underset{\sim}{S}$ before the analysis: see Andrews, Gnanadesikan and Warner [1] for methods for assessing multivariate normality, Gnanadesikan and Kettenring [9] for robust estimates of variances and covariances and Devlin, Gnanadesikan and Kettenring [3] for robust estimates of correlations. The minimization of F with respect to the independent parameters is done by an iterative method which makes use of the first derivatives of F and the information matrix. It converges rapidly from an arbitrary starting point to a local minimum of F. If there are several minima of F there is no guarantee that the method will converge to the absolute minimum. At the minimum of F, the information matrix may be computed and used to compute standard errors for all estimated parameters.

The derivatives of F with respect to the elements of $\underset{\sim}{\Lambda}_y$, $\underset{\sim}{\Lambda}_x$, $\underset{\sim}{B}$, $\underset{\sim}{\Gamma}$, $\underset{\sim}{\Phi}$, $\underset{\sim}{\Psi}$, $\underset{\sim}{\Theta}_\delta$, and $\underset{\sim}{\Theta}_\varepsilon$ may be obtained by matrix differentiation. Writing $\underset{\sim}{A} = \underset{\sim}{B}^{-1}$, $\underset{\sim}{D} = \underset{\sim}{B}^{-1}\underset{\sim}{\Gamma}$, $\underset{\sim}{C} = \underset{\sim}{D}\underset{\sim}{\Phi}\underset{\sim}{D}' + \underset{\sim}{A}\underset{\sim}{\Psi}\underset{\sim}{A}'$ and

$$\underset{\sim}{\Omega} = \begin{pmatrix} \underset{\sim}{\Omega}_{yy} & \underset{\sim}{\Omega}_{yx} \\ \underset{\sim}{\Omega}_{xy} & \underset{\sim}{\Omega}_{xx} \end{pmatrix} = \underset{\sim}{\Sigma}^{-1}(\underset{\sim}{\Sigma} - \underset{\sim}{S})\underset{\sim}{\Sigma}^{-1} \ , \tag{10}$$

the derivatives are, except for a scale factor of 2 in the first four derivatives,

$$\partial F/\partial \Lambda_y = \Omega_{yy}\Lambda_y C + \Omega'_{xy}\Lambda_x \Phi D' \quad , \tag{11}$$

$$\partial F/\partial \Lambda_x = \Omega_{xy}\Lambda_y D\Phi + \Omega_{xx}\Lambda_x \Phi \quad , \tag{12}$$

$$\partial F/\partial B = -A'\Lambda'_y(\Omega_{yy}\Lambda_y C + \Omega'_{xy}\Lambda_x \Phi D') \quad , \tag{13}$$

$$\partial F/\partial \Gamma = A'\Lambda'_y(\Omega_{yy}\Lambda_y D + \Omega'_{xy}\Lambda_x)\Phi \quad , \tag{14}$$

$$\partial F/\partial \Phi = D'\Lambda_y\Omega_{yy}\Lambda_y D + \Lambda'_x\Omega_{xy}\Lambda_y D + D'\Lambda_y\Omega'_{xy}\Lambda_x + \Lambda'_x\Omega_{xx}\Lambda_x \quad , \tag{15}$$

$$\partial F/\partial \Psi = A'\Lambda_y\Omega_{yy}\Lambda_y A \quad , \tag{16}$$

$$\partial F/\partial \Theta_\varepsilon = \Omega_{yy} \quad , \tag{17}$$

$$\partial F/\partial \Theta_\delta = \Omega_{xx} \quad . \tag{18}$$

In these expressions the symmetry of Φ, Ψ, Θ_ε and Θ_δ has not been taken into account. Symmetry may be handled by equality constraints on the off-diagonal elements.

2.4. *Information Matrix*

The elements of S have an asymptotic multinormal distribution with mean matrix Σ . Let ε_{ij} be a typical element of $S - \Sigma$. Then, asymptotically

$$NE(\varepsilon_{gh}\varepsilon_{ij}) = \sigma_{gi}\sigma_{hj} + \sigma_{gj}\sigma_{hi} \, . \tag{19}$$

The asymptotic distribution of the elements of $\Omega = \Sigma^{-1}(\Sigma - S)\Sigma^{-1}$ is multivariate normal with means zero and variances and covariances given by

$$NE(\omega_{\alpha\beta}\omega_{\mu\nu}) = \sigma^{\alpha\mu}\sigma^{\beta\nu} + \sigma^{\alpha\nu}\sigma^{\beta\mu}.$$

This follows immediately by multiplying

$$\omega_{\alpha\beta} = \sum_{gh}\sum\sigma^{\alpha g}(\sigma_{gh} - s_{gh})\sigma^{h\beta} \quad \text{and} \quad \omega_{\mu\nu} = \sum_{ij}\sum\sigma^{\mu i}(\sigma_{ij} - s_{ij})\sigma^{j\nu}$$

and using (19).

The information matrix may be obtained by using the following lemma.

LEMMA. Let the element of Σ be functions of two parameter matrices $M = (\mu_{gh})$ and $N = (\nu_{ij})$ and let $F(M, N) = \frac{1}{2}[\log|\Sigma| + \mathrm{tr}(S\Sigma^{-1})]$ with $\partial F/\partial M = A\Omega B$ and $\partial F/\partial N = C\Omega D$, where A, B, C, D are independent of S. Then we have asymptotically

$$E(\partial^2 F/\partial\mu_{gh}\partial\nu_{ij}) = (\underset{\sim}{A}\underset{\sim}{\Sigma}^{-1}\underset{\sim}{C}')_{gi}(\underset{\sim}{B}'\underset{\sim}{\Sigma}^{-1}\underset{\sim}{D})_{hj} + (\underset{\sim}{A}\underset{\sim}{\Sigma}^{-1}\underset{\sim}{D})_{gj}(\underset{\sim}{B}'\underset{\sim}{\Sigma}^{-1}\underset{\sim}{C}')_{hi}.$$

Proof. Writing $\partial f/\partial\mu_{gh} = a_{g\alpha}\omega_{\alpha\beta}b_{\beta h}$ and $\partial F/\partial\nu_{ij} = c_{i\mu}\omega_{\mu\nu}d_{\nu j}$, where it is assumed that every repeated subscript is to be summed over, we have (cf. Kendall and Stuart, [21] Eq. 18.57)

$$E(\partial^2 F/\partial\mu_{gh}\partial\nu_{ij}) = NE(\partial F/\partial\mu_{gh}\partial F/\partial\nu_{ij}) = NE(a_{g\alpha}\omega_{\alpha\beta}b_{\beta h}c_{i\mu}\omega_{\mu\nu}d_{\nu j})$$

$$= Na_{g\alpha}b_{\beta h}c_{i\mu}d_{\nu j}E(\omega_{\alpha\beta}\omega_{\mu\nu}) = a_{g\alpha}b_{\beta h}c_{i\mu}d_{\nu j}\ (\sigma^{\alpha\mu}\sigma^{\beta\nu} + \sigma^{\alpha\nu}\sigma^{\beta\mu})$$

$$= (a_{g\alpha}\sigma^{\alpha\mu}c_{i\mu})\ (b_{\beta h}\sigma^{\beta\nu}d_{\nu j}) + (a_{g\alpha}\sigma^{\alpha\nu}d_{\nu j})\ (b_{\beta h}\sigma^{\beta\mu}c_{i\mu})$$

$$= (\underset{\sim}{A}\underset{\sim}{\Sigma}^{-1}\underset{\sim}{C}')_{gi}(\underset{\sim}{B}'\underset{\sim}{\Sigma}^{-1}\underset{\sim}{D})_{hj} + (\underset{\sim}{A}\underset{\sim}{\Sigma}^{-1}\underset{\sim}{D})_{gj}(\underset{\sim}{B}'\underset{\sim}{\Sigma}^{-1}\underset{\sim}{C}')_{hi}.$$

It should be noted that the lemma is quite general in that both $\underset{\sim}{M}$ and $\underset{\sim}{N}$ may be row or column vectors or scalars and $\underset{\sim}{M}$ and $\underset{\sim}{N}$ may be identical in which case, of course, $\underset{\sim}{A} \equiv \underset{\sim}{C}$ and $\underset{\sim}{B} \equiv \underset{\sim}{D}$.

The above lemma can be applied repeatedly to compute all the elements of the information matrix. For example, with $\underset{\sim}{\Omega}$ and $\underset{\sim}{\Lambda}$ as before and with

$$\underset{\sim}{T}'[m \times (m + n)] = [\underset{\sim}{A}' \quad \underset{\sim}{0}]$$

$$\underset{\sim}{P}'[m \times (m + n)] = [\underset{\sim}{C} \quad \underset{\sim}{D}\underset{\sim}{\Phi}],$$

$$\partial F/\partial\underset{\sim}{B} = -2\underset{\sim}{T}'\underset{\sim}{\Lambda}'\underset{\sim}{\Omega}\underset{\sim}{\Lambda}\underset{\sim}{P}.$$

2.5. *The Handling of Fixed and Constrained Parameters*

Let $\underset{\sim}{\mu}' = (\mu_1, \mu_2,\ldots, \mu_k)$ be a vector of *all* elements in all eight parameter matrices $\underset{\sim}{\Lambda}_y$, $\underset{\sim}{\Lambda}_x$, $\underset{\sim}{B}$, $\underset{\sim}{\Gamma}$, $\underset{\sim}{\Phi}$, $\underset{\sim}{\Psi}$, $\underset{\sim}{\Theta}_\varepsilon$, and $\underset{\sim}{\Theta}_\delta$, and consider F as a function of $\underset{\sim}{\mu}$. This function is continuous and has continuous derivatives of first and second order except where $\underset{\sim}{\Sigma}$ is singular. The totality of these derivatives is represented by a gradient vector $\partial F/\partial\underset{\sim}{\mu}$ and a symmetric Hessian matrix $\partial^2 F/\partial\underset{\sim}{\mu}\partial\underset{\sim}{\mu}'$. Some $k - r$ of the μ's are fixed. Let the remaining r μ's form a vector $\underset{\sim}{\nu}' = (\nu_1, \nu_2,\ldots, \nu_r)$. Derivatives $\partial F/\partial\underset{\sim}{\nu}$ and $\partial^2 F/\partial\underset{\sim}{\nu}\partial\underset{\sim}{\nu}'$ are obtained from $\partial F/\partial\underset{\sim}{\mu}$ and $\partial^2 F/\partial\underset{\sim}{\mu}\partial\underset{\sim}{\mu}'$ by elimination of the rows and columns corresponding to the fixed $\underset{\sim}{\mu}$'s. Among $\nu_1, \nu_2,\ldots, \nu_r$ there are some distinct parameters $\theta_1, \theta_2,\ldots,\theta_s$, assumed to be identifiable. Let $k_{ig} = 1$ if $\nu_i = \theta_g$ and $k_{ig} = 0$ otherwise and let $\underset{\sim}{K} = (k_{ig})$, $i = 1, 2,\ldots, r$, $g = 1, 2,\ldots, s$. Then we have

$$\partial F/\partial\underset{\sim}{\theta} = \underset{\sim}{K}'\ \partial F/\partial\underset{\sim}{\nu} \qquad \partial^2 F/\partial\underset{\sim}{\theta}\ \partial\underset{\sim}{\theta}' = \underset{\sim}{K}'\ \partial^2 F/\partial\underset{\sim}{\nu}\partial\underset{\sim}{\nu}'\underset{\sim}{K}$$

and

$$E(\partial^2 F/\partial\underset{\sim}{\theta}\partial\underset{\sim}{\theta}') = \underset{\sim}{K}'E(\partial^2 F/\partial\underset{\sim}{\nu}\ \partial\nu')\underset{\sim}{K} \tag{20}$$

The elements of the information matrix on the right-hand side of (20) are obtained as described in the previous section.

2.6. *Basic Minimization Algorithm*

The function $F(\theta)$ may be minimized numerically by Fisher's scoring method or the method of Fletcher and Powell [8] (see also Gruvaeus and Jöreskog [12]).

The minimization starts at an arbitrary starting point $\theta^{(1)}$ and generates successively new points $\theta^{(2)}, \theta^{(3)}, \ldots$, such that $F(\theta^{(s+1)}) < F(\theta^{(s)})$ until convergence is obtained. Let $g^{(s)}$ be the gradient vector $\partial F/\partial \theta$ at $\theta = \theta^{(s)}$ and let $E^{(s)}$ be the information matrix $E(\partial^2 F/\partial\theta\partial\theta')$ evaluated at $\theta = \theta^{(s)}$. Then Fisher's scoring method computes a correction vector by solving the equation system

$$E^{(s)}\delta^{(s)} = g^{(s)}$$

and then computes the new point as

$$\theta^{(s+1)} = \theta^{(s)} - \delta^{(s)}$$

This requires the computation of $E^{(s)}$ and the solution of a symmetric equation system in each iteration and this is often quite time consuming. An alternative is to use the method of Fletcher and Powell, which evaluates only the inverse of $E^{(1)}$ and in subsequent iterations E is improved, using information built up about the function, so that ultimately E converges to an approximation of the inverse of $\partial^2 F/\partial\theta\, \partial\theta'$ at the minimum.

A computer program LISREL based on the Fletcher and Powell method has been written by Jöreskog and Sörbom [20]*.

2.7. *Tests of Hypotheses*

When the maximum likelihood estimates of the parameters have been obtained, the goodness of fit of the model may be tested, in large samples, by the likelihood ratio technique. Let H_0 be the null hypothesis of the model under the given specifications of fixed, constrained, and free parameters. First consider the case when the alternative hypothesis H_1 is that Σ is any positive definite matrix. Then minus twice the logarithm of the likelihood ratio is $(N/2)F_0$ where F_0 is the minimum value of F. If the model holds, this is distributed, in large samples, as χ^2 with

$$d = \frac{1}{2}(p + q)(p + q + 1) - s$$

degrees of freedom, where, as before, s is the total number of independent

* The program is available on a rental basis from International Educational Services, P. O. Box A 3650, Chicago, Illinois 60690.

parameters estimated under H_0.

Let H_0 be any specified hypothesis concerning the parametric structure of the general model which is more restrictive than an alternative hypothesis, H_1, say. In large samples one can then test H_0 against H_1. Let F_0 be the minimum of F under H_0 and let F_1 be the minimum of F under H_1. Then $F_1 < F_0$ and minus twice the logarithm of the likelihood ratio becomes $(N/2)(F_0 - F_1)$. Under H_0 this is distributed approximately as χ^2 with degrees of freedom equal to $s_1 - s_0$, the difference in number of independent parameters estimated under H_1 and H_0. In many situation, it is possible to set up a sequence of hypotheses such that each one is a special case of the preceding and to test these hypotheses sequentially.

In a more exploratory situation the χ^2-goodness-of-fit-values can be used as follows. If a value of χ^2 is obtained, which is large compared to the number of degrees of freedom, the fit may be examined by an inspection of the magnitudes of the first derivatives of F with respect to the fixed parameters. Often such an inspection or the results of analysis will suggest ways to relax the model somewhat by introducing more parameters. The new model usually yields a smaller χ^2. A large drop in χ^2, compared to the difference in degrees of freedom, indicates that the changes made in the model represent a real improvement. On the other hand, a drop in χ^2 close to the difference in number of degrees of freedom indicates that the improvement in fit is obtained by "capitalizing on chance", and the added parameters may not have real significance and meaning.

3. APPLICATIONS

3.1. Models for Longitudinal Data

The characteristic feature of a longitudinal research design is that the same measurements are obtained from the same people at two or more occasions. The purpose of a longitudinal or panel study is to assess the changes that occur between the occasions and to attribute these changes to certain background characteristics and events existing or occurring before the first occasion and/or to various "treatments" that occur after the first occasion. Wiley and Harnischfeger [26] have given an account of the conceptual issues in the attribution of change in educational studies. In the sociological literature there has been a number of articles concerned with the specification of models incorporating causation and measurement errors, and analysis of data from panel studies; see Bohrnstedt [2], Heise [14, 15] , Duncan [5, 6]. Jöreskog and Sörbom [19] have given an outline of statistical models and methods for longitudinal data.

An Example: The stability of alienation

As an example of a longitudinal study, we use some data analyzed in more detail by Wheaton, Muthén, Alwin and Summers [25]. This study was concerned with the stability over time of attitudes such as alienation and the relation to background variables such as education and occupation. Data on attitude scales were collected from 932 persons in two rural regions in Illinois at three points in time: 1966, 1967 and 1971. The variables we use for the present illustration are the *Anomia* subscale and the *Powerlessness* subscale, taken to be indicators of *Alienation*. We use these subscales from 1967 and 1971 only. The background variables are respondent's education (years of schooling completed) and Duncan's Socioeconomic Index (SEI). These are taken to be indicators of respondent's socioeconomic status (SES). We analyze these variables under two different models as shown in Figures 1 A - B, none of which correspond to Wheaton et al's final model. In these figures, observed variables are enclosed in rectangles whereas latent variables are enclosed in ellipses. Residuals (errors in equations) and errors of measurements are included in the diagram but are not enclosed. A one-way arrow pointing from one variable x to another variable y indicates a possible direct causal influence of x on y, whereas a curved two-way arrow between x and y indicates that x and y may correlate without any causal interpretation of this correlation being given. A coefficient associated with an arrow may also be included in the path diagram.

Specification

Let

y_1 = ANOMIA 67 y_3 = ANOMIA 71

y_2 = POWERLESSNESS 67 y_4 = POWERLESSNESS 71

x_1 = EDUCATION x_2 = SEI

ξ = SES η_1 = ALIENATION 67 η_2 = ALIENATION 71

The LISREL specification of model 1A is

$$\begin{pmatrix} y_1 \\ y_2 \\ y_3 \\ y_4 \end{pmatrix} = \begin{bmatrix} 1 & 0 \\ \lambda_1 & 0 \\ 0 & 1 \\ 0 & \lambda_2 \end{bmatrix} \begin{pmatrix} \eta_1 \\ \eta_2 \end{pmatrix} + \begin{pmatrix} \varepsilon_1 \\ \varepsilon_2 \\ \varepsilon_3 \\ \varepsilon_4 \end{pmatrix} \tag{21}$$

$$\begin{pmatrix} x_1 \\ x_2 \end{pmatrix} = \begin{pmatrix} 1 \\ \lambda_3 \end{pmatrix} \xi \quad \begin{pmatrix} \delta_1 \\ \delta_2 \end{pmatrix} \tag{22}$$

$$\begin{pmatrix} 1 & 0 \\ \beta & 1 \end{pmatrix} \begin{pmatrix} \eta_1 \\ \eta_2 \end{pmatrix} = \begin{pmatrix} \gamma_1 \\ \gamma_2 \end{pmatrix} \xi + \begin{pmatrix} \zeta_1 \\ \zeta_2 \end{pmatrix} \tag{23}$$

FIG 1 A

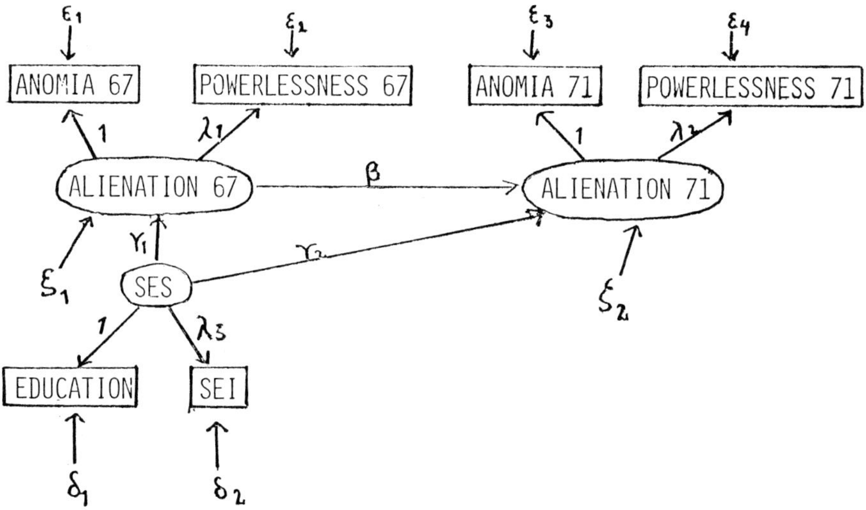

FIG 1 B

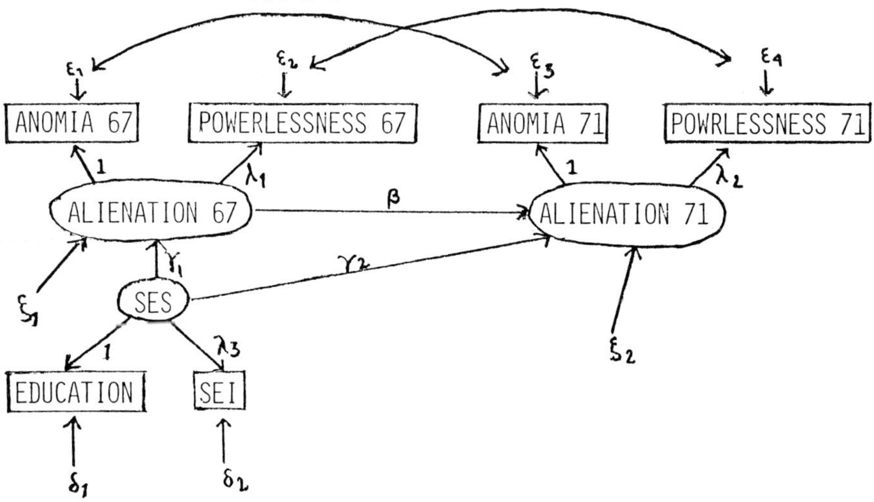

Fig. 1 A - B. Models for the study of Stability of Alienation.

with $\underset{\sim}{\Phi}(1\times 1) = \phi = \text{Var}(\xi)$, $\underset{\sim}{\Psi}(2\times 2) = \text{diag}(\psi_{11}, \psi_{22})$, where $\psi_{ii} = \text{Var}(\zeta_i)$ $i = 1,2$, and $\underset{\sim}{\Theta}_\varepsilon(4\times 4)$ and $\underset{\sim}{\Theta}_\delta(2\times 2)$ diagonal. Here we have assumed that the scales for η_1, η_2 and ξ have been chosen to be the same as for y_1, y_3 and x_1, respectively.

In model 1B the only difference is that the $\underset{\sim}{\Theta}_\varepsilon$ is not diagonal but

$$\underset{\sim}{\Theta}_\varepsilon = \begin{bmatrix} \theta_{11}^{(\varepsilon)} & & & \\ 0 & \theta_{22}^{(\varepsilon)} & & \\ \theta_{31}^{(\varepsilon)} & 0 & \theta_{33}^{(\varepsilon)} & \\ 0 & \theta_{42}^{(\varepsilon)} & 0 & \theta_{44}^{(\varepsilon)} \end{bmatrix},$$

with $\theta_{ii}^{(\varepsilon)} = \text{Var}(\varepsilon_i)$ and $\theta_{ij}^{(\varepsilon)} = \text{Cov}(\varepsilon_i, \varepsilon_j)$.

Identification

We have six observed variables with 21 variances and covariances. Model 1A has 15 parameters (3 λ's, 1 β, 2 γ's, 1 ϕ, 2 ψ's and 6 θ's) so that if all these are identified the model will have 6 degrees of freedom.

The reduced form of (23) is

$$\eta_1 = \gamma_1\xi + \zeta_1$$

$$\eta_2 = (\gamma_2 - \beta\gamma_1)\xi + (\zeta_2 - \beta\zeta_1)$$

$$= \pi\xi + \nu \quad , \text{ say.}$$

We have

$$\text{Cov}(y_1, x_1) = \text{Cov}(\eta_1, x_1) = \gamma_1\phi \tag{24}$$

$$\text{Cov}(y_2, x_1) = \lambda_1\text{Cov}(\eta_1, x_1) = \lambda_1\gamma_1\phi \tag{25}$$

$$\text{Cov}(y_3, x_1) = \text{Cov}(\eta_2, x_1) = \pi\phi \tag{26}$$

$$\text{Cov}(y_4, x_1) = \lambda_2\text{Cov}(\eta_2, x_1) = \lambda_2\pi\phi \ . \tag{27}$$

If we use x_2 instead of x_1 in these equations all four right sides will be multiplied by λ_3. Hence λ_3 is overdetermined since

$$\lambda_3 = \text{Cov}(y_i, x_2) \ / \ \text{Cov}(y_i, x_1) \ , \ i = 1, 2, 3, 4.$$

With λ_3 determined, ϕ is determined by

$$\mathrm{Cov}(x_1, x_2) = \lambda_3\phi \ .$$

With ϕ determined, equations (24) - (27) determine γ_1, λ_1, π and λ_2, respectively. Furthermore

$$\mathrm{Cov}(y_1, y_2) = \lambda_1 \ \mathrm{Var}(\eta_1) = \lambda_1(\gamma_1{}^2\phi + \psi_{11}) \ ,$$

which determines ψ_{11}, and

$$\mathrm{Cov}(y_3, y_4) = \lambda_2 \ \mathrm{Var}(\eta_2) = \lambda_2[\pi^2\phi + \mathrm{Var}(\nu)]$$

which determines

$$\mathrm{Var}(\nu) = \psi_{22} + \beta^2\psi_{11} \ .$$

For given λ_1, λ_2, γ_1, π , ϕ and ψ_{11} the four equations

$$\mathrm{Cov}(y_1, y_3) = \gamma_1\pi\phi - \beta\psi_{11} \ , \qquad (28)$$

$$\mathrm{Cov}(y_1, y_4) = \lambda_2(\gamma_1\pi\phi - \beta\psi_{11}) \ , \qquad (29)$$

$$\mathrm{Cov}(y_2, y_3) = \lambda_1(\gamma_1\pi\phi - \beta\psi_{11}) \ , \qquad (30)$$

$$\mathrm{Cov}(y_2, y_4) = \lambda_1\lambda_2(\gamma_1\pi\phi - \beta\psi_{11}) \ , \qquad (31)$$

show that β is overdetermined. Then, with β determined, $\gamma_2 = \pi + \beta\gamma_1$ and ψ_{22} are obtained. The errors variances $\theta_{ii}^{(\varepsilon)}$ are determined from $\mathrm{Var}(y_i)$, i = 1, 2, 3, 4, and $\theta_{ii}^{(\delta)}$ from $\mathrm{Var}(x_i)$, i = 1, 2. Hence it is clear that model 1A is identified and has six independent restrictions on $\underset{\sim}{\Sigma}$.

In model 1B we have two more parameters namely $\theta_{31}^{(\varepsilon)}$ and $\theta_{42}^{(\varepsilon)}$. These will be added to the right sides of (28) and (31), respectively. However, since (29) and/or (30) can be used to determine β it is clear that also $\theta_{31}^{(\varepsilon)}$ and $\theta_{42}^{(\varepsilon)}$ are determined by (28) and (31) respectively. Hence model 1B is also identified and has four degrees of freedom.

Estimation and testing

To estimate models 1A - B we use the data in TABLE I. The maximum likelihood estimates of the parameters with standard errors in parenthesis are given in Table II. The stability of alienation over time is reflected in the parameter β . The influence of SES on Alienation at the two occasions is significant in model 1A. The coefficient for 1967, γ_1, is -0.614 with a standard error of 0.056 and for 1971, γ_2, it is -0.174 with a standard error equal to 0.054. The negative sign of the γ-coefficients indicates that for high socioeconomic status the alienation is low

TABLE I

Covariance Matrix for the Model of Fig. 1 (n = 932)

y_1	11.834					
y_2	6.947	9.364				
y_3	6.819	5.091	12.532			
y_4	4.783	5.028	7.495	9.986		
x_1	- 3.839	- 3.889	- 3.841	- 3.625	9.610	
x_2	-21.899	-18.831	-21.748	-18.775	35.522	450.288

and vice versa. However, the overall fit of the model 1A is not acceptable; χ^2 with six degrees of freedom equals 71.544. Since the same scales are used at the two occasions, it seems plausible that errors may be correlated for the same variable over time, see e.g. Jöreskog and Sörbom [19]. Thus the model 1B is intuitively more plausible. As can be seen from TABLE II the inclusion of these correlations results in a model with an acceptable overall fit. A test of the hypothesis that both $\theta_{31}^{(\varepsilon)}$ and $\theta_{42}^{(\varepsilon)}$ are zero yields $\chi^2 = 66.774$ with 2 degrees of freedom so that this hypothesis must be rejected.

3.2. Models with Reciprocal Causation

The models in section 3.1 were such that the structural equation system was recursive, i.e. with $\underset{\sim}{B}$ lower triangular. Another characteristic feature was that the latent variables were acting as underlying causes of observed variables. In this section we consider a different type of model namely with reciprocal causation (simultaneity or interdependence). Another feature of this type of models is that latent variables may appear not only as underlying causes but also as effects of observed variables [18].

A model of Duncan, Haller and Portes

Sociologists have often called attention to the way in which one's peers -- e.g. best friends -- influence one's decisions -- e.g. choice of occupation. They have recognized that the relation must be reciprocal -- if my best friend influences my choice, I must influence his. Duncan, Haller and Portes [7] presented a simultaneous equation model of peer influences on occupational choice, using a sample of Michigan high-school students paired with their best friends. A striking result was that the direct path from respondent's choice to friend's choice turned out to be very similar to the direct path from friend's choice to respondent's choice. The authors interpret educational and occupational choice as two indicators of a single latent variable "ambition", and specify reciprocal causation between the ambitions rather than between the choices. This model with simultaneity and errors of measurement is displayed in Figure 2.

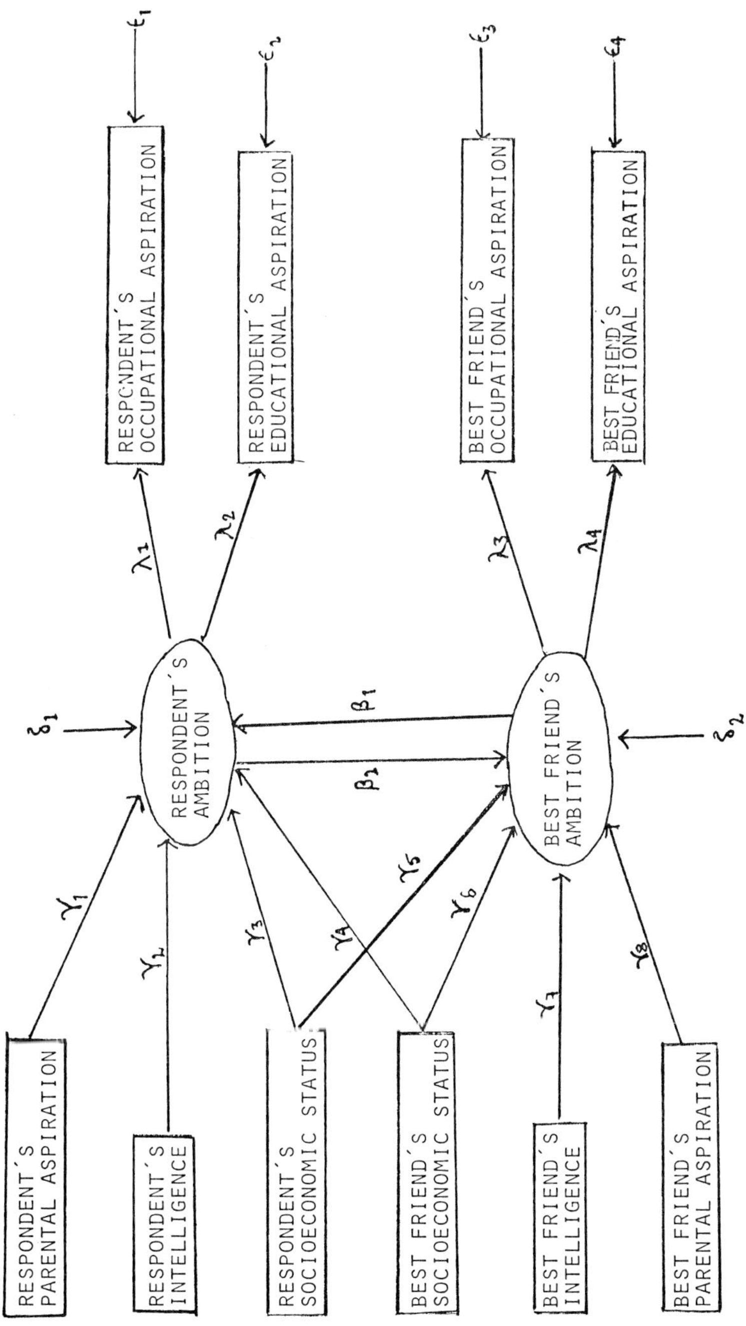

Fig. 2. A model of Duncan, Haller and Portes (1971).

TABLE II

Maximum Likelihood Estimates for the Models in Fig. 1A - B

The standard errors of the estimates are given within parenthesis.

Parameter	Model in Figure 1A	Model in Figure 1B
λ_1	0.888 (.041)	0.979 (.062)
λ_2	0.849 (.040)	0.922 (.059)
λ_3	5.331 (.430)	5.221 (.422)
β	0.705 (.054)	0.607 (.051)
γ_1	- 0.614 (.056)	- 0.575 (.056)
γ_2	- 0.174 (.054)	- 0.227 (.052)
ψ_{11}	5.307 (.473)	4.847 (.468)
ψ_{22}	3.742 (.388)	4.089 (.405)
ϕ	6.663 (.641)	6.803 (.650)
$\theta_{11}^{(\delta)}$	2.947 (.500)	2.807 (.508)
$\theta_{22}^{(\delta)}$	260.911 (18.241)	264.818 (18.155)
$\theta_{11}^{(\varepsilon)}$	4.014 (.343)	4.734 (.454)
$\theta_{22}^{(\varepsilon)}$	3.199 (.271)	2.566 (.404)
$\theta_{33}^{(\varepsilon)}$	3.699 (.373)	4.401 (.516)
$\theta_{44}^{(\varepsilon)}$	3.626 (.292)	3.075 (.435)
$\theta_{31}^{(\varepsilon)}$	- - -	1.624 (.314)
$\theta_{42}^{(\varepsilon)}$	- - -	0.340 (.261)
χ^2	71.544	4.770
d.f.	6	4

Specification

Let

x_1 = RESPONDENT'S PARENTAL ASPIRATION

x_2 = RESPONDENT'S INTELLIGENCE

x_3 = RESPONDENT'S SOCIOECONOMIC STATUS

x_4 = BEST FRIEND'S SOCIOECONOMIC STATUS

x_5 = BEST FRIEND'S INTELLIGENCE

x_6 = BEST FRIEND'S PARENTAL ASPIRATION

y_1 = RESPONDENT'S OCCUPATIONAL ASPIRATION

y_2 = RESPONDENT'S EDUCATIONAL ASPIRATION

y_3 = BEST FRIEND'S OCCUPATIONAL ASPIRATION

y_4 = BEST FRIEND'S EDUCATIONAL ASPIRATION

η_1 = RESPONDENT'S AMBITION

η_2 = BEST FRIEND'S AMBITION

In terms of our general model we take $\xi_i \equiv x_i$, i.e. in equation (3) we take $\underset{\sim}{\Lambda}_x(6 \times 6) = \underset{\sim}{I}$ and $\underset{\sim}{\delta} = \underset{\sim}{0}$.

The structural equations are

$$\begin{pmatrix} 1 & \beta_1 \\ \beta_2 & 1 \end{pmatrix} \begin{pmatrix} \eta_1 \\ \eta_2 \end{pmatrix} = \begin{pmatrix} \gamma_1 & \gamma_2 & \gamma_3 & \gamma_4 & 0 & 0 \\ 0 & 0 & \gamma_5 & \gamma_6 & \gamma_7 & \gamma_8 \end{pmatrix} \begin{pmatrix} \xi_1 \\ \xi_2 \\ \xi_3 \\ \xi_4 \\ \xi_5 \\ \xi_6 \end{pmatrix} + \begin{pmatrix} \zeta_1 \\ \zeta_2 \end{pmatrix}, \tag{32}$$

and the equations relating the η's to the y's are

$$\begin{pmatrix} y_1 \\ y_2 \\ y_3 \\ y_4 \end{pmatrix} = \begin{bmatrix} 1 & 0 \\ \lambda_1 & 0 \\ 0 & 1 \\ 0 & \lambda_2 \end{bmatrix} \begin{pmatrix} \eta_1 \\ \eta_2 \end{pmatrix} + \begin{pmatrix} \varepsilon_1 \\ \varepsilon_2 \\ \varepsilon_3 \\ \varepsilon_4 \end{pmatrix} \tag{33}$$

In $\underset{\sim}{\Lambda}_y$ we have fixed the scales for η_1 and η_2 to be the same as in y_1 and y_3, respectively. Since $\underset{\sim}{\xi} \equiv \underset{\sim}{x}$ and there are no constraints on $\underset{\sim}{\Phi}$, $\underset{\sim}{\Phi} = \underset{\sim}{\Sigma}_{xx}$. The matrix $\underset{\sim}{\Psi}(2 \times 2)$ is

$$\underset{\sim}{\Psi} = \begin{bmatrix} \psi_{11} & \\ \psi_{21} & \psi_{22} \end{bmatrix}$$

with $\psi_{ii} = \text{Var}(\zeta_i)$, $i = 1, 2$ and $\psi_{21} = \text{Cov}(\zeta_1, \zeta_2)$ and the matrix Θ_ε is diagonal with diagonal elements $\theta_{ii}^{(\varepsilon)} = \text{Var}(\varepsilon_i)$.

Since $\xi \equiv x$, the structural equations are equivalent to

$$B\eta = \Gamma x + \zeta$$

with reduced form

$$\eta = B^{-1}\Gamma x + B^{-1}\zeta. \tag{34}$$

Furthermore we have

$$y = \Lambda_y \eta + \varepsilon = \Lambda_y B^{-1}\Gamma x + \Lambda_y B^{-1}\zeta + \varepsilon = \Pi x + z\ , \tag{35}$$

where

$$\Pi = \Sigma_{yx}\Sigma_{xx}^{-1} = \Lambda_y B^{-1}\Gamma$$

and

$$z = \Lambda_y B^{-1}\zeta + \varepsilon$$

Identification

Since Ψ is unconstrained, it follows from (34) that there is a one-to-one correspondence between Ψ and the covariance matrix of η, Ω say, where

$$\Omega = \begin{bmatrix} \omega_{11} & \\ \omega_{21} & \omega_{22} \end{bmatrix} .$$

From the first part of (35) we have

$$\Sigma_{yy} = \Lambda_y \Omega \Lambda_y' + \Theta_\varepsilon = \begin{bmatrix} \omega_{11} + \theta_{11}^{(\varepsilon)} & & & \\ \lambda_1\ \omega_{11} & \lambda_1^2\ \omega_{11} + \theta_{22}^{(\varepsilon)} & & \\ \omega_{21} & \lambda_1\ \omega_{21} & \omega_{22} + \theta_{33}^{(\varepsilon)} & \\ \lambda_2\ \omega_{21} & \lambda_1\ \lambda_2\omega_{21} & \lambda_2\omega_{22} & \lambda_2\omega_{22} + \theta_{44}^{(\varepsilon)} \end{bmatrix} \tag{36}$$

Furthermore we have

$$B^{-1} = (1 - \beta_1\beta_2)^{-1} \begin{pmatrix} 1 & -\beta_1 \\ -\beta_2 & 1 \end{pmatrix}$$

and

$$B^{-1}\Gamma = (1-\beta_1\beta_2)^{-1} \begin{pmatrix} \gamma_1 & \gamma_2 & \gamma_3-\beta_1\gamma_5 & \gamma_4-\beta_1\gamma_6 & -\beta_1\gamma_7 & -\beta_1\gamma_8 \\ -\beta_2\gamma_1 & -\beta_2\gamma_2 & \gamma_5-\beta_2\gamma_3 & \gamma_6-\beta_2\gamma_4 & \gamma_7 & \gamma_8 \end{pmatrix}$$

The first and third rows of Π are identical to the first and second row of $B^{-1}\Gamma = D$, say, respectively. The second row of Π is λ_1 times the first row of D and the fourth row of Π is λ_2 times the second row of D. Hence it is clear that λ_1 and λ_2 are identified and that

$$\lambda_1 = \frac{\pi_{2i}}{\pi_{1i}} \quad \text{and} \quad \lambda_2 = \frac{\pi_{4i}}{\pi_{3i}} , \quad i = 1, 2, \ldots, 6 .$$

Since D are two rows of Π, D is identified. From D it follows that γ_1, γ_2, γ_7, γ_8, β_1 and β_2 are determined as

$$\gamma_1 = d_{11}, \quad \gamma_2 = d_{12}, \quad \gamma_7 = d_{25}, \quad \gamma_8 = d_{26}$$

$$\beta_1 = -(d_{15}/d_{25}) = -(d_{16}/d_{26}) \quad ,$$

$$\beta_2 = -(d_{21}/d_{11}) = -(d_{22}/d_{12}) \quad .$$

γ_3, γ_4, γ_5 and γ_6 are then determined by d_{13}, d_{14}, d_{23}, d_{24}. With λ_1 and λ_2 determined we can now determine ω_{11}, ω_{12} and ω_{22} from the off-diagonal elements of Σ_{yy}. Finally, the $\theta_{ii}^{(\varepsilon)}$, i = 1, 2, 3, 4, can be determined from the diagonal elements of Σ_{yy} .

This analysis shows that all parameters are identified. All together there are 40 parameters (2 β's, 8 γ's, 3 ω's, 2 λ's, 4 $\theta^{(\varepsilon)}$'s and 21 ϕ's). The degrees of freedom will be (1/2) 10 · 11 - 40 = 15.

Analysis of data

To illustrate the analysis we use the correlation matrix published by Duncan, Haller and Portes [7]. This is reproduced here in TABLE III. An overall goodness of fit test gives $\chi^2 = 26.70$ with 15 degrees of freedom which is significant at the 5% level but not at the 2.5% level. A test of hypothesis $\psi_{21} = 0$ gives $\chi^2 = 0.19$ with 1 degree of freedom and a test of $\beta_1 = \beta_2$, given $\psi_{21} = 0$, gives $\chi^2 = 0.06$ with 1 degree of freedom. Hence it is clear that these hypotheses can-

TABLE III

Correlation Matrix for Duncan-Haller-Portes Model (N = 329)

y_1	1.000									
y_2	0.625	1.000								
y_3	0.327	0.367	1.000							
y_4	0.422	0.327	0.640	1.000						
x_1	0.214	0.274	0.112	0.084	1.000					
x_2	0.411	0.404	0.290	0.260	0.184	1.000				
x_3	0.324	0.405	0.305	0.279	0.049	0.222	1.000			
x_4	0.293	0.241	0.411	0.361	0.019	0.186	0.271	1.000		
x_5	0.300	0.286	0.519	0.501	0.078	0.336	0.230	0.295	1.000	
x_6	0.076	0.070	0.278	0.199	0.115	0.102	0.093	-0.044	0.209	1.000

not be rejected. An overall test of goodness of fit of the model with $\psi_{21} = 0$ and $\beta_1 = \beta_2$ gives $\chi^2 = 26.95$ with 17 degrees of freedom. This is not significant at the 5% level. The maximum likelihood estimates with their standard errors are given in TABLE IV. The standardized solution in which η_1 and η_2 are scaled to unit variance is also given in TABLE IV.

4. CONCLUSION

We have described a general approach to handle many models involving latent variables, measurement error and simultaneity. By allowing the parameters of the model to be fixed, free or constrained, great flexibility is obtained in that the general model contains a wide range of specific models useful in the social and behavioral sciences. Using the LISREL program, each model may be estimated by the maximum-likelihood method on the assumption that the observed variables have a multinormal distribution. When the estimates have been obtained, the information matrix may be computed and used to determine standard errors for the estimated parameters. The over-all fit of the model may be assessed by means of a large sample χ^2-test and specific parametric hypotheses may be tested by large sample χ^2-tests. The general approach has been illustrated by means of two examples, one, in which the latent variables are acting as underlying causes of observed variables and which includes a recursive structural equation system and one, in which the latent variables appear both as underlying causes and effects of observed variables and which also includes reciprocal causation.

TABLE IV

Maximum Likelihood Estimates for the Model in Fig. 2 with $\psi_{21} = 0$ and $\beta_1 = \beta_2$.

The standard errors of the estimates are given in parenthesis.

Parameter	Unscaled Solution	Standardized Solution
λ_1	1.000*	0.768
λ_2	1.059 (0.089)	0.813
λ_3	1.000*	0.827
λ_4	0.934 (0.070)	0.772
β_1	- 0.180 (0.039)	- 0.194
β_2	- 0.180 (0.039)	- 0.167
γ_1	0.163 (0.039)	0.213
γ_2	0.253 (0.042)	0.329
γ_3	0.219 (0.042)	0.285
γ_4	0.074 (0.042)	0.096
γ_5	0.077 (0.041)	0.093
γ_6	0.236 (0.041)	0.285
γ_7	0.356 (0.043)	0.431
γ_8	0.163 (0.039)	0.197
ψ_{11}	0.281 (0.046)	0.476
ψ_{22}	0.265 (0.045)	0.387
$\theta_{11}^{(\varepsilon)}$	0.411 (0.051)	0.411
$\theta_{22}^{(\varepsilon)}$	0.339 (0.051)	0.339
$\theta_{33}^{(\varepsilon)}$	0.314 (0.046)	0.314
$\theta_{44}^{(\varepsilon)}$	0.403 (0.046)	0.403

Asterisks denote parameter values fixed by scaling.

REFERENCES

[1] ANDREWS, D. F., GNANADESIKAN, R. and WARNER, J. L. (1973). Methods for assessing multivariate normality. *Multivariate Analysis - III*, (P. R. Krishnaiah, ed.). Academic Press, New York. 95-116.

[2] BOHRNSTEDT, G. W. (1969). Observations on the measurement of change. *Sociological Methodology*. (E. F. Borgatta, ed.). Jossey-Bass, San Francisco.

[3] DEVLIN, S. J., GNANADESIKAN, R. and KETTENRING, J. R. (1975). Robust estimation and outlier detection with correlation coefficients. *Biometrika*, **62**, 531-545.

[4] DUNCAN, O. D. (1966). Path analysis: Sociological examples. *American Journal of Sociology*, **72**, 1-16.

[5] DUNCAN, O. D. (1969). Some linear models for two-wave, two-variable panel analysis. *Psychological Bulletin*, **72**, 177-182.

[6] DUNCAN, O. D. (1972). Unmeasured variables in linear models for panel analysis. *Sociological Methodology*. (H. L. Costner, ed.). Jossey-Bass, San Francisco.

[7] DUNCAN, O. D., HALLER, A. O. and PORTES, A. (1971). Peer influence on aspirations, a reinterpretation. *Causal Models in the Social Sciences*. (H. M. Blalock, Jr., ed.). Aldine-Atherton, Inc.

[8] FLETCHER, R. and POWELL, M. J. D. (1963). A rapidly convergent descent method for minimization. *Comput. J.*, 163-168.

[9] GNANADESIKAN, R. and KETTENRING, J. R. (1972). Robust estimates, residuals, and outlier detection with multiresponse data. *Biometrics*, **28**, 81-124.

[10] GOLDBERGER, A. S. (1971). Econometrics and psychometrics: a survey of communalities. *Psychometrika*, **36**, 83-107.

[11] GOLDBERGER, A. S. (1972). Structural equation methods in the social sciences. *Econometrica*, **40**, 929-1001.

[12] GRUVAEUS, G. T. and JÖRESKOG, K. G. (1970). A computer program for minimizing a function of several variables. *Research Bulletin* 70-14. Educational Testing Service, Princeton, N. J.

[13] HAUSER, R. M. and GOLDBERGER, A. S. (1971). The treatment of unobservable variables in path analysis. *Sociological Methodology*. (H. S. Costner, ed.). Jossey-Bass, San Francisco. 81-117.

[14] HEISE, D. R. (1969). Separation reliability and stability in test-retest correlation. *American Sociological Review*, **34**, 93-101.

[15] HEISE, D. R. (1970). Causal inference from panel data. *Sociological Methodology*. (E. F. Borgatta and G. W. Bohrnstedt, eds.). Jossey-Bass, San Francisco. 3-27.

[16] JÖRESKOG, K. G. (1969). A general approach to confirmatory maximum likelihood analysis. *Psychometrika*, **34**, 183-202.

[17] JÖRESKOG, K. G. (1973). A general method for estimating a linear structural equation system. *Structural Equation Models in the Social Sciences*. (A. S. Goldberger and O. D. Duncan, eds.). Seminar Press, New York. 85-112.

[18] JÖRESKOG, K. G. and GOLDBERGER, A. S. (1975). Estimation of a model with multiple indicators and multiple causes of a single latent variable. *J. Amer. Statist. Assoc.*, **10**, 631-639.

[19] JÖRESKOG, K. G. and SÖRBOM, D. (1977). Statistical models and methods for analysis of longitudinal data. *Latent variables in Socioeconomic models.* (D. J. Aigner and A. S. Goldberger, eds.). North-Holland Publishing Co., Amsterdam.

[20] JÖRESKOG, K. G. and SÖRBOM, D. (1976). LISREL III-Estimation of linear structural equation systems by maximum likelihood methods. International Educational Services, Chicago.

[21] KENDALL, M. G. and STUART, A. (1961). *The Advanced Theory of Statistics. Inference and Relationship.* Griffin, London. Vol. 2.

[22] TORGERSON, W. S. (1958). *Theory and Methods of Scaling.* Wiley, New York.

[23] WERTS, C. E. and LINN, R. L. (1970). Path Analysis: Psychological examples. *Psychological Bulletin* , **74**, 193-212.

[24] WERTS, C. E., JÖRESKOG, K. G. and LINN, R. L. (1973). Identification and estimation in path analysis with unmeasured variables. *American Journal of Sociology* , **78**, 1469-1484.

[25] WHEATON, B., MUTHÉN, B., ALWIN, D. and SUMMERS, G. (1977). Assessing reliability and stability in panel models with multiple indicators. *Sociological Methodology*, (D. R. Heise, ed.).

[26] WILEY, D. E. and HARNISCHFEGER, A. (1973). Post hoc, ergo propter hoc. Problems in the attribution of change. Studies of Educative Processes, Chicago, Report No. 7.

[27] WRIGHT, S. (1934). The method of path coefficients. *Ann. Math. Statist.*, **5**, 161-215.

[28] WRIGHT, S. (1960). Path coefficients and regressions: Alternative or complementary concepts? *Biometrics* , **16**, 189-202.

P.R. Krishnaiah, ed., APPLICATIONS OF STATISTICS

THE FLABBINESS AND INSTABILITY OF REGRESSION ANALYSIS AND COMPUTATIONAL METHODS FOR IMPROVEMENT

HARRIET KAGIWADA

The Rand Corporation, Santa Monica, California

R. KALABA

Department of Economics and Biomedical Engineering,
University of Southern California
Los Angeles, California

K. MEASE

Departments of Biomedical Engineering,
University of Southern California,
Los Angeles, California

Many recent papers have called attention to the flabbiness or instability of regression analysis. In part the problem is a computational one due to the ill-conditioning of the matrix of the normal equations.

It is pointed out that the use of dynamic programming can be of great help numerically. It is suggested that previous regression studies be re-examined using this new approach.

1. INTRODUCTION

One of the most common tools of inquiry in the social sciences is regression analysis [14]. This is due, in part, to the availability of powerful digital computers to do arithmetic calculations and computer programs (SPSS, BMDP, etc.) to guide the calculations and so analyze the data. Yet a flabbiness of the regression coefficients is commonly observed [10]. This instability of the computed coefficients is partly statistical in nature, which is well known. But it is also computational in nature and is so serious that much doubt is cast upon the results of earlier studies that involve "routine regression analysis."

Here we shall examine the origins of the problem and suggest means for overcoming the computational difficulties.

Suppose that an explained variable y is a linear combination of N explaining variables $x_1, x_2, \ldots, x_N$ and a random term ε,

$$y = \sum_{j=1}^{N} x_j \beta_j + \varepsilon. \tag{1.1}$$

Suppose that L observations on $x_1, x_2, \ldots, x_N$ and the corresponding value of y are available. In vector-matrix notation we write

$$y = X\beta + \varepsilon, \tag{1.2}$$

where now y is a column vector with L rows. In ordinary least squares theory we estimate the true vector β by minimizing the sum of squares of residuals S,

$$S = (X\beta - y, X\beta - y), \tag{1.3}$$

Where we have employed the usual inner product notation

$$(a,b) = a^T b, \tag{1.4}$$

and a^T is the transpose of the vector a. It is easy to see that the vector $\hat{\beta}$ which minimizes in Eq. (1.3) satisfies the linear algebraic equations

$$X^T X\hat{\beta} = X^T y, \tag{1.5}$$

which are the so-called normal equations. If

$$\det (X^T X) \neq 0, \tag{1.6}$$

then the exact solution of Eq. (1.5) is given as

$$\hat{\beta} = (X^T X)^{-1} X^T y. \tag{1.7}$$

However even if inequality (1.6) holds, it is by no means a simple task for the computer to determine $\hat{\beta}$ if $\det (X^T X)$ is close to zero. The difficulty is not hard to see. Since

$$(X^T X)^{-1} = \operatorname{adj}(X^T X)/\det(X^T X), \tag{1.8}$$

we may expect the inverse to contain large elements with both positive and negative signs. Carrying out the operations called for in Eq. (1.7) with the limited number of digits carried in each number can lead to answers that are wrong by orders of magnitude in size and which may even have the wrong sign. In econometrics this is referred to as the problem of multicollinearity [14]. It is clearly present when the columns of the matrix X are nearly dependent.

One method for improving the computational situation is provided by ridge regression [2]. In this method we recognize that the computed β should be such that $(X\beta-y, X\beta-y)$ is small, and in addition $\beta^T\beta$ should not be too large. This leads to the problem of determining β so as to minimize

$$S_1 = (X\beta - y, X\beta - y) + \lambda(\beta,\beta), \quad \lambda > 0, \tag{1.9}$$

where λ is a positive parameter. The vector $\hat{\beta}$ which minimizes S_1 satisfies the linear algebraic system

$$(X^T X + \lambda I)\ \hat{\beta} = X^T\ y. \tag{1.10}$$

Adding λ to each element down the main diagonal of the matrix $X^T X$ helps remove the ill-conditioning of the matrix $X^T X$ by increasing each eigenvalue (including the smallest of the matrix $X^T X$) by λ, which increases the value of the determinant. When λ is zero, we have the original ordinary least squares problem which may be difficult to solve numerically; when λ is greater than zero, we have moved away from the original problem, but have an easier task computationally [7]. The approaches which we propose later include ridge regression as a special case.

Let us suppose that we wish to find a vector β so that $X\beta$ is close to y and at the same time β is close to a vector c, where c is a "current best estimate" of the true vector of regression coefficients. The vector c may come from previous calculations, an "equal weights" hypothesis [1], or be set equal to the null vector as in ridge regression. This leads us to the problem of finding the vector β which minimizes the quadratic expression

$$R_N(\beta) = \lambda\ (X\beta - y, X\beta - y)\ +\ \lambda(\beta - c, \beta - c). \tag{1.11}$$

In this case the vector which minimizes is the solution of the linear algebraic system

$$(X^T X + \lambda I)\ \hat{\beta} = X^T\ y + \lambda\ c, \tag{1.12}$$

which shows that $\hat{\beta}$ depends linearly on the observation vector y.

In the following pages we shall indicate how the minimization of $R_N(\beta)$ can be achieved *without* solving the linear algebraic system in Eq. (1.12). This involves use of the sequential methods of dynamic programming [15] and should permit use of smaller values of λ than would be possible using routine methods of solving the linear algebraic system in Eq. (1.12). In addition two methods will be given for characterizing the solution of Eq. (1.5) (which is difficult to obtain computationally) in terms of solutions of Eq. (1.12) (which are relatively easy to obtain using dynamic programming). It is of course important to have both of these methods lead to the same results to have trustworthy computed values. Earlier regression studies may not pass this test. Other approaches are presented in [16].

2. DYNAMIC PROGRAMMING SOLUTION FOR FIXED VALUES OF λ AND C

Our task is to find the vector β which minimizes the quadratic expression

$$R_N(\beta) = (X\beta-y, X\beta-y) + \lambda(\beta-c, \beta-c), \tag{2.1}$$

where β is a column vector with N rows, X is an $L \times N$ rectangular matrix, y is a column vector with L rows, λ is a positive scalar, and c is a column vector with N rows. Consider the more general problem of minimizing the quadratic expression

$$R_M(\beta) = \lambda (\beta_1-c_1)^2 + \lambda(\beta_2-c_2)^2 + \ldots + \lambda_M(\beta_M-c_M)^2 + \sum_{i=1}^{L} (\sum_{j=1}^{M} x_{ij} \ \beta_j - y_i)^2. \tag{2.2}$$

When M = N, we have the original problem; when M = 1, we have the simple problem of minimizing

$$R_1(\beta) = \lambda(\beta_1-c_1)^2 + \sum_{i=1}^{L} (x_{i1} \ \beta_1-y_i)^2. \tag{2.3}$$

Let $x^{(M)}$ be the column vector which is the Mth column of the matrix X,

$$x^{(M)} = \begin{bmatrix} x_{1M} \\ \vdots \\ x_{LM} \end{bmatrix}. \tag{2.4}$$

Now let us define the functions

$$f_M(y) = \min_{\beta} R_M(\beta), \quad M=1,2,\ldots,N, \tag{2.5}$$

so that these functions represent the minimum values of R_M viewed as functions of y for $M=1,2,\ldots,N$.

The principle of optimality [15] enables us to relate the function f_M to the function f_{M-1}. This equation is

$$f_M(y) = \min_{\beta_M} \left[\lambda(\beta_M-c_M)^2 + f_{M-1}(y-\beta_M \ x^{(M)}) \right], \quad M = 2,3,\ldots,N. \tag{2.6}$$

In order to use this recurrence relation we use the fact that f_M is actually quadratic in y and can be written

$$f_M(y) = (y, Q_M y) + 2(p_M, y) + r_M, \tag{2.7}$$

where Q_M is a symmetric L×L matrix, p_M is a vector with L rows, and r_M is a scalar. Some simple but tedious algebraic manipulations then enable us to produce the recurrence relations

$$Q_M = Q_{M-1} - \frac{A^{(M)}}{\lambda + K_M}, \tag{2.8}$$

$$p_M = p_{M-1} - \frac{(\lambda c_M + \rho_M)\ \alpha^{(M)}}{\lambda + K_M}, \tag{2.9}$$

$$r_M = r_{M-1} + \lambda\ c_M^{\ 2} - \frac{(\lambda c_M + \rho_M)^2}{\lambda + K_M}, \tag{2.10}$$

where the auxiliary quantities are given by

$$\alpha^{(M)} = Q_{M-1}\ x^{(M)}, \tag{2.11}$$

$$\rho_M = (p_{M-1}, x^{(M)}), \tag{2.12}$$

$$K_M = (\alpha^{(M)}, x^{(M)}), \tag{2.13}$$

and the matrix $A^{(M)}$ is

$$A^{(M)} = \alpha^{(M)}\ (\alpha^{(M)})^T. \tag{2.14}$$

By taking

$$Q_o = I,\ p_o = 0,\ r_o = 0, \tag{2.15}$$

we can use these recurrence relations to obtain Q_M, p_M and r_M for $M = 1,2,\ldots,N$. The minimizing value of β_N is given by

$$\beta_N = \frac{\lambda c_N + \rho_N + (\alpha^{(N)}, y)}{\lambda + K_N}, \tag{2.16}$$

and all of the elements of the right hand side are known computationally. Similarly β_{N-1}, $\beta_{N-2}, \ldots, \beta_1$ can be determined [3].

The stability and accuracy of such a dynamic programming calculation are well known [14], and successful calculations for a process of this type are reported in [1,3,4]. Much remains to be done with regard to writing of computer programs, testing of the programs, and improvement of the algorithms with regard to accuracy and speed. However, the basic experiments which were done at RAND over a decade ago leave little doubt about the superiority of this approach over routine use of Gaussian elimination, Cholesky decomposition and related methods.

3. SUCCESSIVE APPROXIMATIONS

The vector c in Eq. (2.1) represents a "current approximation" to $\hat{\beta}$, and the vector β which minimizes $R_N(\beta), \beta^{(1)}$, is an improved approximation to $\hat{\beta}$ for a fixed value of $\lambda > 0$. We may now replace c in Eq. (2.1) by $\beta^{(1)}$, use dynamic programming to minimize again, and call the minimizing vector $\beta^{(2)}$. This process may be repeated again and again to produce the sequence of vectors $\beta^{(1)}, \beta^{(2)}, \ldots$. It can be shown that this sequence of vectors converges, that the limit is given by

$$\lim_{n\to\infty} \beta^{(n)} = \hat{\beta} , \tag{3.1}$$

and that the smaller λ is, the faster is the convergence. This points up the importance of using dynamic programming to solve the basic minimization problem, for we expect to be able to use smaller values of λ with it than with conventional methods.

Thus the use of successive approximations provides one method for obtaining $\hat{\beta}$ accurately.

4. ANALYTIC CONTINUATION

Now let us consider a second method for obtaining $\hat{\beta}$ numerically. The vector which minimizes in Eq. (2.1) satisfies the linear algebraic equations

$$(X^TX + \lambda I)\ \beta = X^T y + \lambda c, \tag{4.1}$$

the solution of which may be represented in the form

$$\beta = (X^TX + \lambda I)^{-1} (X^T y + \lambda c). \tag{4.2}$$

From this we see that as a function of λ, β may be written into the form of a convergent power series

$$\beta(\lambda) = \hat{\beta} + v_1 \lambda + v_2 \lambda^2 + \ldots , \tag{4.3}$$

where $\hat{\beta}$ satisfies

$$(X^T X)\hat{\beta} = X^T y , \tag{4.4}$$

and $v_1, v_2, \ldots$ are certain vectors.

We may use this information as follows. For several small values of λ we compute the solutions of the minimization problem in Eq. (2.1). Then by running parabolas through these points for each component of $\beta(\lambda)$ we can estimate $\hat{\beta}$, v_1, and v_2. Or more sophisticated methods could be used. In any event we characterize $\hat{\beta}$ in a second manner as

$$\lim_{\lambda \to 0} \beta(\lambda) = \hat{\beta}, \tag{4.5}$$

which provides a second independent method for obtaining $\hat{\beta}$.

5. NUMERICAL EXPERIMENTS

Numerical experiments have been conducted for studying regression problems via ordinary least squares, ridge, and optimal filtering analyses, and to compare two computational methods of solution, Gaussian elimination and dynamic programming. The test problem is to estimate the column vector β, of dimension 4, in the equation,

$$y \cong X\beta + \varepsilon,$$

where X is a matrix of dimension 25×4, y is a column vector of dimension 25, and ε is a column vector of errors and has dimension 25. We generated the matrix X in this way. One hundred random numbers were produced with zero mean and standard deviation unity. The first 75 numbers were entered into the first three columns of X. For each entry in the fourth column, we took the sum of the corresponding elements in the first three columns plus a constant γ times a random number, thus using the remaining random numbers. The constant γ could be set at values between and including zero to one. The matrix X becomes singular when $\gamma = 0$, so we can regard γ as a kind of a conditioning parameter. The true regression coefficients were chosen to be $\beta_1 = 1$, $\beta_2 = 0$, $\beta_3 = 2$, $\beta_4 = 1$. For the dependent vector y, we produced $X\beta$. The vector ε is zero unless otherwise stated.

In ordinary least squares analysis, the aim is to minimize the quantity

$$S = (X\beta - y,\ X\beta - y), \tag{1.3}$$

by solving the linear algebraic system

$$X^T X\beta = X^T y. \tag{1.5}$$

Both a standard matrix inversion routine based on Gaussian elimination and a dynamic programming were used in our experiments to solve (1.5).

In ridge analysis, the task is to minimize

$$S_1 = (X\beta-y\ ,\ X\beta-y) + \lambda(\beta,\beta) \tag{1.9}$$

for $\lambda > 0$. the linear system to be solved is

$$(X^TX + \lambda I)\beta = X^Ty. \tag{1.10}$$

Again, a standard Gaussian elimination method and dynamic programming were used.

In our optimal filtering analysis we minimize

$$R_N = (X\beta-y,\ X\beta-y) + \lambda(\beta-c,\ \beta-c), \tag{1.11}$$

where c is an initial estimate of the solution. We employed the dynamic programming algorithm of Eqs. (2.8)-(2.16) for solving this minimization problem, though other methods could have been used. When the parameter λ is set equal to zero, we have the ordinary least squares problem. When the vector c is set equal to zero, for $\lambda>0$, then we have the ridge problem.

We now present some numerical results. First, we solved the least squares problem of Eqs. (1.3) and (1.5) by Gaussian elimination and also by dynamic programming. We set $\lambda=0$ in the dynamic programming calculation. We studied the solutions produced as the conditioning parameter γ was decreased. The interesting results are shown in Table I. They cover the range of $\gamma = 10^{-5}$ to 10^{-8}. For the larger values of γ, the problem is apparently easy to solve and both methods reproduce the true value, $\beta_1=1$. As γ is decreased, the estimates erode in accuracy, with the dynamic programming solution being the more accurate. In the interval where γ is just slightly larger than 10^{-8}, a tremendous change in the ill-conditioning takes place and both methods give dismal results. Let us now study this interesting region more closely.

We select the value of the conditioning parameter, $\gamma = 0.4\times10^{-7}$. We perform a ridge analysis, using both the Gaussian elimination and the dynamic programming methods of solution. For values of λ greater than or equal to 10^{-11}, both give basically the same results. See the first line in Table II. Now look down the first column of figures obtained by the Gaussian elimination method. The estimates of β_1 get worse. Next, look down the column of figures obtained in the first approximation of dynamic programming. The estimates improve dramatically.

TABLE I

Numerical Results for Least Squares Analysis

Values of β_1

Conditioning Parameter γ	Gaussian Elimination Method	Dynamic Programming Method
1.0×10^{-5}	1.0000	1.0000
1.0×10^{-6}	.9973	1.0003
1.0×10^{-7}	.8101	1.0220
0.4×10^{-7}	.2387	1.1569
1.0×10^{-8}	- .7241	-0.8721

TABLE II

Numerical Results for Ridge Analysis

$\gamma = 0.4 \times 10^{-7}$

Values of β_1

Ridge Parameter λ	Gaussian Elimination Method	Dynamic Programming	
		1st Approximation	20th Approximation
10^{-11}	0.4994	0.5005	0.5100
10^{-12}	.4952	.5034	.5624
10^{-13}	.4627	.5327	<u>.8197</u>
10^{-14}	.3219	.8502	1.2778

Now look at the third column for results obtained after twenty successive approximations. The underscored entry is much better than that from the first approximation, and better than that obtained via Gaussian elimination. In the calculation of the last entry, we believe that numerical errors have come into play.

It is interesting to see that these values of λ are many orders of magnitude smaller than those commonly encountered in ridge regression studies to date.

6. DISCUSSION

The seriousness of the problem of numerical determination of regression coefficients and the promising palliatives suggested in previous sections suggest that the following studies should be carried out:

1. Develop and test a FØRTRAN IV program to carry out the dynamic programming solution of regression problems.
2. Re-examine selected earlier regression studies with this new tool to see if the earlier results stand up or not.
3. Improve the dynamic programming algorithm by careful study of the recurrence relations.
4. Investigate the statistical properties of the new estimators.
5. Study implications for ANOVA.
6. Develop the theory for generalized least squares estimation.
7. Study the connections with ridge regression.
8. Develop dynamic programming algorithms for other ways of handling prior information about the regression coeffiecients (e.g., that they may all be approximately equal, that they may have known signs, etc.).

REFERENCES

[1] BELLMAN, R., KALABA, R. and LOCKETT, J. (1965). Dynamic programming and ill-conditioned linear systems. *J. Math. Anal. Appl.*, **10**, No. 1, 206-215.

[2] MCDONALD, G. and SCHWING, R. (1971). Instabilities of regression estimates relating air pollution to mortality. *Gen. Motors Res. Lab., Res. Pub.* GMR-1124, Warren, Michigan.

[3] BELLMAN, R., KALABA, R. and LOCKETT, J. (1966). *Numerical Inversion of the Laplace Transform.* Elsevier, New York.

[4] BELLMAN, R., KALABA, R. and LOCKETT, J.(1965). Dynamic programming and ill-conditioned linear systems - II. *J. Math. Anal. Appl.*, **12**,393-400.

[5] PHILLIPS, D. (1962). A technique for the numberical solution of certain integral equations of the first kind. *J. Assoc. Comput. Mach.*, **9**,84-97

[6] TWOMEY, S. (1963). On the numerical solution of Fredholm integral equations of the first kind by the inversion of the linear system produced by quadrature. *J. Assoc. Comput. Mach.*, **10**, 97-101.

[7] TIKHONOV, A.(1963). On the solution of ill-posed problems and the method of regularization. *Dokl. Akad. Nauk SSR*, **151**,501-504 and (1963) **153**,49-52.

[8] FORSYTHE, G. (1953). Solving linear algebraic equations can be interesting. *Bull. Amer. Math. Soc.*, **59**, 299-329.

[9] LEHMAN, R. S. (1962). Dynamic programming and gaussian elimination. *J. Math. Anal. Appl.*, **5**, No. 1, 499-501.

[10] WAINER, H. Estimating coefficients in linear models: it don't make no nevermind. (Unpublished ms.)

[11] DAWES, R. and CORRIGAN, B. (1974). Linear models in decision making. *Psychological Bulletin*, **81**, 95-106.

[12] EINHORN, H. J. Equal weighting in multi-attribute models: a rationale, and example and some extensions. (Unpublished ms.)

[13] WILKS, S. S. (1938). Weighting systems for linear functions of correlated variables when there is no dependent variable. *Psychometrika*, **3**, 23-40.

[14] MALINVAUD, E. (1970). *Statistical Methods of Econometrics.* American Elsevier, N. Y.

[15] BELLMAN, R. and KALABA, R. (1965). *Dynamic Programming and Modern Control Theory.* Academic Press, N. Y.

[16] LAWSON, C. and HANSON, R. (1974). *Solving Least Squares Problems.* Prentice-Hall, Englewood Cliffs.

P.R. Krishnaiah, ed., APPLICATIONS OF STATISTICS

ON HIERARCHICAL CLASSIFIER THEORY AND INTERACTIVE DESIGN*

LAVEEN N. KANAL

Laboratory for Pattern Analysis, Department of Computer Science
University of Maryland, College Park, Maryland

and

L.N.K. Corporation, Silver Spring, Maryland

This paper summarizes aspects of some recent theoretical work on hierarchical classifiers, presents an overview of the conceptual and implementation design of MIPACS - the Maryland Interactive Pattern Analysis and Classification System - and briefly describes its application to the design of hierarchical classifiers for the classification of clouds in satellite imagery.

1. INTRODUCTION

One can consider a number of different paradigms for pattern recognition [1]. At present feature extraction-classification remains the basis of practically oriented machine recognition systems. The implementation of such systems involves several iterations of the design cycle with labeled or unlabeled samples.

Reasons for taking an automated, interactive, graphics-oriented approach to pattern analysis and classification were summarized in a survey and commentary on Interactive Pattern Analysis and Classification Systems (IPACS), which also described a number of representative IPACS then implemented or under development [2]. MIPACS, which is partially implemented at the Laboratory for Pattern Analysis of the University of Maryland, represents our attempt to achieve the capabilities desired in IPACS. In addition to proving its immediate usefulness in pattern analysis and classification studies, MIPACS has stimulated the theoretical investigation of multistage multiclass classification.

The following sections describe certain aspects of multistage multiclass classification and hierarchical classifiers, and present an overview of the conceptual and implementation design of MIPACS. As an example of its applications, a brief description of the use of MIPACS in the design of hierarchical and single stage classifiers for cloud classification is included.

*Research supported in part by the Control and Automation Branch, Engineering Division, National Science Foundation, under grants GK-41602 and GK-43323, and in part by the Mathematics and Information Sciences Directorate, Air Force Office of Scientific Research, Air Force Systems Command, USAF, under grant AFOSR 76-2901.

2. MULTISTAGE DECISION SCHEMES AND HIERARCHICAL CLASSIFIERS

A state-space model for multistage multiclass classification has been formulated. The model is general enough to describe many parametric and nonparametric schemes for pattern classification. Systematic approaches to the design of decision trees in terms of the design of the tree skeleton, the allocation of features to nodes, and the construction of decision policies have provided some new insights on the design of hierarchical classifiers [3, 4, 5]. This section and Section 3 summarize the motivation and some of the results of this work.

Measurement cost, decision complexity and the relationships between measurement complexity, sample size and error probability place an upper bound on the performance of practical classification schemes [4, 6]. The interaction between these factors makes it necessary that one limit the number of features used in the classifier. Since the goodness of the features depends on the form of the classifier with which they will be used, feature selection and classifier design should be closely coupled. Thus distance measures and related evaluation criteria, which attempt to evaluate features without taking into account the particular classifier structure in which the features are to be used cannot be depended upon and are best treated as aids to further exploration [1].

The tradeoffs between performance and complexity underlie practical approaches to classifier design. The design of the classifier consists of estimating parameters of the various discriminant functions that may be employed, and designing the decision logic (which uses the outputs of the discriminant functions to classify a sample) so as to satisfy classifier performance requirements in the field.

All the problems of feature selection and classifier design are accentuated in the multiclass multimodal case. For example, in an M class problem, for a given sample size per class, dimensionality sample size considerations may dictate that no more than n features be used. It is likely that the best set of n features to discriminate between one group of classes is quite different from the best n features for another set of classes. Limiting consideration to some n features to be used in a single step to make an M-way decision may lead to a forced compromise in the choice of the feature set. Splitting the decision making into several stages, with subsets of the entire set of classes being handled at any stage, might be better. Then at each stage of the decision process, the n features best suited for that classification task could be used. This permits the multiclass classification problem to be decomposed into a series of less complex decision problems.

The published work on methods of designing multistage decision schemes in this context may be categorized into three broad groups: (a) conversion of decision tables to optimal decision trees [7, 8, 9, 10, 11]; (b) sequential pattern classification methods [12, 13]; (c) hierarchical classification methods [1, 14,

15, 16, 17, 18, 19]. The algorithms for converting decision tables to optimal decision trees do not address the problems of feature selection and misclassification error. Designing the decision table is itself the difficult part of the process in pattern classification, since this determines the error rate. Sequential schemes are differentiated from hierarchical classifiers in the ordering imposed on the sequence of feature measurements and on the possible class labels. Sequential pattern recognition schemes are characterized by a linear ordering on the features. In most cases no particular order is imposed on the class labels--any class label may be accepted at any stage. Hierarchical classifiers are characterized by the hierarchical ordering imposed both on the features and the class labels. Some classes are rejected from further consideration as candidates for the test sample's label, at each stage of the measurement process.

In [3, 4], it was shown that many parametric and nonparametric classification schemes can be described in terms of a general theoretical model--a state space graph model for multistage multiclass classification. An example graph is shown in Figure 1. A state consists of a measurement set, and a set of possible class labels. An edge in the graph has an associated measurement cost and represents the action of observing a feature or a set of features. A goal state is any state with one or zero (reject) class labels as the set of possible classifications. The cost of decision making, i.e. reaching a goal state, is assumed to be a weighted sum of measurement cost and misclassification risk.

Two types of admissible search strategies are defined. An S-admissible strategy finds the optimal goal node, i.e. category, when the goal risk is a function of only the feature measurements on the paths to that node in the graph. A B-admissible strategy terminates with the optimal solution when the risk depends on all the observations. Bayes optimal strategies are shown to be implementable as a special case of a B-admissible search. Both strategies use bounds on the misclassification risk to direct the search in the state-space. Methods of computing upper and lower bounds on the risk, given a subset of the total measurement set, are derived for various parametric and nonparametric schemes.

A hierarchical classifier is a particular type of state-space graph in that a single path is followed to classify a test sample and at each stage some classes are rejected from further consideration as possible class labels for the sample. The application of optimization methods, such as dynamic programming and branch and bound, in a systematic approach to the design of hierarchical statistical classifiers is investigated in [4, 5] where some new methods of reducing the complexity of the design algorithms, at the expense of some degradation in performance, are developed.

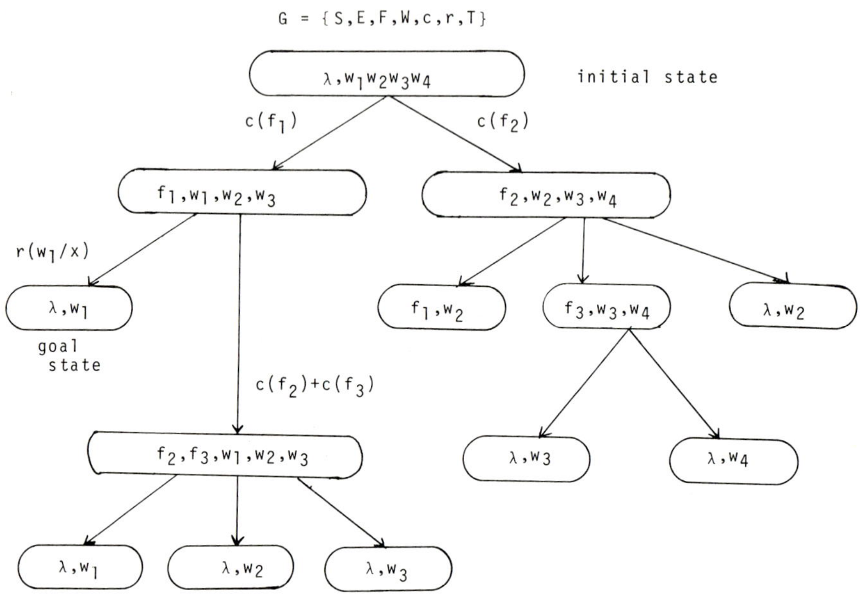

S = Set of states (nodes)
E = Set of edges
c = Cost function on edges
r = Risk function on goal nodes
F = Feature set
W = Class label set
T = Strategy function
s = (Fs,Ws) where $Fs \subseteq F$, $Ws \subseteq W$
λ = Reject class label

Fig. 1. State space model of multistage classification.

3. AUTOMATED OPTIMAL TREE DESIGN

A decision tree T consists of a root node ℓ_1, nonterminal nodes ℓ_i, and terminal nodes. Each terminal node is labeled with a class label. Note that the same class label can occur at more than one terminal node. The sample to be classified is subjected to a sequence of decision rules, i.e. tests, on the path from the root to a terminal node. The choice of the next edge to be traversed, i.e. the next test, may depend on the last test, or more generally on all the measurements taken on the sample.

Two results show that the design of optimal hierarchical classifiers is not a simple task even under highly simplifying assumptions. For statistically independent features at each node of a decision tree, it is easily shown that optimizing the average correct recognition rate at each node does not necessarily lead to an overall optimal tree. For classification in w_i, $i = 1\ldots M$, the overall tree performance, defined by $P_c(T)$, the probability of correct recognition, is given by

$$P_c(T) = \sum_{i=1}^{m} P(w_i)P_c(w_i)$$

where $P(w_i)$ is the a priori probability of class w_i and $P_c(wi)$ is the probability of correctly recognizing class w_i. For statistically independent features $P_c(w_i)$ can be written as the product of the probabilities $P(w_i/\ell_{k_j})$ for correctly recognizing class w_i at nodes ℓ_{k_j} on a path from the root to terminal node w_i. Thus $P_c(w_i) = \prod_j P_c(w_i/\ell_{k_j})$. Letting $W(\ell_{k_j})$ be the set of all terminal nodes below ℓ_{k_j}, the average correct recognition rate $P_c(\ell_{k_j})$ at node ℓ_{k_j} is

$$P_c(\ell_{k_j}) = [\sum_{w_i \varepsilon W(\ell_{k_j})} P(w_i).\ P_c(w_i/\ell_{k_j})]/\sum_{w_i \varepsilon W(\ell_{k_j})} P(w_i)\ .$$

Optimizing $P_c(\ell_{k_j})$ does not necessarily optimize $P_c(T)$.

A second result is that even for statistically independent features, the feature assignment (at each node of a given tree structure) which gives the node decision with the least average error, is not necessarily the overall optimal assignment of features. The conclusion is that, in general, maximizing classification performance at each individual node of a decision tree does not result in an overall optimum decision tree.

The design of an optimal decision tree may be viewed as a search problem involving (a) specifying a tree skeleton, i.e. the form of the tree along with the terminal node labels; (b) assigning features to individual nodes, and (c) specifying decision rules at each node. Finding the optimal decision tree by exhaustive search over all possible tree skeletons, feature assignments to each tree, and decision rules for each assignment, will be infeasible for most problems. Employing dynamic programming and a phased approach to tree design, which takes into account the degree of a priori information available to the designer concerning the class hierarchy, i.e. the tree skeleton, and the features to be used at the nodes, the computational complexity can be considerably reduced [4]. But the 'curse of dimensionality' associated with dynamic programming procedures remains.

Some new suboptimal techniques for tree design allow further reduction. These involve:

(a) clustering decision rules applicable to each node using similarity measures defined for decision vectors, representing each cluster by a prototype, and using dynamic programming to search over the set of prototypes for the optimal rule among the prototype rules;

(b) discarding rules which are not 'efficient'. A rule is efficient at a node if there is no other rule which dominates it. It is proved in [4] that the set of rules d_i^*, i = 1,2...N, which maximize the overall tree performance $P_c(T)$, must be efficient at the corresponding nodes ℓ_i, i = 1,2...N, of the tree. By finding, and then rejecting, rules which are dominated by a given rule d_i, efficient rules are found;

(c) using a branch and bound strategy for assigning features to nodes of a given tree skeleton in a sequential, top down manner.

Investigation of the above suboptimal techniques to automated tree design has thus far been done in limited contexts under simplifying assumptions, primarily to gain the kind of general insights available from theoretical analysis. In practice such techniques are best viewed as subroutines to be utilized in a recursive, interactive approach to the design of hierarchical classifiers. MIPACS was designed to facilitate such an interactive approach.

4. MIPACS

The objectives of MIPACS are: to analyze data sets, with particular interest in contrasting several data sets by applying known statistical, and mathematical tools and by experimenting with various representations of data, particularly graphical representations; and to synthesize classification logics, by using standard decision-theoretic techniques, trial and error, and empirical techniques. MIPACS shares some goals and procedures with well-known statistical packages such as BMD and SPSS, but differs greatly in its emphasis and capability in (a) interaction, (b) data representation, and (c) design of classification logic. The takeoff point for our conceptual design of MIPACS was OLPARS (On Line Pattern Analysis and Recognition System) developed earlier at the U.S. Air Force's Rome Air Development Center.

MIPACS is presently implemented on a DEC GT 40 refresh display connected to a Univac 1106/1108 System. Currently effort is directed at implementing it in a computer environment consisting of a GT 40, a PDP 11/45 and the UNIX operating system.*

MIPACS supports the creation and interpretation of hierarchical classifiers with a flexible choice of features, classes and decision policies at the branch points. For multiclass classifier design, MIPACS currently permits one to (a) structure sets of data in a tree according to a priori knowledge--this is the extensive structure; (b) design a decision tree on the data by selecting and using a number of statistical decision theoretic and graphical-empirical techniques--this is the intensive classification structure; and finally (c) to test the resulting classifier and iterate the entire process. The entire procedure is interactive with graphic output and can be iterated in a short time to give (1) either a satisfying hierarchical decision logic, or (2) the conclusion that the data categories are intensively insufficiently discriminable.

MIPACS uses five different data representations. These are: (a) a universe of raw data objects--arrays of measurements or samples; (b) a corresponding universe of extracted feature vectors; (c) a tree structuring on sets of preclassified data objects (in the case of initially unknown classes, all data is input with a single class label); (d) a decision tree or a tree structure of sets of intensively classified data objects; and (e) a graphic representation of an object or set of objects. For each of the different data representations, there exists a MIPACS file for storing it. The five file types are briefly described next [20].

1. R$ - Raw Vector File (Fig.2)

A file of variable length vectors of measurements taken from the objects. These measurements may have spatial or temporal ordering as in the cases of sampled waveforms or images, but need not. Each vector has an a priori, (extensive) label assigned to it even if there is only one class.

2. F$ - Feature Vector File (Fig.3)

A file of fixed length (d less than 50) vectors of features extracted from the raw vectors of R$. Each features is real valued and the order of features is not important. Each feature vector is labeled the same as the raw data vector from which it was extracted.

3. S$ - Extensive Structure File (Fig.4)

A file which contains the user's a priori structuring of data objects. No data vectors are present--only a tree of sets and statistical information associated with each set. Sets are represented by bit maps which specify exactly which vectors of the feature vector universe belong to the set.

*A version of MIPACS now exists at the Dept. of Medical Informatics of the Free University of Amsterdam on a PDP 11/70 system.

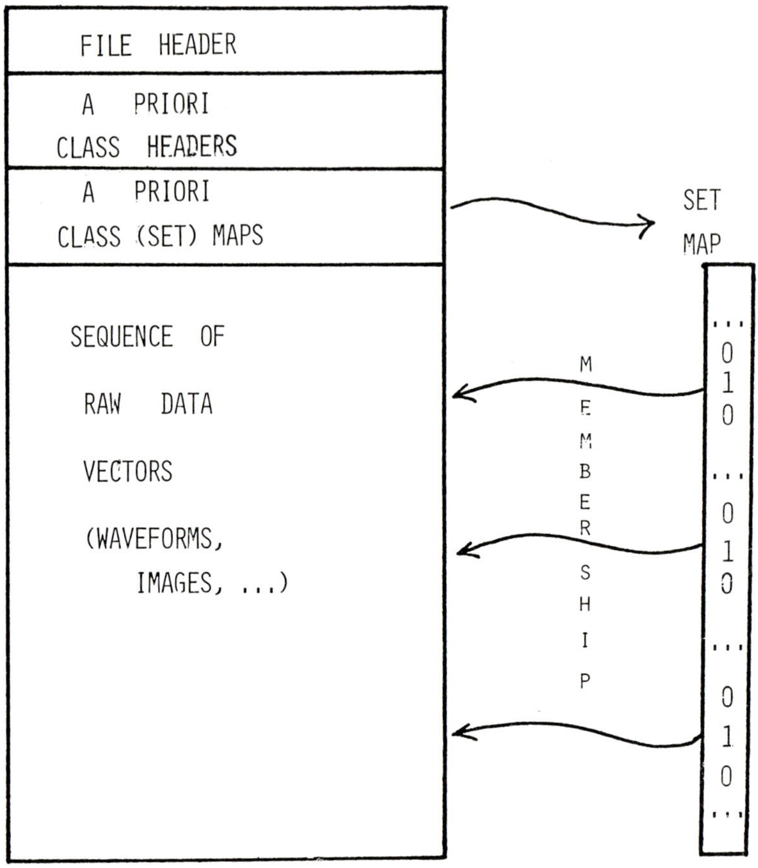

Fig. 2. Raw data file.

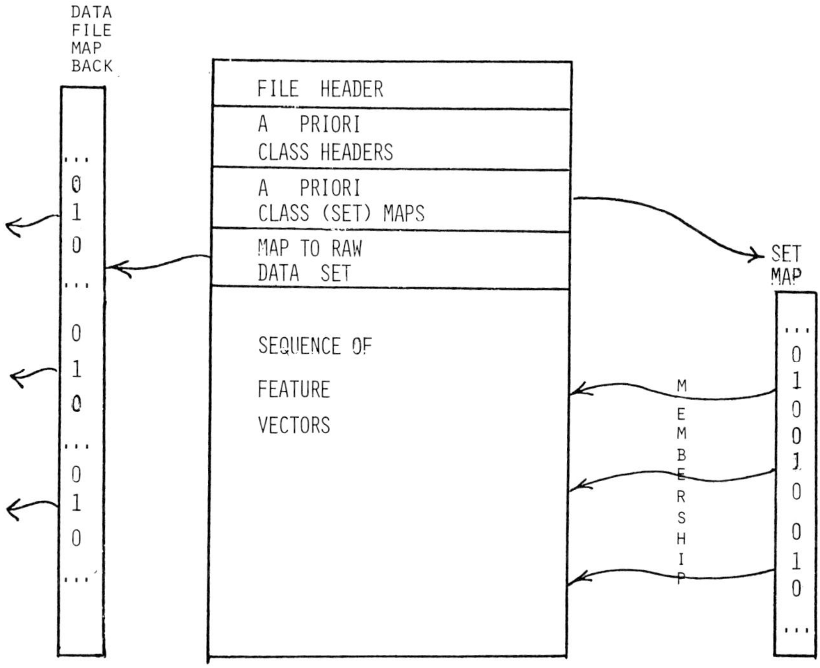

Fig. 3. Feature vector file.

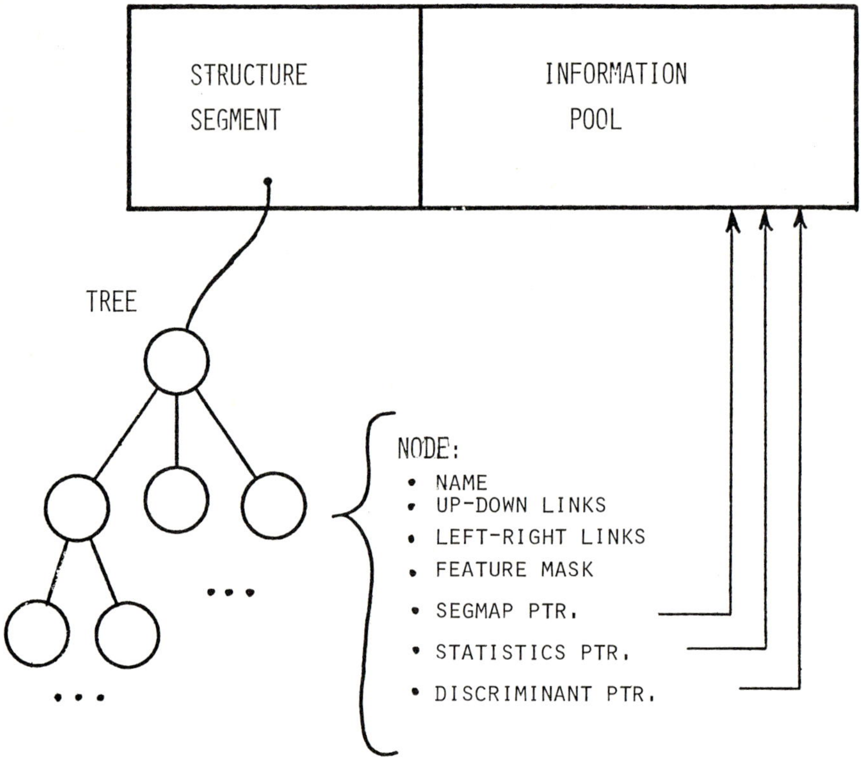

Fig. 4. Structure file.

4. *D$ - Decision File, or Intensive Structure File*

A file which contains a decision tree. At each node of the tree, a classification (branching) function is stored which can be interpreted by MIPACS to classify single vectors. No set maps and no statistics are stored in a decision file since the decision tree does not necessarily relate to any data. However, when MIPACS classifies a universe of feature vectors using the tree, sets of objects will be formed at each non-root node. (i.e. set-maps will be created). Now the file will be called an intensive structure file since it does structure a universe of data objects.

5. *G$ - Graph File*

A file containing an abstract representation of a graph or picture which a MIPACS processor has produced for the user. The purpose of the graph file is to preserve in permanent form a graph which the user may want to save or output to one or more plotting devices.

4.1. Pattern Analysis Using MIPACS

Pattern analysis research using MIPACS consists of transformations of the data, proceeding for example as shown in Figure 5, through the different data representations defined by R$, F$, S$ and D$, all of which may be viewed graphically via G$.

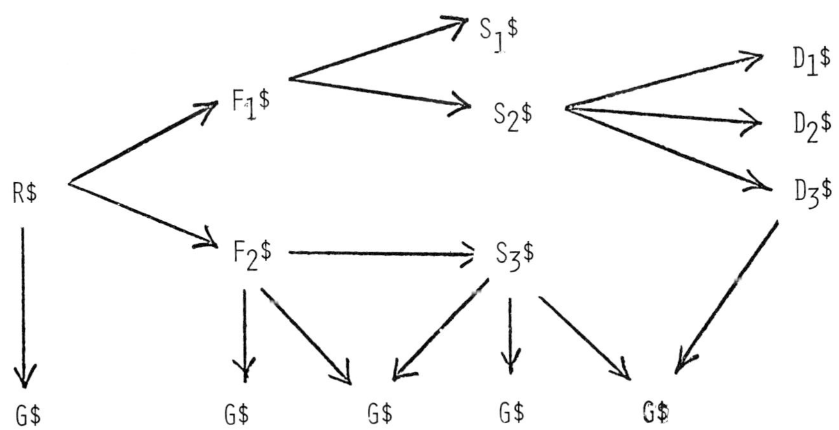

Fig. 5. Pattern analysis research using MIPACS.

The role of MIPACS processors is readily defined in terms of the data transformations they perform. Fig. 6. gives examples of the types of processors--feature extractors, mappers, classifiers, etc. Examples of specific processors might be fourier series expansion as feature extraction, Fisher discriminant as a mapping, etc. A run consists of a series of calls to MIPACS processors, each one calling for a transformation of data from one representation to another, or the inference of a classification function, or the display of a graph. With two exceptions, the result of all processing is saved in MIPACS files on back-up store. This enables independent execution of processors, which in turn implies safety from system crashes, possibility for parallel processing, and smaller core requirements. The exceptional processors are the Command Language Interpreter which manages all interactive user input and the viewing processor which manages output of graph files G$ to the CRT.

4.2. *Inferring and Evaluating Classifiers.*

MIPACS allows one to construct decision trees as follows. The skeleton of the tree can be specified as any legitimate tree structure. Sets of sample feature vectors are assigned to the terminal nodes of the tree by the user. Once terminal sets are assigned, a set is assigned to each node of the tree by successively unioning the sets of successors up the tree. To produce the decision tree from this hierarchy of sets, a branching function must be assigned to each node of the tree.

MIPACS offers several automatic methods and several empirical methods of computing branching functions for a given node of the tree using the training samples derived from the sets at the successor nodes. If there are two successors, 2-class discriminants apply (Figure 7). If there are n successors, multiclass methods must be employed to give an n-way branching function.

For each node of the tree, the user is free to choose the type of branching function he wants to try, i.e. Fisher discriminant, maximum likelihood classification function, etc., subject only to the 2-class or multiclass restriction imposed by the degree of the node. The user must also declare for each branching function what features will be used in the computation of the function. He does this by setting a feature mask at each successor node of the mode to which the branching function will be assigned.

The training sample assigned to any node may also be projected onto one or two space using a variety of mappings, e.g. the optimal discriminant plane, principal components, nonlinear mapping, which are summarized in [2]. The classification surface is then specified graphically on the CRT display.

Another approach is to allow the user the capability of defining procedural discriminants in terms of the features, using a procedural language. The language permits the construction of decision trees, or programs, using the Fortran

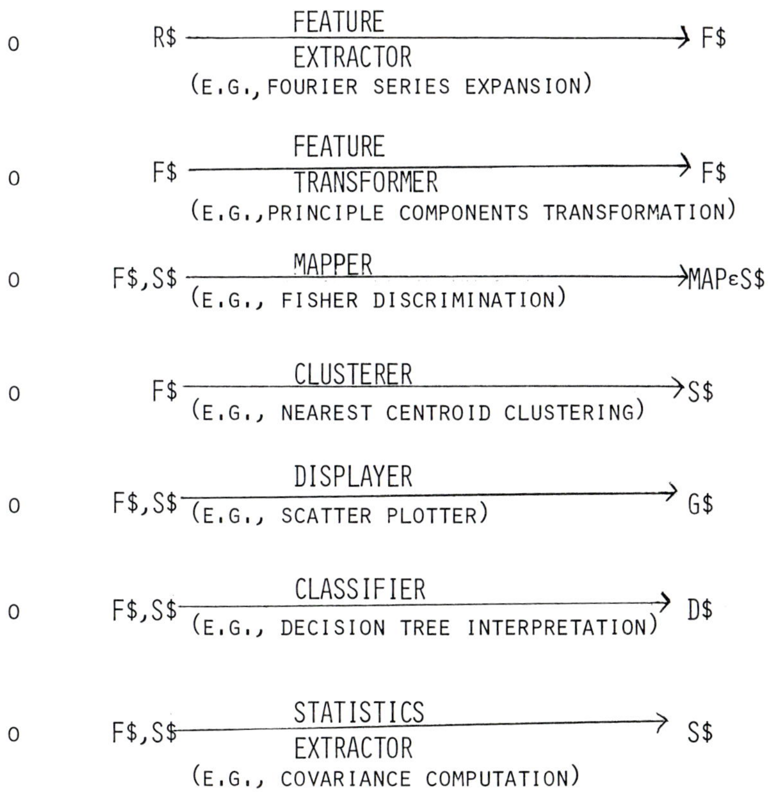

Fig. 6. MIPACS processors defined by data I/O.

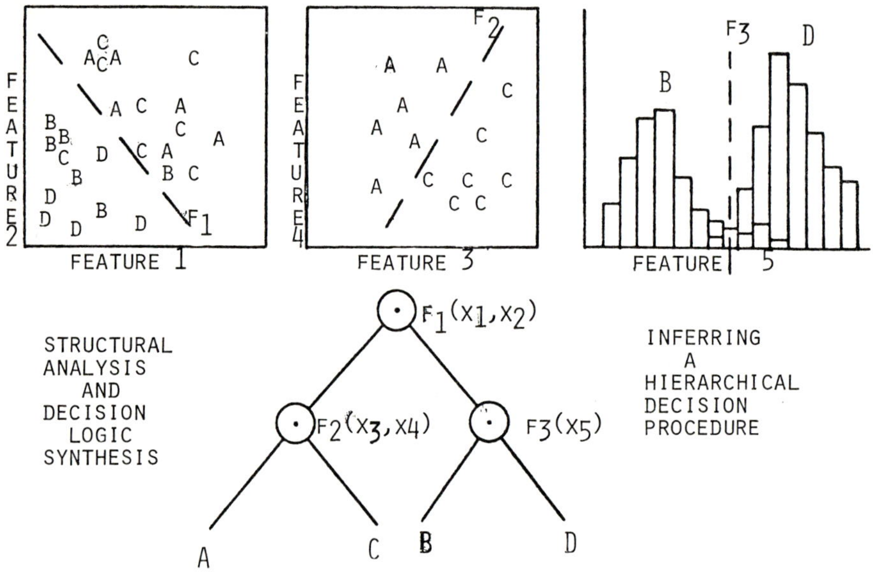

Fig. 7. Inferring a decision tree.

IF statement. Statement labels in such a program are symbolic, not numeric and the consequent of an IF clause can be either a branch or a classification. All variables in arithmetic or boolean expressions in the IF clause must be features in the feature vector. For example, the following procedure defines a 3-way branching function.

```
          IF(F1.GT.F3) GOTO BORC
          CLASS 1
     BORC IF(F2**2,LE.F3/(F4+1).AND.F1.LT.0) CLASS 3
          CLASS 2
```

Once branching functions have been assigned to each non-terminal node of the tree, the tree design is complete. At that time, the sample sets of vectors at the nodes which have been used to construct the branching functions are no longer relevant. The decision tree can now be applied to any set of sample data to create new sets of vectors (specified by setmaps) at the nodes of the tree. A confusion matrix is automatically supplied by computing the intersection between all a priori sets of the test samples and all a posteriori sets of the decision structure.

In the current terminal version of MIPACS, the user has the following options for designing and testing decision trees [21]:

(1) Using the program MIPACS*MIPACS.INPUT or the MIPACS option NEWTRE the user can define an arbitrary tree skeleton and assign sets of samples to its nodes.

(2) Using MIPACS options TREOPS and NEWMSK the user can indicate what features are to be used by the branching functions to be assigned to the decision tree nodes.

(3) Using MIPACS options DESIGN or STEVAL the user can automatically or interactively design branching functions at each non-terminal node of the tree.

(4) Using MIPACS option CLASFY the user can test his decision tree by using it to classify some structured set of data--either the same data used to train the branching functions on the decision tree, or some other set of test data.

5. CLOUD PATTERN CLASSIFICATION WITH MIPACS

A sample run of the terminal version of MIPACS on NOAA-1 geosynchronous satellite VISSR (Visible and Infrared Spin Scan Radiometer) image data, is shown in Display 1. References [22] and [23] contain an extensive report on results of similar experiments using MIPACS. The purpose of these experiments was to design and evaluate single stage and multistage decision trees for classifying sample areas selected from VISSR images into the following categories: (1) cumulonimbus clouds; (2) low clouds (cumulus, cumulus congestus, stratocumulus, and stratus); (3) cirrus clouds; (4) mixed clouds (cirrus with low clouds, cirrus with middle clouds, and cirrus with low and middle clouds).

DISPLAY 1

CLOUD DATA EX. 1 OF 5

```
BEGIN MIPACS
  3/FTEVAL  4/STEVAL  5/DEFINE  6/DESIGN  7/FTRANS
  8/FILOPS
*OPT*
5:

  3/TRESET  4/TREOPS  5/NEWTRE
*OPT*
4:

(FIL)FILNAM
CLDS:

 STRUCTURE OUTPUT,NODE INFO IS;NODE NAME,NVECTS,NFEATS
 TOP      243        47
     LOW        86        47
     OTH       157        47
          MIX        87       47
          RES        70       47
              HI        24        47
              CUM       46        47
```

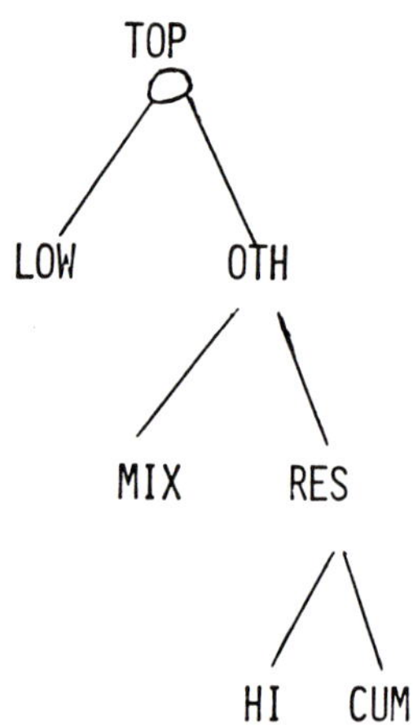

DISPLAY 1 (CONT'D)

CLOUD DATA EX. 2 OF 5

```
3/VIEWTR  4/VIEWND  5/VIEWST  6/VIEWFT  7/NEWSET
8/NEWMSK
*OPT*
8:
(NOD)NODNAM:(LST)IDLIST
LOW:1,2,3,4:
*OPT*
8:
(NOD)NODNAM:(LST)IDLIST
OTH:1,2,3,4:
*OPT*
8:
(NOD)NODNAM:(LST)IDLIST
MIX:2,3:
*OPT*
8:
(NOD)NODNAM:(LST)IDLIST
RES:2,3:
*OPT*
8:
(NOD)NODNAM:(LST)IDLIST
HI:2,3,4:
*OPT*
8:
(NOD)NODNAM:(LST)IDLIST
CUM:2,3,4:
```

DISPLAY 1 (CONT'D)

CLOUD DATA EX. 3 OF 5

```
*OPT*
3:

(NOD)NODNAM
TOP:

  STRUCTURE OUTPUT,NODE INFO IS;NODE NAME,NVECTS,NFEATS
  TOP      243       47
      LOW       86        4
      OTH      157        4
          MIX        87       2
          RES        70       2
               HI        24        3
               CUM       46        3
*OPT*
2:

  3/TRESET  4/TREOPS  5/NEWTRE
*OPT*
2:

  3/FTEVAL  4/STEVAL  5/DEFINE  6/DESIGN  7/FTRANS
  8/FILOPS
```

DISPLAY 1 (CONT'D)

CLOUD DATA EX. 4 OF 5

```
*OPT*
/DESIGN:
  3/FISHER  4/QUADRT  5/MAXLIK  6/VOTING  7/ONERST
  8/CLASFY  9/TRESET 10/TREOPS
*OPT*
/FISHER:

(FIL)FILNAM:(NOD)NODE   :(REA)PROB
CLDS:TOP:-.5:

  FISH. DIR. .32+00 .21+00-.65+00  .66+00
  LIN. THRESHOLD= -.1701+01
  FISHER DISTANCE=  .4105+00
*OPT*
/MAXLIK:

(FIL)FILNAM:(NOD)NODE
CLDS:OTH:

MAX-LIKELIHOOD SCHEME AT NODE OTH FOR 2 CLASSES
*OPT*
/QUADRT:

(FIL)FILNAM:(NOD)NODE  :(REA)PROB
CLDS:RES:0.4:

QUAD. THRESHOLD= .108+03
```

DISPLAY 1 (CONT'D)

CLOUD DATA EX. 5 OF 5

```
*OPT*
/CLASFY:

(FIL)SFILE :(NOD)SNODE :(FIL)DFILE :(NOD)DNODE
CLDS:RES:CLDS:RES:

  APRIORI CLASSES:  2  APOSTERIORI CLASSES:  2
  PRIORI/POST  HI    CUM
   HI          24     0
   CUM          0    46
*OPT*
/CLASFY:

(FIL)SFILE  :(NOD)SNODE  :(FIL)DFILE  :(NOD)DNODE
CLDS:TOP:CLDS:TOP:

  APRIORI CLASSES:  4  APOSTERIORI CLASSES:  4
  PRIOR/POST    LOW   MIX   HI    CUM
   LOW            7    62    0     17
   MIX            1    73    1     12
   HI             3    21    0      0
   CUM            0     5    0     41
```

While MIPACS speeds up the process of designing and evaluating decision tree logics, experience using it continues to reinforce the importance of theory and thought, and interaction with individuals having deep knowledge of the physical processes generating the data.

REFERENCES

[1] KANAL, L. (1974). Patterns in pattern recognition, 1968-1974. *IEEE Trans. Information Theory*, IT-20, 6, 697-722.

[2] KANAL, L. N. (1972). Interactive pattern analysis and classification systems: a survey and commentary. *Proc. IEEE*, 60, 10, 1200-1215.

[3] KANAL, L. N. and KULKARNI, A. V. (1976). Admissible strategies in a state-space model for multistage, multiclass classification. Abstract, *Proc. 1976 International Symposium on Information Theory*. Ronneby, Sweden.

[4] KULKARNI, A. V. (1976). Optimal and heuristic synthesis of hierarchical classifiers. Ph.D. dissertation, University of Maryland, Dept. of Computer Science Technical Report Series, TR-469.

[5] KULKARNI, A. V. and KANAL, L. N. (1976). An optimization approach to hierarchical classifier design. *Proc. 3rd International Joint. Conf. on Pattern Recognition*. San Diego, California. *IEEE Catalog #76CH1140-3C*.

[6] KANAL, L. N. and CHANDRASEKARAN, B. (1968). On dimensionality and sample size in statistical pattern classification. *Proc. National Electronics Conf.*, 2-7, NEC Inc., Oak Brook, Illinois.

[7] KNUTH, D. (1971). Optimum binary search trees. *Acta Informatica*, 1, 14-25.

[8] MEISEL, W. S. and MICHALOPOULOS, D. A. (1973). A partitioning algorithm with application in pattern classification and the optimization of decision trees. *IEEE Trans. Computers*, C-22, 93-103.

[9] POLLACK, S. L., HICKS, H. T. and HARRISON, W. J. (1971). *Decision Tables - Theory and Practice*. Wiley Interscience, New York.

[10] REINWALD, L. T. and SOLAND, R. M. (1966). Conversion of limited entry decision tables to optimal computer programs I - minimum average processing time. *JACM*, 13, 339.

[11] WINSTON, P. (1969). A heuristic program that constructs decision trees. MIT Project MAC, Memo No. 173.

[12] FU, K. S. (1968). *Sequential Methods in Pattern Recognition and Machine Learning*. Academic Press.

[13] WALD, A. (1947). *Sequential Analysis*. Wiley, New York.

[14] FRIEDMAN, J. (1975). A variable metric decision rule for nonparametric classification. SLAC-PUB-1573, SLAC, Stanford, California.

[15] FUKUNAGA, K. and NARENDRA, P. M. (1974). A branch and bound algorithm for computing k-nearest neighbors. *IEEE Trans. Computers*, C-24, 7.

[16] HART, P. (1969). Searching probabilistic decision trees. *AI Group Tech. Note No. 2*, SRI Project 7494, Stanford Research Inst., Stanford, California.

[17] MATTSON, R. L. and DAMMANN, J. E. (1965). A technique for detecting and coding subclasses in pattern recognition problems. *IBM Journal Res. and Dev.*

[18] NADLER, M. (1971). Error and reject rates in a hierarchical pattern recognizer. *IEEE Trans. Computers*, C-20.

[19] WU. C., LANDGREBE, D. and SWAIN, P. (1975). The decision tree approach to classification. TR-75-17, Purdue University.

[20] STOCKMAN, G. C. (1975). Maryland interactive pattern analysis and classification system, Part I: concepts. University of Maryland, Computer Science Technical Report Series, TR-408.

[21] STOCKMAN, G. C. and KULKARNI, A. V. (1976). Maryland interactive pattern analysis and classification system, Part II. user documentation. University of Maryland, Computer Science Technical Report Series, TR-482.

[22] PARIKH, J. A. (1976). Cloud pattern classification from visible and infrared data. University of Maryland, Computer Science Technical Report Series, TR-442.

[23] PARIKH, J. A. (1977). Automatic cloud classification and segmentation. Ph.D. dissertation, Department of Computer Science, University of Maryland.

P.R. Krishnaiah, ed., APPLICATIONS OF STATISTICS

TRANSFORMATIONS YIELDING RELIABILITY MODELS BASED ON INDEPENDENT RANDOM VARIABLES: A SURVEY*

N. LANGBERG
Florida State University, Tallahassee, Florida

F. PROSCHAN
Florida State University, Tallahassee, Florida

A. J. QUINZI
Florida State University, Tallahassee, Florida
and
Temple University, Philadelphia, Pennsylvania

This is an expository paper presenting various ways of transforming dependent models into independent ones and displaying applications in a variety of contexts including reliability modeling, life testing, and nonparametric estimation in the study of competing risks.

1. INTRODUCTION

The central theme of this survey is the transformation of dependent models into independent ones. By "dependent (independent) models" we mean multivariate stochastic models whose joint probability distribution are distributions of dependent (independent) random variables. Each of the transformations discussed here can be used to convert the original dependent model into an independent model which is equivalent (in a specified sense) to the original model. It is the purpose of this paper to present:

(a) some key theorems upon which such transformations are based, and

(b) a variety of applications in reliability and biometry.

We do not give formal proofs of the results presented; these may be found in the original papers cited. Rather, we motivate the key ideas by examining important special cases and several illustrative examples.

2. DISTRIBUTIONS WITH EXPONENTIAL AND PROPORTIONAL HAZARD MINIMA

In this section and the next we describe methods for converting dependent models into independent ones based upon the assumption that the joint distribution of the random variables in the original (dependent) model belongs to a specified family of distributions.

*Research supported by the Air Force Office of Scientific Research under AFOSR Grant 74-2581C.

We begin with some terminology and notation. A *life length* T is a nonnegative random variable such that $\lim_{t\to\infty} P(T > t) = 0$. Suppose that a system consists of n components with random life lengths $T_1,\ldots,T_n$. We say that the system is a *series (parallel)* system if the failure of the system coincides with the earliest (latest) component failure. Thus, the life length of the corresponding series (parallel) system is given by $\min(T_i, 1 \le i \le n)$ [$\max(T_i, 1 \le i \le n)$]. Series and parallel systems are examples of more general systems in reliability known as *coherent* systems [see Birnbaum, Esary, and Saunders (1961) or Barlow and Proschan (1975)].

A random vector $(T_1,\ldots,T_n)$ has *exponential minima* if $\min(T_i, i \in I)$ is exponentially distributed for every nonempty subset I of $\{1,\ldots,n\}$. In reliability terms, a random vector has exponential minima if the life length of every series subsystem (i.e., every series system which may be formed by using a subset of the n components) is exponentially distributed. In particular, individual components (series systems of size one) have exponential life lengths. The (n-dimensional) random vectors T and U are *marginally equivalent in minima* ($T \overset{m}{=} U$, in symbols) if $\min(T_i, i \in I)$ and $\min(U_i, i \in I)$ have the same distribution for each nonempty $I \subset \{1,\ldots,n\}$.

A particularly important multivariate distribution with exponential minima is the multivariate exponential (MVE) of Marshall and Olkin (1967). The classic paper of Marshall and Olkin (1967) and the model derived therein have prompted numerous investigations [see the annotated bibliography of Kotz (1974)]. The following characterization of the MVE is a particularly useful one.

Theorem 2.1. A random vector $(U_1,\ldots,U_n)$ has the (n-dimensional) MVE distribution if and only if there exists a collection H_J, $J \in \mathcal{J}$, $\cup\mathcal{J} = \{1,\ldots,n\}$, of independent exponential random variables such that $U_i = \min(H_J, J \in \mathcal{J}, i \in J)$, $i = 1,\ldots,n$.

Theorem 2.1 is an immediate extension of Theorem 3.2 of Marshall and Olkin (1967). To simplify notation we shall adopt the following conventions throughout the remainder of this paper. Let $\mathcal{J}$ denote the collection of all nonempty subsets of $\{1, \ldots, n\}$. Whenever an element $\{i_1,\ldots,i_m\}$ of $\mathcal{J}$ appears as a subscript, as, for example, in $H_{\{i_1,\ldots,i_m\}}$, we shall often write instead $H_{i_1 \ldots i_m}$. We say that two random variables X and Y are stochastically equal and write $X \overset{st}{=} Y$, if X and Y have the same probability distribution. Unless otherwise indicated, all random vectors are assumed to be n-dimensional.

Esary and Marshall (1974) prove the following:

Theorem 2.2. Suppose that a random vector T has exponential minima. Then there exists a random vector U with the MVE distribution of Marshall and Olkin such that $T \overset{m}{=} U$.

Theorem 2.2 can be used to obtain a consistent estimator for system reliability as the following example illustrates:

Example 2.3. Suppose that an estimate of system reliability for an arbitrary *coherent* system of n components is desired prior to manufacture of the system. Suppose also that the only available failure data, however, is for n-component *parallel* systems whose component life lengths have the same joint distribution as those of the given system. If component life lengths have exponential minima, then by Theorem 2.2 there is a random vector U with the MVE distribution such that $T \overset{m}{=} U$. Consistent estimators for the parameters of the MVE, given the failure data from parallel systems as above, have been obtained by Proschan and Sullo (1976). Since the reliability of the system can be expressed as a continuous function of survival probabilities $P[\min(T_i, i \in I) > t]$, $I \in \mathscr{I}$ [see Esary and Marshall (1970)], we can replace $P[\min(T_i, i \in I) > t]$ by an estimator for $P[\min(U_i, i \in I) > t]$ given by Proschan and Sullo (1976) and thus obtain a consistent estimator for system reliability.

In view of Theorem 2.1, we can state Theorem 2.2 in the following equivalent form.

Theorem 2.3. Suppose that a random vector T has exponential minima. Then there exists a collection $\{H_J, J \in \mathscr{J}\}$, $\cup \mathscr{J} = \{1,\ldots,n\}$, of independent exponential random variables such that for each $I \in \mathscr{I}$,

$$\min(T_i, i \in I) \overset{st}{=} \min(H_J, J \in \mathscr{J}, J \cap I \neq \emptyset).$$

Consider, for example, a *series* system whose component life lengths $T_1,\ldots,T_n$ have exponential minima and are mutually dependent. If we view the independent random variables $\{H_J\}$ of Theorem 2.3 as the component life lengths in a new series system, then, in effect, Theorem 2.3 allows us to transform a dependent model into an independent one, while *preserving* the life distribution of the original system. Under different conditions we shall see in Section 4 how to transform a dependent model into an independent one, while preserving not only system life length, but also the probabilities of occurrence of certain "failure patterns". We shall also see that such transformations from dependent to independent models are not only of interest in their own right, but also have important statistical applications to the theory of competing risks and to statistical life testing, in general.

Esary and Marshall (1974) establish existence only in their proof of Theorem 2.3 by using the special nature of coherent systems in reliability theory. The proof of Theorem 2.3 given by Langberg, Proschan and Quinzi (1977a) is considerably more elementary and specifies *explicitly* the distributions of the independent random variables $\{H_J\}$ as follows. Suppose that $P[\min(T_i, i \in I) > t] = \exp(-\mu_I t)$, $I \in \mathscr{I}$, for some collection $\{\mu_I, I \in \mathscr{I}\}$ of positive constants. Then the random variable H_J in Theorem 2.3 is exponentially distributed with parameter λ_J given by

$$\lambda_J = (-1)^{\#(J)-1}\Big(\mu_{1 \cdots n} - \sum_{i \in J} \mu_{\overline{\{i\}}} + \sum_{\substack{i_1, i_2 \subset J \\ i_1 < i_2}} \mu_{\overline{\{i_1, i_2\}}} - \cdots + (-1)^{\#(J)} \mu_{\bar{J}}\Big), \tag{2.1}$$

where $\#(J)$ is the cardinality of J and $\bar{J}$ is the complement of J in $\{1,\ldots,n\}$. Formula (2.1) provides an *explicit* solution for the parameters $\{\lambda_J\}$ in terms of the known constants $\{\mu_I\}$. Formula (2.1) also indicates ways of testing the validity of the assumption of exponential minima as the following examples illustrate.

First, suppose it is known *a priori* that, due to the structure of a particular system, it is impossible for the components in some subset J (generally, some collection $\{J_1,\ldots,J_m\}$ of subsets) to fail simultaneously. This is equivalent to assuming that the corresponding parameter $\lambda_J = 0$. If the corresponding linear combination of (known) constants $\{\mu_I\}$ given by (2.1) does *not* yield $\lambda_J = 0$, then the assumption of exponential minima must be wrong. Similarly, if some combination of the $\{\mu_I\}$ yields a λ_J which is negative, then the assumption of exponential minima is likewise incorrect.

More generally, formula (2.1) indicates a heuristic method for testing the statistical hypothesis of exponential minima. For example, consider a two-component system with component life lengths T_1 and T_2. If $P[\min(T_i, i \in I) > t] = \exp(-\mu_I t)$, $I \in \mathcal{J}$, then by (2.1),

$$\lambda_1 = \mu_{12} - \mu_2 \geq 0$$

$$\lambda_2 = \mu_{12} - \mu_1 \geq 0$$

$$\lambda_{12} = \mu_1 + \mu_2 - \mu_{12} \geq 0.$$

Consequently, we would expect that *estimates* for the μ_I's, together with an allowance for random error, would satisfy a similar set of inequalities. If not, we would tend to doubt the hypothesis of exponential minima.

Employing the same technique of proof used to prove Theorem 2.3, Langberg, Proschan and Quinzi (1977a) [hereafter referred to as LPQ (1977a)] obtain a generalization of Theorem 2.3. First we introduce some terminology. The *hazard function* R associated with the distribution function F of a nonnegative random variable is the function $R(t) = -\log[1 - F(t)]$, $t \geq 0$. The (multivariate distribution of the) nonnegative random vector T has *proportional hazard minima* if there exists a collection $\{\mu_I, I \in \mathcal{J}\}$ of positive constants such that $P[\min(T_i, i \in I) > t] = \exp[-\mu_I R(t)]$, $I \in \mathcal{J}$, where $R(\cdot)$ is a hazard function, i.e., a nonnegative, nondecreasing function satisfying $R(0) = 0$ and $R(\infty) = \infty$. LPQ (1977a) prove the following:

Theorem 2.4. Suppose that a random vector T has proportional hazard minima with hazard function R. Suppose further that R is continuous at $t_0 = \sup\{t: R(t) = 0\}$. Then there exists a collection $\{H_J, J \in \mathcal{J}\}$, $\cup \mathcal{J} = \{1,\ldots,n\}$, of independent random variables with hazard functions proportional to $R(\cdot)$ such that for each $I \in \mathcal{I}$,

$$\min(T_i, i \in I) \stackrel{st}{=} \min(H_J, J \in \mathcal{J}, J \cap I \neq \emptyset).$$

Furthermore, the constants of proportionality $\{\lambda_J, J \in \mathcal{J}\}$ are given by (2.1).

Remark 2.5. Note that in the special case $R(t) = t$, $t \geq 0$, the conclusion of Theorem 2.4 holds by Theorem 2.3.

The assumption of proportional hazard minima holds, for example, when the random vector T has minima with the Weibull distribution having a fixed scale parameter.

3. ADDITIVE FAMILIES OF DISTRIBUTIONS

Again using the same technique of proof as in the proof of Theorem 2.3, LPQ (1977a) obtain a result for additive families of distributions which is analogous to Theorem 2.3 except that "sum" plays the role of "minimum".

Let $\mathcal{F} = \{F_\theta, \theta \in \Theta\}$ be a family of distributions parameterized by θ, and let g be a binary operation on the set of real numbers. The collection $\mathcal{F}$ is said to form an *additive family with respect to g* if for every $\theta_1, \theta_2 \in \Theta$: X_1 and X_2 are independent random variables with respective distributions F_{θ_1} and F_{θ_2} implies that $g(X_1,X_2)$ has distribution $F_{\theta_1+\theta_2}$. In Theorem 2.3, the family of interest is the additive family of exponential distributions, $F_\theta(x) = 1 - \exp(-\theta x)$, with respect to $g(x_1,x_2) = \min(x_1,x_2)$. The following result applies to additive families $\{F_\theta, \theta \in \Theta\}$ with respect to $g(x_1,x_2) = x_1 + x_2$, where $\int x dF_\theta(x) = \theta$.

Theorem 3.1. For each $I \in \mathcal{I}$, let the random variable T_I have distribution $F_{\mu_I} \in \mathcal{F} = \{F_\mu, \mu \in \mathcal{M}\}$, where $\mathcal{F}$ is an additive family of distributions with respect to $g(x_1,x_2) = x_1 + x_2$ satisfying $\int x dF_\mu(x) = \mu$. Then there exists a collection $\{S_J, J \in \mathcal{J}\}$ of *independent* random variables such that the distribution of S_J belongs to $\mathcal{F}$, $J \in \mathcal{J}$, and for each $I \in \mathcal{I}$,

$$T_I \stackrel{st}{=} \sum_{J \in \mathcal{J}: J \cap I \neq \emptyset} S_J.$$

Furthermore, the mean λ_J of the random variable S_J is given by (2.1).

Examples of families satisfying the hypothesis of Theorem 3.1 are the Poisson family with mean μ and the gamma family with mean μ and unit scale parameter. A sample application in reliability follows.

Example 3.2. Suppose that an n-component system is exposed to shocks which are not necessarily fatal. For each $I \in \mathcal{J}$, a shock of type I simultaneously affects all components exclusively in subset I. For example, the shock pattern for a two-component system might be exhibited as in Figure 3.1 below:

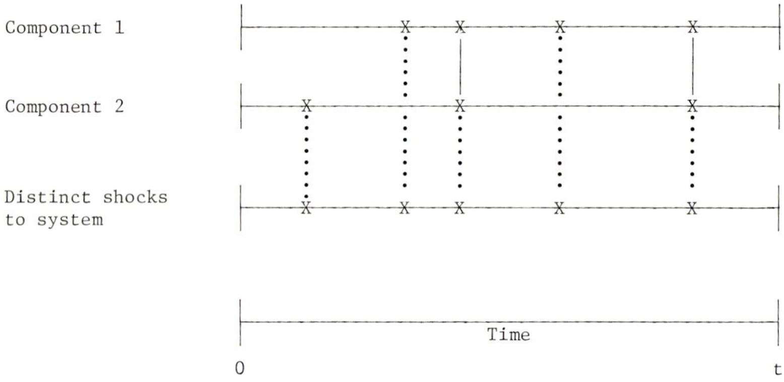

Fig. 3.1.

Figure 3.1 indicates that a total of 5 distinct shocks occurred in the interval $[0, t]$: 2 shocks affected component 1 alone, 1 shock affected component 2 alone, and 2 shocks affected both components simultaneously. Let $N_I(t)$ be the number of shocks in the interval $[0, t]$ which are simultaneously received by the components exclusively in subset I. In Figure 3.1, $N_1(t) = 2$, $N_2(t) = 1$, $N_{12}(t) = 2$. Let $K_I(t)$ be the number of *distinct* (in time) shocks in the interval $[0, t]$ which are received by the components in subset I. In Figure 3.1, $K_1(t) = 4$, $K_2(t) = 3$, $K_{12}(t) = 5$. Note that, in general,

$$K_I(t) = \sum_{J \cap I \neq \emptyset} N_J(t),$$

so that the processes $\{K_I(t), t \geq 0\}$, $I \in \mathcal{J}$, are generally dependent. If $\{K_I(t), t \geq 0\}$ is a Poisson process with intensity $\mu_I > 0$, $I \in \mathcal{J}$, then we conclude from Theorem 3.1 that there exists a collection $\{\{N_J^*(t), t \geq 0\}, J \in \mathcal{J}\}$ of *independent* Poisson processes such that for every $I \in \mathcal{J}$,

$$K_I(t) \overset{st}{=} \sum_{J \in \mathcal{J}: J \cap I \neq \emptyset} N_J^*(t).$$

Furthermore, the intensity λ_I of the process $\{N_I^*(t), t \geq 0\}$ is given by (2.1).

4. PRESERVING SYSTEM LIFE LENGTH AND FAILURE PATTERNS

It is desirable to have methods for converting dependent models into independent ones which *preserve* essential features of the original (dependent) model. For example, in the case of a series system whose component life lengths have exponential minima, Theorem 2.3 allows us to convert a dependent model into an independent one, while preserving the system life length, i.e., the *minimum* of the component life lengths. In this section we show how, under more general conditions, it is possible to convert a dependent model into an independent one, while preserving features of the original model in addition to system life length.

Consider an arbitrary series system of n components. In many practical applications we are able to observe:

(1) the time at which the system fails, and

(2) the identity of the component or set of components which fails.

Note that although we use the language of reliability theory (series system, component, etc.), the general model has application in a variety of contexts. For example, in population mortality studies, the data on each subject includes (1) the age at death and (2) the cause of death. Suppose that an individual dies due to one of n possible causes. An individual, in this context, can be viewed as an n-component series system who dies due to the occurrence of one or more of the n possible causes. As another example, suppose a personnel study is undertaken to determine the departure patterns of employees in a large company. The data on each employee might consist of (1) the length of stay, i.e., the time from arrival to termination and (2) the reason for termination. In general, one could imagine any model where observations include (1) the time at which a particular event occurs and (2) the identity of the cause or set of causes (among a finite number) which results in the occurrence of the event. Moreover, one or more of the "causes" might be identified with the withdrawal of a unit from observation, resulting in censored or truncated data. For convenience we continue to employ the language of reliability theory although the results are applicable in many other situations.

In this section we show how it is possible to replace a series system of dependent components by a series system of *independent* components while simultaneously preserving

(1)' the distribution of the time to system failure, and

(2)' the probability of occurrence of each failure pattern.

By "failure pattern" we mean, in the case of a series system, the failure of a set of components whose simultaneous failure causes (i.e., coincides with) the failure of the system.

We begin with some terminology and notation. If T is the vector of component life lengths in an n-component system, we say that *failure pattern I occurs*, and write $\xi(T) = I$, if the simultaneous failures of the components exclusively in subset I coincide with the failure of the system. Let S and T be the vectors of component life lengths of two systems with respective life lengths S and T. We say that the systems are *equivalent in life lengths and patterns* ($S \overset{LP}{=} T$, in symbols) if

$$P(S > t,\ \xi(S) = I) = P(T > t,\ \xi(T) = I)$$

for every $t \geq 0$ and every $I \in \mathcal{J}$.

Miller (1977) proves the following existence result:

Theorem 4.1. Let T_i be the life length of component i, $i = 1,\ldots,n$, and let T be the life length of the corresponding series system. Assume that the functions

$$F(t, I) = P(T \leq t,\ \xi(T) = I),\ I \in \mathcal{J}, \tag{4.1}$$

have no common discontinuities and that $P(T_i = T_j) = 0$ for $i \neq j$. Then there exists a vector S of independent random variables such that $T \overset{LP}{=} S$, and at least one of the S_i is a life length. The distributions of $S_1,\ldots,S_n$ are uniquely determined on $\{t\colon \bar{F}(t) > 0\}$, where $\bar{F}(t) = P(T > t)$, $t \geq 0$.

We can paraphrase Theorem 4.1 as follows. Under the given hypothesis, the original (dependent) system with life length $T = \min(T_i,\ 1 \leq i \leq n)$ can be replaced by a system with life length $S = \min(S_i,\ 1 \leq i \leq n)$, where the S_i's are *independent* random variables in such a way that $S \overset{LP}{=} T$. Tsiatis (1975) proves a similar result in the context of competing risk theory by assuming that the joint distribution of $T_1,\ldots,T_n$ has continuous partial derivatives. It is noteworthy that the nature of the dependence in the original model is unspecified in Theorem 4.1, i.e., the original components might be dependent in any way whatsoever. LPQ (1977b) show that the assumption of no common discontinuities in Theorem 4.1 is a *necessary* as well as a sufficient condition for the replacement of a dependent model by an independent one. Moreover, we provide *explicit expressions* for the appropriate distributions in the independent model.

Before presenting the LPQ (1977b) result, we motivate the theorem as follows. Let $T = \min(T_i,\ 1 \leq i \leq n)$, where $T_1,\ldots,T_n$ are the (dependent) component life lengths. Let $\bar{F}(t, I)$ be the joint probability that the system survives beyond time t and failure pattern I occurs. For example, if $n = 3$, $\bar{F}(t, \{1,2\}) = P(T > t,\ T_1 = T_2 < T_3)$ so that ties are possible. The problem as posed in LPQ (1977b) can be stated as follows. Given the vector T of (dependent) life lengths, determine a random vector S such that $S \overset{LP}{=} T$, where $S_1,\ldots,S_n$ are expressible in terms of *independent* random variables. By "expressible" we mean that the S_i are either themselves independent random variables or else can be expressed as functions of independent random variables. The solution is found by letting $S_1,\ldots,S_n$ be the life lengths of components in a theoretical n-component series system, where the components are exposed to shocks according to the following shock model.

Each component fails if it receives a shock. Independent sources of shock are present in the environment-one source for each $I \in \mathcal{I}$. A shock from source I simultaneously kills the components exclusively in subset I. Let H_I denote the time (measured from the origin) until a shock from source I occurs. Then $S_i = \min(H_I,\ I \in \mathcal{I},\ i \in I)$, $1 \leq i \leq n$, and $S = H$, where $S = \min(S_i,\ 1 \leq i \leq n)$ and $H = \min(H_I,\ I \in \mathcal{I})$. Define

$$\xi^*(H) = \begin{cases} I, & \text{if } H_I < H_J \text{ for each } J \neq I \\ \emptyset, & \text{otherwise.} \end{cases}$$

To allow for simultaneous failures among the components in the original system, we permit the dimension of the vector H to be greater than or equal to that of T. Generally, if T has dimension n, then the vector H of times until shock has dimension (at most) $2^n - 1$. The subscripts on the components of H are understood to be ordered lexicographically. It follows that $P(S > t,\ \xi(S) = I) = P(T > t,\ \xi(T) = I)$ if and only if

$$P(H > t,\ \xi^*(H) = I) = P(T > t,\ \xi(T) = I) \tag{4.2}$$

for each $t \geq 0$ and each $I \in \mathcal{I}$. If (4.2) holds for *every* subset I of $\{1,\ldots,n\}$, we write $H \overset{LP}{=} T$. The problem will be solved if we determine *independent* random variables H_I, $I \in \mathcal{I}$, such that (4.2) holds for every $t \geq 0$ and every subset I of $\{1,\ldots,n\}$. LPQ (1977b) prove the following:

Theorem 4.2. Let $T = \min(T_i,\ 1 \leq i \leq n)$ be the life length of an n-component series system, where T_i is the life length of component i, $i = 1,\ldots,n$. Define $\bar{F}(t,I) = P(T > t,\ \xi(T) = I)$ and $F(t, I) = P(T \leq t,\ \xi(T) = I)$, $I \in \mathcal{I}$. Let $\bar{F}(t) = P(T > t)$ and $\alpha(F) = \sup\{x:\ \bar{F}(x) > 0\}$. Then the following statements hold:

(i) A necessary and sufficient condition for the existence of a set of independent random variables H_I, $I \in \mathcal{I}$ which satisfy $H \overset{LP}{=} T$, where $H = \min(H_I, I \in \mathcal{I})$, is that the functions $F(\cdot, I)$, $I \in \mathcal{I}$, have no common discontinuities in the interval $[0, \alpha(F))$.

(ii) The random variables $\{H_I,\ I \in \mathcal{I}\}$ in (i) have corresponding survival probabilities $\{\bar{G}_I(\cdot),\ I \in \mathcal{I}\}$ which are uniquely determined on the interval $[0, \alpha(F))$ as follows:

$$\bar{G}_I(t) = \exp\{-\int_0^t [dF^C(\cdot, I)/\bar{F}]\} \cdot \prod_{a_j(I) \leq t} [\bar{F}(a_j(I))/\bar{F}(a_j(I)^-)],\quad 0 \leq t < \alpha(F), \tag{4.3}$$

where $F^C(\cdot, I)$ is the continuous part of $F(\cdot, I)$, $\{a_j(I)\}_j$ is the set of discontinuities of $F(\cdot, I)$, $I \in \mathcal{I}$, and the product over an empty set is defined as unity.

The following generalization of Theorem 4.2 for arbitrary (not necessarily coherent) systems also holds:

Theorem 4.3. Let T_i denote the life length of component i, i = 1,...,n, in an arbitrary n-component system with life length T. Define $F(\cdot, I)$ and $\alpha(F)$ as in Theorem 4.2. Then (i) and (ii) of Theorem 4.2 hold.

Example 4.4. Suppose that the vector (T_1, T_2) has a bivariate distribution with survival probability:

$$\bar{F}(t_1, t_2) = (1 + t_1 + t_2)^{-1}, \quad t_1 \geq 0,\ t_2 \geq 0.$$

Note that T_1 and T_2 are mutually dependent. If T_1 and T_2 are the component life lengths in a two-component series system, we may conclude from (4.3) of Theorem 4.2 that the original system is equivalent in life length and failure patterns to a system involving *independent* times H_1 and H_2 until shock, where

$$\bar{G}_i(t) = P(H_i > t) = (1 + 2t)^{-\frac{1}{2}}, \quad i = 1,2, t \geq 0.$$

5. APPLICATIONS TO THE THEORY OF COMPETING RISKS, LIFE TESTING, AND CENSORED DATA PROBLEMS

The reader will note that in every model thus far considered, we have obtained explicit expressions for the appropriate distributions in the independent model, whereas existence alone is proven in other approaches. Thus, our results are not only more general but also more readily applicable, especially when explicit solutions are called for. The probabilistic results obtained are, of course, of interest in their own right since they facilitate the analysis of various dependent models. However, a significant statistical payoff is also derived from our approach. In this section we show how our probabilistic solution to the conversion problem of Theorem 4.2 can be used to unify the nonparametric approach to estimation problems in competing risk theory, life testing, and certain incomplete data problems.

The theory of competing risks derives its name from the fact that, during a person's lifetime, he is exposed to several risks of death (various fatal diseases, accidents, etc.) which can be viewed as "competing" for his life. A series system of r dependent components with life length $T = \min(T_i, 1 \leq i \leq r)$ such that failure pattern I occurs $[\xi(T) = I]$ becomes, in the terminology of competing risks, an individual with life length $T = \min(T_i, 1 \leq i \leq r)$ exposed to r dependent risks of death, where T_i is the age at death if risk i were the only risk present in the environment, $1 \leq i \leq r$, and ξ is the cause of death, i.e., the subset I of $\{1,\ldots,r\}$ such that $T = T_i$ for each $i \in I$ and $T \neq T_i$ for each $i \notin I$. When death results from a single cause, then ξ is the index i for which $T = T_i$. [In an incomplete or censored data problem, one of the random variables T_j represents the time at which an individual becomes "unobservable" for a reason other than death, while the remaining variables typically represent various causes of death.] The biomedical researcher is interested in making inferences about *unobservable* quantities (viz., the random variables $T_1,\ldots,T_r$) by using data from *observable* quantities - in this case, the lifetime T and cause ξ. In showing how Theorem 4.2 may be applied, we focus on the following question: How can we estimate the marginal survival probabilities corresponding to a given risk (or combination of risks) operating along without competition from the other risks? That is, how can we estimate the $2^r - 1$ survival probabilities (so-called "net probabilities") $\bar{M}_J(t) = P[\min(T_j,\ j \in J) > t],\ J \subset \{1,\ldots,r\}$?

Throughout the remainder of this paper let $(T_{1i},\ldots,T_{ri})$, $i = 1,\ldots,n$, represent a random sample of size n from the joint distribution of the nonnegative random variables $T_1,\ldots,T_r$. To conform to the usual notation, we reserve 'n' for sample size. Thus, let $\mathcal{J}$ now denote the collection of all nonempty subsets of the set $\{1,\ldots,r\}$ of risks. We adopt all of the previous notation subject to the substitution of 'r' for 'n'. For each distribution F, let $\bar{F}(\equiv 1 - F)$ be the corresponding survival function. For each $I \in \mathcal{J}$, let $M_I(t) = P(T_I \leq t)$, where $T_I = \min(T_i,\ i \in I)$. In the competing risk model, only the following are observed:

$$T_{0i} = \min(T_{1i},\ldots,T_{ri}),\ 1 \leq i \leq n,$$

and

$$\xi_i,\ 1 \leq i \leq n,$$

where $\xi_i = I$ if and only if $T = T_i$ for each $i \in I$ and $T \neq T_i$ for each $i \notin I$. Let $0 \equiv T'_{00} \leq T'_{01} \leq \cdots \leq T'_{0n} \equiv T_{max}$ denote the ordered values of the observations $T_{01},\ldots,T_{0n}$.

Consider now the following assumptions:

(A1) The risks (i.e., the random variables $T_1,\ldots,T_r$) are mutually independent.

(A2) No ties are possible, i.e., $P(T_i = T_j) = 0$ for each $i \neq j$, $i,j = 1,\ldots,r$.

(A3) The distributions of $T_1,\ldots,T_r$ have no common discontinuities.

(A4) The random variables $T_1,\ldots,T_r$ have a joint distribution which is absolutely continuous.

Note that Assumptions (A2) and (A3) together imply that death results from a single cause.

Assuming (A1), (A2), and (A3), Peterson (1975) examines the following estimator for $\bar{M}_J$:

$$\hat{\bar{M}}_J(t) = \prod_R [(n - R)/(n - R + 1)], \tag{5.1}$$

where the product is over the ranks R of those observations T'_{0i}, $1 \leq i \leq n$, such that $T'_{0i} \leq t < T_{max}$ and T'_{0i} corresponds to a death from *at least one* cause in subset J. If $T_{max} = T_{ji}$ for some $j \in J$ and some i, $1 \leq i \leq n$, then $\hat{\bar{M}}_J(t)$ is defined to be zero for $t > T_{max}$. Otherwise, $\hat{\bar{M}}_J(t)$ is undefined for $t > T_{max}$.

The estimator (5.1) is a generalization of the well-known product-limit estimator for a survival probability proposed by Kaplan and Meier (1958). If $T_{0i} = T_{ji}$ for each i, $1 \leq i \leq n$, and some fixed j, $1 \leq j \leq r$, then (5.1) reduces to a step function with jumps of height 1/n at each T_{0i}, *thus yielding the usual empirical estimate* of $\bar{M}_j(t)$. Assuming (A1), (A2), and (A3), Peterson (1975) shows that the estimator (5.1) is maximum likelihood, (weakly) consistent, and, regarded as a process in t, converges to a normal process.

In the remainder of this section *we drop the assumption (A1) of independent risks*. How then can we estimate the functions $\bar{M}_I(t)$, $I \in \mathcal{I}$? Note that formula (4.3) of Theorem 4.2 exhibits a relationship between distributions of *observable* quantities [viz., the survival probability $\bar{F}(t)$ and the functions $F(t, i)$, $1 \leq i \leq r$] and distributions associated with the theoretical random variables H_I, $I \in \mathcal{I}$, which are *unobservable*. Replacing $\bar{F}(t)$ and $\bar{F}(t, i)$ in (4.3) by their empirical counterparts thus allows us to *estimate* the distributions G_i, $1 \leq i \leq r$, associated with the unobservable random variables H_i, $1 \leq i \leq r$. Unfortunately, the distributions G_i, $1 \leq i \leq r$, are, in general, *different* from the marginal distributions M_i, $1 \leq i \leq r$, which we seek to estimate. The natural question then is how to relate the unobservable (but estimable) functions G_i, $1 \leq i \leq r$, to the marginal distributions M_i, $1 \leq i \leq r$. More generally, how can we relate the functions M_I to the survival probabilities $\bar{G}_I$ given by (4.3)?

Assuming no ties (A2) and no common discontinuities among the marginal distributions (A3), Peterson (1975) finds necessary and sufficient conditions for a relationship to exist between the functions $\bar{G}_I$ and $\bar{M}_I$. Dropping the assumption (A2) of no ties and weakening the assumption (A3) of no common discontinuities, LPQ (1977c) prove the following:

Theorem 5.1. Assume that the functions $F(\cdot, I)$, $I \in \mathcal{I}$, in Theorem 4.2 have no common discontinuities. Let $I \in \mathcal{I}$. Then for each $t \in [0, \alpha(F))$,

$$\bar{M}_I(t) = \prod_{J \cap I \neq \emptyset} \bar{G}_J(t) \tag{5.2}$$

if and only if the following two conditions hold:

$$\bar{M}_I(a)/\bar{M}_I(a^-) = \begin{cases} \bar{F}(a)/\bar{F}(a^-), & a \in D(F(\cdot, \mathcal{I}_I)) \\ 1 & , \text{otherwise}, \end{cases} \tag{5.3a}$$

and

$$P(T_{\bar{I}} \geq t \mid T_I = t) = P(T_{\bar{I}} > t \mid T_I > t), \tag{5.3b}$$

where $\bar{G}_J(t)$ is given by (4.3) and $D(F(\cdot, \mathcal{I}_I))$ is the set of discontinuities of the function $F(t, \mathcal{I}_I) = P(T > t, \xi(T) \in J, J \cap I \neq \emptyset)$.

Desu and Narula (1977) arrive at a condition similar to (5.3b) when the assumption of absolute continuity (A4) and hence also the assumption of no ties (A2) hold. We remark that the assumption of no ties (A2) does *not* hold, e.g., in models where failures or deaths from simultaneous causes can occur. An important family of multivariate distributions for which assumption (A2) *fails* is the family of multivariate exponential distributions of Marshall and Olkin (1967). We illustrate with an example.

Example 5.2. For simplicity, suppose that the random vector (T_1, T_2) has the Marshall-Olkin bivariate exponential distribution with survival probability:

$$P(T_1 > t_1, T_2 > t_2) = \exp[-\lambda_1 t_1 - \lambda_2 t_2 - \lambda_{12} \max(t_1, t_2)],$$

for $t_1, t_2 \geq 0$ and $\lambda_1, \lambda_2, \lambda_{12} > 0$. Since the marginal distributions M_1 and M_2 are continuous, condition (5.3a) of Theorem 5.1 holds trivially. Condition (5.3b) with $I = \{1\}$ states that

$$P(T_2 \geq t | T_1 = t) = P(T_2 > t | T_1 > t).$$

An easy computation shows that $P(T_2 \geq t | T_1 = t) = P(T_2 > t | T_1 > t) = \exp(-\lambda_2 t)$. Thus, Theorem 5.1 may be applied when the joint distribution belongs to the family of Marshall-Olkin MVE distributions, whereas other approaches to the estimation problem do not apply here since the assumption of no ties (A2) fails to hold.

Formula (4.3), via Theorem 5.1, can now be used to suggest estimators for the functions M_I, $I \in \mathcal{I}$, in the important practical cases when independence fails to hold and when ties are allowed. Suppose that the joint distribution of $T_1,\ldots,T_r$ satisfies (5.3 a,b). For each $i = 1,\ldots,n$, only the following are observed:

$$T_{0i} = \min(T_{1i},\ldots,T_{ri}),$$

and

$$\xi_i,$$

where $\xi_i = J$ if and only if $T_{0i} = T_{ji}$ for *each* $j \in J$ and $T_{0i} \neq T_{ji}$ for each $j \notin J$. In accordance with Theorem 5.1, we estimate $\bar{M}_I(t)$ for $t \leq T_{max}$ by

$$\tilde{\bar{M}}_I(t) = \prod_{J \cap I \neq \emptyset} \tilde{\bar{G}}_J(t),$$

where $\tilde{\bar{G}}_J$ is the function resulting from (4.3) by replacing $\bar{F}$ and $F(\cdot, I)$ by their empirical counterparts, $I \in \mathcal{I}$. In analogy with (5.1), the resulting statistic can be expressed as follows:

$$\tilde{\bar{M}}_I(t) = \prod_R [(n - R)/(n - R + 1)], \tag{5.4}$$

where the product is over the ranks R of those observations $T'_{0,R}$, $1 \leq R \leq n$, such that $T'_{0,R} \leq t \leq T_{max}$ and $T'_{0,R}$ corresponds to a death from the *simultaneous* causes $j \in J$ such that $J \cap I \neq \emptyset$. If for some i, $T_{max} = T_{ji}$ for each $j \in J$ with $J \cap I \neq \emptyset$, then $\tilde{\bar{M}}_I(t)$ is defined to be zero for $t > T_{max}$. Otherwise, $\tilde{\bar{M}}_I(t)$ is undefined for $t > T_{max}$.

Optimality properties of the estimator (5.4) readily suggest themselves because of its resemblance to the (generalized) Kaplan-Meier estimator (5.1) and its reliance upon empirical distributions. Moreover, there is evidence that formula (4.3) can be used in a similar way to estimate other quantities of interest to the biomedical researcher.

REFERENCES

[1] BARLOW, R. E. and PROSCHAN, F. (1975). *Statistical Theory of Reliability and Life Testing: Probability Models*. Holt, Rinehart and Winston, New York.

[2] BIRNBAUM, Z. W., ESARY, J. D. and SAUNDERS, S. C. (1961). Multicomponent systems and structures, and their reliability. *Technometrics*, 3, 55-77.

[3] DESU, M. M. and NARULA, S. C. (1977). Reliability estimation under competing causes of failure. *The Theory and Applications of Reliability with Emphasis on Bayesian and Nonparametric Methods*, (I. Shimi and C. P. Tsokos, eds.). Academic Press. In press.

[4] ESARY, J. D. and MARSHALL, A. W. (1970). Coherent life functions. *SIAM J. Appl. Math.*, 18, 810-814.

[5] ESARY, J. D. and MARSHALL, A. W. (1974). Multivariate distributions with exponential minimums. *Ann. Statist.*, 2, 84-98.

[6] KAPLAN, E. L. and MEIER, P. (1958). Nonparametric estimation from incomplete observations. *J. Amer. Statist. Assoc.*, 53, 457-481.

[7] KOTZ, S. (1974). Annotated and abstracted bibliography on multivariate statistical models. Temple University Technical Report.

[8] LANGBERG, N., PROSCHAN, F. and QUINZI, A. J. (1977a). Converting dependent models into independent ones, with applications in reliability. *The Theory and Applications of Reliability, with Emphasis on Bayesian and Nonparametric Methods*, (I. Shimi and C. P. Tsokos, eds.). Academic Press. In press.

[9] LANGBERG, N., PROSCHAN, F. and QUINZI, A. J. (1977b). Converting dependent models into independent ones, preserving essential features. Submitted for publication.

[10] LANGBERG, N., PROSCHAN, F. and QUINZI, A. J. (1977c). Dependent competing risks. In process.

[11] MARSHALL, A. W. and OLKIN, I. (1967). A multivariate exponential distribution. *J. Amer. Statist. Assoc.*, 62, 30-44.

[12] MILLER, D. R. (1977). A note on independence of multivariate lifetimes in competing risks models. *Ann. Statist.* In press.

[13] PETERSON, A. V., JR. (1975). Nonparametric estimation in the competing risks problem. Stanford University Technical Report No. 13.

[14] PROSCHAN, F. and SULLO, P. (1976). Estimating the parameters of a multivariate exponential distribution. *J. Amer. Statist. Assoc.*, **71**, 465-472.

[15] TSIATIS, A. (1975). A nonidentifiability aspect of the problem of competing risks. *Proc. Nat. Acad. Sci., USA.* **72**.

P.R. Krishnaiah, ed., APPLICATIONS OF STATISTICS

STATISTICAL PROBLEMS IN MENTAL TESTING

FREDERIC M. LORD

Educational Testing Service
Princeton, New Jersey

Many problems of mental testing are problems in applied statistics. Before discussing some of these, I am going to try to give a brief outline of some of the statistical theory of mental testing. My main emphasis will be on the so-called classical theory, although I shall mention some newer developments also.

The average man on the street is puzzled as to why there should be any theory of mental testing. A student answers the questions on a test. He gets a certain number of these right. His knowledge of the material is clearly represented by his number of right answers. What need is there for any kind of statistical theorizing?

The need for a theory begins to become apparent when we consider: if the same person had been tested at a different time with a distinct but parallel form of the same test, he would very likely have gotten a different score. The discrepancy is by no means negligible. If a person were administered several parallel forms of a 50-item test, the standard deviation of his number of right answers over these tests would typically be more than three score points. This amount of error is best appreciated by comparing it with the standard deviation of number-right scores across people. For a typical 50-item test, this latter might be about six score points.

What sort of theory is needed to cope with such a situation? An effective theory states that a person's scores on parallel test forms differ because of errors of measurement.

There are several possible sources of errors of measurement. The difference between parallel test forms is one source. Changes in the person tested from moment to moment is a second source. Even if a person were tested repeatedly with the *same* form of a test, his score would be expected to vary. If a person does not know the answer to an item, he often picks among available choices by a more or less random process. He might even toss a coin to determine his choice. Random, or partially random, guessing is clearly one source of error in the test score.

If an observed test score contains an error of measurement, this implies that the difference between the test score and the error of measurement is something of special interest. We call it the *true score*. This is a technical term with a defined meaning.

Some psychologists are unwilling to deal with things they cannot observe directly. For this reason, they raise objections to the notion of a true score. To a statistician, however, the true score is simply a parameter describing the person. The statistician usually does not expect to observe a parameter directly. He is happy to be able to estimate its value.

The true-score parameter, along with others, in principle determines the frequency distribution of scores that would be observed if we could repeatedly test a given person without the effect of fatigue, learning, memory, etc. The true score *may* be defined as the expected value of the observed score.

Much of mental test theory is concerned with estimating what would have happened if there had been no errors of measurement. Certainly, this is what the tester would like to know.

The theory of errors in mental test theory is often quite different from the theory of errors in other disciplines. The classical theory of errors, widely used in astronomy and physics, deals with a situation where repeated measurements are made on a single object of interest. Frequently the observations are found to be normally distributed. In such situations, theory and application are rather straightforward.

At an opposite extreme we have the problems of econometrics. It is recognized, for example, that the unemployment statistics and figures for the gross national product contain errors of measurement. Yet there is no way to obtain two independent measures of these quantities at a given time. This leads to many solved and unsolved problems in the estimation of functional or structural relationships.

Mental test theory lies between these extremes. Clearly we cannot repeatedly test the same individual without changing him in the process. Classical test theory has dealt with this problem, for example, by scoring the odd-numbered items of the test and the even-numbered items of the test separately. The difference between the scores on the two half-tests is considered to be due to errors of measurement. Here we have only two measures on each individual. However, if we have a large number of individuals, it is possible to develop a useful theory.

Classical test theory begins with the equation $X = T + E$, which states that the observed score X is a sum of a true score and an error of measurement. It is usual to start with three so-called assumptions. First, that the errors of measurement are unbiased--the expected error of measurement is zero. Second that the true score and error are uncorrelated in a group of individuals. Third, if we have two different tests, that the errors of measurement on the two tests are uncorrelated with each other and with both true scores.

On careful examination, it appears that there is no possible way that the first two assumptions can be falsified by any set of data. It would seem better, therefore, to consider these *not* as assumptions but as part of the definition of what is meant by true score and by error.

Classical mental test theory is based on the reasonable assumption that we can build parallel forms of a test, at least to an acceptable approximation. There are various possible definitions of parallel. A simple one for our purpose here is to say that the tests are parallel if every person has the same true score on both tests, and if the distribution of the errors of measurement is the same for both tests. The odd-numbered items and the even-numbered items often provide tests that are reasonably parallel by this definition.

Classical test theory is concerned exclusively with first and second moments and functions of them. The model from which it starts is the simplest possible linear model. You will have no trouble visualizing how means, variances, and covariances of quantities of interest may be estimated by analysis of variance components or by other simple procedures.

Let me quickly go over some basic formulas. If X and X' are parallel forms of a test, their covariance $\sigma_{XX'}$ in any group is equal to the true score variance σ_T^2. This equation (equation 3, Appendix) provides a practical way to estimate the true-score variance from actual test data.

Clearly $\sigma_E^2 = \sigma_X^2 - \sigma_T^2$ (equation 4). This gives a practical estimate of the variance of the errors of measurement in any group.

A quantity of great interest in classical test theory is the correlation ρ_{XT} between observed score and true score. It is easily shown that in the population this is automatically equal to the square root of the correlation between parallel test forms $\rho_{XT}^2 = \rho_{XX'}$ (equation 5). This quantity is called the *reliability coefficient* of the test.

Statisticians have sometimes pointed out that mental testers frequently use correlation coefficients when they should use regression coefficients. It is true that mental test workers become so accustomed to using correlation coefficients that it may not occur to them to use a regression coefficient when the problem really requires it. The reason for the extensive use of correlation coefficients is the following. Test scores as they are usually used often have no meaningful origin or unit of measurement. Number-right scores have a certain absolute meaning for a person familiar with the test administered; but for published tests, number-right scores are usually transformed, linearly or nonlinearly, to some scale with an arbitrary origin and unit of measurement--for example, to a scale with a mean of 500 and a standard deviation of 100. Once this has been done, there is no longer any reason to use regression coefficients, rather than correlation coefficients, in studying a relationship between test scores.

Why do publishers use transformed scores rather than raw scores? If simple number-right scores were used, small differences between test forms would make scores not comparable from form to form.

Formulas (6) to (9) from classical test theory are listed in the handout to indicate the kind of formulas obtainable.

Darrell Bock [2] has already described to you one of the newer, more interesting approaches to mental test theory. I will mention briefly another recent model called the binomial error model. This model deals with tests that are *randomly parallel*, rather than strictly parallel as in the classical model. Two tests are called *randomly parallel* if the items in each are a random sample from the same pool of items. Under these conditions, the number-right score observed for a person has a binomial distribution, the binomial parameter being the proportion-correct true score (the proportion of items in the entire pool that the person would answer correctly).

When such a mathematical model holds, it is no longer necessary to build parallel tests. All quantities of interest can be estimated by using known properties of the binomial distribution.

An objection to this model is that we do not ordinarily build tests in this fashion. This model could be extended to deal with stratified sampling of items, which *does* approximate a common procedure of test construction. Even without the rather severe complications introduced by stratification, the binomial error model still yields results that are very useful for many purposes.

A few formulas obtained from the binomial error model are listed in the handout. Equation (10) gives the standard error of the number-right score of a single individual on an n-item test. The standard error is due to sampling of test items. An unbiased estimate of this sampling error is given in equation (11). This is very readily calculated from the individual's number-right score X_a. For a whole group of individuals, the variance of the errors of measurement is estimated without bias by (12).

A special case is of considerable interest. If the regression of true score on observed score is linear, then the marginal frequency distribution of observed score is necessarily a negative hypergeometric distribution, as indicated by (13), where the subscripted parentheses denote ascending factorials. When the regression of true score on observed score is linear, the frequency distribution of true score is either a beta distribution or some other distribution having the same moments up through order n.

Regardless of the shape of the regression of true scores on observed score, the true-score moments are readily obtained from the observed-score moments without any assumptions beyond the binomial model itself. The relationship is given by equation (15). The quantity on the left is the true-score moment. The numerator on the right denotes the factorial observed-score moment. The denominator is a descending factorial. Thus, the first n moments of the true-score distribution are readily estimated from the factorial moments of any sample observed-score distribution. Moments of order higher than n cannot be estimated from the data.

With this background, let me go on to mention eight important unsolved or imperfectly solved problems in mental test theory. I believe all these problems are well within the reach of current statistical theory and are ready for useful solution.

1. Mental test theory seems a very appropriate place for application of Bayesian statistics. It is usually natural to act as if the people we have tested were a random sample from some population. If we have tested people like these before, we may have a very good idea about the shape of the distribution of ability or other trait in this population.

A main concern of test theory is to estimate what would have happened if there had been no errors of measurement. It is thus natural for a Bayesian to make use of a prior distribution of *true* scores. If we know the frequency distribution of the errors of measurement, and we also have a prior distribution for the true scores, then it should theoretically be possible to solve just about any reasonable question in a wide area of test theory.

It does not seem plausible, as a first approximation, to assume a normal distribution either for the true scores or for the errors of measurement. If we use the binomial error model, an obvious and convenient prior for a distribution of true score is the beta distribution given by equation (14). If one has only a few tens or hundreds of examinees in the sample, I have found that equation (14) usually is satisfactory. It usually gives a satisfactory fit to most sets of data.

With a sample size of two or three thousand, as is commonly available for most nationally known tests, it is usually possible to show that the beta prior distribution does not fit the data. Equation (14) shows a *two*-parameter beta distribution. Even with a *four*-parameter beta distribution, it is frequently impossible to fit the data.

Given the binomial error model (or a generalized binomial error model), it is not hard to demonstrate a lack of fit by using equation (15), or otherwise. Of course, this could indicate that the model itself was wrong, not the prior distribution. It is usually possible, however, to find a prior distribution which, taken together with the binomial error model, will fit the data. This is commonly true even when we increase the number of observations into the tens of thousands and even into the hundreds of thousands. We can still find priors that imply a distribution of observed scores that differs insignificantly from the actual sample distribution, as determined by a chi-square test. But these priors are not beta distributions. They have no simple form. Usually they require fitting more than four parameters from the data.

My own conclusion is that the use of *subjective* priors in mental test theory will probably be limited to situations where the number of people tested is not large. When we have a thousand or more examinees, the prior distribution should be estimated from the data, rather than determined subjectively.

Equation (15) strongly suggests that this can be done. At first glance, it might appear that equation (15) solves the problem. However, there are certain restrictive conditions that moments must satisfy if they are the actual moments of a frequency distribution. Estimated moments (estimated by equation (15)) will rarely satisfy these conditions.

One method is currently in experimental use for estimating true-score distributions from observed-score distributions. I have discussed this method in Patil [9]. Other and better methods for dealing with this interesting problem will surely be developed.

2. Another area, a rapidly growing area in which there are many unsolved problems, is the area of matrix sampling. If the State of California wishes to estimate the mean level of performance of seventh-grade students in mathematics, it may start with a pool of, say, 100 seventh-grade-level mathematics achievement items. It is too expensive in pupil time to give all these items to all the seventh-grade pupils in California. It is inefficient simply to do the obvious thing and give all 100 items to a random sample of pupils. The efficient procedure is to administer different small random samples of items drawn from the 100-item pool to different samples of students. This is matrix sampling. This is a situation where it is appropriate to use the theory of randomly parallel tests.

It is clearly possible to estimate mean pupil performance by this approach. Formulas for the standard error of the resulting estimate of the mean are available. This is not nearly enough, however, to guide a practical person setting up such a program towards finding an optimally cost-effective answer to the many questions that arise. How large should the pool of items be? How many different samples of items should be drawn from the pool? How large should each sample of items be, in view of the various costs involved? To how many students should each sample of items be administered?

The State of California would not only wish to estimate the mean performance of the students, but also the standard deviation of their level of performance. This can also be done. Standard error formulas for the resulting estimates can be obtained, but they are very lengthy and complex. In my opinion, not only the numerical work of evaluating standard errors, but also the *mathematical* work of deriving the standard error formulas, should be done by computers. Some computer programs for this purpose have been written by Dayhoff [5] and by Carney [3, 4].

When both means and standard deviation are to be estimated, the problem of cost-effective design of a testing program becomes complicated. Yet such programs must be designed, and much money is spent on them. Important work on these problems has been done by Madow and by others. Much more work remains to be done.

3. It would be helpful if small sample frequency distributions could be found for various statistics commonly used in mental test theory. A commonly used quantity is the one on the right-hand side of equation (8). Under certain

normality assumptions, Kristof [7] has shown that the obvious sample estimator of this function has a distribution that is closely related to an F distribution.

It would be very helpful to have a small sample frequency distribution for some convenient sample estimator of the quantity on the right-hand side of equation (9). This formula is the one commonly used for determining whether or not two tests measure the same psychological dimension except for errors of measurement and except for possible differences in the origin and unit of measurement of the score scale. Clearly it is important in mental testing to estimate the correlation between true scores on two different tests, yet nothing is known about the small sample distribution of the statistics involved.

4. A related and in some ways more serious problem arises in work with partial correlations. As shown by equation (9), the effect of errors of measurement on a zero-order correlation is simply to reduce the correlation below what it would be if there were no errors. The effect of errors of measurement on a partial correlation may be much more serious. It would not be at all unusual for the sign of the partial correlation to be changed by the introduction of errors of measurement. If the test reliability coefficients are available, equation (9) can be used as a basis for correcting a partial correlation for this effect of the errors of measurement. The resulting formula for the corrected partial is fairly simple. When this formula is applied to sample correlations, in certain cases the resulting estimate will have a very large sampling error. Little is known about the large-sample distribution of the corrected partial; nothing is known about its small sample distribution.

5. A not unrelated problem appears in analysis of covariance. Particularly in the field of education, it is common to attempt to use analysis of covariance to compare groups while supposedly "correcting" for the effects of a covariate. We wish to know how the group given the special training would have turned out if they had started at the same level of initial aptitude as the control group. There are various problems that may prevent answering such a question. The problem of concern here is that we do not know the initial aptitude of the students, we merely have an unreliable measure of this. Statistical adjustments to match the treatment group and the control group on an unreliable measure can give very different answers than if matching were done on a reliable measure. An attempt to deal with this problem appears in Lord [8], also in Stroud [11] and in DeGracie and Fuller [6]. This problem affects a large number of analyses that are continually being carried out and used as a basis for practical decisions. It is important that some convenient method for dealing with this be devised.

6. A closely related problem is the problem of using an unreliable measure in such a way as actually to match groups on true score--that is, to select two subgroups that will have identical true-score distributions. Since the distribution of true scores can be estimated from the distribution of observed scores, it

should be possible to achieve this. I do not believe it has ever been attempted in practice, however.

7. Suppose a physicist has many different instruments in his laboratory, yet he is unable to determine whether or not two different instruments measure the same physical variable. This seems like an absurd state of affairs, yet the psychometrician is in exactly this position.

We do have *asymptotic* procedures in certain cases for determining whether or not two tests differ only in origin and unit of measurement and in errors of measurement. We have no small-sample procedure for investigating this.

When two different test publishers build two different tests of reading comprehension, even when the tests measure the same psychological dimension, it is very likely that the tests differ in more than origin, scale, and errors of measurement. At best, it is likely that the true scores on one test represent a *non*-linear transformation of the true scores on the other test. How can we determine whether there is a perfect nonlinear relation between the true scores, as opposed to the alternative that there is no one-to-one relationship between the true scores. If there is a perfect nonlinear relationship, we can still say in a sense that the tests are measuring the same dimension. If there is no one-to-one relationship between true scores, the tests are measuring different dimensions. We need a statistical test to investigate this, but no good significance test is available. Thus there is no good way to determine whether two tests that claim to measure that same trait in fact do so.

8. Let me mention one final problem. The major purpose of mental testing may be thought of as estimating an ability parameter that describes individual students. Statistical procedures are usually evaluated by considering their bias and their sample variance. As discussed earlier, however, there is often no uniquely proper scale for measuring ability. If θ^* is a nonlinear transformation of θ, θ^* may be just as good a measure of ability as θ. How do we evaluate estimators of the ability parameter of individuals in such a situation? The notion of an unbiased estimator becomes meaningless. Even the sampling variance of the estimator is useless, because the unit of measurement in which it is expressed has no unique merit. What is required is that the sample estimators insofar as possible should rank the individuals in order of their true ability, whether θ or θ^*. How does one evaluate and compare different sample estimators for such a purpose? A useful way of dealing with this problem is provided by an information statistic discussed by Birnbaum [1]. A related information statistic has been discussed by Rao [10 pp. 270-271]. There still does not seem to be a widely known and well-unified theory for dealing with this kind of problem, however.

APPENDIX

$$X = T + E \tag{1}$$

If X' is parallel to X ,

$$X' = T + E' \tag{2}$$

$$\sigma_T^2 = \sigma_{XX'}^2 = \sigma_X^2 \rho_{XX'} \tag{3}$$

$$\sigma_E^2 = \sigma_X^2 - \sigma_T^2 = \sigma_X^2 - \sigma_{XX'} = \sigma_X^2(1 - \rho_{XX'}) \tag{4}$$

$$\rho_{XT}^2 = \rho_{XX'} \tag{5}$$

$$\rho_{XT}^2 = \sigma_T^2/\sigma_X^2 = 1 - \sigma_E^2/\sigma_X^2 \tag{6}$$

$$\sigma_E^2 = \frac{1}{2}\,\varepsilon(X - X')^2 = \sigma_{E'}^2 \tag{7}$$

If X is divided into parallel half tests x and x' so that $X \equiv x + x'$,

$$\rho_{XX'} = \frac{2\rho_{xx'}}{1 + \rho_{xx'}} \quad . \tag{8}$$

For two different tests X and Y ,

$$\rho(T_X, T_Y) = \frac{\rho_{XY}}{\sqrt{\rho_{XX'}\rho_{YY'}}} \quad . \tag{9}$$

Under the binomial error model, the standard error of the number-right score X_a of a single individual on an n -item test, due to sampling of test items, is

$$\sigma^2(X_a|T_a) = \sigma^2(E_a) = T_a(n - T_a)/n \quad . \tag{10}$$

An unbiased estimate of this is

$$\hat{\sigma}^2(X_a|T_a) = \hat{\sigma}^2(E_a) = X_a(n - X_a)/(n - 1) \quad . \tag{11}$$

-A2-

The error variance over a group of individuals is estimated without bias by

$$\hat{\sigma}_E^2 = \frac{1}{n-1}[\bar{x}(n-\bar{x}) - s_x^2] \ . \tag{12}$$

If the regression of T on X is linear, then the (marginal) frequency distribution of X is necessarily a negative hypergeometric distribution:

$$f(X) \propto \frac{(-n)_X\ (a)_X}{(-b)_X\ X!} \quad \text{where} \quad (a)_N \equiv a(a+1)\ \dots\ (a+N-1)\ . \tag{13}$$

In this case, the frequency distribution of T is the beta distribution

$$g(T) \propto (T/n)^{a-1}(1-T/n)^{b-n} \tag{14}$$

(or, equally well, any other distribution having the same first n moments).

Regardless of the linearity of the regression,

$$\mu_r' = M_{[r]}/n^{[r]} \quad , \qquad r = 1,2,\dots, n \quad . \tag{15}$$

REFERENCES

[1] BIRNBAUM, A. (1968). Some latent trait models and their use in inferring an examinee's ability. In *Statistical Theories of Mental Test Scores* (F. M. Lord and M. R. Novick), pp. 453-479. Addison-Wesley, Reading, Massachusetts.

[2] BOCK, R. D. (1976). Characterizing a latent trait distribution. *Proceedings of the Symposium on Applications of Statistics*, Dayton, Ohio.

[3] CARNEY, E. J. (1970). Multiplication of polykays using ordered partitions. *Ann. Math. Statist.*, **41**, 1749-1752.

[4] CARNEY, E. J. (1972). Polykays and ordered partitions--some computing problems. University of Rhode Island, Kingston, Tech. Rept. 72-105.

[5] DAYHOFF, E. (1966). Generalized polykays, an extension of simple polykays and bipolykays. *Ann. Math. Statist.*, **37**, 226-241.

[6] DEGRACIE, J. S. and FULLER, W. (1972). Estimation of the slope and analysis of covariance when the concomitant variable is measured with error. *J. Amer. Statist. Assoc.*, **67**, 930-937.

[7] KRISTOF, W. (1963). Statistical inferences about the error variance. *Psychometrika*, **28**, 129-143.

[8] LORD, F. M. (1960). Large-sample covariance analysis when the control variable is fallible. *J. Amer. Statist. Assoc.*, **55**, 307-321.

[9] PATIL, G. P. (Ed.) (1970). *Random Counts in Scientific Work*, **2**. *Random Counts in Biomedical and Social Sciences*. The Pennsylvania State University Press, University Park, Pennsylvania.

[10] RAO, C. R. (1965). *Linear Statistical Inference and Its Applications*. Wiley, New York.

[11] STROUD, T. W. F. (1972). Comparing conditional means and variances in a regression model with measurement errors of known variances. *J. Amer. Statist. Assoc.*, 67, 407-412.

P.R. Krishnaiah, ed., APPLICATIONS OF STATISTICS

SELF-SELECTIVITY PROBLEMS IN ECONOMETRIC MODELS

G. S. MADDALA

Department of Economics
University of Florida*
Gainesville, Florida

1. INTRODUCTION

There are many problems in economics where the data that we have are generated by the choices made by individuals. This poses some estimation problems which we will discuss in this paper. The following are some illustrative examples of situations where these problems occur:

(a) Returns to college education: If we are given data on incomes of a sample of individuals, some of whom have college education and others not, we have to take account in our analysis the fact that those who have college education are those that chose to go to college and those that do not have college education are those that have chosen (for some reason) not to go to college. This is what we call "self-selectivity". A naive and commonly used way of analyzing these differences is to define a dummy variable

$$D = 1 \text{ if the individual goes to college}$$
$$= 0 \text{ otherwise}$$

and estimate an earnings function with D as an extra explanatory variable. This, however, is not a satisfactory solution to the problem.

(b) Effects of unions on wages: Suppose we have data on wages of unionized and non-unionized workers and we have to estimate the effects of unions on wages. Again a naive way is to add a dummy variable representing unionization in an equation determining the wages. This does not take into account the "self-selectivity" problem - that some workers chose to join the union and others did not. Of course, here there are other problems like restriction of entry or union selectivity in addition to the self-selectivity of the workers. There may also be some employer-selectivity that makes employers more selective in their choice of employees if the employees are unionized.

(c) Demand for housing: In studies of demand for housing it is customary to analyze the demand for owner-occupied housing and rental housing separately, or else to regress housing expenditures (imputed rent for owner-occupied) on other explanatory variables and a dummy variable defined as:

*Financial support from the NSF under grant SOC-76-04356 to the University of Florida is gratefully acknowledged.

D = 1 if owner-occupied housing
= 0 otherwise

Again, there is a self-selectivity problem here - some individuals choose to own houses and others choose to rent. In the estimated demand functions this choice has to be taken into account.

(d) Returns to migration: This problem is similar to the problem of returns to college education. Some individuals choose to migrate, others not. There is thus the self-selectivity problem.

(e) Effects of Employment Agencies on Post-unemployment Wages: Suppose we have data on post-unemployment wages of a number of workers - some of whom have sought the help of employment agencies and others have not sought any such help. To estimate the effect of the employment agencies on post unemployment wages, one naive method would again be using a dummy variable as an explanatory variable - the dummy variable defined to be 1 if the individual seeks help from the employment agency and defined to be 0 otherwise. This method does not give a satisfactory answer because there is a self-selectivity problem here - those who went to the employment agency are precisely the ones that expected that they would benefit from the agency and those that did not go are precisely the ones that expected no benefit from the agency.

(f) Labor force participation of married women: Here again the problem is that some women choose to join the labor force and some do not. This self-selectivity problem has been analyzed by Gronau (1974), Heckman (1974), Lewis (1974) and Nelson (1975). The mean wage of working women is not a correct measure of what the mean wage of women would have been if all women went to work.

There are many other problems one can enumerate. But the examples mentioned above are sufficiently illustrative of the type of problems commonly encountered. In the following sections we will outline the estimation problems for such models.

2. A SIMPLE MODEL WITH NO EXPLANATORY VARIABLES

We will first present a simple model with no explanatory variables (similar to the model considered by Lewis (1974)). This will highlight both the problems and the essentials of the methods of estimation. In the next section we will introduce explanatory variables and further complications in the model.

Let us consider example (a) on returns to college education discussed in the previous section. Suppose that we are given data on the variable y (income) for n individuals. n_1 of them have had college education and n_2 of them have not. We are interested in estimating the average increment in income due to going to college.[1] One simple solution is to regress y on a dummy variable D defined as

[1]For simplicity we are avoiding the problems of costs of going to college, foregone earnings, etc.

$D = 1$ if the individual has college education

$= 0$ otherwise

The coefficient of D will measure the average increase in income due to going to college. If $\bar{y}_1$ is the mean income of those with college education and $\bar{y}_2$ the mean income of those without college education then this coefficient is nothing but $\bar{y}_1 - \bar{y}_2$.

At first glance it appears that this is all that can be done with the data. However, the estimate $\bar{y}_1 - \bar{y}_2$ can be criticized on grounds that it does not take into account the fact that those who went to college did so precisely because they expected their incomes to be higher than otherwise and those who did not go, chose not to go to college precisely because they did not expect their incomes to be higher by doing so. Thus there is a selectivity bias in the estimate $\bar{y}_1 - \bar{y}_2$. We will now show that by adding a few plausible assumptions we can do something more with the same data.

Suppose that all the individuals have the same opportunities available to them and that an individual goes to college if he thinks his income will be higher by having college education. Otherwise he does not.

Let y_1^* be the expected income with college education

y_2^* be the expected income without college education

The individual goes to college if $y_1^* > y_2^*$

The dummy variable now is

$D = 1$ if $y_1^* > y_2^*$

$= 0$ otherwise.

However, the expected incomes y_1^* and y_2^* are not observable. We will denote the realized or observed values by y_1 and y_2 and we observe y_1 for only those individuals that go to college and y_2 for those individuals that do not. We will now add the further assumption that expectations are right (individuals do not make mistakes) in the sense that

$$y_1 > y_2 \text{ iff } y_1^* > y_2^*$$

Thus, the definition of the dummy variable D is now

$$\begin{aligned} D &= 1 \text{ if } y_1 > y_2 \\ &= 0 \text{ otherwise} \end{aligned} \tag{1}$$

Also, we observe y_1 iff $D = 1$. We observe y_2 otherwise.

Suppose that $y_1 = \mu_1 + u_1$

$y_2 = \mu_1 + u_2$

and u_1 and u_2 have a bivariate normal distribution with means zero and a covariance matrix

$$\begin{pmatrix} \sigma_1^2 & \sigma_{12} \\ \sigma_{12} & \sigma_2^2 \end{pmatrix}$$

Our objective is to estimate μ_1, μ_2, σ_1^2, σ_2^2 and σ_{12}. The difference $\mu_1 - \mu_2$ now measures the average increase in income due to college education. There are two assumptions being made in the above formulation. First, that people choose. Second, that they are right in their decisions. The first assumption is very reasonable. The second assumption can be defended on the grounds that it is hard to formulate a thoery of systematic mistakes. The argument is similar to the one made in defense of rational expectation models.

There is one interesting feature of this model that is worth mentioning here. Let λ be the probability that an individual goes to college.

Then

$$\begin{aligned} \lambda &= \text{Prob.} \quad (y_2 < y_1) \\ &= \text{Prob.} \quad (u_2 - u_1 < \mu_1 - \mu_2) \end{aligned} \tag{2}$$

Since u_1 and u_2 are normally distributed, $u_2 - u_1$ is also normally distributed with variance $\sigma^2 = \sigma_1^2 + \sigma_2^2 - 2\sigma_{12}$.

Define

$$u = \frac{u_2 - u_1}{\sigma} \tag{3}$$

$$Z = \frac{\mu_1 - \mu_2}{\sigma} \tag{4}$$

Then

$$\lambda = \text{Prob } (u < Z) = \int_{-\infty}^{Z} \frac{1}{\sqrt{2\pi}} \exp\left(-\frac{t^2}{2}\right) dt \tag{5}$$

An estimate of λ is $\frac{n_1}{n_1+n_2}$. Given this number (the proportions of individuals in the sample having college education) we can estimate Z by looking at the cumulative normal distribution. Thus if $n_1 = n_2$ then the estimate of Z is zero and $\hat{\mu}_1 = \hat{\mu}_2$. Thus, the incremental returns to college education are zero. This result is in marked contrast to the result we would obtain if we regarded the dummy variable D as exogenous. For instance, if the sample means are $\bar{y}_1$ = \$14,000 and $\bar{y}_2$ = \$10,000, the naive estimate of $\mu_1 - \mu_2$ is \$4,000. The model considered here estimates $\mu_1 - \mu_2$ as zero if $n_1 = n_2$. This result is a consequence of the fact that we assume the sample to be a random sample and that everyone in the sample has an opportunity to choose whether or not to have college education. Given that the choice has been 50-50, the model attributed the difference of \$4,000 entirely to self-selectivity.

The estimation of the parameters in this model proceeds as follows:

Noting that $\frac{n_1}{n_1+n_2}$ is an estimate of λ, we use equation (5) to get an estimate of $\frac{\mu_1-\mu_2}{\sigma}$.

Next we note that u_1 and u have a joint normal distribution with mean vector zero and a covariance matrix

$$\begin{pmatrix} \sigma_1^2 & \sigma_{1u} \\ \sigma_{1u} & 1 \end{pmatrix}$$

where

$$\sigma_{1u} = (\sigma_{12} - \sigma_1^2)/\sigma$$

and that[1].

$$E(u_1 | u<Z) = -\sigma_{1u}\frac{f(Z)}{F(Z)} = \theta_1 \text{ (say)} \tag{6}$$

$$\begin{aligned} E(u_1^2 | u<Z) &= \sigma_1^2 - \sigma_{1u}^2 Z \frac{f(Z)}{F(Z)} \\ &= \sigma_1^2 + \theta_1 Z\sigma_{1u} \end{aligned} \tag{7}$$

where f(Z) and F(Z) are respectively the density function and the distribution function of the standard normal. Hence for the subsample on y_1 that we have, we get

$$\begin{aligned} E(y_1 | D=1) &= \mu_1 + E(u_1 | u<Z) \\ &= \mu_1 + \theta_1 \end{aligned} \tag{8}$$

and

$$\text{Var}(y_1 | D=1) = \text{var}(u_1 | u<Z) = \sigma_1^2 + \theta_1 Z\sigma_{1u} - \theta_1^2 . \tag{9}$$

Similarly, denoting by σ_{2u} the covariance between u_2 and u we get

[1]These are well-known formulae for moments of truncated normal. See, for instance, Johnson and Kotz (1972), pp. 112-113.

$$\sigma_{2u} = (\sigma_2^2 - \sigma_{12})/\sigma$$

and

$$E(u_2|u>Z) = \sigma_{2u} \frac{f(Z)}{1-F(Z)} = \theta_2 \text{ (say)} \tag{10}$$

$$E(u_2^2|u>Z) = \sigma_2^2 + \sigma_{2u}^2 Z \frac{f(Z)}{1-f(Z)}$$

$$= \sigma_2^2 + \theta_2 Z\sigma_{2u} \tag{11}$$

Hence for the subsample on y_2 that we have, we get

$$E(y_2|D=0) = \mu_2 + E(u_2|u>Z)$$

$$= \mu_2 + \theta_2 \tag{12}$$

and

$$\text{Var } (y_2|D=0) = \text{Var}(u_2|u>Z) = \sigma_2^2 + \theta_2 Z\sigma_{2u} - \theta_2^2 \tag{13}$$

we have to solve for the five parameters μ_1, μ_2, σ_1^2, σ_2^2 and σ_{12} from the five equations (5), (8), (9), (12) and (13) using the five sample statistics:

$\frac{n_1}{n_1+n_2}$: the proportion making the choice

$\bar{y}_1$, $\bar{y}_2$: the sample means of y_1 and y_2

s_1^2, s_2^2 : the sample variances of y_1 and y_2.

This solution is not an easy task because it involves some nonlinearities. But there are some interesting conclusions that can be drawn from these equations. Firstly, we might want to know whether $\bar{y}_1$ and $\bar{y}_2$ are respectively upward or downward biased estimates of μ_1 and μ_2. In equation (8) since

$$\theta_1 = \left(\frac{\sigma_1^2 - \sigma_{12}}{\sigma}\right) \frac{f(Z)}{F(Z)}$$

we can easily see that $\bar{y}_1$ will be an upward biased estimate of μ_1 iff $\sigma_1^2 > \sigma_{12}$. Similarly, in equations (12) since $\theta_2 = \left(\frac{\sigma_2^2 - \sigma_{12}}{\sigma}\right) \frac{f(Z)}{1-F(Z)}$, $\bar{y}_2$ will be an upward biased estimate of μ_2 iff $\sigma_2^2 > \sigma_{12}$. Thus, in the special case when the correlation between the residuals is zero or negative, *both* the sample means are upward biased estimates of the true means μ_1 and μ_2.

If σ_{12} is positive but $\sigma_1^2 > \sigma_{12} > \sigma_2^2$ then $\bar{y}_1$ is an upward biased estimate of μ_1 and $\bar{y}_2$ is a downward biased estimate of μ_2 and hence $\bar{y}_1 - \bar{y}_2$ is an upward biased estimate of the true difference $\mu_1 - \mu_2$. Similarly if $\sigma_2^2 > \sigma_{12} > \sigma_1^2$ then $\bar{y}_1 - \bar{y}_2$ is a downward biased estimate of $\mu_1 - \mu_2$. In other cases the bias in $\bar{y}_1 - \bar{y}_2$ is not unambiguous.

Also, from equation (9) we see that the conditional variance of y_1 will be greater than σ_1^2 iff $\theta_1 Z\sigma_{1u} > \theta_1^2$. Similarly, from equation (13) we see that the conditional variance of y_2 will be greater than σ_2^2 iff $\theta_2 Z\sigma_{2u} > \theta_2^2$. Since θ_1 and σ_{1u} are of opposite signs and θ_2 and σ_{2u} are of the same signs, we get the result that if $Z>0$, the conditional variance of y_1 is unambiguously less than σ_1^2, and that if $Z<0$, the conditional variance of y_2 is unambiguously less than σ_2^2.

It might be instructive to look at the magnitude of the biases for a few special cases. Consider the following:

Case (i) $\mu_1 = \$13{,}000$, $\mu_2 = \$10{,}000$

$\sigma_1 = \$\ 1{,}500$, $\sigma_2 = \$\ 2{,}000$

$\zeta = -.05$ (is the correlation between u_1 and u_2)

We get

$$\sigma \simeq \$3{,}040,\ Z = \frac{\mu_1 - \mu_2}{\sigma} \simeq 1.0$$

$$f(Z) = .242,\ F(Z) = .841$$

Hence

$$\theta_1 = \frac{\sigma_1(\sigma_1 - \zeta\sigma_2)}{\sigma} \frac{f(Z)}{F(Z)} \simeq \$357$$

$$\theta_2 = \frac{\sigma_2(\sigma_2 - \zeta\sigma_1)}{\sigma} \frac{f(Z)}{1-F(Z)} \simeq \$2{,}750$$

Also if $S_1^2 = \text{Var}\,(y_1|y_1>y_2)$ and $S_2^2 = V(y_2|y_2>y_1)$ we get from (9) and (13)

$$S_1 = \$1{,}300 \text{ and } S_2 = \$1{,}220$$

Thus, both the sample means will be upward biased estimates of the true means and both the sample variances will be downward biased estimates of the true variances. Also note that $\bar{y}_1 - \bar{y}_2$ *underestimates* the difference $\mu_1 - \mu_2$ by about 80 percent in this case. The biases are smaller in magnitude for the first group partly because 84% of the individuals in the sample are in that group.

Case (ii) Consider case (i) with $\sigma_1 = \$1{,}000$ and $\sigma_2 = \$3{,}000$. We now get $\sigma = \$3{,}600$, $Z \simeq .85$, $f(Z) = .278$, $F(Z) = .802$. Hence $\theta_1 \simeq \$240$ and $\theta_2 \simeq \$4{,}050$. Thus the conditional mean of y_2($14,500) is greater than the conditional mean of y_1 ($13,240) even though $\mu_1 > \mu_2$.

Case (iii) $\mu_1 = \$13,000$, $\mu_2 = \$10,000$, $\sigma_1 = \$3,000$, $\sigma_2 = \$1,00$ and $\zeta=0.5$. We get $\sigma = \$2,650$, $Z \simeq 1.15$, $f(Z) = .206$, $F(Z) = .875$, $\theta_1 = \$640$ and $\theta_2 = -\$300$. In this case $\bar{y}_1 - \bar{y}_2$ overestimates the true difference $\mu_1 - \mu_2$ by about 30 percent ($940).

These examples illustrate that the direction of biases can be either way. Also, it is possible for the sample mean $\bar{y}_2$ to be higher than $\bar{y}_1$ (on the average) even if $\mu_1 > \mu_2$. The sign of $\mu_1 - \mu_2$ is determined entirely by the proportion $\frac{n_1}{n_1+n_2}$ (through equation (5)) and not by $\bar{y}_1 - \bar{y}_2$.

The estimate of μ_1 we get from the preceding analysis gives us the mean income if everyone in the sample went to college and the estimate of μ_2 gives us the mean income if everyone decided not to go to college. We might also be interested in estimating what the mean incomes of the two groups would have been if they made the opposite choice. This can be easily calculated as follows:

Let

$$\delta_1 = E(y_1|y_1<y_2)$$

$$\delta_2 = E(y_2|y_1>y_2)$$

Then δ_1 is what the mean income of those who did not go to college would have been if they did. Similarly, δ_2 is what the mean income of those who went to college would have been if they did not.

Since

$$E(y_1) = P(y_1>y_2)\ E(y_1|y_1>y_2) + P(y_1<y_2)\ E(y_1|y_1<y_2)$$

we get

$$\mu_1 = \lambda(\mu_1 + \theta_1) + (1-\lambda)\delta_1$$

Similarly

$$\mu_2 = \lambda\ \delta_2 + (1-\lambda)\ (\mu_2 + \theta_2)$$

Hence

$$\delta_1 = \mu_1 - \left(\frac{\lambda}{1-\lambda}\right)\theta_1 \qquad (14)$$

and

$$\delta_2 = \mu_2 - \left(\frac{1-\lambda}{\lambda}\right)\theta_2 \qquad (15)$$

Note that $\lambda = F(Z)$.

For the numbers considered in case (i) we get

$$\delta_1 = \$11,125 \text{ and } \delta_2 = \$9,476.$$

Note that in this case the actual mean income of the group who did not go to college is \$12,750 and it is \$13,360 for the group that went to college. By definition δ_1 will always be less than $\mu_2 + \theta_2$ and δ_2 will always be less than $\mu_1 + \theta_1$.

The above model can easily be extended to the case where there are explanatory variables determining both y_1 and y_2. This will be discussed in the next section.

There is a lot of discussion on returns to college education and whether college matters (see Solmon and Taubman (1973), Taubman and Wales (1972) and the references in them). However, in all these studies the usual regressions are dummy variable regressions (with perhaps some explanatory variables and IQ scores added) and the estimates of returns to college education are all of the form $\bar{y}_1 - \bar{y}_2$ in the above discussion. This is not a correct measure because of the self-selectivity involved in the data. Also, often a question is asked as to what the average income of those who did not go to college would have been if they did (see, for instance Mundel (1972)). Again, the answers usually given to such questions are not the right ones. The correct procedure would be to calculate parameters like δ_1 and δ_2 in equations (14) and (15). The parameters of interest in all these problems are μ_1, μ_2, σ_1^2, σ_2^2, σ_{12}, δ_1 and δ_2 - not $\bar{y}_1 - \bar{y}_2$.

3. MODELS WITH EXPLANATORY VARIABLES

The model considered in the previous section can be written as:

$$y_{1t} = \beta_1' X_{1t} - \varepsilon_{1t} \tag{16}$$

$$y_{2t} = \beta_2' X_{2t} + \varepsilon_{2t} \tag{17}$$

Where X_{1t} and X_{2t} are any vectors of exogenous explanatory variables (with or without overlapping elements) and the observed value of the dependent variable y_t is defined as

$$\begin{aligned} y_t &= y_{1t} \quad \text{iff} \quad y_{1t} > y_{2t} \\ &= y_{2t} \quad \text{iff} \quad y_{1t} < y_{2t} \end{aligned} \tag{18}$$

A model similar to this was considered by Maddala and Nelson (1974). The model (Model II in the paper) is a disequilibrium model of the following form:

$$D_t = \beta_1' X_{1t} + \varepsilon_{1t} \quad \text{(Demand function)} \tag{19}$$

$$S_t = \beta_2' X_{2t} + \varepsilon_{2t} \quad \text{(Supply function)} \tag{20}$$

Since in periods of excess demand, the quantity transacted is equal to the supply and in periods of excess supply, the quantity transacted is equal to the demand, we find that Q_t, the observed quantity is

$$Q_t = \text{Min} \quad (D_t, S_t) \tag{21}$$

Also, we assume that we know for each observation whether Q_t is on the demand function or on the supply function (i.e. the sample separation is available). We also assume that ε_{1t} and ε_{2t} have a joint normal distribution with means zero and a covariance matrix

$$\Sigma = \begin{pmatrix} \sigma_1^2 & \sigma_{12} \\ & \sigma_2^2 \end{pmatrix} \tag{22}$$

When

$$Q_t = D_t \text{ we get } \varepsilon_{1t} = Q_t - \beta_1' X_{1t}$$

$$\text{and } \varepsilon_{2t} > Q_t - \beta_2' X_{2t} \text{ since } S_t > Q_t$$

Similarly, when $Q_t = S_t$ we get

$$\varepsilon_{2t} = Q_t - \beta_2' X_{2t}$$

and

$$\varepsilon_{1t} > Q_t - \beta_1' X_{1t} \text{ since } D_t > Q_t$$

Hence the likelihood function for the estimation of the parameters is

$$L(\beta_1, \beta_2, \sigma_1^2, \sigma_{12}, \sigma_2^2) =$$

$$\prod_t \left[\int_{g_{2t}}^{\infty} f(g_{1t}, \varepsilon_{2t})\, d\varepsilon_{2t}\right]^{I_t} \left[\int_{g_{1t}}^{\infty} f(\varepsilon_{1t}, g_{2t})\, d\varepsilon_{1t}\right]^{1-I_t} \tag{23}$$

where

$$I_t = 1 \quad \text{iff } Q_t = D_t$$

$$= 0 \quad \text{if } Q_t = S_t$$

$$g_{1t} = Q_t - \beta_1' X_{1t}$$

$$g_{2t} = Q_t - \beta_2' X_{2t}$$

and $f(\varepsilon_1, \varepsilon_2)$ is the joint distribution of ε_1 and ε_2.
The maximum likelihood estimates are obtained by maximizing (23) with respect to all the parameters.

The model considered in equations (16)-(18) is the same as the disequilibrium model in equations (19)-(21) except that the 'Min' in equation (21) has to be replaced by a 'Max'. This will reverse some inequalities and change the limits of integrations in (23).

The condition $y_{1t} - y_{2t} \geq 0$ in (18) can be written as $\beta_1' X_{1t} - \beta_2' X_{2t} \geq \varepsilon_{2t} - \varepsilon_{1t}$. In general, the condition determining the sample separation may involve other exogenous variables and error terms. Hence we can write it as

$$\alpha' Z_t \geq \varepsilon_t \tag{24}$$

where Z_t is a vector of exogenous variables including some or all of X_{1t} and X_{2t}. Since α is estimable only up to a scale factor, without any loss of generality we will assume ε_t to have a unit variance. Thus, the model can now be written as

$$y_t = \beta_1' X_{1t} + \varepsilon_{1t} \quad \text{iff} \quad \alpha' Z_t \geq \varepsilon_t \tag{25}$$

$$y_t = \beta_2' X_{2t} + \varepsilon_{2t} \quad \text{iff} \quad \alpha' Z_t < \varepsilon_t \tag{26}$$

The model is like a switching regression model considered by Goldfelt and Quandt (1973) but since ε_t is correlated with ε_{1t} and ε_{2t} we can call it a switching regression model with endogenous switching, as suggested in Maddala and Nelson (1975). We will assume that ε_{1t}, ε_{2t} and ε_t have a trivariate normal distribution with mean vector zero and covariance matrix Σ (which can be singular)

$$\Sigma = \begin{pmatrix} \sigma_1^2 & \sigma_{12} & \sigma_{1\varepsilon} \\ & \sigma_2^2 & \sigma_{2\varepsilon} \\ & & 1 \end{pmatrix} \tag{27}$$

The likelihood function for this model is

$$L(\beta_1, \beta_2, \gamma, \sigma_1^2, \sigma_2^2, \sigma_{1\varepsilon}, \sigma_{2\varepsilon})$$

$$= \Pi \left[\int_{-\infty}^{\alpha' Z_t} g(y_t - \beta_1' X_{1t}, \varepsilon_t)\, d\varepsilon_t \right]^{I_t} \left[\int_{\alpha' Z_t}^{\infty} f(y_t - \beta_2' X_{2t}, \varepsilon_t)\, d\varepsilon_t \right]^{1-I_t} \tag{28}$$

where

$$I_t = 1 \text{ if } \gamma' Z_t \geq \varepsilon_t$$
$$= 0 \text{ otherwise}$$

and g and f are the bivariate normal density functions of ε_{1t}, ε_t and ε_{2t}, ε_t respectively. Note that σ_{12} does not occur at all in this expression and thus σ_{12} is not estimable. It is only $\sigma_{1\varepsilon}$ and $\sigma_{2\varepsilon}$ that are estimable. In the special case $\varepsilon_t = \frac{\varepsilon_{1t}-\varepsilon_{2t}}{\sigma}$ where $\sigma^2 = \mathrm{Var}(\varepsilon_{1t} - \varepsilon_{2t})$, it can be easily verified that from the consistent estimates of σ_1^2, σ_2^2, $\sigma_{1\varepsilon}$ and $\sigma_{2\varepsilon}$, we can get a consistent estimate of σ_{12}. This is the case with the model considered in equations (19-21) for which the likelihood function is given in (23).

The maximization of the likelihood functions (23) and (25) may be cumbersome in practice. Lee (1975) discusses a simple two-stage estimation method for this model which runs as follows: Define the dichotomous variable

$$\begin{aligned} I_t &= 1 \quad \text{iff} \quad \gamma' Z_t \geq \varepsilon_t \\ &= 0 \text{ otherwise} \end{aligned} \tag{29}$$

Then using the probit method we can get consistent estimates of γ which will be used in further analysis:

We define

$$u_{1t} = f(\gamma' Z_t) \;/\; F(\gamma' Z_t)$$

and (30)

$$u_{2t} = f(\gamma' Z_t) \;/\; (1-F(\gamma' Z_t))$$

where $f(\cdot)$ and $F(\cdot)$ are the density function and the distribution function of the standard normal. Then we have the following relations (similar to (6), (7), (10) and (11) earlier)

$$\begin{aligned} E(\varepsilon_{1t} | I_t = 1) &= -\sigma_{1\varepsilon}\, u_{1t} \\ E(\varepsilon_{1t}^2 | I_t = 1) &= \sigma_1^2 - \sigma_{1\varepsilon}^2\, (\gamma' Z_t)\, u_{1t} \end{aligned} \tag{31}$$

$$\begin{aligned} E(\varepsilon_{2t} | I_t = 0) &= \sigma_{2\varepsilon}\, u_{2t} \\ E(\varepsilon_{2t}^2 | I_t = 0) &= \sigma_2^2 + \sigma_{2\varepsilon}^2\, (\gamma' Z_t)\, u_{2t} \end{aligned} \tag{32}$$

Hence

$$E(y_t | I_t = 1) = \beta_1' X_{1t} - \sigma_{1\varepsilon}\, u_{1t} \tag{33}$$

$$E(y_t | I_t = 0) = \beta_2' X_{2t} + \sigma_{2\varepsilon}\, u_{2t} \tag{34}$$

These equations can be written as

$$y_t = \beta_1' X_{1t} - \sigma_{1\varepsilon}\, u_{1t} + \eta_{1t} \quad \text{for} \quad I_t = 1 \tag{35}$$

$$y_t = \beta_2' X_{2t} + \sigma_{2\varepsilon} u_{2t} + \eta_{2t} \quad \text{for} \quad I_t = 0 \tag{36}$$

where

$$E(\eta_{1t}|I_t = 1) = E(\eta_{2t}|I_t = 0) = 0$$

$$\text{Var}\ (\eta_{1t}|I_t = 1) = \sigma_1^2 - \sigma_{1\varepsilon}^2\, u_{1t}\, (\gamma' Z_t + u_{1t}) \tag{37}$$

$$\text{Var}\ (\eta_{2t}|I_t = 0) = \sigma_2^2 + \sigma_{2\varepsilon}^2\, u_{2t}\, (\gamma' Z_t - u_{2t}) \tag{38}$$

The estimation procedure is the following. Using data as I_t, get a consistent $\hat{\gamma}$ of γ using the probit method. Then construct $\hat{u}_{1t}$, $\hat{u}_{2t}$ by substituting $\hat{\gamma}$ for γ in (30). Next estimate (35) and (36) by ordinary least squares by regressing y_t on X_{1t} and $\hat{u}_{1t}$ for the subset of observations for which $I_t = 1$ and y_t on X_{2t} and $\hat{u}_{2t}$ for the subset of observations for which $I_t = 0$, to obtain consistent estimates $\hat{\beta}_1$, $\hat{\beta}_2$ of β_1 and β_2. Next compute

$$\hat{\varepsilon}_{1t} = y_t - \hat{\beta}_1' X_{1t} \quad \text{for the set } I_t = 1 \tag{39}$$

$$\hat{\varepsilon}_{2t} = y_t - \hat{\beta}_2' X_{2t} \quad \text{for the set } I_t = 0. \tag{40}$$

and regress $\hat{\varepsilon}_{1t}^2$ on $(\hat{\gamma}' Z_t)\, u_{1t}$ and $\hat{\varepsilon}_{2t}^2$ on $(\hat{\gamma}' Z_t)\, \hat{u}_{2t}$ to get consistent estimates of σ_1^2, $\sigma_{1\varepsilon}$ and σ_2^2, $\sigma_{2\varepsilon}$ respectively (see equations (31) and (32)).

Thus, we get consistent estimates of all the parameters (except σ_{12} which is not estimable). These estimates can be used either as final estimates or as starting values for a one-step iteration in the computation of maximum likelihood estimates. They can also be used in the computation of weighted least squares estimates from (35) and (36). Note that the residuals in equations (35) and (36) do not have constant variances. Their variances are given by (37) and (38) which can be estimated from the above mentioned initial consistent estimates. Using these estimated variances, one can use a weighted least squares procedure to get more efficient estimates of the parameters in (35) and (36).

One can also combine equations (35) and (36) and write the unconditional expectation of y_t as:

$$\begin{aligned} E(y_t) &= E(y_t|I_t = 1)\ .\ P(I_t = 1) + E(y_t|I_t = 0).\ P(I_t = 0) \\ &= \beta_1' X_{1t} F_t + \beta_2' X_{2t} (1-F_t) + (\sigma_{2\varepsilon} - \sigma_{1\varepsilon}) f_t \end{aligned} \tag{41}$$

where

$$f_t = f(\gamma' Z_t) \text{ and } F_t = F(\gamma' Z_t)$$

Thus, regressing y_t or $X_{1t}F_t$, $X_{2t}(1-F_t)$ and f_t we get consistent estimates of β_1, β_2 and $(\sigma_{2\varepsilon} - \sigma_{1\varepsilon})$. Then using (39), (40), (31) and (32) we get consistent estimates of σ_1^2, σ_2^2, $\sigma_{1\varepsilon}$ and $\sigma_{2\varepsilon}$. If we write

$$y_t = E(y_t) + W_t$$

where $E(y_t)$ is given by (41), then $E(W_t) = 0$ and $\mathrm{Var}(W_t)$ can be evaluated once we have consistent estimates of all the parameters. Hence, again one can estimate the parameters in (41) by a weighted least squares method.

In some cases the self-selectivity may be such that y_t is observed only if $I_t = 1$. i.e. the observations y_t are defined as

$$y_t = y_{1t} \quad \text{iff } y_{1t} \geq y_{2t}$$

$$= 0 \text{ otherwise.}$$

This is the model considered by Nelson (1975). The model is one of labor force participation of married women.

y_{1t} is the market wage

y_{2t} is the reservation wage

The person is in the labor force if $y_{1t} \geq y_{2t}$. In this case the actual wage observed is $y_t = y_{1t}$. If $y_{1t} < y_{2t}$ then all we observe is that the person is not in the labor force. Note that y_{2t} is never observed. This introduces some identification problems that do not exist in the earlier model given by (16), (17), and (18). Specifically, we now have to impose the condition that either $\sigma_{12} = 0$ or that there is at least one explanatory variable in X_{1t} that does not occur in X_{2t}. To see this note that in (24) we now have

$$\gamma' Z_t = \frac{\beta_1' X_{1t} - \beta_2' X_{2t}}{\sigma} \quad \text{and} \quad \varepsilon_t = \frac{\varepsilon_{2t} - \varepsilon_{1t}}{\sigma}$$

where

$$\sigma^2 = \sigma_1^2 + \sigma_2^2 - 2\,\sigma_{12}$$

Thus, from the probit regression model (based on the dichotomous variable I_t) we get consistent estimates of $\frac{\beta_{1j}}{\sigma}$ and $\frac{\beta_{2j}}{\sigma}$ for the elements of β_1 and β_2 corresponding to non-overlapping variables in X_1 and X_2, and $\frac{\beta_{1k} - \beta_{2k}}{\sigma}$ corresponding to the common variables in X_1 and X_2. Next from equation (35) we can get consistent estimates of β_1 and $\sigma_{1\varepsilon} = \left(\frac{\sigma_{12} - \sigma_1^2}{\sigma}\right)$. Since we now have estimates for all the elements of β_1, if there is at least one variable in X_1 not included in X_2, then from the estimate of $\frac{\beta_{1j}}{\sigma}$ corresponding to this variable, we now get a consistent estimate of σ and hence consistent estimates of all the elements of β_2. Next we use equations (39) and (31) as before to get a consistent estimate of σ_1^2 and given estimates of σ, σ_1^2 and $\sigma_{1\varepsilon}$ we can now get an estimate of σ_{12} and finally from an estimate of σ^2 we can get an estimate of σ_2^2. Thus, all parameters are estimable.

Alternatively, if $\sigma_{12} = 0$ we have $\sigma^2 = \sigma_1^2 + \sigma_2^2$ and $\sigma_{1\varepsilon} = -\frac{\sigma_1^2}{\sigma}$. From the estimates of $\frac{\beta_1 - \beta_2}{\sigma}$ obtained from the probit regression, β_1 and $\frac{\sigma_1}{\sigma}$ from equation (35) and σ_1^2 from equations (39) and (31) we can get estimates of all the parameters. An alternative discussion of these conditions for identification in this model can be found in Nelson (1976). Finally, the likelihood function for this model is

$$L(\beta_1, \beta_2, \sigma_1^2, \sigma_{12}).$$

$$= \prod_t \left[\int_{-\infty}^{g_{2t}} f(g_{1t}, \varepsilon_{2t})\, d\varepsilon_{2t}\right]^{I_t} (1-F_t)^{1-I_t} \tag{43}$$

where

$$g_{1t} = y_t - \beta_1' X_{1t}$$

$$g_{2t} = y_t - \beta_2' X_{2t}$$

$$F_t = F\left(\frac{\beta_1' X_{1t} - \beta_2' X_{2t}}{\sigma}\right)$$

In his empirical work Nelson (1976) obtained the maximum likelihood estimates by maximizing the likelihood function (43). He imposed the condition $\sigma_{12} = 0$ since all the variables for X_1 were also included in X_2 in his model.

4. CONCLUSION

The paper discusses methods for analyzing models involving self-selection. Applications of these methods can be found in Lee (1976), Nelson (1976) and Trost (1976). Lee analyzes the effect of unions on wages. Here the self-selectivity is the decision of whether or not to join the union. Nelson analyzes the problem of labor force participation of women. Here the decision is whether or not to join the labor force. Trost analyzes demand for housing. Here the decision is whether to own or rent. Further discussion of these problems and the empirical results can be found in these papers and also in Maddala and Nelson (1976).

REFERENCES

[1] GOLDFELD, S. M. and QUANDT, R. E. (1973). The estimation of structural shifts by switching regressions. *Annals of Social and Economic Measurement*, 2, 4, 475-485.

[2] GRONAU, R. (1974). Wage comparisons: a selectivity bias. *Journal of Political Economy*, 82, 6, 1119-1143.

[3] HECKMAN, J. (1974). Shadow prices, market wages and labor supply. *Econometrica*, 42, 4, 679-694.

[4] JOHNSON, N. L. and KOTZ, S. (1974). Distributions in statistics: continuous multivariate distributions. Wiley & Sons, New York.

[5] LEE, LUNG-FEI (1975). Estimation of some limited dependent variable models by two stage methods. Discussion paper, University of Rochester, Sept. 1975.

[6] LEE, LUNG-FEI (1976). Unionism and wage rates: A simultaneous equations model with qualitative and limited dependent variables. Discussion paper, University of Florida, Aug. 1976.

[7] LEWIS, H. G. (1974). Comments on selectivity biases in wage comparisons. *Journal of Political Economy*, 82, 6, 1145-1155.

[8] MADDALA, G. S. and NELSON, F. D. (1974). Maximum likelihood methods for models of markets in disequilibrium. *Econometrica*, 42, 6, 1013-1030.

[9] MADDALA, G. S. and NELSON, F. D. (1975). Switching regression models with exogenous and endogenous switching. *Proc. Business and Economic Statistics Section, American Statistical Association, 1975.* 423-426.

[10] MADDALA, G. S. and NELSON, F. D. (1976). Limited dependent variable models in economics (manuscript).

[11] MUNDEL, D. S. (1972). Federal aid to higher education: An analysis of federal subsidies to undergraduate education. The Economics of Federal Subsidy Programs. Submitted to the Joint Economic Committee, U. S. Congress, 92nd Congress, 2nd Session.

[12] NELSON, F. D. (1975). Estimation of economic relationships with censored, truncated and limited dependent variables. Unpublished Ph.D. Thesis, University of Rochester, 1975.

[13] NELSON, F. D. (1976). Censored regression models with unobserved, stochastic censoring thresholds. Social Science Working Paper No. 115, March 1976, California Institute of Technology.

[14] SOLMON, L. C. and TAUBMAN(1973). *Does College Matter?* Academic Press, New York.

[15] TAUBMAN, P. J. and WALES, T. (1974). *Higher Education and Earnings.* McGraw-Hill, New York.

[16] TROST, R. (1976). Demand for housing - a model based on inter-related choices between owning and renting. Discussion Paper, University of Florida, Sept. 1976.

P.R. Krishnaiah, ed., APPLICATIONS OF STATISTICS

SOME APPLICATIONS OF COMPLEX DISTRIBUTIONS TO THE TESTING OF STATISTICAL HYPOTHESES

KENNETH S. MILLER
Riverside Research Institute, New York, New York

If ϕ is a most powerful (MP) or a uniformly most powerful (UMP) level α test of a statistical hypothesis testing problem described in terms of a test statistic q, then the power function β of the test may be expressed as an integral of the density function F of q, say $\beta = \int_k^\infty F(q)dq$, where the threshold k is determined by the level α. The usefulness of this formula for β is greatly enhanced if F is known explicitly. It is to this problem that we address our efforts; that is, our main concern is to determine explicit formulas for density functions of test statistics for a variety of hypothesis testing problems. We consider a number of MP and UMP tests -- as well as tests invariant with respect to various groups of transformations -- and show how to calculate their frequency functions in a tractable form. Both the single cell and multiple cell cases receive attention. We also indicate some practical areas to which complex distribution theory may be applied, and give a brief summary of the closed form density functions associated with the second order statistics of the bivariate complex Gaussian distribution.

1. INTRODUCTION

If $\mathcal{X} = \{x(t) \mid t \in T\}$ is a real wide-sense stationary stochastic process with covariance function R, then the power spectrum of $\mathcal{X}$ is real, nonnegative, and symmetric about zero. In studies of the statistics of ionized media it has been observed that the spectrum of the underlying random process is not symmetric about zero, but around some nonzero frequency, say β, where β may be interpreted as mean frequency [26], [27], [29]. This particular physical phenomenon, among others, then leads to complex stochastic processes, and in particular, complex Gaussian processes [2], [3], [7], [20], [22]. Now the existence of a complex density function is a useful tool, say, for example, for calculating likelihood ratios and distributions of certain sufficient statistics. In this paper we shall survey some properties of complex multivariate Gaussian density functions, and in particular, adumbrate some applications to hypothesis testing theory.

The concept of a complex Gaussian density function apparently was first introduced by Wooding [31]. (See also [3], [22], [24].) We recall the definition. Suppose $\underset{\sim}{H}$ is an n × n Hermitian matrix and $\underset{\sim}{z}$ an n-dimensional complex vector ($\underset{\sim}{z}'$ its transpose, $\bar{\underset{\sim}{z}}$ its conjugate). Then $q = \bar{\underset{\sim}{z}}'\underset{\sim}{H}\underset{\sim}{z}$ is a real Hermitian form.

Thus if $\underset{\sim}{R}$ is a positive definite Hermitian matrix and $\underset{\sim}{c}$ a complex vector, then $(\bar{\underset{\sim}{z}} - \bar{\underset{\sim}{c}})'\underset{\sim}{R}^{-1}(\underset{\sim}{z} - \underset{\sim}{c})$ is positive, and

$$f(\underset{\sim}{z}) = \frac{1}{\pi^n \det \underset{\sim}{R}} e^{-(\bar{\underset{\sim}{z}} - \bar{\underset{\sim}{c}})'\underset{\sim}{R}^{-1}(\underset{\sim}{z} - \underset{\sim}{c})}$$

has area one,

$$\int_{-\infty}^{\infty} f(\underset{\sim}{z})\, d\underset{\sim}{z} = 1 .$$

That is, f has the properties of a density function. If $\underset{\sim}{z}$ is a complex n-dimensional random vector, then we shall say that $\underset{\sim}{z}$ is normal $(\underset{\sim}{c},\underset{\sim}{R})$ in n complex dimensions if its density function is f. (There are other approaches.) in particular we have

$$\mathscr{E}\underset{\sim}{z} = \underset{\sim}{c}$$

$$\mathscr{E}(\underset{\sim}{z}-\underset{\sim}{c})(\bar{\underset{\sim}{z}}-\bar{\underset{\sim}{c}})' = \underset{\sim}{R}$$

$$\mathscr{E}(\underset{\sim}{z}-\underset{\sim}{c})(\underset{\sim}{z}-\underset{\sim}{c})' = \underset{\sim}{0} .$$

2. DISTRIBUTIONS ASSOCIATED WITH THE COMPLEX GAUSSIAN

Let $\mathscr{Z} = \{z(t) \mid -\infty < t < \infty\}$ be a complex mean zero stationary Gaussian process, and let $\underset{\sim}{z}'_n = \{z_n(t_1), z_n(t_2)\}$, $1 \leqq n \leqq N$, be N independent two-dimensional vector observations on $\mathscr{Z}$. If $\psi(h) = \mathscr{E}z(t+h)\bar{z}(t)$ is the covariance function of $\mathscr{Z}$, then the covariance matrix of the vector $\underset{\sim}{z}_n$ is

$$\underset{\sim}{\Psi} = \mathscr{E}\underset{\sim}{z}_n\bar{\underset{\sim}{z}}'_n = \begin{bmatrix} \psi(0) & \psi(h) \\ \bar{\psi}(h) & \psi(0) \end{bmatrix} , \quad h = t_1 - t_2 \neq 0 .$$

Thus

$$\hat{\psi}(h) = \frac{1}{N} \sum_{n=1}^{N} z_n(t_1)\bar{z}_n(t_2)$$

is an unbiased estimate of $\psi(h)$, $h \neq 0$, and

$$\hat{\psi}(0) = \frac{1}{2N} \sum_{n=1}^{N} |\underset{\sim}{z}_n|^2$$

is an unbiased estimate of $\psi(0)$. In many statistical problems $\hat{\psi}(h)$ and $\hat{\psi}(0)$ are sufficient statistics.

More generally, and in simpler notation, let $\underset{\sim}{z}' = \{z,\zeta\}$ be a two-dimensional complex random vector normal $(\underset{\sim}{0},\underset{\sim}{R})$ where $\underset{\sim}{R}$ is positive definite. We write

$$\underset{\sim}{R}^{-1} = \begin{bmatrix} \gamma_{11} & \gamma_{12} \\ \bar{\gamma}_{12} & \gamma_{22} \end{bmatrix} , \quad \gamma_{12} = \gamma_1 + i\gamma_2 .$$

Note that we do not assume the variance (diagonal) terms of $\underset{\sim}{R}$ to be equal. If $\underset{\sim}{z}_1,\ldots,\underset{\sim}{z}_N$ are independent observations on $\underset{\sim}{z}$, then we may define the second order statistics

$$u = \sum_{n=1}^{N} z_n \bar{\zeta}_n \equiv u_1 + iu_2$$

$$w_1 = \sum_{n=1}^{N} |z_n|^2$$

$$w_2 = \sum_{n=1}^{N} |\zeta_n|^2$$

and

$$w = \tfrac{1}{2}(w_1 + w_2) .$$

The joint density function of u_1, u_2, w_1, w_2 is

$$j(u_1,u_2,w_1,w_2) = \frac{(w_1w_2 - |u|^2)^{N-2}}{\pi(N-1)!(N-2)!(\det \underset{\sim}{R})^N} \times e^{-2[\gamma_1u_1 + \gamma_2u_2 + \frac{1}{2}(\gamma_{11}w_1+\gamma_{22}w_2)]} \qquad (1)$$

for w_1, $w_2 > 0$, $w_1w_2 > |u|^2$, $-\infty < u_1$, $u_2 < \infty$; and $j = 0$ otherwise. Equation (1) is obtained most readily by first finding the characteristic function of u_1, u_2, w_1, w_2 and then computing its Fourier transform. Of course, one recognizes (1) as the bivariate complex Wishart distribution. That is,

$$j(\underset{\sim}{\Omega}) = \frac{(\det\underset{\sim}{\Omega})^{N-2}}{\pi(N-1)!(N-2)!(\det \underset{\sim}{R})^N} e^{-\mathrm{tr}\underset{\sim}{\Omega}\underset{\sim}{R}^{-1}}$$

where

$$\underset{\sim}{\Omega} = \begin{bmatrix} w_1 & u \\ \bar{u} & w_2 \end{bmatrix} \equiv \sum_{n=1}^{N} \underset{\sim}{z}_n \bar{\underset{\sim}{z}}'_n .$$

Goodman [7] has generalized (1) to the p-variate case.

The temptation to find the density functions of various combinations of u_1, u_2, w_1, w_2 is overwhelming. Many of these distributions appear in (a not generally available publication) by Goodman [6] and in [22], [23], [24]. A modest knowledge of special functions and a modicum of manipulative ability yield:

Variate	*Form of Density Function*
u_1, u_2, w	$I_{\frac{1}{2}(2N-3)}$
$\lvert u\rvert, w_1, w_2$	I_o
w_1, w_2	I_{N-1}
u_1, u_2	K_{N-1}
$\frac{u_1}{\sqrt{w_1 w_2}}, \frac{u_2}{\sqrt{w_1 w_2}}$	${}_2F_1$ = elementary
$\frac{u_1}{w}, \frac{u_2}{w}$	${}_2F_1$
u_1, w	I_{N-1}
$\lvert u\rvert, w$	$I_o \times I_{\frac{1}{2}(2N-3)}$
$\frac{\lvert u\rvert}{\sqrt{w_1 w_2}}$	${}_2F_1$
$\frac{\lvert u\rvert}{w}$	P_{2N-1} (if $\gamma_{11} = \gamma_{22}$)
u_1	$K_{N-\frac{1}{2}}$
u_2	$K_{N-\frac{1}{2}}$
$\lvert u\rvert$	$I_o \times K_{N-1}$
$\frac{u_2}{u_1}$	${}_2F_1$ = elementary

Variate	*Form of Density Function*
$\arctan \frac{u_2}{u_1}$	$\int K_{N-1}$ = elementary
w_1	Elementary
w_2	Elementary
w	$I_{N-\frac{1}{2}}$
$w_1 w_2 - \lvert u\rvert^2$	K_1
$w^2 - \lvert u\rvert^2$	$I_{\frac{1}{2}(2N-3)}$

In the above Table, I_ν is the modified Bessel function of the first kind and order ν, K_μ is the modified Bessel function of the second kind and order μ, ${}_2F_1$ is the hypergeometric function, and P_{2N-1} is the Legendre polynomial of degree 2N-1. Recall that if m is an integer, $I_{m-\frac{1}{2}}$ and $K_{m-\frac{1}{2}}$ may be written as finite sums of elementary functions.

Now suppose $\underset{\sim}{z}$ is normal $(\underset{\sim}{c},\underset{\sim}{R})$, that is, has non-zero mean. Then the complex first order statistics are

$$X = \frac{1}{N} \sum_{n=1}^{N} z_n \equiv X_1 + iX_2$$

$$Y = \frac{1}{N} \sum_{n=1}^{N} \zeta_n \equiv Y_1 + iY_2$$

and the second order statistics are

$$U = \frac{1}{N} \sum_{n=1}^{N} (z_n - X)(\overline{\zeta}_n - \overline{Y}) \equiv U_1 + iU_2$$

$$W_1 = \frac{1}{N} \sum_{n=1}^{N} |z_n - X|^2$$

$$W_2 = \frac{1}{N} \sum_{n=1}^{N} |\zeta_n - Y|^2$$

$$W = \tfrac{1}{2}(W_1 + W_2) \quad .$$

One may show that $\underset{\sim}{x}' = \{X,Y\}$ and $\underset{\sim}{V}' = \{U_1,U_2,W_1,W_2\}$ are independent. The complex random vector $\underset{\sim}{x}$ is normal $(\underset{\sim}{c}, N^{-1}\underset{\sim}{R})$ and the joint density function of $\underset{\sim}{V}$ is

$$J(U_1,U_2,W_1,W_2) = \frac{N^{2N-2}\,(W_1W_2 - |U|^2)^{N-3}}{\pi(N-2)!(N-3)!(\det \underset{\sim}{R})^{N-1}} \times e^{-2N[\gamma_1U_1 + \gamma_2U_2 + \frac{1}{2}(\gamma_{11}W_1+\gamma_{22}W_2)]}$$

for W_1, $W_2 > 0$, $W_1W_2 > |U|^2$, $-\infty < U_1$, $U_2 < \infty$, and is zero otherwise. Note that J is the same as j [see (1)] (with one less degree of freedom). Hence, if we replace the lower case u's and w's in the above Table by corresponding capital letters, then their density functions are trivially available. One also observes that the joint density function of X_2/X_1 and Y_2/Y_1 is a hypergeometric function -- which may be written as an elementary function.

3. SOME SIMPLE APPLICATIONS TO HYPOTHESIS TESTING

There are many applications of complex distributions to statistical problems. For example, we cite [1-16], [19-23], [26-29]. However, I shall confine my attention to certain problems in hypothesis testing theory.

In testing statistical hypotheses, one invariably forms the likelihood ratio and chooses a statistic q which is a monotone function of this ratio. If F is the density function of q, then the power function is of the form

$$\beta = \int_k^\infty F(q)\,dq .$$

The usefulness of this formula is greatly enhanced if the frequency function of q is known *explicitly*. This is the specific type of problem to which we shall address our attention.

If radar observations are made on an ionized media, then the wake echoes are the sum of a large number of scattering returns from a continuum of uncorrelated "turbules" in motion. Over an appropriate time interval, these statistics remain stationary. Thus by the central limit theorem, it is not unreasonable to represent the wake echoes by a complex stationary Gaussian process. The optimality of the likelihood ratio tests rests on this assumption. We shall consider a number of hypothesis testing problems, and their test statistics, and then look into the distributions of these statistics.

Suppose we wish to distinguish between two classes of ionized media. One method is to compare spectra, or alternatively, covariances. If, for example, the covariance function of a stationary complex Gaussian process is of the form

$$\psi(h) = a\, e^{-\frac{1}{2}(2\pi bh)^2}\, e^{i2\pi ch}$$

(which is a reasonable model) then a may be interpreted as the total power, b as the spread, and c as the mean frequency.

Let us now suppose that our echoes are N two-dimensional independent complex Gaussian vectors, $\underset{\sim}{z}_1,\ldots,\underset{\sim}{z}_N$, each normal $(\underset{\sim}{0},\underset{\sim}{R})$ where $\underset{\sim}{R}$ is positive definite, but not necessarily stationary. Then it is not unreasonable to consider the problem of testing (at level α) the simple hypothesis

$$\mathcal{H}_o \quad : \quad \underset{\sim}{R} = \underset{\sim}{R}_o \tag{2a}$$

against the simple alternative

$$\mathcal{H}_1 \quad : \quad \underset{\sim}{R} = \underset{\sim}{R}_1 \quad . \tag{2b}$$

Forming the likelihood ratio leads to the test statistic

$$q = \frac{1}{N}\sum_{n=1}^{N} \overline{\underset{\sim}{z}}'_n(\underset{\sim}{R}_o^{-1} - \underset{\sim}{R}_1^{-1})\underset{\sim}{z}_n \tag{3}$$

(a monotone increasing function of the likelihood ratio). Thus a test function ϕ is

$$\phi(\underset{\sim}{z}) = \begin{cases} 1 \ , & q > k \\ 0 \ , & q < k \end{cases}$$

where the threshold k is determined by the level α. If F_o is the density function of q under the null hypothesis $\mathcal{H}_o$, then

$$\alpha = \mathcal{E}_o\phi(\underset{\sim}{z}) = \int_k^\infty F_o(q)\, dq \ ,$$

and if F_1 is the density function of q under the alternative hypothesis $\mathcal{H}_1$, then the power β is given by

$$\beta = \mathcal{E}_1\phi(\underset{\sim}{z}) = \int_k^\infty F_1(q)\, dq \ .$$

That is, ϕ is a MP level α test of (2).

In many problems the total power is *not* a distinguishing feature of either hypothesis. Thus if $\underset{\sim}{z}$ is normal $(\underset{\sim}{c},\underset{\sim}{R})$ where

$$\underset{\sim}{R} = \begin{bmatrix} R_{11} & R_{12} \\ \bar{R}_{12} & R_{22} \end{bmatrix}$$

we would like to test

$$\begin{array}{lll} \mathcal{H}_o : & |\lambda| \leq |\lambda_o| & \\ & & \text{any } R_{11}, R_{22} > 0, \quad \text{any } \underset{\sim}{c}, \quad \text{any } \angle\lambda \\ \mathcal{H}_1 : & |\lambda| > |\lambda_o| & \end{array} \tag{4}$$

where λ is the correlation coefficient. For this problem

$$Q = \frac{|U|}{\sqrt{W_1 W_2}} \tag{5}$$

is a UMP invariant test statistic of (4) with respect to some group G. (The statistics U, W_1, W_2 are defined in Section 2.) If $R_o \equiv R_{11} = R_{22}$, that is the stationary case,

$$Q = \frac{|U|}{W} \tag{6}$$

is a UMP invariant test statistic. One also has that in the mean zero stationary case,

$$q = \frac{\sum_{n=1}^{N} \bar{\underset{\sim}{z}}'_n \underset{\sim}{R}_o^{-1} \underset{\sim}{z}_n}{\sum_{n=1}^{N} \bar{\underset{\sim}{z}}'_n \underset{\sim}{R}_1^{-1} \underset{\sim}{z}_n} \tag{7}$$

is a MP invariant test statistic of

$$\begin{array}{lll} \mathcal{H}_o : & \lambda = \lambda_o & \\ & & \text{any } R_o > 0 \\ \mathcal{H}_1 : & \lambda = \lambda_1 & . \end{array}$$

We mention in passing that if z is a one-dimensional complex random variable normal (c,R), and $z_1,\ldots,z_N$ are N independent observations on z, then with

$$Z = \frac{1}{N} \sum_{n=1}^{N} z_n$$

and

$$W = \frac{1}{N} \sum_{n=1}^{N} |z_n - Z|^2 ,$$

the statistic

$$q = |Z| \tag{8}$$

leads to a UMP invariant test of

$$\mathcal{H}_o : \quad |c| \leq |c_o| \qquad \angle c \text{ arbitrary, R known}$$

$$\mathcal{H}_1 : \quad |c| > |c_o| ,$$

and if

$$\rho = \frac{|c|}{\sqrt{R}} ,$$

then

$$q = \frac{|Z|}{\sqrt{W}} \tag{9}$$

leads to a UMP invariant test of

$$\mathcal{H}_o : \quad \rho \leq \rho_o \qquad \text{any } R > 0, \ \angle c \text{ arbitrary}$$

$$\mathcal{H}_1 : \quad \rho > \rho_o .$$

Now as we mentioned earlier, our goal is to find the frequency functions of (3), (5), (6), (7), (8), (9). If we write (3) in the form

$$q = \frac{1}{N} \sum_{n=1}^{N} \bar{z}_n' G z_n$$

where

$$G = R_o^{-1} - R_1^{-1} ,$$

then while it is true that R_o and R_1 are positive definite, the matrix G may be definite, indefinite or semidefinite. By diagonalizing G one may show that q is of the form $I_{N-\frac{1}{2}}$ if G is definite, of the form $K_{N-\frac{1}{2}}$ if G is indefinite, and exponential if G is semidefinite. (See [22], [23].)

For (5) and (6) we have seen that the density function of Q is the hypergeometric function. To find the frequency function of (7), the simultaneous diagonalization of $\underset{\sim}{R}_o^{-1}$ and $\underset{\sim}{R}_1^{-1}$ leads to the beta distribution. The density function of (8) is well-known (an I_o Bessel function) and it is easy to show that the frequency function of (9) is a confluent hypergeometric function. These and other problems are discussed in detail in [23].

4. THE MULTIPLE CELL PROBLEM

If observations are made on an ionized media at various downstream distances corresponding to k = 1,2,...,K, then at each cell k we have N independent samples. In the simplest (one-dimensional) complex case, let $z^{(k)}$, $1 \leq k \leq K$, be K independent complex scalar variates normal $(c^{(k)}, R^{(k)})$ where $R^{(k)} > 0$ and the $c^{(k)}$ are assumed known. Let $z_1^{(k)}, \ldots, z_N^{(k)}$ be N independent observations on $z^{(k)}$, $1 \leq k \leq K$. Based on the NK observations $z_n^{(k)}$, $1 \leq n \leq N$, $1 \leq k \leq K$, we wish to test, at level α, the simple null hypothesis

$$\mathcal{H}_o \quad : \quad \underset{\sim}{R} = \underset{\sim}{R}_o \tag{10a}$$

against the simple alternative

$$\mathcal{H}_1 \quad : \quad \underset{\sim}{R} = \underset{\sim}{R}_1 \tag{10b}$$

where

$$\underset{\sim}{R} = ||\delta_{jk} R^{(k)}|| \quad .$$

The usual likelihood ratio arguments lead to the test statistic

$$q = \sum_{k=1}^{K} \sum_{n=1}^{N} |z_n^{(k)} - c^{(k)}|^2 (R_o^{(k)^{-1}} - R_1^{(k)^{-1}})$$

$$= \sum_{n=1}^{N} (\bar{\underset{\sim}{z}}_n - \bar{\underset{\sim}{c}})' (\underset{\sim}{R}_o^{-1} - \underset{\sim}{R}_1^{-1}) (\underset{\sim}{z}_n - \underset{\sim}{c}) \tag{11}$$

where $\underset{\sim}{z}_n' = \{z_n^{(1)}, \ldots, z_n^{(K)}\}$ and $\underset{\sim}{c}' = \{c^{(1)}, \ldots, c^{(K)}\}$. If we find the characteristic function of q, expand in partial fractions, invert and rearrange terms, we may find the density function F of q (an elementary function). If N = 1, then except for a scalar factor, F is of the form

$$\begin{vmatrix} e^{-\gamma_1 q} D_1 & \cdot & \cdot & \cdot & \cdot & e^{-\gamma_J q} D_J \\ 1 & \cdot & \cdot & \cdot & \cdot & 1 \\ \gamma_1 & \cdot & \cdot & \cdot & \cdot & \gamma_J \\ \cdot & \cdot & \cdot & \cdot & \cdot & \cdot \\ \gamma_1^{J-2} & \cdot & \cdot & \cdot & \cdot & \gamma_J^{J-2} \end{vmatrix}, \quad q > 0 \tag{12}$$

where

$$\gamma_k^{-1} = R^{(k)}(R_o^{(k)^{-1}} - R_1^{(k)^{-1}})$$

and $\gamma_k > 0$, $1 \leq k \leq J \leq K$ while

$$D_k = \prod_{j=J+1}^{K} (\gamma_j - \gamma_k)^{-1}, \quad 1 \leq k \leq J .$$

A similar expression, valid for $q < 0$, holds for the set $\gamma_k < 0$, $J+1 \leq k \leq K$.

Now suppose $\underset{\sim}{z}^{(k)}$, $1 \leq k \leq K$, are K independent two-dimensional complex random vectors normal $(\underset{\sim}{0}, \underset{\sim}{R}^{(k)})$ where the $\underset{\sim}{R}^{(k)}$ are positive definite, and let $\underset{\sim}{z}_1^{(k)}, \ldots, \underset{\sim}{z}_N^{(k)}$ be N independent observations on $\underset{\sim}{z}^{(k)}$. Based on the NK observations $\underset{\sim}{z}_n^{(k)}$, we wish to test the simple null hypothesis

$$\mathcal{H}_o \quad : \quad \underset{\sim}{R} = \underset{\sim}{R}_o \tag{13a}$$

against the simple alternative

$$\mathcal{H}_1 \quad : \quad \underset{\sim}{R} = \underset{\sim}{R}_1 \tag{13b}$$

where

$$\underset{\sim}{R} = ||\delta_{jk} \underset{\sim}{R}^{(k)}|| .$$

This leads to the test statistic

$$q = \sum_{k=1}^{K} \sum_{n=1}^{N} \bar{\underset{\sim}{z}}_n^{(k)'} (\underset{\sim}{R}_o^{(k)^{-1}} - \underset{\sim}{R}_1^{(k)^{-1}}) \underset{\sim}{z}_n^{(k)} . \tag{14}$$

Another analysis similar to that used in the preceding problem (but more laborious) leads to an explicit expression for the density function of q. In particular, if $N = 1$ and all the $R_o^{(k)^{-1}} - R_1^{(k)^{-1}}$ are indefinite, then the density function is of the same form as (12).

Now we shall consider various invariant tests that are independent of the total power. In the one-dimensional case we may write

$$\begin{aligned} \mathcal{H}_o &: \quad P = P_o \\ & \qquad\qquad\qquad \text{any } R^{(1)} > 0 \\ \mathcal{H}_1 &: \quad P = P_1 \end{aligned} \tag{15}$$

where

$$P = \frac{1}{R^{(1)}} \left|\left|\delta_{jk} R^{(k)}\right|\right| \quad , \; 2 \leq k \leq K .$$

This leads to

$$q = \frac{\sum_{k=1}^{K} \sum_{n=1}^{N} |z_n^{(k)} - c^{(k)}|^2 R_o^{(k)^{-1}}}{\sum_{k=1}^{K} \sum_{n=1}^{N} |z_n^{(k)} - c^{(k)}|^2 R_1^{(k)^{-1}}} \tag{16}$$

as a MP invariant test of (15). Arguments similar to those used above lead to the density function of q. In particular, if $N = 1$, the frequency function is similar to (12) with J replaced by K, γ_k replaced by

$$\lambda_k = \frac{R_1^{(k)}}{R_o^{(k)}} ,$$

and $e^{-\gamma_k q} D_k$ replaced by

$$(\lambda_k - q)^{K-2} \, U_{\lambda_k}$$

where

$$U_\lambda = \begin{cases} 1 \quad , & \lambda > q \\ 0 \quad , & \lambda < q . \end{cases}$$

In the bivariate case we have the option of testing

$$\begin{aligned} \mathcal{H}_o &: \quad \underset{\sim}{P} = \underset{\sim}{P}_o \qquad \text{any } R^{(k)} > 0 \\ \mathcal{H}_1 &: \quad \underset{\sim}{P} = \underset{\sim}{P}_1 \end{aligned} \tag{17}$$

where

$$\underset{\sim}{P} = ||\delta_{jk} R^{(k)^{-1}} \underset{\sim}{R}^{(k)}||$$

(and the $R^{(k)}$ are the diagonal elements of $\underset{\sim}{R}^{(k)}$) or

$$\begin{aligned} \mathcal{H}_o &: \quad \underset{\sim}{Q} = \underset{\sim}{Q}_o \qquad \text{any } R^{(1)} > 0 \\ \mathcal{H}_1 &: \quad \underset{\sim}{Q} = \underset{\sim}{Q}_1 \end{aligned} \tag{18}$$

where

$$\underset{\sim}{Q} = ||\delta_{jk} R^{(1)^{-1}} \underset{\sim}{R}^{(k)}|| .$$

For the problem of (17)

$$q = \prod_{k=1}^{K} \frac{\sum_{n=1}^{N} \bar{\underset{\sim}{z}}_n^{(k)'} \underset{\sim}{R}_o^{(k)^{-1}} \underset{\sim}{z}_n^{(k)}}{\sum_{n=1}^{N} \bar{\underset{\sim}{z}}_n^{(k)'} \underset{\sim}{R}_1^{(k)^{-1}} \underset{\sim}{z}_n^{(k)}} . \tag{19}$$

It appears that one can find the density function of (19) in tractable form only for special values of K and N. For example, for N = 1 and K arbitrary, and under the assumption that the correlation coefficients are independent of the cell k, then the frequency function of q may be written in terms of

$$S_K(y) = \int_0^{\infty} \left(\frac{\sin\sigma}{\sigma}\right)^K \cos y\sigma \, d\sigma$$

where y is a linear function of ln q. Of course $S_K(y)$ may be exhibited as a piece-wise (K-1) *th* degree polynomial in y. For the problem of (18),

$$q = \frac{\sum_{k=1}^{K} \sum_{n=1}^{N} \bar{\underset{\sim}{z}}_n^{(k)'} \underset{\sim}{R}_o^{(k)^{-1}} \underset{\sim}{z}_n^{(k)}}{\sum_{k=1}^{K} \sum_{n=1}^{N} \bar{\underset{\sim}{z}}_n^{(k)'} \underset{\sim}{R}_1^{(k)^{-1}} \underset{\sim}{z}_n^{(k)}} \tag{20}$$

is a MP invariant test statistic. With an appropriate identification of parameters, its density function is identical with the density function of the statistic of (16). All the above problems, and others, are treated in detail in [23].

REFERENCES

[1] ARENS, R. (1957). Complex processes for envelopes of normal noise. *Trans. I.R.E.*, IT-3, 204-207.

[2] CRAMÉR, H. and LEADBETTER, M. R. (1967). *Stationary and Related Stochastic Processes*. John Wiley & Sons, Inc., New York.

[3] DOOB, J. L. (1953). *Stochastic Processes*. John Wiley & Sons, Inc., New York.

[4] DUGUNDJI, J. (1958). Envelopes and pre-envelopes of real waveforms. *Trans. I.R.E.*, IT-4, 53-57.

[5] GIRI, N. (1965). On the complex analogues of T^2 - and R^2 - tests. *Ann. Math. Statist.*, 36, 664-670.

[6] GOODMAN, N. R. (1957). On the joint estimation of the spectra, co-spectrum and quadrature spectrum of a two-dimensional stationary Gaussian process. *Scientific Paper No.* 10, Engineering Statistics Laboratory, College of Engineering, New York University, New York, New York.

[7] GOODMAN, N. R. (1963). Statistical analysis based on a certain multivariate complex Gaussian distribution (An introduction). *Ann. Math. Statist.*, 34, 152-177.

[8] GRETTENBERG, T. L. (1965). A representation theorem for complex normal processes. *Trans. I.E.E.E.*, IT-11, 305-306.

[9] GUPTA, R. P. (1964). Some extensions of the Wishart and multivariate t-distributions in the complex case. *J. Indian Statist. Assoc.*, 2, 131-136.

[10] KABE, D. G. (1966). Complex analogues of some classical noncentral multivariate distributions. *Austral. J. Statist.*, 8, 99-103.

[11] KABE, D. G. (1966). Complex analogues of some results in classical normal multivariate regression theory. *Austral. J. Statist.*, 8, 87-91.

[12] KABE, D. G. (1966). On the distribution of the complex analog of Rao's U statistic. *J. Indian Statist. Assoc.*, 4, 189-194.

[13] KAILATH, T. (1966). The complex envelope of white noise. *Trans. I.E.E.E.*, IT-12, 397-398.

[14] KELLY, E. J., REED, I. S. and ROOT, W. L. (1960). The detection of radar echoes in noise I. *J. Soc. Indust. Appl. Math.*, 8, 309-341.

[15] KELLY, E. J., REED, I. S. and ROOT, W. L. (1960). The detection of radar echoes in noise II. *J. Soc. Indust. Appl. Math.*, 8, 481-507.

[16] KHATRI, C. G. (1965). Classical statistical analysis based on a certain multivariate complex Gaussian distribution. *Ann. Math. Statist.*, 36, 98-114.

[17] KINSMAN, B. (1965). *Wind Waves*. Prentice-Hall, New Jersey.

[18] LEHMANN, E. L. (1959). *Testing Statistical Hypotheses*. John Wiley & Sons, Inc., New York.

[19] LOWENTHAL, S. and ARSENAULT, H. (1970). Image formation for coherent diffuse objects: statistical properties. *J. Opt. Soc. Amer.*, 60, 1478-1483.

[20] MILLER, K. S. (1969). Complex Gaussian processes. *SIAM Rev.*, 11, 544-567.

[21] MILLER, K. S. (1975). Complex random fields. *Inform. Sciences*, 9, 185-225.

[22] MILLER, K. S. (1974). *Complex Stochastic Processes.* Addison-Wesley, Reading, Massachusetts.

[23] MILLER, K. S. (1977). *Hypothesis Testing with Complex Distributions.* (To be published).

[24] MILLER, K. S. (1975). *Multivariate Distributions.* R. E. Krieger Pub. Co., Huntington, New York.

[25] MILLER, K. S. (1973). Serial correlation and spectral moment statistics. *Austral. J. Statist.*, 15, 180-184.

[26] MILLER, K. S. and ROCHWARGER, M. M. (1972). A covariance approach to spectral moment estimation. *Trans. I.E.E.E.*, IT-18, 588-596.

[27] MILLER, K. S. and ROCHWARGER, M. M. (1970). Estimation of spectral moments of time series. *Biometrika*, 57, 513-517.

[28] MILLER, K. S. and ROCHWARGER, M. M. (1976). Hypothesis testing of complex covariance matrices. *Trans. I.E.E.E.*, IT-22, 26-33.

[29] MILLER, K. S. and ROCHWARGER, M. M. (1970). On estimating spectral moments in the presence of colored noise. *Trans. I.E.E.E.*, IT-16, 303-309.

[30] PHILLIPS, O. M. (1966). *The Dynamics of the Upper Ocean.* Cambridge University Press, Cambridge.

[31] WOODING, R. A. (1956). The multivariate distribution of complex normal variables. *Biometrika*, 43, 212-215.

[32] YAO, K. (1973). A representation theorem and its applications to spherically-invariant random processes. *Trans. I.E.E.E.*, IT-19, 600-608.

P.R. Krishnaiah, ed., APPLICATIONS OF STATISTICS

THE WEIGHTED DISTRIBUTIONS: A SURVEY OF THEIR APPLICATIONS

G. P. PATIL AND C. RADHAKRISHNA RAO

The Pennsylvania State University
University Park, Pennsylvania

The Indian Statistical Institute
New Delhi, India

1. INTRODUCTION

The concept of weighted distributions has been introduced and formalized recently, see for example, Rao (1965). Although the situations that involve weighted distributions seem to occur frequently in various fields, the underlying concept of weighted distributions as a major stochastic concept does not seem to have been widely recognized. This paper is intended to provide the needed focus to the notion of weighted distributions and the inference problems associated with them. To this end, some interesting and varied applications from a number of disciplines have been indentified. While they may look quite unrelated at the first glance, they are all indeed natural examples of the single underlying theme of weighted distributions.

To begin with, the paper defines and illustrates the concept of weighted distributions. Situations leading to weighted discrete distributions are discussed first. These relate to the analysis of family data, the problem of family size and alcoholism, the aerial survey involving visibility bias in wildlife ecology, and a discrete renewal system. The problems generating weighted continuous distributions refer to computer generation of random variables using rejection technique, renewal theory, continuous waiting time paradox, cell kinetics and early disease detection, forest products research, particle size distributions by thin section methods and low density traffic streams. Lastly, the weight size distributions used in small particle physics and sedimentology are recognized as the weighted distributions. So also are recognized the moment distributions in economics. It should be worthwhile to comment here that while the earlier examples involve weighted distributions essentially because of built in probability sampling at some stage in the problem, the last two examples can not be explained on the probability sampling basis. These involve a distinctly different recording mechanism responsible for generating a class of weighted distributions.

2. NOTATION AND TERMINOLOGY

Consider a natural mechanism generating a random variable X with *pdf*

$f(x;\ \theta)$ where $\theta \in \Omega$, the parameter space. For drawing a random sample of observations on X, we have to use a method of selection which gives the same chance of including in the sample any observation produced by the original mechanism. But in practice it may so happen that the relative chances of inclusion of two observations x and y are $w(x)\colon\ w(y)$ where $w(\cdot)$ is non-negative valued function. Then the recorded X to be denoted by X_w has the *pdf*

$$f^w(x;\ \theta) = \frac{w(x)f(x;\ \theta)}{\omega} \tag{2.1}$$

where $\omega = E[w(X)] = \int w(x)f(x;\ \theta)dx$ or $= \Sigma w(x)f(x;\ \theta)$ depending on whether X is continuous or discrete. Further, if $0 \leq w(x) \leq 1$, ω is the probability of including an observed value in the sample.

The distribution defined by (2.1) is called a weighted distribution with weight function w(x) which can be arbitrary. When an investigator collects a sample of observations produced by nature, according to a certain model, the original distribution may not thus be reproduced. The main interest in any investigation is, however, to determine the characteristics of the original distribution. Further, it also becomes important to assess the nature and amount of distortion caused in the determination of these characteristics in case the change in the underlying distribution due to sampling bias is ignored.

The following examples may help illustrate a few general situations involving "non-response" responsible for generating weighted distributions.

(i) *Truncation:* The distribution of a random variable truncated to a set T is a weighted distribution with weight function $w(x) = 1$ for $x \in T$ and zero elsewhere.

(ii) *Missing Data:* If a proportion $1-w(x)$, $o \leq w(x) \leq 1$, out of the natural frequency of x is missing, that is if the probability of x missing is $1-w(x)$, the pdf to use for the analysis of the observed data is with the weight function w(x).

(iii) *Damaged Observations:* Consider a damage model where an observation $X = x$ is reduced to y by a destructive process with *pdf* $d(y|x)$ (See Rao (1965). Then the probability that the observation $X = x$ is undamaged is $d(x|x)$, and the distribution of the undamaged observation is the weighted distribution with $w(x) = d(x|x)$. For example, under binomial survival model, $d(x|x) = \theta^x$, $0 < \theta < 1$. An investigator recording only undamaged observations will need to work with a corresponding weighted distribution.

The following table gives various weight functions used in literature that we have come across.

We note that the weight functions in the table are all monotone functions, either increasing or decreasing.

The assumptions of the implications of the relationships between the original distribution of x and the weighted distribution obtained using some weight function $w(x)$ can generate interesting and useful characterization results. See, for example, Rao, Rao-Rubin (1964), Patil-Ord (1975), Gupta (1975), and sections 7 and 8 of this paper.

A TABLE OF WEIGHT FUNCTIONS USED

No	$x \geq 0$	$w(x)$	references
1	General	x	most of the references in this paper
2	discrete	x^{α}, $\alpha > 0$	Rao (1965), Patil-Ord (1975)
3	continuous	x^{α}, $\alpha > 0$	Brown (1972), Patil-Ord (1975)
4	discrete	$1-(1-\beta)^{x}$, $0 < \beta < 1$	Haldane (1938), Rao (1965) Neel-Schull (1966), Cook-Martin (1974)
5	discrete	$x + 1$	This paper
6	discrete	$x(x-1)\ldots(x-r+1)$	Patil-Ord (1975), Gupta (1975)
7	discrete	ϕ^{x}, $0 < \phi < 1$	Rao-Rubin (1964), Rao (1965), Kemp (1973)
8	continuous	$e^{\omega x}$	Patil-Ord (1975)

PART I: DISCRETE MODELS AND SIZE-BIASED PROBABILITY SAMPLING

3. THE ANALYSIS OF FAMILY DATA

Various demographic and social studies involve family size and sex ratio as important factors which have some bearing on the main study. The following examples show as to how the weighted distributions arise here as a result of the size-biased probability sampling. This discussion is based on Rao (1965) and Neel and Schull (1966).

3.1. Sex Ratio and Weighted Binomials: The following data relate to brothers and sisters in families of 104 boys who were admitted to a postgraduate course at the Indian Statistical Institute.

Let us assume that in families of given size n, the probability of a family with x boys coming into our record is proportional to x. Also, suppose that the number of boys follow a binomial distribution with probability parameter π. Then

$$f(x;\ \pi) = \binom{n}{x} \pi^x (1-\pi)^{n-x},$$
$$w(x) = x,\ E\{w(X)\} = \omega = n\pi,$$
$$f^w(x;\pi) = \binom{n-1}{x-1} \pi^{x-1} (1-\pi)^{n-x},$$
$$E[X_w - 1] = (n-1)\ \pi,$$
$$E[X_w/n] = \pi + \frac{1-\pi}{n} > \pi,$$
$$E[(X_w-1)/(n-1)] = \pi.$$

If k boys representing families of size $n_1, n_2, \ldots, n_k$ report $x_1, x_2, \ldots, x_k$ boys, $E[(\Sigma X_i - k)/(\Sigma n_i - k)] = \pi$, and an unbiased estimate of π is $\tilde{\pi} = (\Sigma x_i - k))/(\Sigma n_i - k) = (414 - 104)/(726 - 104) = 310/622 \approx \frac{1}{2}$. Whereas if one wrongly treats x_i's as observations on k randomly drawn families with at least one boy, i.e., as arising from a truncated binomial, then the estimate of π will have a serious upward bias.

TABLE III.I

based on Rao (1965), p. 325

Family Size	1	2	3	4	5	6	7	8	9	10	11	12	13	15	Total
No. of Families	1	6	6	13	12	7	14	11	12	8	6	5	2	1	104
Brothers:	1	8	12	34	34	29	59	50	54	46	32	31	16	8	414
Sisters:	0	4	6	18	26	13	39	38	54	34	34	29	10	7	312

3.2. Geometric Distribution as Weighted Log Series for Family Size:

A geometric distribution is sometimes found to provide a good fit to an observed distribution of family size, specially when data are obtained from children attending a school. But this may be the effect of sampling with families of large size having a higher chance of being recorded with w(x) proportional to x, and the actual distribution may well be logarithmic series (Rao, 1965). What we have is:

$$f(x;\ \theta) = \frac{1}{x}\ \theta^x/[-\log(1-\theta)] = \alpha\theta^x/x \quad x = 1,2,\ldots$$

$$w(x) = x,\ E[w(X)] = \omega = \alpha\theta/(1-\theta) \tag{3.2}$$
$$f^w(x;\ \theta) = (1-\theta)\theta^{x-1} \qquad x = 1,2,\ldots$$

3.3. A Study of Albinism and Weighted Binomials: Fisher (1934), Haldane (1938), Rao (1965).

If we wish to study the distribution of albino children in families capable of producing such children, we may contact a large number of families and ascertain from each family the number of albinos. The families with at least one albino child provide a truncated binomial distribution and the probability of a child being an albino can be estimated from such a distribution. But this method of investigation is wasteful as the proportion of abnormal families is investigated. A convenient method in such a case is first to discover an albino child and and through it obtain the information about the family to which it belongs. But such a procedure may not give equal chance to all families in which albinos have occurred. The exact chance for a family with x albinos is that of detecting at least one of its albino children, which may be a function of x. The weight functions of the following form have been studied.

$$w_1(x) = 1 - (1-\beta)^x, \; 0 \leq \beta \leq 1$$
$$w_2(x) = x^\alpha \qquad (3.3)$$

Rao (1965) observes that $\alpha = 1/2$ seems to provide a good fit.

3.4. Analysis of Family Data in Human Heredity: Neel and Schull (1966), p. 211-229.

A primary object in collecting family data is to compare the proportion of affected children actually observed with some theoretical proportion based upon the type of mating and the suspected mode of inheritance. To this end, we may select families at random without reference to the type of offspring produced and study only the families which contain at least one affected child. If we are interested in a rare inherited trait, then a random selection of all families will necessarily lead to a preponderance of families in the data which could yield no information, since the majority of families would not possess the rare gene. Ascertainment through affected individuals may then be the only reasonable way of going about the data collection.

Neel and Schull consider binomial distributions with weight function

$$w(x) = 1 - (1-\beta)^x, \; 0 \leq \beta \leq 1 \qquad (3.4)$$

and estimate the sighting probability β to be 80/100, and use this estimate further in estimating the desired common probability parameter of the binomial distributions.

4. ALCOHOLISM AND FAMILY SIZE

In a social psychological study relating to alcoholism Smart (1963) tested various hypotheses on the basis of data obtained from 242 alcoholics treated in three alcoholism clinics in Ontario, Canada. One of the variables recorded is the size of the family to which an alcoholic belongs. What is the appropriate theoretical distribution with which the observed distribution can be compared?

It is clear that the method of ascertainment is such that it gives greater chance to families of larger size being represented in our sample. Then the question arises as what is the appropriate weight function to be applied to the actual distribution of family size in the population to make it comparable to the observed distribution.

Let p_n be the proportion of families of size n in the population, and suppose that each individual has a probability k, independently of the others, of becoming an alcoholic. Then the distribution of family size among families having at least one alcoholic is

$$p_n^* = \frac{[1-(1-k)^n]\, p_n}{\Sigma[1-(1-k)^n]p_n} \tag{4.1}$$

which is the limit as $k \longrightarrow 0$, is of the form

$$p_n^* = \frac{n\, p_n}{\Sigma n\, p_n} . \tag{4.2}$$

It may be noted that the probability of a family being of size n and having r alcoholics is

$$p_{nr} = \binom{n}{r} k^r (1-k)^{n-r}\, p_n . \tag{4.3}$$

Suppose our sampling mechanism is such that a family with r alcoholics gets a chance proportional to r of being selected. Then the probability for a family of size n with r alcoholics coming into our sample is proportional to

$$p_{nr}^* = r\, p_{nr} \tag{4.4}$$

and the corresponding probability for a family of size n is

$$p_n^* = \frac{\Sigma r p_{nr}}{\Sigma\Sigma r\, p_{nr}} = \frac{n\, p_n}{\Sigma\, n\, p_n} \tag{4.5}$$

which is the same as (4.2). Thus the method of selection of a family with probability proportional to the number of alcoholics does not alter the distribution of family size among families with at least one alcoholic in the population.

Sprott (1964), following the work of Smart (1963), used the 1931 Canadian Census data for finding p_n and fitted the weighted distribution (4.5) to the observed distribution of family size. The observed and expected values are given in the Table.

TABLE

Frequency Distribution of Family Size as Observed on the 242 Alcoholics and Expected on the Assumed Hypothesis

Family Size	1	2	3	4	5	6	7	8	9	TOTAL
Observed	21	32	40	47	29	23	20	11	19	242
Expected	34.2	51.4	47.3	37.6	26.4	17.9	12.1	7.4	8.2	242

The value of χ^2 goodness of fit is 39.1 which is high for 8 degrees of freedom, so that the weight function $w(n) = n$ in (4.5) does not explain the observed data. There is still excess of observed families with larger size over the expected values indicating the possibility that the weight function is of a higher order of magnitude than the family size, which may have some sociological significance.

5. AERIAL SURVEY AND VISIBILITY BIAS

Visibility bias is a recognized problem in aerial survey techniques for estimating wildlife population density. This source of error is generally conceded to be the main cause of inaccurate aerial census data and depending on various contributing environmental factors can produce severely biased population density estimates.

The visibility bias is present because of the failure to observe some animals. Cook and Martin (1974) have presented a model for quadrat sampling of randomly occurring groups whose size follows a single parameter power series distribution when there is a probability $\beta > 0$ of missing single animals.

The sampling model is based on three main assumptions:

(a) that animals occur within quadrats in groups of varying size,

(b) that each animal has a probability β of being observed, and

(c) that, conditional on observing at least one member of a group, the entire group is observed with certainty.

For the purposes of the present discussion, the main point of interest being to estimate the mean group size μ, let the random group size X have *pdf* $f(x;\ \theta)$. It may be easily verified that the recorded group size X_w has pdf

$$f^w(x;\ \theta) = \frac{w(x;\ \beta)f(x;\ \theta)}{\sum_x w(x;\ \beta)f(x;\ \theta)} \tag{5.1}$$

which, for small β, can be approximated by $f^*(x;\ \theta) = xf(x;\ \theta)/\mu$, where $\mu = \Sigma xf(x;\ \theta)$, the mean group size.

Based on the data consisting of g group sizes observed, $x_1, x_2, \ldots, x_g$, the sample mean $\bar{x}$ overestimates μ as its expectation is $\mu + \sigma^2/\mu$ where σ^2 is the variance of X. We consider the following two models:

(i) If we take X-1 ~ Poisson (θ), that is,

$$f(x;\ \theta) = e^{-\theta}\theta^{x-1}/(x-1)!, \tag{5.2}$$

then the estimating equation for the maximum likelihood estimate of μ based on the corresponding $f^*(x;\ \theta)$ is

$$\bar{x}-1 = \hat{\mu} - \frac{1}{\hat{\mu}}. \tag{5.3}$$

(ii) if we take X ~ Zero-truncated Poisson (θ), that is,

$$f(x;\ \theta) = e^{-\theta}\,\frac{\theta^x}{x!}\,/(1-e^{-\theta}), \tag{5.4}$$

then the likelihood equation based on the corresponding $f^*(x;\ \theta)$ is

$$(\bar{x}-1)/(1-e^{-(\bar{x}-1)}) = \hat{\mu}. \tag{5.5}$$

Cook and Martin (1974) report the following aerial moose census data collected over northeast of Minnesota in 1970. The number of groups observed is g = 113.

Group size x:	1	2	3	4	5	6
Observed pf:	.398	.407	.133	.044	.009	.009

We obtain $\bar{x} = 1.84$. Based on the model for X-1 ~ Poisson (θ), the maximum likelihood estimate of μ is 1,505, whereas for X ~ zero-truncated Poisson (θ), the maximum likelihood estimate of μ is 1.478.

6. FOREST DISEASE AND LINE TRANSECT SAMPLING

Consider a forest stand of trees. Suppose that an investigator is interested in estimating the proportion of diseased trees in the stand using a line transect for sampling. For logistic reasons, the investigator uses forest paths to get to the line transect and records the run length x of healthy trees on the transect across the path intersection. The following drawing may provide a schematic picture.

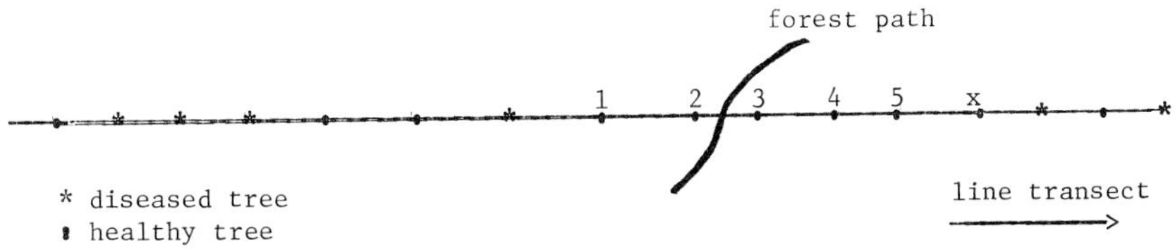

If L represents the random distance between successive trees on the line transect, the probability that the investigator records a healthy run length x is proportional to $w(x) = E[L_1 + L_2 + \ldots + L_{x+1}] = (x+1)E[L]$. Without loss of generality, one may take $E[L] = 1$ and $w(x) = x+1$.

Now, if $X \sim$ Geometric (θ), that is,

$$f(x;\ \theta) = (1-\theta)^x\theta, \quad x = 0,1,2,\ldots$$

$$E[X] = \mu = (1-\theta)/\theta, \tag{6.1}$$

$$E[X+1] = 1/\theta.$$

Also,

$$f_w(x;\ \theta) = (x+1)(1-\theta)^x\theta^2\ ,$$

$$E[X_w] = 2\cdot(1-\theta)/\theta = 2E[X], \tag{6.2}$$

$$E[\frac{1}{X_w+1}] = \theta.$$

Note that $X_w \sim$ Negative Binomial $(2,\ \theta)$. To estimate μ on the basis of x_1, x_2, ..., x_n observed at intersections of n paths with the transect, the best unbiased estimate for μ is $\frac{1}{2}\ \bar{x}$ and *not* $\bar{x}$ as seen from (6.2).

Further, based on these n observations on X_w, the best unbiased estimate for θ is given by $1/(x+1)$. However, a random sample on X does not provide any unbiased estimate for θ.

7. DISCRETE WAITING TIME PARADOX

Feller (1966) states continuous waiting time paradox, and this is discussed in this paper in Part II on continuous models. A discrete version of it arises in the context of the line transect sampling on one side of the forest paths for the purposes of estimating incidence as discussed in the previous section. Using the terminology there, let the run length between diseased trees follow a geometric process with mean length unity, that is, let $p = \frac{1}{2}$, and $E[X] = q/p = 1$ be the expected length of the healthy patch on the line transect *across* a forest path. If V denotes the healthy patch length on the line transect on one side of a forest path, a question arises as to what is $E[V]$? Is $E[V] = 1/2$, because the intersection of the forest path with the line transect has a random location relative to the healthy patch? Or, is $E[V] = 1$. The initial reasoning leading to $E[V] = 1/2$ can be modified by observing that what is observed is not V, but its size-biased version V_w for which $E[V_w] = 1/2\ E[X_w] = 1/2 \cdot 2 = 1$.

More situations involving discrete waiting time paradox may be identified. Uppuluri and Patil (1976) discuss exact discrete analog of Feller's continuous waiting time paradox in connection with inferences about rare events.

Now, we raise two further questions. Is the discrete waiting paradox an exclusive property of the geometric process? Or, of the weight function $w(x) = x+1$? We have the following partial answers when X has a power series distribution (for definitions, see Patil and Joshi (1968)).

Theorem: Assume that X has a power series distribution with series function $f(\theta) = \Sigma a(x)\theta^x$. Let the weight function be $w(x) = x+1$ for the weighted version X_w of X. Then $E[X_w] = 2E[X]$ if and only if $f(\theta) = (1-\theta)^{-1}$, in which case X has the geometric distribution with parameter $1-\theta$.

Proof: We know that

$$E[X] = \theta \frac{d}{d\theta} [\log f(\theta)].$$

Further, we observe that X_w can be shown to have the power series distribution with series function $f_w(\theta) = \frac{d}{d\theta}[\theta f(\theta)]$. The solution of

$$\frac{d}{d\theta} [\log f_w(\theta)] = 2 \frac{d}{d\theta} [\log f(\theta)]$$

gives $f(\theta) = (1-\theta)^{-1}$.
The proof for 'if' part is obvious.

Theorem: Assume that X has a power series distribution with series function $f(\theta) = \Sigma a(x)\theta^x$. Let $[f(\theta)]^k = \Sigma b(x, k)\theta^x$. Let $w(x) \geq 0$, and the corresponding weighted version of X be X_w. Then $E[X_w] = kE[X]$ if and only if $w(x) = b(x, k)/a(x)$.

In particular, if X has a geometric distribution, $E[X_w] = 2E[X]$ if and only if $w(x) = x + 1$.

Proof: We observe that X_w has power series distribution with series function $f_w(\theta) = \Sigma w(x)a(x)\theta^x$. The rest of the proof is straightforward.

PART II. CONTINUOUS MODELS AND SIZE-BIASED PROBABILITY SAMPLING

8. RENEWAL THEORY AND ITS APPLICATIONS

Cox (1962) considers the following problem: Suppose that we have a number of independent realizations of the same renewal process, for example a number of components of the same type in use on different machines. Suppose that to investigate the distribution of failure-time a survey is made at time t to obtain the ages of the components currently in use. The distribution of the observations will be that of U_t, the backward recurrence time.

Cox obtains the limiting pdf of U_t as $t \to \infty$ to be $\bar{F}(x)/\mu$, where $\bar{F}(x) = 1 - F(x)$ with $F(x) = \int_0^x f(x)dx$, the distribution function of the life length of a component with expected mean life length μ. He also obtains the same result alternatively using the weighted-distribution-argument resultant from what he calls length-biased sampling as follows: Consider for any renewal process a recurrence-time R defined in the following way. First, we take a sampling point chosen at random over a very long time interval. Then R is defined as the time measured from the sampling point forward to the next renewal. Entirely the same properties would hold for the time measured from the sampling point back to the previous renewal.

It is clear that one samples from a population of failure-times distributed according to f(x) while the probability of selection of any individual in the population is proportional to its length x. If X_w denotes the failure time of the component in whose life the sampling point falls, then X_w has the *pdf* $f_w(x) = xf(x)/\mu$. Conditionally on $X_w = x_w$, the *pdf* of R is rectangular over $(0, x_w)$, which leads to its marginal *pdf* at x to be $\int_x^\infty \frac{1}{x_w} \cdot f_w(x)dx = \bar{F}(x)/\mu$.

Cox (1969) gives the following example for length-biased sampling. An idealized model of a textile yarn is an assembly of parallel fibers, with *pdf* of fiber length f(x). The fiber left-ends are arranged at random along a line. Take a particular cross-section of the yarn, that is, a particular point on the line, and consider the fibers that intersect this cross-section. This is length-biased sampling; the *pdf* of length of the fibers is $xf(x)/\mu$. For discussions on similar problems, reference may be made to papers by Coleman (1972), Daniels (1942), Palmer (1948), Kriens (1963), and Moran (1966, 1969).

It may be instructive to record a few more examples.

(1) Morrison (1973): Size-biased sampling has implications for certain survey research data that is gathered at an arbitrary point in time. When you ask a person, "When did you last purchase Product A?", or "When did you last attend a baseball game?", etc., you are obtaining particular outcomes of a random variable U_t where U_t is the time elapsed since the last purchase until the randomly selected interview time t.

(2) Robson (1975): In assessing the extent of utilization of the national parks and other recreational facilities, an investigator asks an individual present on such a location as to since when he has been there. The data recorded on a number of such surveyed individuals have the size-biased feature.

9. WAITING TIME PARADOX

Feller, p. 11-13, introduces the waiting time paradox as follows: "Buses arrive in accordance with a Poisson process, the expected time between consecutive buses being α^{-1} (=1, say). I arrive at an epoch t say noontime sharp. What is the expectation $E[W_t]$ of my waiting time W_t for the next bus? Two contradictory answers stand to reason:

(a) The lack of memory of the Poisson process implies that $E[W_t]$ should be independent of t, that is, $E[W_t] = E[W_0] = 1$.

(b) The epoch of my arrival is 'chosen at random' in the interval between two consecutive buses, and for reasons of symmetry $E[W_t] = \frac{1}{2}$.

Let X be the inter-arrival time between two consecutive buses. It is given that X has the *pdf* $f(x) = e^{-x}$ of a standard exponential with $E[X] = 1$. Then it can be shown that the inter-arrival time X* between two consecutive buses that cover my epoch t has *pdf* $f^*(x) = xe^{-x}$, so that $E[X^*] = 2$.

It is clear that $E[W_t] = E[E[W_t|X^*]] = E[\frac{1}{2} X^*] = \frac{1}{2} \cdot 2 = 1$. We note that using X instead of X* in the preceding equation leads to the wrong answer of $\frac{1}{2}$ discussed in (b) above.

Now, in what follows, we prove a theorem that characterizes exponential distribution within the setup of the continuous waiting time paradox resulting from $E[X^*] = 2E[X]$.

Theorem: Let the inter-arrival times have a distribution belonging to the linear exponential family defined by the *pdf* $f(x) = \exp\{\theta x + a(x) - b(\theta)\}$, where $E[X] = b'(\theta) < \infty$. Let X* have *pdf* $f^*(x) = xf(x)/E[X]$. Then X is exponential if and only if $E[X^*] = 2E[X]$.

Proof: From Theorem, $E[X^*] = E[X] + V(X)/E[X]$, and therefore $E[X^*] = 2E[X]$ implies $V(X) = \{E[X]\}^2$, which relation characterizes the exponential distribution within linear exponential family, as proved in Wani and Patil [10].

Now for the exponential distribution within the setup of the continuous waiting time paradox resulting from $E[X^*] = 2E[X]$, we characterize the weight function to be $w(x) = x$, as follows.

Theorem: Let X have *pdf* $f(x) = \lambda e^{-\lambda x}$. Let X* have *pdf* $f^*(x) = \frac{w(x)f(x)}{E[w(X)]}$, where $w(x) > 0$ with $E[w(X)] < \infty$. Then $E[X^*] = 2E[X]$ if and only if $w(x) = x$.

Proof: The condition that $E[X^*] = 2E[X]$ in this simplifies to

$$\frac{\int xw(x)e^{-\lambda x}dx}{\int w(x)e^{-\lambda x}dx} = \frac{2}{\lambda} .$$

We note that LHS represents the mean value function for a linear exponential family with exponent parameter λ. Following Patil and Shorrock [7], the mean value function has to be of the gamma distribution with index parameter 2, implying that $w(x) = x$.

Before we conclude this section, we note that Kotz and Johnson (1974) characterize the exponential distribution to be the distribution of the inter-arrival time X by requiring the waiting time W_t to be identically distributed like X, when t is chosen at random.

10. CELL CYCLE ANALYSIS AND PULSE LABELING

The following is based on Takahashi (1966) and Zelen (1974). The idealized cell cycle model for a proliferating cell consists of four phases. After mitosis, a cell which is destined to proliferate enters a (i) pre-DNA synthesis phase called G_1, (ii) DNA-synthesis or S-phase, (iii) post-DNA synthesis or G_2 phase, and finally (iv) mitosis during which time cell division occurs.

The cell cycle of a population of cells may be studied by exposing the cells to pulse labeling. Only those cells that are in S-phase get labeled. Further, when a labeled cell divides, the label is generally passed on to the daughter cells.

Periodically, the samples of cells from the cell population are observed under a microscope. The information on the proportion of cells labeled and the proportion of labeled mitotic cells can be used to estimate important characteristics of the kinetics of the cell cycle. Of particular interest is the mean duration of the S-phase and the mean cell cycle time.

A difficulty is in the interpretation of the experimental data. A diversity of opinions have been expressed in this connection. The data of Defendi and Manson (1963) or of Tolani (1965), if evaluated by Quastler's revised method (1963) indicate nearly twice as large values for mean S-phase durations. Should Quastler's original method (1959) be true, van't Hof's estimates (1965) for mean S-phase duration must be halved. A standardization of interpreting method is therefore needed to render the published data comparable on a common basis.

As Zelen observes, it is not generally realized that the radioactive labeling of cells is actually choosing a sample of cells by a length-biased sampling procedure. Cells with longer S-phase durations will have a higher probability of being labeled. Hence, the parameters obtained from labeling experiments are different from those of the target population of cells under investigation. It is both instructive and interesting to see that the following analysis using a limiting size-biased sampling argument can explain the puzzle of double and half mentioned above for the mean S-phase duration.

Let X and T denote the durations of the S-phase and the cell cycle. Let their joint *pdf* be f(x, t). Then, the observed x and t have the *pdf* given by

$$f^{w}(x, t) = \frac{xf(x,t)}{E[X]} \qquad 0 \leq x \leq t \leq \infty.$$

It follows that $E(X_w) = E(X)[1+C^2]$ and $E(T_w) = E(T) + \frac{Cov(X,T)}{E(X)}$ where C is the coefficient of variation of X. Assuming that the coefficient of variation of X is unity which is the case when X is exponential, we get

$$E[X_w] = 2E[X] > E[X].$$

Further, if X and T-X are uncorrelated,

$$E[T_w] = E[T] + E[X] > E[T].$$

Thus under the assumptions of X being exponential, and with X and T-X being uncorrelated, the mean duration of the S-phase at the label point will be larger than the population mean by a factor of two and the cell cycle length will have a bias of E[X].

11. EFFICACY OF EARLY SCREENING FOR DISEASE

Currently there is increasing emphasis on early detection programs which identify individuals who are unaware that they have a particular chronic disease. The hope is that earlier detection and treatment will result in an enhanced therapeutic benefit to the individual. Zelen (1974) provides an illuminating discussion of certain aspects of the problems involved. The following is taken from this discussion.

Consider an individual to be in one of three disease states, S_0: a disease free state; S_p: a pre-clinical state; S_c: a clinical state. In the state S_0, individuals do not have the disease, or have a form of the disease which cannot be found by the early detection program. The definition of the pre-clinical state is that the individual has the disease, but is unaware of this condition.

The early detection program is capable of identifying the individual as being in S_p. The clinical state is defined as one where the disease has become clinical and has been diagnosed.

Now consider a population of individuals having a particular chronic disease. There will be a probability distribution governing the duration of the pre-clinical disease. At a particular point in time, this population is examined under an early detection program, and those in S_p are identified. Thus an early detection program identifies individuals by a size-biased scheme. Hence, if a tumor is slow growing, the pre-clinical state is long, then the clinical course of the disease will tend to be long, resulting in longer survival of the individual. Thus, regardless of whether the therapy is enhanced by early detection, individuals so found will tend to live longer than the general population of individuals having this same disease. Consequently, evidence of longer survival for an earlier detected group of cases relative to a control group is not valid scientific evidence of the effectiveness of the screening program.

Under suitable assumptions, it can be shown that

$$E[R_w] = \frac{1}{2}\{E[T] + V(T)/E[T]\} \tag{11.1}$$

where T is the duration of S_p, and R is the true lead time gained by earlier diagnosis because of screening or the duration of the pre-clinical state until screening. R_w is the observed R under size-bias. The time gained by early diagnosis cannot be observed directly. Once the disease is diagnosed, its course is interrupted by treatment. However, the mean lead time can be estimated indirectly by

$$E[R_w] = E[T] - (A_c - A_p), \tag{11.2}$$

where A_p is the mean age of an individual when detected in S_p by screening, and A_c is the mean age at transition into S_c if not screened, (Zelen and Feinleib (1960)).

If T is exponential, (11.1) and (11.2) imply $A_p = A_c$, which says that the mean age of those detected early as well as of those who are routinely diagnosed without the benefit of an early detection program are identical. This apparent paradox arises because of two compensating features. Although those diagnosed by an early detection procedure are found earlier, the size-biased feature of the detection shows that the individuals tend to spend longer time in S_p. Furthermore, if the coefficient of variation T exceeds one, the mean age of individuals detected in an early detection program will even be higher than for those detected through routine medical care.

12. FOREST PRODUCTS RESEARCH

The present discussion of size-biased sampling in forest products research is based on Warren (1975). The general problem here concerns identifying a distribution from data collected by a method in which the probability of an individual being included in the sample is a function of that individual's size. A particular instance arises in examining the size of wood cells. A convenient method consists of measuring the size (cross-sectional area) of only those cells selected by a set of random points on a microscopic field. The probability of a cell being selected is then proportional to its cross-sectional area size, in which case $w(x) = x$. In effect, we have two distributions: (1) the underlying distribution of cell size about which we wish to make inferences, and (2) the distribution generated by our sampling mechanism.

The situations where size-biased sampling is conveniently employed commonly exhibit large numbers of small individuals, with declining numbers of large individuals. Under certain environmental conditions, the diameters of trees may have a distribution of this form, and the forest then be economically surveyed by an analogous system.

13. PARTICLE SIZE DISTRIBUTIONS BY THIN SECTION METHODS

Particle size statistics is an important subject area in several scientific fields such as physics, geology, agricultural engineering, etc. The following discussion is primarily based on Kendall and Moran (1963).

Consider a problem of determining the distribution of the sizes of particles embedded in an opaque medium from the measurement of the figures formed by their intersections with a random plane, or from the segments formed by their intersections with a random line. For example, Tallis (1970) considers a problem as follows: A large block of Swiss cheese was thinly sliced and one hundred slices were drawn at random with replacement. Diameters of each hole appearing in the slices were suitably obtained. It was required to estimate the distribution function of the diameters of the spherical airspaces in the cheese.

If R denotes the diameter of a sphere with *pdf* $f(r)$, the probability of this sphere being intersected by a random plane is proportional to r, and therefore the *pdf* of the diameter of a sphere cut by a random plane is

$$f_w(r) = rd(r) / \int_0^\infty rf(r)dr = rf(r)/\mu. \tag{13.1}$$

Actually, the circular section of the sphere with the plane is what is observed and its diameter x recorded. One obtains its density as

$$\phi(x) = x \int_x^\infty (\frac{1}{r} f_w(r)/\sqrt{(r^2-x^2)}\, dr, \tag{13.2}$$

and gets the solution for f(r) as

$$f(r) = -\frac{2r\mu}{\pi} \int_{x^2}^\infty (r^2-x^2)^{-1/2} \frac{d}{dx} (x^{-1}\phi(x))dx. \tag{13.3}$$

In his pioneering paper, Krumbein (1935) applied the above theory to the analysis of thin sections of sediments in petrography, assuming the particles to be spheres. His original theory neglected the fact that larger spheres have a larger probability of being included in the section. Later work by Greenman (1951), Rosenfelt et al. (1953), Packham (1955) also use the incorrect equation by wrongly using f(r) where $f_w(r)$ given by (13.1) needs to be used. Interestingly, however, Krumbein obtained a closer fit to observed values with his simpler and incorrect theory when examining two samples of sand boty by this cross-section method and by actual measurement of the particle size after separation.

We note that if a random line (linear probe) was used instead of a random plane (planar probe), the diameter of the sphere so detected will the *pdf*

$$f_w(r) = r^2 f(r)/\int_0^\infty r^2 f(r)dr = r^2 f(r)/\mu_2'. \tag{13.4}$$

Further, if a sphere of volume v with *pdf* f(v) is selected by a sampling mechanism with probability proportional to its surface area, the *pdf* of the observed v will be

$$f_w(v) = v^{2/3} f(v)/\int_0^\infty v^{2/3} f(v)dv. \tag{13.5}$$

14. AERIAL SURVEY IN TRAFFIC RESEARCH

We have discussed an aerial survey problem in wildlife ecology in Section 5. In quite a different way, a different kind of observational bias enters in the aerial survey of low density traffic streams. This discussion is based on Brown (1972).

Assume that vehicles enter the highway according to a non-homogeneous Poisson process of intensity λ. The vehicles choose velocities at random from a distribution of V with *pdf* f(v). Assume that $E[1/V] < \infty$.

It turns out that the velocity of a vehicle in $[a,b]$ at time t, as $t \to \infty$, has its limiting *pdf* as

$$f_w(v) = \frac{1}{v} f(v) / \int_0^\infty \frac{1}{v} f(v)dv \tag{14.1}$$

for all [a,b]. Moreover, the velocities of vehicles which lie in [a,b] at time t are independent. Thus if one records the velocities of vehicles in [a,b] at time t (for example through aerial photographs) for t large, he will be observing a random sample of Poisson size drawn from the *pdf* $f_w(V)$ and not the *pdf* $f(v)$.

The main effect of this is that he will tend to observe an unduly high proportion of slower vehicles. This can be seen directly by comparing the two distribution functions corresponding to $f(v)$ and $f_w(v)$. And this is intuitively plausible since slower vehicles take longer to traverse [a,b] and therefore tend to be around when the interval is samples. The problem, then, is to construct an estimate of the distribution function of $f(v)$ based on a sample of Poisson size from the distribution function of $f_w(v)$ defined by (14.1).

15. REJECTION TECHNIQUE IN RANDOM VARIABLES GENERATION

In this section we show that the rejection technique used in the generation of random variables is simply a computer counterpart of the concept of weighted distributions. This discussion is based on Patil-Boswell-Friday (1975).

For, suppose we can generate a r.v. Y with pdf $f_Y(y)$, and we want to generate a r.v. X with pdf $f_X(x)$. The rejection technique consists of two parts: generate an observation on Y (=y, say), and accept y to be the realization of X with probability w(y). Proceed until a realization of Y is accepted. Clearly, in order for this method to work, Y has to assume every value that X can assume. Further,

$$f_X(x) = \frac{w(x) f_Y(x)}{\omega}$$

where ω is the probability that a random generation of Y is accepted. Thus, the acceptance probability w(x) is determined by

$$w(x) = \omega \cdot f_X(x)/f_Y(x)$$

implying that w(x) is proportional to the likelihood ratio of r.v. X on r.v. Y. Further, w(x) and ω are maximized by choosing $1/\omega = \sup f_X(x)/f_Y(x)$, which if bounded, the method works. If not bounded, $w(x) > 1$ for some x, and is no longer a probability for such x. What can be done in such a case is an interesting problem. The following approach may have some use in certain situations.

Let X have *pdf* f(x). Let the weight function be arbitrary, that is, $0 < w(x) < \infty$ with $E[w(X)] < \infty$. Let X_w have the *pdf* $f^w(x) = w(x)f(x)/E[w(X)]$.

If X_w has to be computer-generated from available X, X_w may be realized by using the formulation of

$$f_2(x) = w_1(x)f_1(x)/E_1[w_1(X)]$$

and

$$f^w(x) = w_2(x)f_2(x)/E_2[w_1(X)]$$

where $f_1(x) = f(x)$, $w_1(x) = w(x)/n(x) \leq 1$, and $w_2(x) = n(x) = \lceil w(x) \rceil$, the least integer not less than w(x).

Thus, the available x corresponding to *pdf* f(x) is selected with probability $w_1(x)$, and the selected x is recorded $w_2(x)$ times. If necessary, one may computer-generate x also, and repeat the procedure a large number of times. The recorded values then constitute a simulated population of values for X_w. A random sample of desired size on X_w can be drawn as usual from this population of values of X_w.

PART III. WEIGHTED DISTRIBUTIONS AS MASS-SIZE DISTRIBUTIONS IN GEOLOGY AND ECONOMICS

16. WEIGHT-SIZE DISTRIBUTIONS IN A BREAKAGE PROCESS

Problems relating to small particle statistics continue to be important in physics, sedimentology, atmospheric science and air environmental research. This brief discussion here is based on Schultz (1975).

The distribution of particles formed in a physical breakage process is usually analyzed by mass rather than frequency. Consider, for example, a sieve analysis of crushed stone after a standard drum test, where the weights, or masses, of all particles retained by sieves of various sizes as shown in the following histogram.

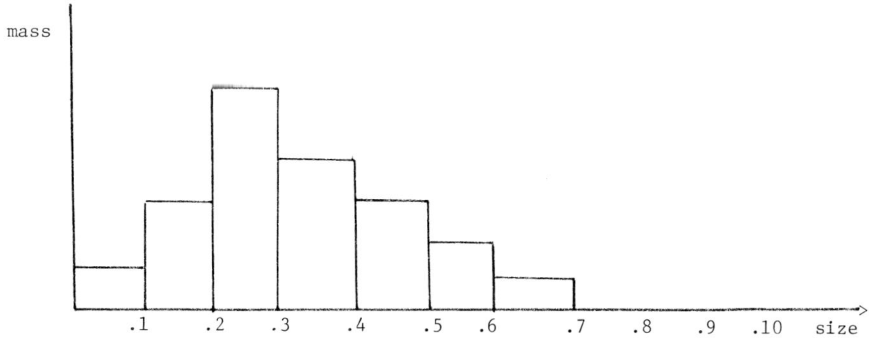

With reference to results such as the above, Krumbein and Pettijohn (see Herdan (1960), p. 291) remarked, "There is one conspicuous manner in which the statistical data of coarse disperse systems differ from the conventional statistical data ... frequency in coarse disperse systems is usually expressed by weight instead of by number. No complete investigations of this ... have been made and the problem of weight versus number is still largely unsolved."

Let X be the diameter size with *pdf* $f(x; \theta)$ with $\mu_j' = E(x^j)$. It is interesting to note that the mass-size density is nothing but $f_w(x; \theta) = x^3 f(x; \theta)/\mu_3'$, the weighted *pdf* with weight function $w(x) = x^3$ of the random variable X_w, say. One may wish to refer to Herdan (1960) for several interesting features and problems related to the weight-size distributions. One problem is to estimate $\mu = E[X]$. It is easy to verify that μ has an alternative interpretation in terms of X_w given by $\mu = H(X_w^3)/H(X_w^2)$, where H stands for the harmonic mean. This is suggested to be an estimating equation for μ in applied literature. It is clear that the observed diameter mean will be estimating μ_4'/μ_3' and not μ in general.

17. MOMENT DISTRIBUTIONS IN ECONOMICS

In the economics literature, weighted distributions with weight functions $w(x) = x^j$ are used. They have been called as moment distributions. The following discussion is based on Hart (1975), Klein (1962), and Ord (1975).

Income distributions at different points of time or location are frequently compared and ranked by some measure of inequality, with low values being preferred to higher values. An income distribution usually follows a typical skew pattern. A main descriptive feature of such a distribution is its degree of inequality. Still another type of tabulation and graph discussed below brings out the degree of inequality.

The Lorenz curve associated with a distribution gives a joint cumulation, both of the frequencies and of the variables being distributed. For example, the distribution of income may be cumulated to show the per cent of income received by the bottom tenth class of spending units, by the bottom two tenth classes, and so on. In other words, if $X \sim F(x)$, where X is the income of an individual having distribution function $F(x)$ and if $F^*(x) = \int_0^x x d\, F(x)/E\,[X]$, the Lorenz curve is the graph of $F^*(x)$ on $F(x)$. Clearly, $F^*(x) \leq F(x)$.

The diagonal line is the curve of equal distribution. The departure of the actual curve from the line of perfect equality shows the degree of inequality. A merit of the Lorenz curve technique is that it enables us to compare distributions in dissimilar units (currencies).

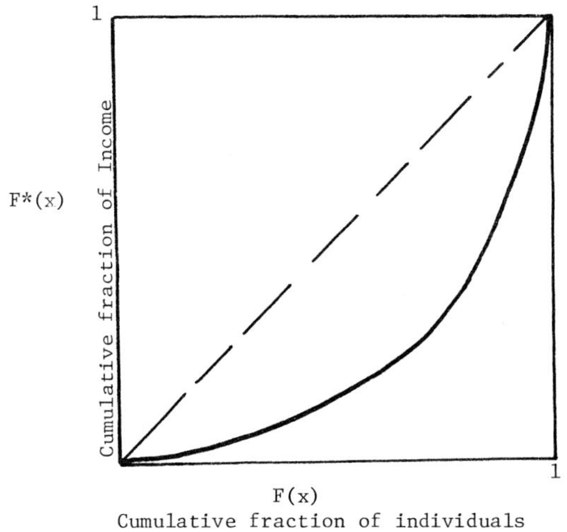

REFERENCES

[1] BERGERUD, A. T. and MANUEL, F. (1969). Aerial census of moose. *J. Wildlife Management*, 33, 910-916.

[2] BLUMENTHAL, S. (1967). Proportional sampling in life length studies. *Technometrics*, 9, 205-218.

[3] BROWN, M. (1972). Low density traffic streams. *Advances in Appl. Probability*, 4, 177-192.

[4] COLEMAN, R. (1972). Sampling procedures for the lengths of random straight lines. *Biometrika*, 59, 415-426.

[5] COOK, R. D. and MARTIN, F. B. (1974). A model for quadrat sampling with "visibility bias". *J. Amer. Statist. Assoc.*, 69, 345-349.

[6] COX, D. R. (1962). *Renewal Theory*. Frome and Long: Butler & Tanner, Ltd.

[7] COX, D. R. (1969). Some sampling problems in technology. In *New Developments in Survey Sampling*, (Johnson and Smith, Eds.). 289-316. University of North Carolina Press.

[8] DANIELS, H. E. (1942). A new technique for the analysis of fibre length distribution in wool. *J. Text. Inst.*, 33, 137-150.

[9] DAVID, H. A. (1973). Waiting time paradoxes and order statistics. *J. Amer. Statist. Assoc.*, 68, 743-745.

[10] EVANS, C. D., TROYER, W. A. and LENSINK, C. J. (1966). Aerial census of moose by quadrat sampling units. *J. Wildlife Management*, 30, 767-776.

[11] FELLER, W. (1966). *Introduction to Probability Theory and Applications*, 2, 10-14. Wiley & Sons, New York.

[12] FISHER, R. A. (1934). The effect of methods of ascertainment upon the estimation of frequencies. *Ann. Eugen.*, 6, 13-25.

[13] GUPTA, R. C. (1975). Some characterizations of discrete distributions by properties of their moment distributions. *Comm. Statist.*, 4, 761-765.

[14] HALDANE, J. B. S. (1938). The estimation of the frequency of recessive conditions in man. *Ann. Eugen.*, **7**, 255-262. London.

[15] HART, P. E. (1975). Moment distributions in economics: an exposition. *J. Roy. Statist. Soc. Ser. A.*, **138**, 423-434.

[16] HART, P. E. (1976). The comparative statistics and dynamics of income distributions. *J. Roy. Statist. Soc. Ser. A.*, **139**, 108-125.

[17] HERDAN, G. (1960). *Small Particle Statistics.* Elsevier, Amsterdam.

[18] KEMP, C. D. (1973). An elementary ambiguity in accident theory. *Accid. Anal. of Prev.*, **5**, 371-373.

[19] KEMP, C. D. (1975). Models for visibility bias in quadrat sampling. Statistics Reports and Reprints, No. 22, University of Bradford.

[20] KENDALL, M. G. and MORAN, P. A. P. (1963). *Geometrical Probability*. Charles Griffin & Co., Ltd., London.

[21] KOTZ, S. and JOHNSON, N. L. (1974). A characterization of exponential distributions by a waiting time property. *Comm. Statist.*, **3**, 257-258.

[22] KRIENS, J. (1963). The procedures suggested by de Wolff and van Heerden for random sampling in auditing. *Statist. Neerlandica* , **17**, 215-231.

[23] MORAN, P. A. P. (1966). Measuring the length of a curve. *Biometrika*, **53**, 359-364.

[24] MORAN, P. A. P. (1969). A second note on recent research in geometrical probability. *Advances in Appl. Probability* , **1**, 73-89.

[25] MORRISON, D. G. (1973). Some results for waiting times with an application to survey data. *The American Statistician*, **27**, 226-227.

[26] NEEL and SCHULL. (1966). *Human Heredity*, 211-227. University of Chicago Press, Chicago.

[27] NOLL, K. E. and PILAT, M. J. (1971). Size distribution of atmospheric giant particles. *Atmospheric Environment*, **5**, 527-540.

[28] ORD, K. (1975). Statistical models for personal income distributions. In *Statistical Distributions in Scientific Work* (G. P. Patil, S. Kotz and K. Ord, eds.), **2**, 151-158.

[29] PALMER, R. C. (1948). The dye sampling method of measuring fibre length distribution. *J. Text. Inst.*, **39**, 8-22.

[30] PALMER, R. C. and DANIELS, H. E. (1947). The sampling problem in single fibre testing. *J. Text. Inst.*, **38**, 94-100.

[31] PATIL, G. P., BOSWELL, M. T. and FRIDAY, D. S. (1975). Chance mechanisms in computer generation of random variables. In *Statistical Distributions in Scientific Work* (G. P. Patil, S. Kotz and K. Ord, eds.).

[32] PATIL, G. P. and JOSHI, S. W. (1968). *A Dictionary and Bibliography of Discrete Distributions.* Oliver and Boyd and Hafner Publishing Company, Edinburgh and New York.

[33] PATIL, G. P. and ORD, J. K. (1975). On size-biased sampling and related form-invariant weighted distributions. *Sankhyā* (To appear).

[34] PROROK, P. C. (1976). The theory of periodic screening I: lead time and proportion detected. *Advances in Appl. Probability* , **8**, 127-143.

[35] RAO, C. R. (1965). On discrete distributions arising out of methods of ascertainment. In *Classical and Contagious Discrete Distributions*, (G. P. Patil, ed.), 320-332. Statistical Publishing Society, Calcutta. Also reprinted in *Sankhyā A*, **27**, 311-324.

[36] RAO, C. R. and RUBIN, H. (1964). On a characterization of the Poisson distribution. *Sankhya*, 26, 295-298.

[37] ROBSON, D. S. (1975). A person communication.

[38] SCHEAFFER, R. L. (1972). Size-biased sampling. *Technometrics*, 14, 635-644.

[39] SCHOTZ, W. E. and ZELEN, M. (1971). Effect of length sampling bias on labelei mitotic index waves. *J. Theoret. Biol.*, 32, 383-404.

[40] SCHULTZ, D. M. (1975). Mass-size distributions: a review and a proposed new model. In *Statistical Distributions in Scientific Work* (G. P. Patil, S. Kotz and K. Ord, eds.), 275-288.

[41] SCHULTZ, D. M. and CROUSE, C. F. (1973). Random splittings: a model for a mass-size distribution. *S. African Statist. J.*, 7, 143-152.

[42] SIMON, R. (1975). Length biased sampling in the estimation of antigen frequencies. National Cancer Institute. A Preliminary Paper.

[43] SMART, R. J. (1963). Alcoholism, birth order, and family size. *J. Abnorm. Soc. Psychology*, 66, 17-23.

[44] SPROTT, D. A. (1964). Use of chi square. *J. Abnormal and Social Psychology*, 69, 101-103.

[45] TAKAHASHI, M. (1966). Theoretical basis for cell cycle analysis. II. Further studies on labeled mitosis wave method. *J. Theoret. Biol.*, 18, 195-209.

[46] TAKAHASHI, M. (1968). Theoretical basis for cell cycle analysis. I. Labeled mitosis wave method. *J. Theoret. Biol.*, 13, 202-211.

[47] TALLIS, G. M. (1970). Estimating the distribution of spherical and elliptical bodies in conglomerates from plane sections. *Biometrics*, 26, 87-104.

[48] TRUCCO, E. and BROCKWELL, P. J. (1968). Percentage labeled mitosis curves in exponentially growing cell populations. *J. Theoret. Biol.*, 20, 321-337.

[49] UPPULURI, V. R. R. and PATIL, S. A. (1976). Inferences about rare events. Oak Ridge National Laboratory. Technical Report. 1-6.

[50] VAN BELLE, G. and SCHNEIDERMAN, M. (1973). Some statistical aspects of environmental pollution and protection. *Int. Statist. Rev.*, 41, 315-331.

[51] WALLACH, R. V. and SICHEL, H. S. (1963). The measurement of mechanical strength and abradability of coke by means of statistical parameters. *J. Inst. Fuel.*, 36, 421-435.

[52] WARREN, W. G. (1975). Statistical distributions in forestry and forest products research. In *Statistical Distributions in Scientific Work*, (G. P. Patil, S. Kotz and K. Ord, eds.), 2, 369-384.

[53] WATSON, G. S. (1971). Estimating functionals of particle size distributions. *Biometrika*, 58, 483-490.

[54] ZELEN, M. (1971). Problems in the early detection of disease and the finding of faults. *Bull. Int. Statist. Inst.*, 38, 649-661.

[55] ZELEN, M. (1974). Problems in cell kinetics and the early detection of disease. *Reliability and Biometry*. *SIAM*, 701-726.

P.R. Krishnaiah, ed., APPLICATIONS OF STATISTICS

AN INVESTIGATION OF THE PROPERTIES OF SOME TWO SAMPLE TWO-STAGE DESIGNS

DALE PRESTON

Division of Biostatistics, School of Public Health
University of California, Los Angeles, California

ROBERT M. ELASHOFF

Department of Biomathematics, School of Medicine
University of California, Los Angeles, California

1. INTRODUCTION

Two-stage experimental designs for testing hypotheses about means of normal distributions, or about binomial proportions, have been discussed for many years (e.g., Dodge and Romig (1929)). Recently, there has been some interest in designing one sample two-stage experiments for testing such hypotheses; see, for example, Hald (1975), Elashoff and Beal (1976), Beal and Elashoff (1975). The work to be discussed in this paper is concerned with the initial investigation of two sample two-stage experimental designs for the comparison of binomial proportions.

The specific experimental situation being considered is a simplified version of a program to screen chemicals for carcinogenicity, such as is done by the National Cancer Institute's Carcinogenesis Bioassay Program. In particular, we are concerned with an assay in which there are only two groups: a control group (non-exposed), and an experimental group (exposed). One is interested in making a decision as to whether or not the chemical being considered is a tumorigen, solely on the basis of the proportion of animals in each group which develop tumors, either at a specific site or at a group of sites.

This paper is organized in this way: Sections 2 and 3 present the statistical models, including the hypotheses of interest, and the operating characteristic for the basic type of designs being considered. Next, Section 4 contains a discussion of an algorithm which has been developed and implemented to find approximately optimal designs. Section 5 presents some numerical results for a number of these designs under several conditions. Finally, these initial results are reviewed and possible extensions of this current work are indicated.

2. THE BASIC STATISTICAL MODEL

In deciding whether or not a chemical is a tumorigenic agent on the basis of tumor incidence rates, one is interested in testing the hypothesis H_0: $p_c = p_e$ against H_1: $p_c < p_e$, where p_c is the proportion of control animals developing tumors at the specified site and p_e is the corresponding proportion in the experimental group. Clearly, this is a composite hypothesis, since it is highly unlikely that the investigator would know the value of p_c in advance; if it were known, the problem reduces to a one sample problem of the type considered by Elashoff and Beal [1976]. However, it is reasonable to assume that in the situation being considered, there will be some knowledge (perhaps quite a bit) of the possible values of p_c prior to the experiment.

An alternative parameterization is used throughout most of the work being discussed in this report. This alternative parameterization is

$$\tau = \ln \frac{p_c}{1-p_c} \qquad \text{the log odds for the control group}$$

and

$$\Psi = \ln\left(\frac{p_c}{1-p_c}\right)\left(\frac{1-p_e}{p_e}\right) \qquad \text{the log odds ratio.}$$

In terms of this alternative parameterization, the hypotheses of interest are

$$H_0' \colon \Psi = 0$$

and

$$H_1' \colon \Psi < 0$$

where τ is clearly a nuisance parameter.

In general, at the first stage of a two sample two-stage experiment, n_{11} control animals and n_{12} experimental animals are observed. Based on the number of successes in the two groups (s_{11}, s_{12}), one decides whether to accept H_0, accept H_1, or continue to a second stage. If the decision is made to go to a second stage, one then observes n_{21} and n_{22} animals. Following the second stage, one must decide to accept one of the two hypotheses. This decision can be based upon $(s_{11}+s_{21}, s_{12}+s_{22})$, or, as will be the case for all designs considered in this paper, solely on the results of the second stage (s_{21}, s_{22}). Designs of this latter type will be called independent two-stage designs (ITSDs).

A two sample ITSD, $\underset{\sim}{d}$, then involves the specification of the four sample sizes, $n_{11}, n_{12}, n_{21}, n_{22}$. For stage one, a partition of the sample space $S_1 = \{(s_{11}, s_{12}) \colon 0 \le s_{11} \le n_{11}, 0 \le s_{12} \le n_{12}\}$ into three mutually exclusive and exhaustive subsets A_1, C_1, and R_1, must be given such that if

$$(s_{11}, s_{12}) \in A_1, \text{ then } H_0 \text{ is accepted;}$$

and if

$$(s_{11}, s_{12}) \in R_1, \text{ then } H_1 \text{ is accepted;}$$

or if

$(s_{11}, s_{12}) \in C_1$, then the experiment is continued to stage two.

At stage two, the analogous sample space, S_2, must be partitioned into two mutually exclusive and exhaustive sets, A_2 and R_2, such that if

$$(s_{21}, s_{22}) \in A_2, \text{ then } H_0 \text{ is accepted;}$$

and if

$$(s_{21}, s_{22}) \in R_2, \text{ then } H_1 \text{ is accepted.}$$

Figure 1 gives a stylized picture of the form of sets A_1, C_1, and R_1 as an aid to understanding these partitions.

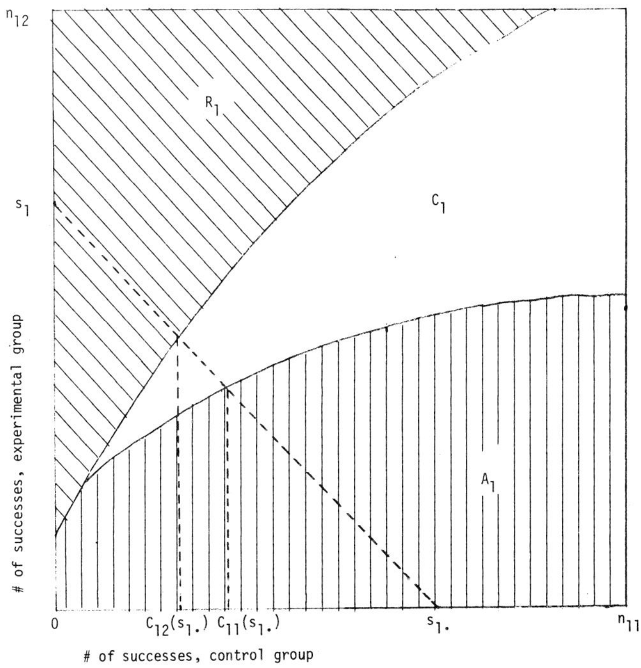

Fig. 1. A graphical representation of the first stage acceptance, rejection and continuation regions for a two sample two-stage design.

In the designs which are considered, the sets A_1, R_1 and C_1 can be defined by specifying constants $C_{11}(s_{1.})$ and $C_{12}(s_{1.})$ for each marginal total $s_{1.} = s_{11}+s_{12}$. The sets A_1, R_1, C_1 are

$$(s_{11}, s_{12}) \in A_1 \qquad \text{if } s_{11} > C_{11}(s_{1\cdot}) ,$$

$$(s_{11}, s_{12}) \in R_1 \qquad \text{if } s_{11} \leq C_{12}(s_{1\cdot}) ,$$

$$(s_{11}, s_{12}) \in C_1 \qquad \text{if } C_{12}(s_{1\cdot}) < s_{11} \leq C_{11}(s_{1\cdot}).$$

Similarly, A_2 and R_2 can be expressed by giving a constant $C_{22}(s_{2\cdot})$ such that

$$(s_{21}, s_{22}) \in A_2 \qquad \text{whenever } s_{21} \geq C_{22}(s_{2\cdot}) ,$$

$$(s_{21}, s_{22}) \in R_2 \qquad \text{whenever } S_{21} < C_{22}(s_{2\cdot}).$$

The designs which will be discussed in this paper involve constraints on the error probabilities, i.e., $\alpha = \Pr(\text{accept } H_1 \mid H_0)$ and $\beta = \Pr(\text{accept } H_0 \mid H_1)$. Thus, it will be necessary to consider the operating characteristic (OC) of an ITSD. The OC is a function of Ψ and τ which gives the probability of accepting H_0, given the design. This can be expressed by the rather formidable looking equation

$$OC(\tau,\Psi) = \sum_{s_{1\cdot}=0}^{n_{11}+n_{12}} (1+e^{\tau})^{-n_{11}}(1+e^{\tau-\Psi})^{-n_{12}} e^{(\tau-\Psi)s_{1\cdot}} \Big[\sum_{s_{11}=C(s_{1\cdot})+1}^{F_1} \binom{n_{11}}{s_{11}}\binom{n_{12}}{s_{1\cdot}-s_{11}} e^{\Psi s_{11}} \Big]$$
$$+ \sum_{s_{11}=C_{12}(s_{1\cdot})+1}^{C_{11}(s_{1\cdot})} \binom{n_{11}}{s_{11}}\binom{n_{12}}{s_{1\cdot}-s_{11}} e^{\Psi s_{11}} \tag{2.1}$$
$$\Big(\sum_{s_{2\cdot}=0}^{n_{21}+n_{22}} (1+e^{\tau})^{-n_{21}}(1+e^{\tau-\Psi})^{-n_{22}} e^{(\tau-\Psi)s_{2\cdot}} \Big) \cdot$$
$$\Big(\sum_{s_{21}=C_{22}(s_{2\cdot})}^{F_2} \binom{n_{21}}{s_{21}}\binom{n_{22}}{s_{2\cdot}-s_{21}} e^{\Psi s_{21}} \Big),$$

with $F_i = \min(n_{i1}, s_{i1}-n_{i1})$, $i = 1,2$. For all designs considered in this investigation, $n_{11} = n_{12} = n_1$, and $n_{21} = n_{22} = n_2$. All of this can be stated in a more easily understood form as

$$OC(\tau,\Psi) = \Pr\{(s_{11},s_{12}) \in A_1)\}+\Pr\{(s_{11},s_{12}) \in C_1\}\cdot\Pr\{(s_{21},s_{22}) \in A_2\}.$$

For a given complete specification wherein one specifies τ, which is the same under both H_0 and H_1, and the value of Ψ_1 under H_1, we have

$$\alpha = \Pr(\text{decide } \Psi = \Psi_1 \mid \Psi = 0,\ \tau) = 1\text{-}OC(0,\tau) = \alpha_1+\lambda_0\alpha_2$$

and

$$\beta = \Pr(\text{decide } \Psi = 0 \mid \Psi = \Psi_1,\ \tau) = OC(\Psi_1,\tau) = \beta_1+\lambda_1\beta_2,$$

where α_1, α_2, β_1 and β_2 are the first and second stage error probabilities, and λ_0 is the probability of continuation to a second stage under H_0, given $\underset{\sim}{d}$ and λ_1, the corresponding probability under H_1. It will be required that

$$\alpha \leq \alpha^* \text{ and } \beta \leq \beta^*. \tag{2.2}$$

The goal of this work is to choose a two-stage design from among those of the type being considered which minimizes a reasonable loss function. The loss function which has been used throughout this investigation is based on the expected sample size. The expected sample size (ASN) can be written as

$$ASN(\tau,\Psi,\underset{\sim}{d}) = n_1+\lambda(\tau,\Psi,\underset{\sim}{d})n_2, \tag{2.3}$$

where n_1 and n_2 are the sample sizes (in each group) at stage 1 and 2 respectively, $\lambda(\tau,\Psi,d)$ is the probability of continuation, given τ,Ψ, and this design $\underset{\sim}{d} = (n_1,n_2,A_1,C_1,A_2)$. The basic loss function is

$$L_2(\underset{\sim}{d}) = (1-\pi)ASN(\tau,0,\underset{\sim}{d})+\pi ASN(\tau,\Psi_1,\underset{\sim}{d}), \tag{2.4}$$

with $\pi \in (0,1)$ reflecting one's prior belief about the probability of H_1 being true. The optimal design, $\underset{\sim}{d}^*$, is chosen such that (2.4) is minimized subject to the constraints in (2.2).

In sum, one observes n_1 animals in each group and performs a Fisher exact test of nominal size α_1^*. If the pair (s_{11}, s_{12}) is in the critical region for this test, we reject H_0. The traditional acceptance region is partitioned into two parts, C_1 and A_1; if $(s_{11}, s_{12}) \in C_1$ we continue to a second stage; otherwise, we accept H_0 and terminate the experiment.

At the second stage, n_2 animals are used in each group and a standard Fisher exact test of nominal size α_2^* is performed. Given n_1, n_2, α_1^* and α_2^*, the size of the continuation region (i.e., the probability of a second stage under H_0), $\lambda_0(n_1,n_2,\alpha_1^*,\alpha_2^*)(=\lambda_0)$ can be found by

$$\lambda_0 = (\alpha^*-\alpha_1)/\alpha_2 \tag{2.5}$$

where α_1 and α_2 are the actual first and second stage α levels. This follows from the fact that $\alpha = \alpha_1+\lambda\alpha_2$. Thus, given α^*, α_1^*, α_2^*, n_1 and n_2, the choice of a design is reduced to a question of choosing a continuation region of the necessary size.

The designs obtained as a solution to this minimization problem are quite different from designs based upon the asymptotic normality of the arc sine transformation of the observed tumor incidence rate. Also, the designs make use of the Fisher exact test and, thus, the data analysis of animal bioassay experiments is essentially the same as the current accepted methods.

3. AN ALTERNATIVE STATISTICAL MODEL

The preceding statistical model specified p_c exactly (relative to the choice of design) and placed a two-point distribution on p_e or Ψ. The assumption that p_c is known exactly is unrealistic. We now weaken this assumption by requiring only a prior distribution on p_c or τ be available from previous studies; except for this change the model and procedure in Section 2 is assumed here. We fit a normal distribution with mean μ and variance σ^2 to the tumor incidence rates (or τ)(all sites exclusive of the reproductive organs) using data on controls from the data base. Lindley (1964) discusses the relationship between the beta prior on p_c and a normal prior on τ as proposed here.

To set up the design problem, additional quantities must be defined. We need to work with expected values of the first and second stage error probabilities, and of the continuation probabilities under H_0 and H_1. These expected values will be denoted by α_1^E, α_2^E, β_1^E, β_2^E, λ_0^E and λ_1^E. As an example of just what these quantities are, consider β_1^E,

$$\beta_1^E = E(\Pr((s_{11},s_{12}) \in A_1) \mid \Psi,\underset{\sim}{d})$$

$$= \int_{-\infty}^{\infty} \sum_{s_{1\cdot}=0}^{2n_1} [(1+e^{\tau})^{-n_1}(1+e^{\tau-\Psi})^{-n_1} e^{(\tau-\Psi)s_{1\cdot}} (\sum_{s_{11}=C_{11}(s_{1\cdot})+1}^{F_1} \binom{n_1}{s_{11}}\binom{n_1}{s_{1\cdot}-s_{11}} e^{\Psi s_{11}})] dF(\tau)$$

$$= \sum_{s_{1\cdot}=0}^{2n_1} [(\sum_{s_{11}=C_{11}(s_{1\cdot})+1}^{F_1} \binom{n_1}{s_{11}}\binom{n_1}{s_{1\cdot}-s_{11}} e^{\Psi s_{11}}) \cdot$$

$$\int_{-\infty}^{\infty} (1+e^{\tau})^{-n_1}(1+e^{\tau-\Psi})^{-n_1} e^{(\tau-\Psi)s_{1\cdot}} dF(\tau)].$$

where $F(\tau)$ is the prior distribution function of τ.

The other quantities are defined similarly. The overall α (β) level of a two-stage design will be defined by $\alpha^E = \alpha_1^E+\lambda_0^E\alpha_2^E$ ($\beta^E = \beta_1^E+\lambda_1^E\beta_2^E$). It should be pointed out that α^E and β^E are not the expected value of α and β with respect to the prior $F(\tau)$, since this expectation involves the expected value of a product of two functions of τ. For example, $E(\alpha) = E(\alpha_1+\lambda_0\alpha_2) = E(\alpha_1)+E(\lambda_0\alpha_2)$, but $E(\lambda_0\alpha_2) \neq E(\lambda_0)E(\alpha_2)$ in general.

Given n_1, n_2, α_1^*, α_2^* and α^*, one can compute α_1^E and α_2^E and use the relationship $\lambda_0^E = (\alpha^*-\alpha_1^E)/\alpha_2^E$ to find the size of the critical region which is needed to achieve the desired α^E level. Thus, as in Section 2, the choice of a design reduces to a question of how to choose the continuation region. The optimal design is then that design which minimizes $E(L_2(\underset{\sim}{d}))$ where the expectation is taken with respect to the density on τ, $F(\tau)$, and $L_2(\underset{\sim}{d})$ is as defined in equation 2.3, using expected ASN's. It can easily be seen that $E(ASN(\tau,\Psi,\underset{\sim}{d})) = n_1+\lambda^E(\Psi,\underset{\sim}{d})n_2$ so that

$$E(L_2(\underset{\sim}{d})) = n_1+(\lambda_0^E\,(1-\pi)+\lambda_1^E\cdot\pi)n_2. \tag{3.2}$$

4. THE MINIMIZATION ALGORITHM

In order to find optimal ITSD's for the problems in Sections 2 and 3, it has been necessary to develop a procedure to choose which points to include in the continuation region for a given n_1, n_2, α_1^*, α_2^* to achieve the desired α-level and to then search through the possible n_1, n_2 combinations for those sample sizes which minimize the loss function. The procedure which has been developed makes no attempt to find the best choice of α_1^* and α_2^*; after the preliminary formulations in Sections 2 and 3 are refined, "optimal" procedures will be investigated. The program performs a restricted search for the optimal design, conditional on the first and second stage nominal test sizes.

It was felt that the continuation region should be made up initially of points for which the probabilities under H_0 and H_1 are approximately equal. This criterion is equivalent to choosing those points for C_1 which have

$$L = \frac{P_{H_0}(s_{11},s_{12})}{P_{H_1}(s_{11},s_{12})} \tag{4.1}$$

close to one, where $P_{H_i}(s_{11},s_{12}) = \Pr\{(s_{11},s_{12}) \mid H_i\}$, $i = 0,1$.

Basically, the algorithm for finding a design, given n_1, n_2, α_1^*, α_2^* and α, involves first finding the Fisher exact critical region of size α_i^* for n_i, $i = 1,2$. Once this is done, one can compute α_1, α_2 and the required λ_0^*. After this, the absolute value of Q_i is computed for each i, $i = 0,1,2,\ldots,2n_1$, where

$$Q_i = \left|\frac{P_{H_0}(C_{11}(i),i-C_{11}(i))}{P_{H_1}(C_{11}(i),i-C_{11}(i))} -1\right| \tag{4.2}$$

These values are sorted into ascending order and the point with the smallest value, Q_j, is added to the continuation region. This changes the value of $C_{11}(j)$ and hence Q_j. The new value for Q_j is computed and sorted into its proper position. The process is then repeated with another point being added to C_1. This continues until the addition of one more point to C_1 would cause λ_0 to exceed λ_0^*.

Once this has been done, the values of β and $L_2(\underset{\sim}{d})$ can be computed. At this point, it is necessary to decide on what design to consider next; i.e., what values of n_1 and/or n_2. This is accomplished in a two-stage procedure. First, n_1 is held fixed and n_2 is varied so as to minimize $L_2(\underset{\sim}{d})$, subject to the constraint $\alpha \leq \alpha^*$ and $\beta \leq \beta^*$. The first constraint is met automatically as a consequence of the procedure used to choose $\underset{\sim}{d}$. If $\beta > \beta^*$, then n_2 is increased; if $\beta < \beta^*$, n_2 is decreased. Once the optimal n_2 is found for a given n_1, the first stage sample size is decreased. The optimal design for this new n_1 is found, and if the loss is reduced, n_1 is again decreased. If the loss increases, the step size is cut in half and the sample size increased by this amount. In this way the program finds an approximately optimal solution.

There are two ways in which the true optimal solution could be missed. First, if the initial estimate of n_1 is too small, it is possible to miss the true optimal design, since once an optimal design is found (and the first design must be optimal in this case), only smaller values of n_1 are considered. Second, if the step size is not 1, it is possible to decrease the loss but skip over a value of n_1 which leads to a greater reduction in loss. It appears from the cases which we have seen that the loss is a relatively flat function in the area of the minimum, and hence the design found can be assumed to be quite close to the true optimal design. If the optimal design has n_1 equal to the initial estimate, one should however, rerun the program with a larger initial estimate for n_1 - clearly this estimate need be no larger than the value of $L_2(\underset{\sim}{d})$ for the design found to be best on the first trial.

Once this procedure had been implemented, it was noticed that one could often greatly reduce the sample sizes necessary, and hence the loss, L_2, by a slight modification of the procedure for choosing C_1. The modification of the procedure is to give preference to those points with a likelihood ratio less than 1. If all such points have been included in C_1 and additional points are needed, those points with $P_{H_0}/P_{H_1} > 1$ are considered. The reason for this improvement is that points with $P_{H_0}/P_{H_1} < 1$ tend to have smaller values for P_{H_0} and hence their inclusion in C_1 has little effect on λ_0 but can result in a great increase in the power of the first stage.

It should be noted that for the design formulation of Section 3, the quantity L that was used is not E(L), but is instead defined as the ratio of the expected probabilities. This procedure is analogous to that used to define the overall α and β levels.

The method used to choose the continuation region for the Section 3 design is closely related to the Bayesian theory for testing a sharp null hypothesis. Dickey and Leintz (1970) consider the situation in which one is interested in hypotheses of the form $H: \eta = \eta_H$ versus an alternative $\overline{H}: \eta \neq \eta_H$, concerning the partitioned parameter vector $\theta = (\eta, \xi) \in \Theta$. They present several forms for the weighted likelihood ratio, the ratio of the posterior to the prior odds, in particular proving that when the prior distribution of θ has a certain form the weighted likelihood ratio is equivalent to Savage's Density Ratio. The relevant areas of their development will be outlined briefly in order to make the connection between the weighted likelihood ratio and the ratio of expected probabilities used in this work.

The Bayesian approach to testing a sharp hypothesis considers the manner in which the odds, 0, are modified upon the observation of a data vector, D. The prior odds of H are defined

$$0(H) = P(H)/P(\overline{H})$$

where $P(K)$, $K = H,\overline{H}$, denotes the probability of event K. The posterior odds are just

$$0(H \mid D) = P(H \mid D)/P(\overline{H} \mid D)$$

which is (trivally) equal to $L \cdot 0(H)$ where $L = 0(H \mid D)/0(H)$ and L is called the weighted likelihood ratio.

Let the data be distributed according to the density function $f(D \mid \theta)$ and let $P(\theta \mid H)$ and $P(\theta \mid \overline{H})$ be prior distributions for θ under H and $\overline{H}$ respectively then the weighted likelihood of D given H and $\overline{H}$ is

$$\Phi(D \mid K) = \int f(D \mid \theta) dP(\theta \mid K) \qquad K = H,\overline{H}$$

and the unconditional expected density is

$$\Phi(D) = P(H)\Phi(D \mid H)+P(\overline{H})\Phi(D \mid \overline{H}).$$

Then we have, upon application of Bayes Theorem,

$$P(K \mid D) = P(K)\Phi(D \mid K)/\Phi(D) \qquad K = H,\overline{H}.$$

Thus

$$L = [P(H \mid D)/P(\overline{H} \mid D) \; / \; P(H)/P(\overline{H})]$$

$$= \Phi(D \mid H)/\Phi(D \mid \overline{H})$$

which is just the ratio of expected densities under the null and alternative hypotheses.

We are interested in the parameter vector $\theta = (\Psi,\tau)$ with $\theta \in \Theta$ $\{\Psi \leq 0, -\infty < \tau < \infty\}$. The sharp null hypothesis is H: $\Psi = 0$ and the alternative is $\overline{H}$: $\Psi < 0$. However, for the development of the designs considered in this paper we have effectively reduced the parameter space to $\Theta' = \{\Psi = 0, \Psi^*(<0), -\infty < \tau < \infty\}$. Thus the sharp null hypothesis and its alternative are:

$$H: \quad \Psi = 0$$

and

$$\overline{H}: \quad \Psi = \Psi^*.$$

An experiment leads to an outcome $D = (s_{11}, s_{12})$ with

$$f(D \mid \theta) = \binom{n}{s_{11}}\binom{n}{s_{12}} e^{\tau s_1 - \Psi s_{12}} / [(1+e^{\tau})^n (1+e^{\tau-\Psi})^n]. \tag{4.3}$$

The prior distribution for θ is just a normal distribution for τ with known mean and variance with $\Psi = 0$ under H and $\Psi = \Psi^*$ under $\overline{H}$. Thus the weighted likelihood of an outcome is given by

$$\Phi(D \mid \Psi) = \int \binom{n}{s_{11}}\binom{n}{s_{12}} e^{\tau s_1 \cdot -\Psi s_{12}} \cdot e^{-\frac{1}{2}(\tau-\mu)^2/\sigma^2} \div ([(1+e^{\tau})^n (1+e^{\tau-\Psi})^n]\sqrt{2\pi\sigma})\, d\tau \tag{4.4}$$

and the weighted likelihood ratio is just

$$L = \Phi(D \mid \Psi = 0)/\Phi(D \mid \Psi = \Psi^*)$$

which is identical to the ratio of expected probabilities under the null and alternative hypotheses which is used in the Section 3 formulation.

The equivalence of the weighted likelihood ratio and the ratio of expected probabilities provide a justification for the modification of the algorithm which is mentioned above. When the weighted likelihood ratio is less than 1 the data tends to favor the alternative hypothesis and hence we should be more willing to continue in these cases.

5. RESULTS

A FORTRAN program has been written to carry out the search for optimal designs, using the approach outlined above. In this search, only values of n_1 and n_2 less than 200 are allowed. Binomial coefficients are computed using a table of log factorials, which is stored in core. For the formulation in Section 3, the only density considered so far is the normal. The integrations necessary for these designs are carried out using an algorithm for Gauss-Legendre-Simpson's product type two point integration (Boland (1971)).

A preliminary numerical investigation of a number of two sample ITSD's has been carried out, and the results are presented in this section. These analyses were carried out using the procedures outlined in the previous section, except that the risk function was generalized to include a penalty for going to the second stage. The risk function is

$$L_2^* = N_1 + r\cdot(\pi\lambda_1 + (1-\pi)\lambda_0)\cdot n_2, \tag{5.1}$$

where the parameter, r, can be interpreted as representing a relative increase in the cost per experimental animal between the two stages. It was also noted by Elashoff and Beal (1976) that by varying r one could control the first stage error probabilities, α_1 and β_1. The effect of varying r on the class of designs being considered here will be discussed following the presentation of some results.

The performance of two sample ITSD's will be assessed by the quantity

$$S = \frac{L_1 - L_2^*}{L_1} \times 10^2, \tag{5.2}$$

where L_1 is the one-stage sample size needed to satisfy the statistical constraints. For all exact designs described below, L_1 was computed using the arc sine transformation.

$$L_1 = \frac{(z_\alpha + z_\beta)^2}{2(\arcsin\sqrt{p_0} - \arcsin\sqrt{p_1})^2}; \tag{5.3}$$

where z_α and z_β are the standard normal deviates corresponding to α and β. The information provided in TABLES I and III is not sufficient to specify the design since one clearly needs to know which points are in A_1, C_1, R_1, A_2 and R_2. For each optimal design a listing of these sets is produced and a plot made. Figures 2 and 3 give the plots for the two cases: one each for the Section 2 and Section 3 designs.

TABLE I presents the first and second stage sample sizes, the continuation probabilities under H_0 and H_1, and the savings of a number of ITSD's, when H_0 and H_1 are specified exactly (p_c known precisely). TABLE II is a summary of the error rates, both overall and at stage one and stage two. TABLE III is the analog to TABLE I for the Section 3 designs. The columns labeled μ and σ^2 contain the mean and variance of the (normal) prior on τ.

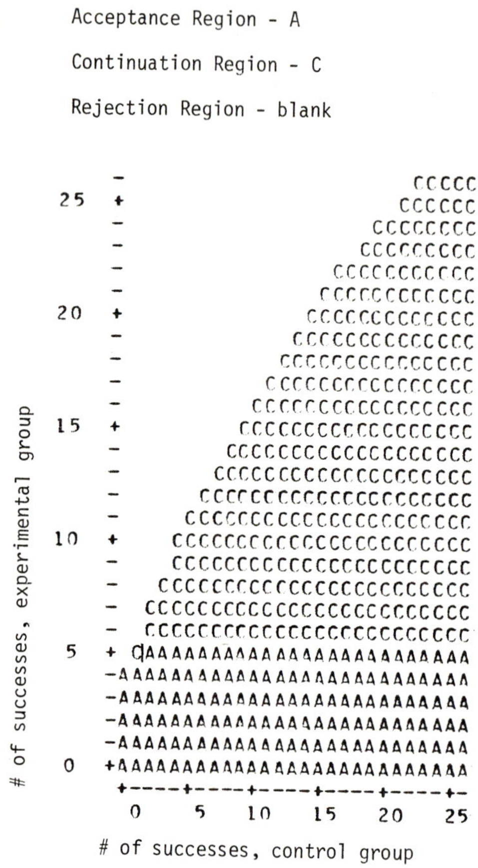

Fig. 2. Two sample two-stage design (p_c known precisely)
$p_c = .15$, $p_c = .30$, $\alpha^* = .05$, $\beta^* = .20$

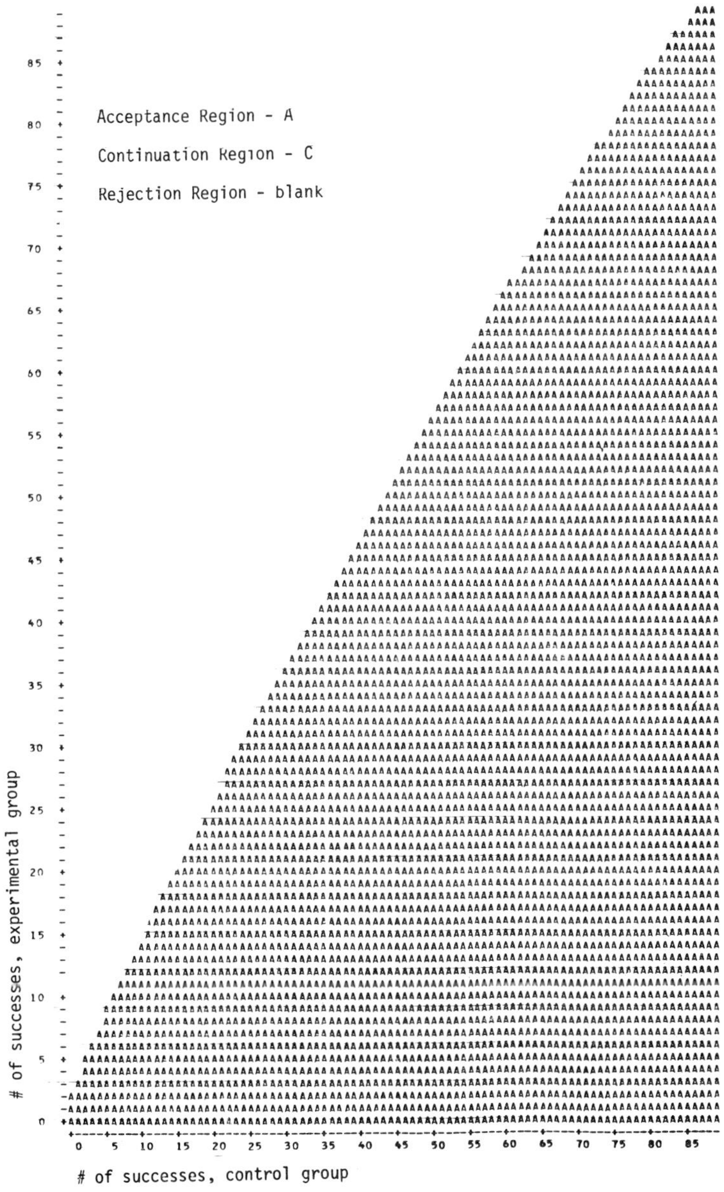

Fig. 2. (continued)

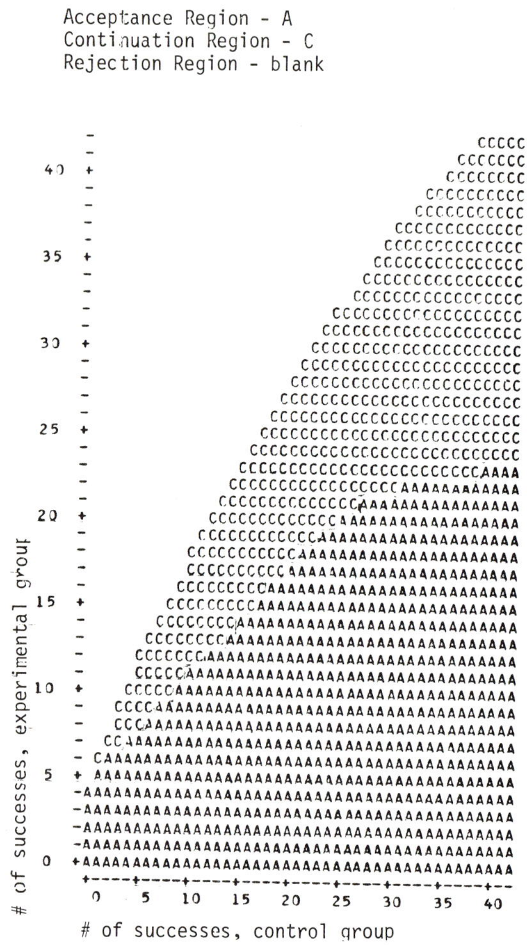

Fig. 3. Two sample two-stage design (normal prior on τ)
$\Psi = -0.8873$, $\mu = -1.7346$, $\sigma = 0.40$, $p_c = .15\text{-}.30$, $\alpha^* = .05$, $\beta^* = .20$

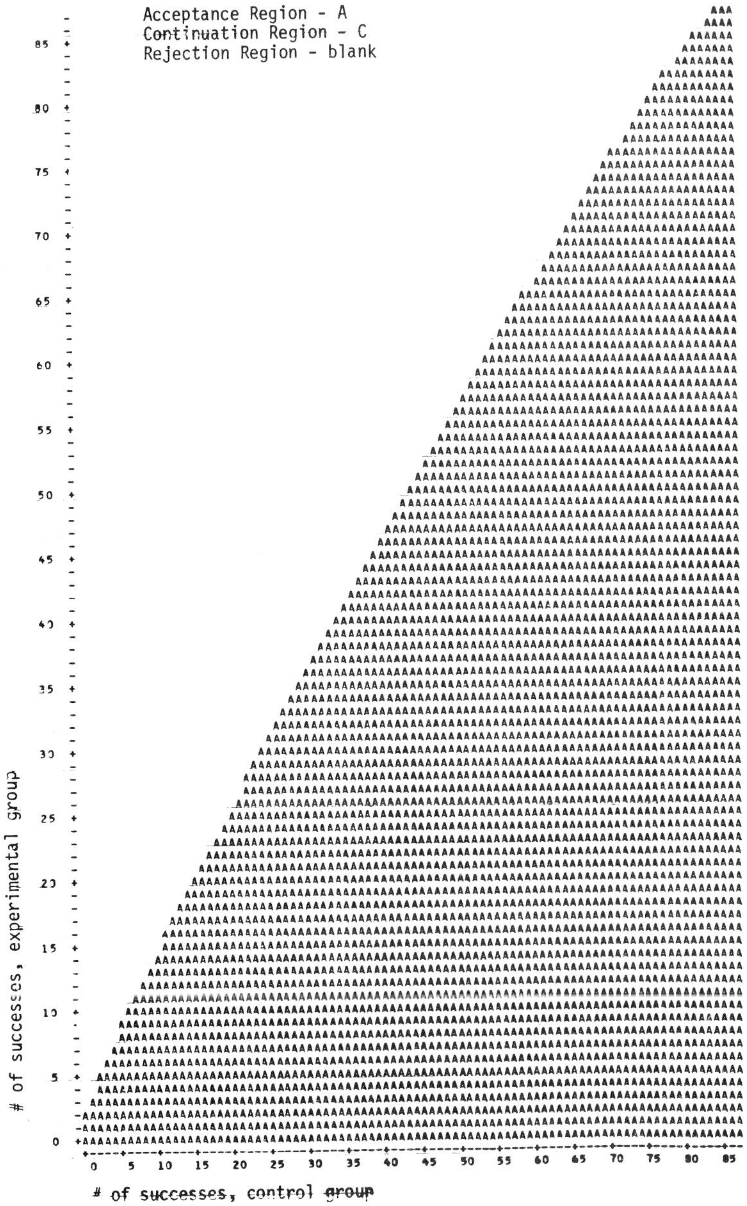

Fig. 3. (continued)

TABLE I

Summary of Two Sample Two-Stage Designs

(p_c known precisely)

α^*	β^*	α_1^*	α_2^*	p_c	p_e	π	r	n_1	n_2	L_2^*	λ_0	λ_1	S
.05	.20	.05	.20	.15	.30	.10	1.00	26	89	45.06	.1738	.5777	52
		.05	.20	.15	.30	.10	1.50	29	89	54.59	.1738	.5777	42
		.05	.20	.15	.30	.50	1.00	36	62	56.05	.1933	.5827	40
		.05	.10	.15	.30	.10	1.00	22	83	60.62	.4370	.7202	35
		.05	.20	.05	.15	.10	1.00	23	94	57.05	.3188	.7531	45
		.05	.20	.10	.25	.10	1.00	18	64	39.92	.3025	.7022	47
		.07	.20	.10	.25	.10	1.00	24	55	36.39	.1892	.5495	52
		.05	.20	.50	.75	.10	1.00	19	25	23.52	.1426	.5232	48
.05	.10	.05	.20	.15	.30	.10	1.00	42	109	63.25	.1581	.5273	53
		.05	.10	.15	.30	.10	1.00	38	106	77.12	.3442	.5932	40
		.05	.10	.10	.25	.10	1.00	40	77	77.51	.4811	.5409	26
		.05	.10	.05	.15	.10	1.00	40	117	122.2	.7016	.7011	15

TABLE II

Summary of Conditional Error Probabilities for Two Sample Two-Stage Designs

(p_c known precisely)

α^*	β^*	α_1^*	α_2^*	p_c	p_e	π	r	α	α_1	α_2	β	β_1	β_2
.05	.20	.05	.20	.15	.30	.10	1.00	.0499	.0247	.1452	.1995	.1553	.0765
		.05	.20	.15	.30	.10	1.50	.0499	.0247	.1452	.1995	.1553	.0765
		.05	.20	.15	.30	.50	1.00	.0477	.0209	.1387	.1984	.1053	.1597
		.05	.10	.15	.30	.10	1.00	.0498	.0202	.0676	.1975	.0616	.1887
		.05	.20	.05	.15	.10	1.00	.0401	.0180	.1202	.1994	.1204	.1049
		.05	.20	.10	.25	.10	1.00	.0472	.0062	.1355	.1981	.1209	.1099
		.07	.20	.10	.25	.10	1.00	.0493	.0251	.1282	.1999	.1150	.1544
		.05	.20	.50	.75	.10	1.00	.0495	.0265	.1611	.1488	.0932	.2018
.05	.10	.05	.20	.15	.30	.10	1.00	.0495	.0254	.1521	.0995	.0750	.0463
		.05	.10	.15	.30	.10	1.00	.0494	.0248	.0711	.0989	.0324	.1115
		.05	.10	.10	.25	.10	1.00	.0482	.0194	.0599	.0946	.0089	.1538
		.05	.10	.05	.15	.10	1.00	.0468	.0073	.0563	.0997	.0079	.1279

TABLE III

Summary of Two Sample Two-Stage Designs +

(prior on τ)

α^*	β^*	α_1^*	α_2^*	Ψ	μ	σ	$\bar{p}_c^{\,a}$	$\bar{p}_e^{\,b}$	π	r	n_1	n_2	EL_2	λ_0^E	λ_1^E	S
.05	.20	.05	.20	-0.8873	-1.7346	0.40	.15	.30	.10	1.00	70	48	77.47	.1441	.2602	21+
		.05	.20	-0.8873	-1.7346	0.40	.15	.30	.10	1.50	74	27	81.49	.1726	.2951	17+
		.05	.20	-0.8873	-1.7346	0.40	.15	.30	.50	1.00	69	46	78.73	.1522	.2710	20+
		.05	.20	-0.8873	-1.7346	0.40	.15	.30	.10	1.00	42	87	59.65	.1757	.4472	39
		.05	.20	-0.8873	-1.7346	0.40	.15	.30	.10	1.50	47	74	68.00	.1645	.4109	31
		.05	.20	-0.8873	-1.7346	0.40	.15	.30	.50	1.00	47	74	68.29	.1645	.4109	30
		.05	.10	-0.8873	-1.7346	0.40	.15	.30	.10	1.00	36	78	68.87	.4032	.5856	30
		.05	.20	-1.2098	-2.9444	0.40	.05	.15	.10	1.00	26	98	65.87	.3741	.7012	34
		.05	.20	-1.0986	-2.1972	0.40	.10	.25	.10	1.00	32	52	48.14	.2827	.5594	42
		.05	.20	-1.0986	0.00	0.40	.50	.75	.10	1.00	29	21	33.41	.1852	.4353	34
.05	.10	.05	.10	-0.8873	-1.7346	0.40	.15	.30	.10	1.00	52	131	93.58	.3021	.4553	31

+Computed using unmodified algorithm.

a value of p_c corresponding to the mean of τ

b value of p_e corresponding to $\Psi-\mu$

TABLE IV

Summary of Conditional Error Probabilities for Two Sample Two-Stage Designs

(normal prior on τ)

α^*	β^*	α_1^*	α_2^*	Ψ	μ	σ	$\bar{p}_c^{\,a}-\bar{p}_e^{\,b}$	π	r	α^E	α_1^E	α_2^E	β^E	β_1^E	β_2^E
.05	.20	.05	.20	-0.8373	-1.7346	0.40	.15-.30	.10	1.00	.0494	.0306	.1301	.1983	.1329	.2516
		.05	.20	-0.8373	-1.7346	0.40	.15-.30	.10	1.50	.0499	.0314	.1073	.1999	.0717	.4345
		.05	.20	-0.8373	-1.7346	0.40	.15-.30	.50	1.00	.0500	.0305	.1282	.1996	.1275	.2658
		.05	.20	-0.8373	-1.7346	0.40	.15-.30	.10	1.00	.0498	.0245	.1441	.2000	.1570	.0960
		.05	.20	-0.8373	-1.7346	0.40	.15-.30	.10	1.50	.0497	.0266	.1407	.1992	.1454	.1309
		.05	.20	-0.8873	-1.7346	0.40	.15-.30	.50	1.00	.0497	.0266	.1407	.1992	.1454	.1309
		.05	.10	-0.8873	-1.7346	0.40	.15-.30	.10	1.00	.0491	.0227	.0655	.2000	.0682	.2249
		.05	.20	-1.2098	-2.9444	0.40	.05-.15	.10	1.00	.0492	.0043	.1201	.1992	.1213	.1111
		.05	.20	-1.0986	-2.1972	0.40	.10-.25	.10	1.00	.0489	.0143	.1224	.1990	.0982	.1801
		.05	.20	-1.0986	0.00	0.40	.50-.75	.10	1.00	.0499	.0248	.1353	.1947	.0656	.2965
.05	.10	.05	.10	-0.8873	-1.7346	0.40	.15-.30	.10	1.00	.0499	.0278	.0732	.0997	.0642	.0780

a value of p_c corresponding to the mean of τ

b value of p_e corresponding to $\Psi-\mu$

TABLE V

A Typical Search

$p_c = .10$, $p_e = .25$, = .10, C = 1.00, $\lambda = .05$, * = .10,
$^*_1 = .05$, $^*_2 = .10$

n_1	n_2		β	L_2
60	100	.0486	.0345	91.44
60	84	.0494	.0507	89.19
60	68	.0498	.0749	86.30
60	52	.0494	.1118	82.49
60	60	.0495	.0951	84.43
60	56	.0490	.1001	82.80
60	58	.0498	.0955	83.61
60	57	.0493	.0972	83.20
44	57	.0495	.1389	74.75
44	73	.0487	.0980	78.35
44	65	.0490	.1178	76.66
44	69	.0499	.1045	78.67
28	73	.0491	.1361	75.28
28	89	.0487	.0925	80.84
28	81	.0473	.1129	76.09
28	85	.0480	.1017	78.47
36	73	.0483	.1144	75.70
36	89	.0487	.0765	81.29
36	81	.0499	.0924	80.05
36	77	.0491	.1028	77.87
36	79	.0495	.1075	78.96
36	78	.0493	.1001	78.41
40	73	.0404	.1035	77.80
40	89	.0476	.0715	79.94
40	81	.0489	.0851	79.46
40	77	.0482	.0946	77.51+
40	75	.0498	.0981	78.84
40	74	.0496	.1008	78.32
38	77	.0494	.0973	78.48
38	61	.0484	.1462	72.80
38	69	.0486	.1159	75.29
38	73	.0495	.1079	77.45
38	75	.0499	.1023	78.53
39	77	.0498	.0951	79.25
39	61	.0498	.1422	74.60
39	69	.0495	.1126	76.43
39	73	.0500	.1050	78.20

+Best design

TABLE IV is the analog of TABLE II for Section 3 designs. TABLE V is an outline of a typical search.

Before discussing the results obtained this far, it might be constructive to consider the form of the normal prior on τ in terms of p_c. This can be seen from TABLE VI, which gives values of p_c such that $\ln (p_c/1-p_c) = F^{-1}(q)$, for q and 1-q, with q = .005, .05, .125, .25, .50 for two values of μ and $\sigma^2 = 0.40$.

TABLE VI

Description of Normal Prior on in Terms of p_c

cumulative probability \ μ	$\mu = -1.7346$ value of p_c	$\mu = -2.944$ value of p_c
.005	.059	.018
.05	.084	.027
.125	.100	.032
.25	.119	.039
.50	.150	.050
.75	.188	.064
.875	.218	.077
.95	.254	.092
.995	.331	.128

The use of a normal prior for the log odds ratio has been considered by a number of authors. Lindley (1964) suggests its usefulness in the analysis of contingency tables, and notes that it is closely related to the "conjugate" class of beta prior distributions for the proportion itself.

6. DISCUSSION

It is difficult to reach any broad conclusions on the basis of the limited number of designs which have been examined so far. However, a number of interesting relationships do appear and are worthy of discussion.

TABLE I indicates that it is possible to achieve savings of almost 50% over one stage designs with the appropriate choice of α_1^* and α_2^* in the cases considered here. It is interesting to note that a poor choice of α_2^* can have a dramatic effect on the savings achieved. For example, when $\alpha^* = .05$ and $\beta^* = .20$, for $p_c = .15$ and $p_e = .30$, changing α_2^* from .20 to .10 causes the savings to decrease from 52% to 35%. A similar result holds when $\beta^* = .10$.

TABLES I and II also indicate that for fixed α^* and β^*, the value of α_1^* should be increased as p_c decreases. The effect of a choice of α_1^* is demonstrated rather markedly for the examples when $\alpha^* = .05$ and $\beta^* = .10$. It should be noted that for p_c fixed, α_1^* cannot be increased too much, for if it is too large, the actual first stage α-level, α_1, will exceed α^*, and there will be no design of the type we are working with. If α_2^* is chosen to be too large, then λ_0 will be very small and the optimal design will approach a one-stage design; i.e., the savings will decrease. At this time we have no guidelines for the choice of α_1^* and α_2^* other than the rather heuristic approach illustrated in TABLE I.

The effects of changing r, the second stage penalty, and π, the prior probability of H_1, are indicated by two examples in TABLES I and III. Changing r from 1.00 to 1.50 resulted in a decreased savings, as must be the case, but the design chosen was identical in both cases. The reduced savings is due solely to the increased cost for stage 2. Clearly, if r becomes large enough (greater than L_1), then the optimal two-stage design must change. For example, if r is larger than the minimum value of L_1, the second stage should never be taken.

Increasing the prior probability from .1 to .5 resulted in an increase in the loss and thus a fairly substantial decrease in the savings (from 52% to 40%) in the case considered here. The effect of changing α_1^* and α_2^* has not been investigated, but it is felt that this will probably not make too large a difference in these cases.

The results in TABLES III and IV for the Section 3 design are less extensive than those for Section 2 designs. However, except for the first three cases in these tables, the results appear quite parallel to those where p_c is precisely known. The low savings in these 3 cases, along with the large values of n_1 (which are obviously related) indicate the need for the modification of the algorithm for choosing points in C_1, mentioned in Section 4. The difference in the form of C_1 in these cases can be seen by comparing Figures 3 and 4. The remaining comments are not meant to refer to these poor cases.

As noted above, the Section 3 designs seem to follow patterns similar to the Section 2 designs as the various parameters are changed. The general effect of introducing some prior variability in τ seems to be a slight increase in the first stage sample size. The effect on the second stage sample size is not so consistent, but appears to be less marked than the effect on stage 1. The savings tend to be about 10 to 15% less than those achieved by the Section 2 designs.

The comments about the effect of changes in α_1^* and α_2^* made for the Section 2 designs seem to apply here, but at this time we have not yet looked at enough of these designs to reach any firm conclusions. Increasing the second stage costs did result in a change in the design vector as well as a decrease in savings for the one example considered. The effect of changing π was also analogous to the effect in the Section 2 design.

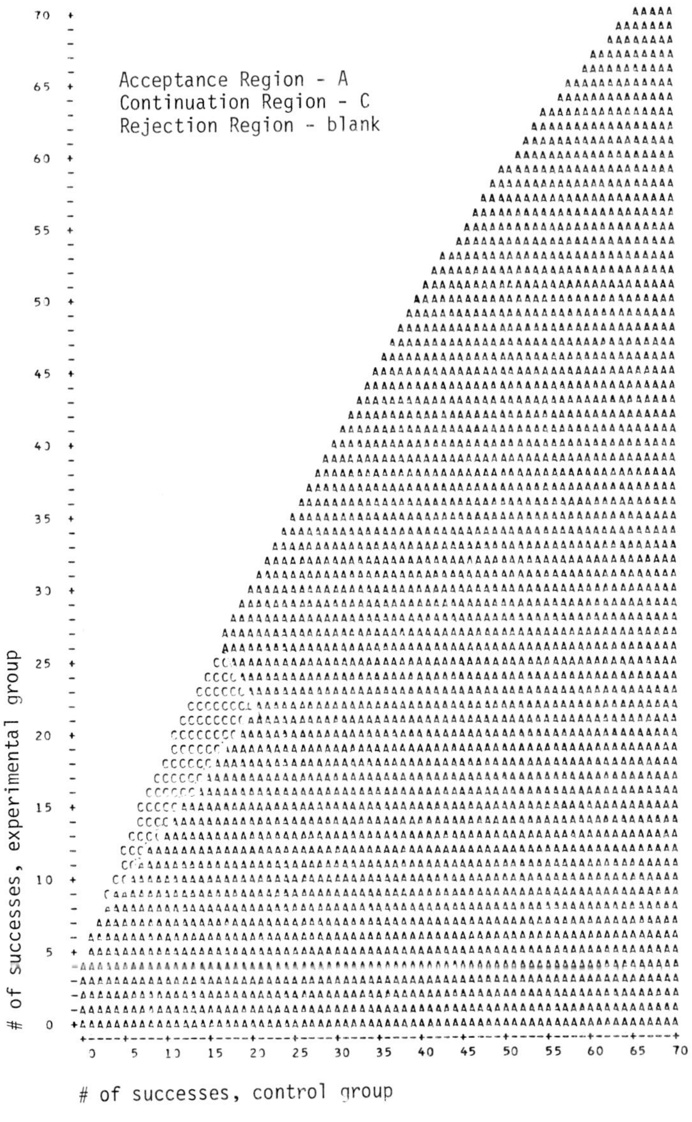

Fig. 4. Two sample two-stage design - unmodified algorithm (normal prior on τ)
$\Psi = -0.8873$, $\mu = -1.7346$, $\sigma = 0.40$, $p_c = .15-.30$, $\alpha^* = .05$, $\beta^* = .20$

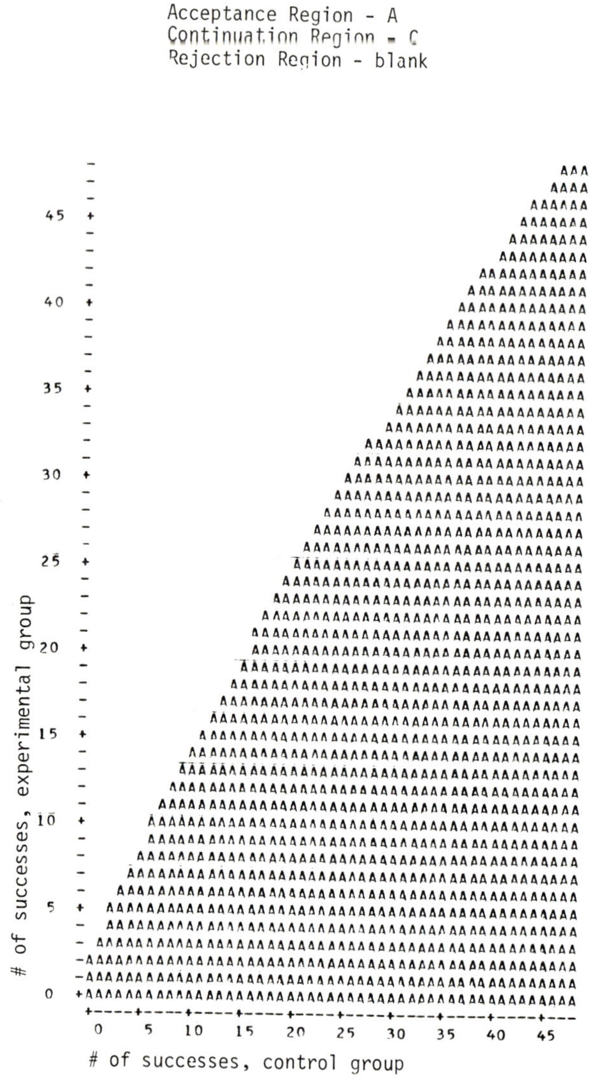

Fig. 4. (continued)

In order better to compare the performance of the effect of not knowing p_c precisely, it would be interesting to conduct a study of its robustness. Such an investigation is being planned. In this investigation, the performance of various optimal designs will be compared for a number of choices of p_c and p_e. It is felt that designs based upon a non-degenerate prior distribution on p_c (Section 3) should be more robust to errors in assumptions than Section 2 designs.

Another interesting comparison which should be made is with the arc sine transformation approach to the problem which was suggested by Elashoff and Beal (1976). They state that the designs suggested by them will achieve the same degree of savings as the comparable one sample design. An initial comparison of the two approaches suggests that the savings achieved by the designs presented in this paper can be much larger than the savings described in their work.

One further point should be noted about the designs discussed here. The ASN's for the values of p_c in the range of interest in tumorigen assays tend to be about 50 in the case when p_c is known and 70 to 80 in the Section 3 model, for $\alpha^* = .05$ and $\beta^* = .20$. This indicated that an assay carried out with two groups of 50 animals in each group is not very powerful, even when the difference between p_c and p_e is quite large. However, it should be noted that when a study is carried out with multiple dose groups, the total number of animals used is increased and the usual one-stage design will be more powerful than in the simple two-stage case considered here.

7. SUMMARY

A class of two sample two-stage designs for use with binary data have been proposed. An algorithm has been developed and programmed to find approximately optimal designs within this class, subject to constraints on the probabilities of false positives and false negatives. A preliminary numerical investigation has been conducted, and the results indicate that these designs can achieve savings of 30% to 50% over one-stage designs subject to the same constraints. This investigation is continuing.

In addition to future work which has been mentioned previously, a number of other possible extensions are worthy of consideration. One possibility is to move into the area of more truly Bayesian approaches to the design problem. This would involve the formulation of appropriate prior distributions for all the parameters and making the decision about acceptance, continuation or rejection on the basis of the posterior distribution of the parameters. Hopefully such an approach would enable one to look at more realistic situations in which multiple dose levels are compared to a control group. Another area of research within the current class of designs might be the investigation of alternative loss formulations. As an example, one might consider a loss which consists of expected costs of misclassification and average sample size. We hope to pursue some or all of these ideas, as well as others, in the near future.

An additional topic of interest is the development of practical two-stage designs and the comparison of the performance of these designs with current one-stage procedures. In particular this involves the study of error rates for various one- and two-stage designs. Initial investigations of this have been completed and results will be forthcoming.

ACKNOWLEDGEMENTS

Computing assistance was obtained from the Health Sciences Computing Facility, UCLA, supported by NIH Special Research Resources Grant RR-3.

REFERENCES

[1] BEAL, S. and ELASHOFF, R. M. (1975). A minimax and Λ-minimax approach to double sampling. Report for Contract NIH-NCI-E-2180.

[2] BOLAND, R. (1971). Algorithm 438: Product type two-point Gauss-Legendre-Simpson's integration (D1). *Collected Algorithms from CACM.*

[3] DICKEY, J. M. and LIENTZ, B. M. (1970). The weighted likelihood ratio, sharp hypothesis about chances, the order of a Markov Chain. *Ann. Math. Statist.*, **41**, 214-226.

[4] DODGE, H. F. and ROMIG, H. G. (1929). A method of sampling inspection. *Bell System Technical Journal*, **8**, 613-31.

[5] ELASHOFF, R. and BEAL, S. (1976). Two-stage screening designs applied to chemical screening with binary data. *Ann. Rev. Biophys. Bioeng.*, **5**, 561-87.

[6] HALD, A. (1975). Optimum double sampling tests of given strength. *J. Amer. Statist. Assoc.*, **70**, 451-56.

[7] LINDLEY, D. (1964). The Bayesian analysis of contingency tables. *Ann. Math. Statist.*, **35**, 1622-43.

P.R. Krishnaiah, ed., APPLICATIONS OF STATISTICS

PATTERN COMPARISON PROBLEMS

J. M. RICHARDSON

Science Center, Rockwell International
Thousand Oaks, California

A class of problems is considered in which the input is composed of two patterns (typically pictorial) representing the same physical reality and the output decision is composed of at least an estimate of the transformation required to bring one pattern into congruence (or best near-congruence) with the other. A number of models of such problems are presented and the solutions of two problems are given detailed discussion.

1. INTRODUCTION

In this paper we consider a set of problems in which the common objective is to determine the relation between two observed patterns. Our decision algorithms are based upon stochastic mathematical models involving strong *a priori* statistical biases on the relation between the patterns but with relatively weak biases on the patterns individually. Although our general approach can be applied to patterns of any kind our explicit discussions in later sections deal exclusively with pictorial patterns since the majority of interesting comparison problems pertain to patterns of this type.

In the following discussion we give a generic model of the comparison of pictorial patterns of which all of our later examples are special cases. For the sake of brevity we give a detailed discussion of the solution of two problems: (a) estimation of the rate of transformation of an image and (b) matching of images where one involves both geometrical and gray - scale transformations of the other. Other examples are formulated but their solutions are not presented. Detailed discussion of the solutions of these cases are given in several forthcoming papers [1].

2. PRELIMINARIES

We first consider some preliminary matters before getting into the main core of our problem in later sections.

We assume, as usual, that the pictorial patterns are presented in the xy-plane in which a general position is denoted by the vector

$$\underline{r} = \begin{pmatrix} x \\ y \end{pmatrix} \tag{2.1}$$

The domain $\mathcal{D}$, corresponding to the field of view, is typically a rectangular one defined by the inequalities

$$-\frac{1}{2} L_x \leq x \leq \frac{1}{2} L_x$$

$$-\frac{1}{2} L_y \leq y \leq \frac{1}{2} L_y$$

However, in many cases we will find it expedient to consider the infinite domain $\mathcal{D}_\infty$ (corresponding to $L_x = L_y = \infty$) with the pictorial patterns becoming sterile at large distances from the origin. This is a rather artificial way of circumventing problems associated with boundaries.

A pictorial pattern involving only shades of gray is represented by a scalar function of $\underline{r}$. A pair of possible observed patterns is denoted by $\rho_1(\underline{r})$ and $\rho_2(r)$. These patterns are represented by stochastic models. For the sake of clarity the *actually* observed patterns are designated by a tilde placed over the above symbols.

It is important to consider transformations of pictorial patterns. The most important type of transformation is geometrical. We begin the treatment of such transformations by the consideration of transformations of points, i.e.

$$\underline{r} \rightarrow T\underline{r} \tag{2.2}$$

where the symbol T denotes the particular transformation applied to $\underline{r}$. Examples of such transformations are:

(a) Translation

$$\underline{r} \rightarrow \underline{r} + \underline{a} \tag{2.3}$$

(b) Rotation

$$\underline{r} \rightarrow \underline{\underline{R}}(\theta) \cdot \underline{r} \tag{2.4}$$

where

$$\underline{\underline{R}}(\theta) = \begin{pmatrix} \cos & -\sin\theta \\ \sin & \cos\theta \end{pmatrix} \tag{2.5}$$

(c) Magnification

$$\underline{r} \rightarrow \xi\underline{r} \tag{2.6}$$

and finally

(d) Affine Transformation

$$r \rightarrow \underline{\underline{A}} \cdot \underline{r} + \underline{a} \tag{2.7}$$

The geometrical transformation of a pictorial pattern $\rho(\underline{r})$ is defined by

$$\rho(\underline{r}) \rightarrow \mathcal{T}\rho(\underline{r}) \triangleq |J(T^{-1}\underline{r}/\underline{r})|^{\nu}\rho(T^{-1}\underline{r}) \tag{2.8}$$

where

$$J(\underline{r}'/\underline{r}) = \text{Det}(\partial\underline{r}'/\partial\underline{r}) = \begin{vmatrix} \partial x'/\partial x & \partial x'/\partial y \\ \partial y'/\partial x & \partial y'/\partial y \end{vmatrix} \tag{2.9}$$

is the Jacobian of the transformation from $\underline{r}$ to $\underline{r}'$. In the case of area-preserving transformations the magnitude of the Jacobian is unity and hence the exponent ν in (2.8) is irrelevant. However, in the case of transformations not preserving area the choice of the exponent has some significance. With the infinite domain $\mathcal{D}_\infty$ the exponent ν implies that the integral

$$\int_{\mathcal{D}_\infty} d\underline{r}\ [\mathcal{T}(d)\ \rho\ (\underline{r})]^{1/2} \tag{2.10}$$

is invariant, i.e. is independent of the vector parameter α.

In the above examples the parameter α has the following meanings: in (a) $\alpha = \underline{a}$, in (b) $\alpha = \theta$, in (c) $\alpha = \xi$ and in (d) $\alpha = (\underline{\underline{A}}, \underline{a})$ (i.e. the four elements of $\underline{\underline{A}}$ plus the two components of $\underline{a}$).

If α is defined so that $\alpha = o$ corresponds to the identity transformation, i.e. $\mathcal{T}(o) = \mathcal{I}$, then for α sufficiently small we can write

$$\mathcal{T}(\alpha) = \ell + \Sigma\alpha_m\ \mathcal{U}_m \tag{2.11}$$

where the α_m are the components of α and the $\mathcal{U}_m$ are the corresponding infinitesimal generators.

3. GENERAL STATEMENT OF THE PROBLEM

In the present section we discuss a generic model embracing all of the image comparison problems under consideration here. As stated before, we confine our attention to pictorial patterns since these yield the comparison problems of greatest interest. A universal objective is to determine the geometrical transformation of one pattern that will yield the best match with another. In special cases, the estimation of other parameters may be desired.

The generic model is defined by the expression

$$\rho_n(\underline{r}) = \mathcal{T}(\alpha_n)\ F(\{\sigma_n(\underline{r})\}, \underline{r}, \lambda_n) + \nu_n(\underline{r}),\ n = 1,2, \tag{3.1}$$

where $\rho_1(\underline{r})$ and $\rho_2(\underline{r})$ are the possible observed images with $\underline{r} \in \mathcal{D}_\infty$.

The other symbols are defined below:

$\mathcal{T}(\alpha_n)$ - geometrical transformation,

$F(\{\sigma_n(\underline{r})\}, \underline{r}, \lambda_n)$ - distorted untransformed image,

$\sigma_n(r)$ - ideal or intrinsic pattern
λ_n - vector parameter,
$\nu_n(\underline{r})$ - various other random effects.

The symbol $\{\sigma_n(\underline{r})\}$ denotes that F depends upon the continuous set of values of $\sigma_n(\underline{r})$ for all $\underline{r} \in \mathcal{D}_\infty$.

The random process $\nu_n(\underline{r})$ is assumed to be spatially stationary and Gaussian with the mean and covariance given by

$$E\,\nu_n(\underline{r}) = 0, \tag{3.2}$$

$$E\,\nu_n(\underline{r})\,\nu_{n'}(\underline{r}') = \delta_{nn'}\,C_\nu\,(\underline{r}-\underline{r}'), \tag{3.3}$$

respectively. The intrinsic or ideal patterns $\sigma_1(\underline{r})$ and $\sigma_2(\underline{r})$ are random processes assumed to be independent of $\nu_1(\underline{r})$ and $\nu_2(\underline{r})$, but with strong mutual correlation. The remaining detailed properties of σ_1 and σ_2 will be specified for each individual case. The nature of the random variables λ_1 and λ_2 will also be specified for each case.

A modification of (3.1) which easier to analyze is one in which the geometrical transformation is applied to both F and ν_n, i.e.

$$\rho_n(\underline{r}) = \mathcal{T}(\alpha_n)\,[F + \nu_n] \tag{3.4}$$

Our attention in later sections will be denoted almost exclusively to this modified form.

4. ESTIMATION OF THE RATE OF TRANSFORMATION OF AN IMAGE

An interesting practical problem associated with the continuous observation of scenery from a moving platform is the estimation of the rate at which the focal-plane image translates, magnifies, rotates, etc. In the following discussion we consider a somewhat simplified model of the problem in which the scenery is regarded as flat. The problems associated with a finite field of view are circumvented by treating the domain of xy-space as infinite but with the artificial condition that the image (input) become sterile at appropriate distances from the origin.

Let us consider the model defined by the expression

$$\rho_n(\underline{r}) = \mathcal{T}(\alpha_n)\,u_n(\underline{r}) \qquad n = 1,2, \tag{4.1}$$

where $\rho_1(\underline{r})$ and $\rho_2(\underline{r})$ are possible observations of a time-evolving image at the times t_1 and t_2, respectively. The transformation parameters α_1 and α_2, referred to the above two times, define geometrical transformations of an earlier "initial" image. The function $u_n(\underline{r})$, the untransformed image at time t_n is given by

$$u_n(\underline{r}) = \mu(\underline{r}) + \nu_n(\underline{r}) \tag{4.2}$$

where $\mu(\underline{r})$ is the part of $u_n(\underline{r})$ that is perfectly correlated for the two times t_1 and t_2 (this corresponds to the assumption $\sigma_1 = \sigma_2 = \mu$ in the generic model). The remaining part of $u_n(\underline{r})$, namely $\nu_n(\underline{r})$, we assume is completely uncorrelated for the two times.

The functions $\mu(\underline{r})$, $\nu_1(\underline{r})$ and $\nu_2(\underline{r})$ are assumed (in an *a priori* sense) to be spatially stationary Gaussian random processes with the properties

$$\begin{cases} E\,\mu\,(\underline{r}) = 0, \\ E\,\nu_n(\underline{r}) = 0, \\ E\,\mu(\underline{r})\,\mu(\underline{r}^1) = C_\mu(\underline{r} - \underline{r}^1), \\ E\,\mu(\underline{r})\;\nu_n(\underline{r}^1) = 0, \\ E\,\nu_n(\underline{r})\,\nu_n^1\,(\underline{r}^1) = \delta_{nn}1\;C_\nu(\underline{r} - \underline{r}^1) \end{cases} \tag{4.3}$$

It is clear that the above covariance expressions appropriately represent the assumptions stated in the last paragraph. It is assumed that the transformation parameters α_1 and α_2 are random variables that are statistically independent of the above random processes and that are otherwise defined by the *a priori* probability density $P(\alpha_1,\alpha_2)$.

The observationally conditioned probability density of α_1 and α_2 is now given by the expression

$$\begin{aligned} &\log P\,(\alpha_1,\alpha_2|\{\tilde{\rho}_1,\,\tilde{\rho}_2\}) \\ &= -\frac{1}{2}\sum_n \int d\underline{r}\;\mathcal{T}_n^{-1}\;\tilde{\rho}_n(\underline{r})\;\mathcal{C}^{-1}\;\mathcal{T}_n^{-1}\;\tilde{\rho}_r\;(\underline{r}) \\ &\qquad + \log P\,(\alpha_1,\alpha_2) + \text{Const.} \end{aligned} \tag{4.4}$$

where $\tilde{\rho}_1$ and $\tilde{\rho}_2$ represent the actual observations of the time-evolving image and where B is a normalization constant. In the integral $\mathcal{T}_n^{-1}$ is the reciprocal of the operator $\mathcal{T}_n$, which in turn is a shorthand notation for $\mathcal{T}(\alpha_n)$, and $\mathcal{C}^{-1}$ is the reciprocal of the covariance operator $\mathcal{C}$ defined by the expression

$$\mathcal{C} f_n(\underline{r}) = \sum_{n} \int d\underline{r}^1 C_{nn^1}(\underline{r} - \underline{r}^1) f_{n^1}(\underline{r}^1) \tag{4.5}$$

where

$$C_{nn^1}(\underline{r} - \underline{r}^1) = C_\mu(\underline{r} - \underline{r}^1) + \delta_{nn^1} C_\nu(\underline{r} - \underline{r}^1) \tag{4.6}$$

The covariance functions C_r and C_ν, have already been defined in (4.3). The integral on $\underline{r}^1$ is assumed to span the infinite domain $\mathcal{D}_\infty$. The actual observations $\tilde{\rho}_1$ and $\tilde{\rho}_2$ are assumed to have the property that the integrals exist.

In the subsequent discussion it will be useful to introduce the covariance operators $\mathcal{C}_\mu$ and $\mathcal{C}_\nu$ defined by

$$\mathcal{C}_\mu f(\underline{r}) = \int d\underline{r}^1 C_\mu(\underline{r} - \underline{r}^1) f(\underline{r}^1), \tag{4.7}$$

and

$$\mathcal{C}_\nu f(\underline{r}) = \int d\underline{r}^1 C_\nu(\underline{r} - \underline{r}^1) f(\underline{r}^1) \tag{4.8}$$

We now make the further simplifying assumptions:

$$\mathcal{C}_\mu >> \mathcal{C}_\nu \tag{4.9}$$

i.e. $\mathcal{C}_\mu - \gamma \mathcal{C}_\nu$ is positive definite even when the scalar multiplier is a large positive real number,

$$\mathcal{C}_\nu = \mathcal{C}_\nu^o \mathcal{I}, \tag{4.10}$$

i.e. the random processes are assumed white, and finally that

$$\mathcal{T}(\alpha_2) = \mathcal{T}(\epsilon)\ \mathcal{T}(\alpha_1) \tag{4.11}$$

where the vector transformation parameter ϵ $(= \alpha_2 - \alpha_1)$ is small in the sense that $\mathcal{T}(\epsilon)$ can be regarded as an infinitesimal transformation to a sufficiently high degree of approximation. We can then write

$$\mathcal{T}^{-1}(\epsilon) = \mathcal{I} - \epsilon^T \mathcal{U} \tag{4.12}$$

where the components of the vector operator $\mathcal{U}$ are the infinitesimal generators corresponding to the components of the vector transformation parameter ϵ. The observationally conditioned probability density of α_1 and α_2, formerly given by (4.4), now reduces to the form

$$\log P(\alpha_1, \alpha_2 | \{\tilde{\rho}_1, \tilde{\rho}_2\})$$
$$= -\frac{1}{2} || \tilde{\rho}_1 - \tilde{\rho}_2 - \epsilon^T \mathcal{U}\tilde{\rho}_2||_\nu^2 + \log P(\alpha_1, \alpha_2) + B, \tag{4.13}$$

where

$$|| v ||_\nu = (v,v)_\nu^{\frac{1}{2}} , \tag{4.14}$$

and

$$(v,w)_\nu = \int d\underline{r}\ v(\underline{r})\ \mathcal{C}_\nu^{-1}\ w(\underline{r}), \tag{4.15}$$

for any two arbitrary functions v and w belonging to L_2.

Regarding α_1 as fixed and the probability density $P(\alpha_1,\alpha_2) = P(\alpha_1,\alpha_1 + \epsilon)$ as practically constant with respect to ϵ compared the other ϵ-dependent terms in (4.13), we find the most probable value of ϵ to be given by

$$\hat{\epsilon} = (\mathcal{U}\tilde{\rho}_2, \mathcal{U}^T\tilde{\rho}_2)^{-1} (\mathcal{U}\tilde{\rho}_2, \tilde{\rho}_1 - \tilde{\rho}_2)_\nu, \tag{4.16}$$

where it is understood that $(\mathcal{U}\tilde{\rho}_2, \mathcal{U}^T\tilde{\rho}_2)_\nu$ is a matrix whose elements are $(\mathcal{U}_m \tilde{\rho}_2, \mathcal{U}_{m'} \tilde{\rho}_2)_\nu$.

In the case in which

$$\tilde{\rho}_2 = \tilde{\rho}_1 + \frac{\partial\tilde{\rho}}{\partial t} dt, \tag{4.17}$$

and

$$\epsilon = dt \frac{d\alpha}{dt} \tag{4.18}$$

we can write

$$\frac{d\alpha}{dt} = (\mathcal{U}\tilde{\rho}, \mathcal{U}^T\tilde{\rho})_\nu^{-1} (\mathcal{U}\tilde{\rho}, \frac{\partial}{\partial t} \tilde{\rho})_\nu. \tag{4.19}$$

as the estimate of the rate of change of α . With the noise-free test pattern

$$\frac{\partial}{\partial t} \tilde{\rho} = (\mathcal{U}^T\tilde{\rho}) \frac{d\alpha}{dt} , \tag{4.20}$$

we then obtain

$$\frac{d\hat{\alpha}}{dt} = \frac{d\alpha}{dt} \tag{4.21}$$

thereby providing a verification of our estimate for the noise-free case.

In the case where the infinitesimal generators commute, i.e. where

$$\mathcal{U}_m \mathcal{U}_{m^1} - \mathcal{U}_{m^1} \mathcal{U}_m = 0,\ m,\ m^1 = 1,\ldots, M \tag{4.22}$$

it is possible to introduce some significant simplifications into the treatment. Since commuting operators have common eigenfunctions we can write

$$\mathcal{U}\psi_q = \lambda_q \psi_q \tag{4.23}$$

where $\psi_q = \psi_q(\underline{r})$ is a scalar function of $\underline{r}$ and where λ_q is an M-dimensional vector given by

$$\lambda_q = \begin{pmatrix} \lambda_{q1} \\ \cdot \\ \cdot \\ \cdot \\ \lambda_{qM} \end{pmatrix} \tag{4.24}$$

If the set of eigenfunctions is denumerable, we can impose the orthonormality condition

$$\int d\underline{r}\ \psi_q^* \ \psi_{q^1} = \delta_{qq^1} \tag{4.25}$$

otherwise the eigenfunctions must be "normalized to a δ-function." In the subsequent discussion we will treat the eigenfunctions as though they are denumerable, keeping in mind that the continuous case can be obtained by suitable limiting processes.

The observed image, now regarded as an explicit function of time, can be represented in terms of the eigenfunctions according to the expression

$$\tilde{\rho}(\underline{r}, t) = \sum_q \tilde{\rho}_q(t)\ \psi_q(\underline{r}) \tag{4.26}$$

where the time-dependent coefficients are given by

$$\tilde{\rho}_q(t) = \int d\underline{r}\ \psi_q^*(\underline{r})\ \tilde{\rho}\ (\underline{r}, t). \tag{4.27}$$

Using (2.23) we find that the operation by $\mathcal{U}$ yields

$$\tilde{\rho}\ (\underline{r}, t) = \sum_q \lambda_q\ \tilde{\rho}_q\ \psi_q(\underline{r}) \tag{4.28}$$

and thus the estimate (4.19) reduces to

$$\frac{d\hat{\alpha}}{dt} = \left[\sum_q |\tilde{\rho}_q|^2 \lambda_q^* \lambda_q^T\right]^{-1} \sum_q \lambda_q^* \tilde{\rho}_q \frac{d}{dt}\tilde{\rho}_q. \tag{4.29}$$

This result has the interesting property that, in the case of a noiseless test pattern given by (4.20), each term in the denominator, i.e. $|\rho_q|^2\lambda_q^*\lambda_q^T$, is proportional to the corresponding term in the numerator, i.e. $\lambda_q^* \tilde{\rho}_q^* \frac{\partial}{\partial t} \tilde{\rho}_q$, where the factor of proportionality is independent of q. This means that, in the case of the noiseless input, the sum over q could be replaced by a sum over a subset of S of q. In fact the subset S could consist of only one value of q provided that $\rho_q \neq o$. In the case in which the input is only slightly noisy a larger, but still small, subset S must be used (provided that not all of the ρ_q vanish for $q \in S$) and for a much noisier input a much more extensive subset S must be used. In any case we obtain the approximate estimate

$$\left(\frac{d\hat{\alpha}}{dt}\right)_S \cong \left[\sum_{q}{}_S |\tilde{\rho}_q|^2 \lambda_q^* \lambda_q^T\right]^{-1} \mathrm{Re} \sum_q{}_S \lambda_q^* \tilde{\rho}_q \frac{d}{dt} \tilde{\rho}_q \qquad (4.30)$$

This result can be implemented in a particularly simple and economical manner. We need only extract a finite set of $\tilde{\rho}_q(t)$'s from the input image $\tilde{\rho}(\underline{r},t)$ using (4.25). These constitute a set of properties of the input image that is adequate for the purpose of estimating $d\alpha/dt$ in accordance with (4.28). This represents a considerable simplification relative to the implementation problems associated with the exact estimate (4.17). However, we hasten to emphasize the simplification is achieved only for the case of commuting infinitesimal generators, a condition seriously limiting the range of applicability of (4.28).

5. MORE COMPLEX COMPARISON PROBLEMS

In this section we consider a set of more complex comparison problems involving the following features:

(a) geometrical and gray-scale transformations,

(b) nonlinear gray-scale transformations,

(c) images produced by different physical systems and

(d) images subject to different illumination directions.

Because of the requirement of brevity a detailed solution is given only for case (a).

(a) *Geometrical and Gray-Scale Transformations*

Here we consider a pattern comparison problem that is somewhat similar to the one treated in the last section. One difference lies in the fact that in the present case the two images are not necessarily closely similar, i.e. they do not represent observations of a single time-evolving image at slightly different times. Another difference is that both geometrical and gray-scale transformations are involved and these are mutually independent.

We consider a model defined by the expressions

$$\rho_1(\underline{r}) = \mathcal{T}(\alpha_1)\,[\lambda_1\, \sigma\,(\underline{r}) + \nu_1\,(\underline{r})], \qquad (5.1a)$$

$$\rho_2(\underline{r}) = \sigma(\underline{r}) + \nu_2(\underline{r}), \tag{5.1b}$$

where $\mathcal{T}(\alpha_1)$ is a geometrical transformation of the type discussed in Section 2, $\sigma(\underline{r})$ is the ideal image and λ_1 is the multiplier involved in the gray scale transformation. The random processes $\nu_1(\underline{r})$ and $\nu_2(\underline{r})$, representing both measurement noise and random intrinsic variations from the ideal image $\sigma(\underline{r})$, are assumed to be spatially stationary and Guassian with the mean and covariance defined by (3.2) and (3.3). The *a priori* statistical properties of α_1, λ_1 and $\sigma(\underline{r})$ are assumed independent of $\nu_1(\underline{r})$ and $\nu_2(r)$ and will be specified later. An additive term in the gray-scale transformation can be represented by the random uniform part of $\nu_1(\underline{r})$.

The above model implies an observationally conditioned probability density of α_1, λ_1 and $\sigma(\underline{r})$ given by

$$\begin{aligned}&\log P\,(\alpha_1, \lambda_1, \{\sigma(\underline{r})\}|\{\tilde{\rho}_1(\underline{r}), \tilde{\rho}_2(\underline{r})\})\\ &= -\tfrac{1}{2}\,||\,\mathcal{T}^{-1}(\alpha_1)\,\tilde{\rho}_1(\underline{r}) - \lambda_1\sigma(\underline{r})||_\nu^2\\ &\quad - \tfrac{1}{2}\,||\,\tilde{\rho}_2(\underline{r}) - \sigma(\underline{r})||_\nu^2\\ &\quad + \log P\,(\alpha_1, \lambda_1, \{\sigma(\underline{r})\}) + \text{const.},\end{aligned} \tag{5.2}$$

where the norm $||\cdot||_\nu$ is defined by (4.14) and (4.15). Since the above probability is a functional of $\sigma(\underline{r})$ it cannot be normalized and we must therefore regard it as a relative probability. Let us assume that $\log P\,(\alpha_1, \lambda_1, \{\sigma(\underline{r})\})$ is a functional of $\sigma(\underline{r})$ that is slowly varying compared with the quadratic functionals occurring earlier on the r.h. side of (5.2). The maximization on $\sigma(\underline{r})$ yields

$$\begin{aligned}&\text{most prob. } \sigma(\underline{r}) = \sigma^*(\underline{r})\\ &= (1+\lambda_1^2)^{-1}\,[\lambda_1\mathcal{T}^{-1}(\alpha_1)\,\tilde{\rho}_1 + \tilde{\rho}_2]\end{aligned} \tag{5.3}$$

Substituting σ^* for σ in (5.2) yields the result

$$\begin{aligned}&\log P\,(\alpha_1, \lambda_1, \{\sigma^*(\underline{r})\}|\{\tilde{\rho}_1(\underline{r}), \tilde{\rho}_2(\underline{r})\})\\ &= -\tfrac{1}{2}\,(1+\lambda_1^2)^{-1}\,[||\,\tilde{\rho}_1\,||_\nu^2 + \lambda_1^2\,||\,\tilde{\rho}_2\,||_\nu^2\\ &\quad - 2\,\lambda_1(\mathcal{T}^{-1}(\alpha_1)\,\tilde{\rho}_1, \tilde{\rho}_2)] + \text{const.}\end{aligned} \tag{5.4}$$

In the above result we have made an assumption on the exponent of the Jacobian in the definition of $\mathcal{T}(\alpha_1)$ in (2.8) expressed by the condition

$$|| \mathcal{T}^{-1}(\alpha_1)\,\tilde{\rho}_1 ||_\nu = || \tilde{\rho}_1 ||_\nu \,. \qquad (5.5)$$

In the cases where the Jacobian is independent of the position $\underline{r}$, the above condition represents no loss of generality because of the inclusion of the gray-scale transformation.

The result (5.4) has the convenient property that the maximization on α_1 can be accomplished with a result that is dependent only on the sign of λ_1. Thus the most probable value of α_1, namely α_1^*, is obtained by maximizing $(\mathcal{T}^{-1}(\alpha_1)\,\tilde{\rho}_1, \tilde{\rho}_2)_\nu$ if λ_1 is positive and minimizing this quantity if λ_1 is negative.

Substitution of α_1^* for α_1 in (5.4) followed by maximization on λ_1 yields the most probable value of λ_1 given by

$$\lambda_1^* = 1/2\gamma \cdot \left(\beta + \sqrt{\beta^2 + 4\gamma^2}\right) \qquad (5.6)$$

where

$$\gamma = (\mathcal{T}^{-1}(\alpha_1^*)\,\tilde{\rho}_1, \tilde{\rho}_2)_\nu \qquad (5.7)$$

and

$$\beta = || \tilde{\rho}_1 ||_\nu \, || \tilde{\rho}_2 ||_\nu^{-1} - || \tilde{\rho}_2 ||_\nu \, || \tilde{\rho}_1 ||_\nu^{-1} . \qquad (5.8)$$

We note that λ_1^* must have the same sign as γ. Therefore, in the determination of α_1^*, the assumed sign of λ_1 must be consistent with the resulting sign of γ. If there are two cases in which consistency is obtained, we must select the one for which $(1 + \lambda_1^{*2})^{-1}\,\lambda_1^*\,\gamma$ has the larger value.

(b) *Nonlinear Gray-Scale Transformation*

In the present and later subsections we consider several additional examples of pattern comparison problems. For the sake of brevity we will give only the problem statements and not the solutions.

The first example is a simple generalization of the problem treated above in Subsection (a), namely the replacement of the linear gray-scale transformation by a nonlinear one. The appropriate model is defined by the following expressions

$$\rho_1(\underline{r}) = \mathcal{T}(\alpha_1)\, g\,(\sigma(\underline{r}) + \nu_1(\underline{r})) \qquad (5.9a)$$

$$\rho_2(\underline{r}) = \sigma(\underline{r}) + \nu_2(\underline{r}) \qquad (5.9b)$$

where $\sigma(\underline{r})$ is the ideal image, $\mathcal{T}(\alpha_1)$ is a geometrical transformation of the same type as above and $g(\cdot)$ is a monotone increasing function which in general in nonlinear. The random processes $\nu_1(\underline{r})$ and $\nu_2(\underline{r})$ are the same as before. The *a priori* statistical properties of $\sigma(\underline{r})$, α_1 and $g(\cdot)$ are specified in general terms.

(c) *Images Produced by Different Physical Systems*

Here we consider a somewhat extrinsic model defined by the expressions

$$\rho_1(\underline{r}) = \mathcal{T}(\alpha_1)\,[\sigma_1(\underline{r}) + \nu_1(\underline{r})]\,, \tag{5.10a}$$

$$\rho_2(\underline{r}) = \sigma_2(\underline{r}) + \nu_2(\underline{r}), \tag{5.10b}$$

where $\mathcal{T}(\alpha_1)$, as before, is a geometrical transformation, $\sigma_1(\underline{r})$ and $\sigma_2(\underline{r})$ are the ideal images and $\nu_1(\underline{r})$ and $\nu_2(\underline{r})$ are the observation noises with statistical properties the same as before. We assume that α_1 is a random variable statistically independent of ν_1 and ν_2 and we further assume that σ_1 and σ_2 are random processes statistically independent of the previous quantities. In particular we assume that the *a priori* relative probability density in the function space of σ_1 and σ_2 is given by an expression of the form

$$\begin{aligned}\log P\,(\{\sigma_1,\sigma_2\}) &= Q_1(\{\sigma_1\}) + Q_2\,(\{\sigma_2\}) \\ &\quad + (\sigma_1, B\,\sigma_2)_\nu\end{aligned} \tag{5.11}$$

where B is a real symmetric positive definite operator. This operator is chosen to represent the kinds of correlation that are expected to exist between σ_1 and σ_2 in view of the underlying physical processes.

(d) *Images Subject to Different Illumination Directions*

Here we consider the case in which the same physical system is used to produce two observed images of a specified scene at vertical incidence but with the position of the sun different in the two cases. We assume for the sake of simplicity that the slopes of the elevation function are small and that the cosine law of scattering of the incident light is valid everywhere. The appropriate model is given by the expression

$$\begin{aligned}\rho_n(\underline{r}) &= \kappa\,\mathcal{T}(\alpha_1)\{\,R\,(\underline{r})\,[\gamma_n - \underline{\beta}_n\cdot\nabla\,h\,(\underline{r})] \\ &\quad + \nu_n\,(\underline{r})\}\,,\quad n = 1,2,\end{aligned} \tag{5.12}$$

where $\mathcal{T}(\alpha_n)$ is the usual geometrical transformation, κ is an optical system parameter, R $(\underline{r})$ is the cosine law reflectivity and the $\nu_n(\underline{r})$ are the observation noises with the same statistical properties as before. The geometrical parameters γ_n and $\underline{\beta}_n$ are defined by

$$\gamma_n = \vec{e}_{sn}\cdot\vec{e}_z \tag{5.13}$$

$$\beta_n = \vec{e}_{sn} \cdot (\vec{e}_x \vec{e}_x + \vec{e}_y \vec{e}_y) \tag{5.14}$$

where $\vec{e}_{sn}$ is the unit vector pointing toward the sun in case n. The quantities $\vec{e}_x, \vec{e}_y \vec{e}_z$ are the unit vectors pointing in the x, y and z directions, respectively, and ∇ is the gradient operator. *A priori* statistical assumptions concerning R and h can be specified in general. However, it is reasonable to assume that the sun directions $\vec{e}_{s1}$ and $\vec{e}_{s2}$ are known *a priori* with negligible error.

6. DISCUSSION

In the above discussion we have presented a rather sparse sampling of pattern comparison problems. It is not difficult to extend the list of such problems and to improve the simulation of reality in the associated models. However, even with the limited set of cases considered here, some general properties of the solutions (including those discussed in Ref. 1) become evident. Each solution involves a spatial filtering operation characteristic of the model followed by a cross-correlation operation. Further operations of a simple algebraic character may be included in special cases.

A few final comments are appropriate concerning possible directive of model improvement. The first such comment, applicable to all of the models considered above, pertains to the replacement of the somewhat artificial infinite field of view $\mathcal{D}_\infty$ by the finite fields of view $\mathcal{D}_1$ and $\mathcal{D}_2$ for the pictorial patterns ρ_1 and ρ_2, respectively. Another direction of improvement is the more realistic modeling of the relation between the ideal patterns σ_1 and σ_2 when these are different from each other. Clearly, a long list of additional model improvements can be formulated. However, in doing so one must attempt to maintain a judicious balance between the degree of reality of the model and tractability of the solution.

REFERENCES

[1] RICHARDSON, J. M. Image matching I. Geometrical and gray-scale transformations; Image matching II. Effect of different physical imaging systems; Image matching III. Effect of different illumination directions. (All to appear in *Info. Sciences*.

P.R. Krishnaiah, ed., APPLICATIONS OF STATISTICS

DECONVOLUTION OF TIME SERIES AS APPLIED TO SPEECH

ENDERS A. ROBINSON
Lincoln, Massachusetts

The vocal tract shape and the excitation function that drives the vocal tract can be obtained from the human speech signal by deconvolution. The speech-producing acoustic tube model of the vocal tract is a deep-source model. If the corresponding surface-source model is constructed, then the sound waves in this model are the physical embodiments of the recursive method of solving the normal equations for the deconvolution operator.

1. INTRODUCTION

Human speech is produced as the result of passing air from the lungs through the vocal tract. The vocal tract is a non-uniform acoustic tube that extends from the glottis to the lips. Each sound produced at the lips is the resultant of two factors, namely (1) the type of wave motion used to excite the vocal tract and (2) the shape of the vocal tract. For example, in order to produce a voiced sound, the excitation function is a series of periodic pulses generated by the vocal cords. In order to produce an unvoiced sound, the excitation function is white random noise generated by turbulent air. These are the two main types of excitation functions that are inputs into the vocal tract at the glottis. The various sounds that are outputs from the vocal tract at the lips are produced by changing the shape of the vocal tract by moving the lips, jaw, tongue, and velum.

A particular sound at the lips represents a segment of a stationary time series that is produced by a given type of excitation function and a given shape of the vocal tract. More specifically, a particular sound is equal to the convolution of its excitation function with the impulse response function of the vocal tract shape. We can make this statement because the vocal tract acts like a linear filter in producing speech.

The problem of deconvolution is one of decomposing the observed output sound into its excitation function and its vocal-tract impulse response. As it turns out, the actual shape of the excitation function is not important, so only the gross features of the excitation function need to be saved, such as (1) its r.m.s. amplitude, (2) knowledge of whether it is a pulse train (in case of a voiced sound) or whether it is random (in case of an unvoiced sound) and (3) the period of the pulses (in case of a voiced sound). That is, the excitation function can be characterized by three parameters, namely (1) r.m.s. value, (2) a binary voiced-unvoiced parameter, and (3) in case of a voiced sound, the pitch period. It also turns out that the vocal-tract impulse response can be characterized by as few as 12 numbers. Thus the result of the deconvolution of a particular sound can be as few as 15 parameters, namely 3 excitation parameters and 12 vocal-tract parameters. Since the sound in question would be represented by a time series of several hundreds or thousands of observations, and since deconvolution would replace these observations by 15 parameters characterizing that sound, this approach results in a considerable reduction of the data. In fact, a reduction of 50 to 1 seems to be a goal in speech analysis that is within reach. Another goal is the identification of the parameters representing a sound with the phonetic symbol for that sound, so that an automatic typewriter could be constructed that would type out printed text from the spoken word.

Finally the results of the deconvolution of speech can be evaluated by reconvolution. For each sound an excitation function is generated from its parameters. This excitation function is convolved with the vocal-tract impulse response function to produce the output time-series segment for that sound. This convolution process is called reconvolution because we are not using the original excitation function but a reconstructed one which agrees with the original in its gross features only. The result of this process is the reconvolved sound. By attaching together all these reconvolved sound segments, we obtain the reconvolved speech signal which can be compared to the original speech signal. Experimental listening tests with various auditors establish that the quality of the reconvolved speech is very close to that of the original speech for a wide range of spoken messages and for many different speakers both male and female.

X-Ray techniques have been used as a direct method of determining the vocal tract shape. Wakita (1973) gave a mathematical method for extracting the vocal-tract shape by showing the relationship of the acoustic tube model to the recursive method given in Robinson (1967) for solving the normal equations. In this paper we show that the vocal tract shape and the excitation function can be obtained from the speech signal by deconvolution. We make use of a constructive proof that relates the speech-producing acoustic tube model, which is a deep-source model, to the corresponding surface-source acoustic tube model. In this way we can physically construct the recursive method for solving the normal equations from the acoustic properties of the system, and can physically obtain the autocorrelation, the prediction error operators, and the reflection coefficients in terms of the model.

Because of the limitation of space, we have been forced to make some of our derivations in this paper quite brief. For background material on the model and the polynomial manipulations, the reader is referred to Robinson (1967, Chapter 3).

2. ACOUSTIC TUBE MODEL

The acoustic tube model is an approximation to the vocal tract consisting of a set of interconnected cylindrical sections piled on top of each other. See Figure 1. Each section may have a different cross-sectional area but has the same height. The cross-sectional areas vary according to the sound being uttered. For any given sound let the cross-sectional area of the nth section be denoted by S_n, where n runs from 1 to N. We let section 1 be the closest to the lips, and let section N be the one closest to the glottis. For the purposes of description, we speak of the lips as being at the top, and the glottis as being at the bottom, of the vertically standing tube made up of the N sections.

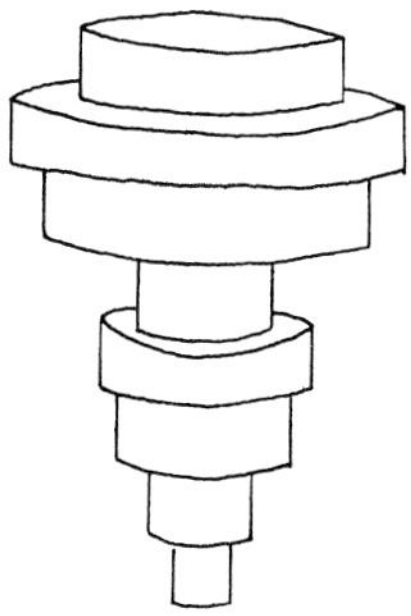

Fig. 1

The acoustic tube can support travelling wave-motion from the glottis to the lips, which we call upgoing waves, and also travelling wave-motion from the lips to the glottis, which we call downgoing waves. At each interface between two adjacent sections of the acoustic tube, a travelling wave will be partially reflected and partially transmitted, the division of energy between the reflected and transmitted portions being governed by the reflection coefficient associated with that interface. Let n denote the interface between sections n and n+1, and let c_n denote the refelction coefficient of this interface. The interpretation of this reflection coefficient is as follows. Let us use the convention that all wave motion is measured in physical units that are proportional to the square-root of energy. Thus the square of the amplitude of a wave is in terms of energy. If a downgoing impulse of unit amplitude is incident on the given interface, then a reflected impulse and a transmitted impulse are produced. See Figure 2.

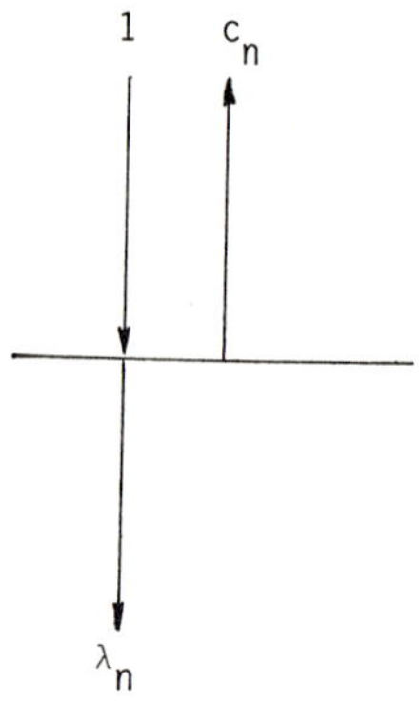

Fig. 2

The reflected impulse has amplitude c_n and the transmitted impulse has amplitude λ_n. By the law of the conservation of energy, the input energy is equal to the output energy, namely the sum of the reflected energy and the transmitted energy, that is

$$1^2 = c_n^2 + \lambda_n^2 .$$

Solving for λ_n we have

$$\lambda_n = \sqrt{1 - c_n^2}$$

The constant λ_n is called the (square-root energy) transmission coefficient. In the above example we considered a downgoing impulse. If instead we consider an upgoing impulse striking the same interface, the reflection coefficient becomes $-c_n$ and the transmission coefficient remains the same. See Figure 3.

Let us now consider the operation of the acoustic tube model in producing speech. The source is the excitation function which represents an upgoing waveform that enters at the lower interface N, namely at the glottis. As a result, speech production represents a deep-source acoustic tube model. This deep source gives rise to upgoing and downgoing wave motion in each section, due to the reflections and transmissions at each interface. In other words, the excited vocal tract represents a reverberating system. When we reach interface 0 (namely, the lips) we meet a *boundary condition*, because the lips are connected to the open air which has infinite cross-sectional area. As a result there is perfect reflection at the lips. If we assume that our wave motion is related to pressure, then

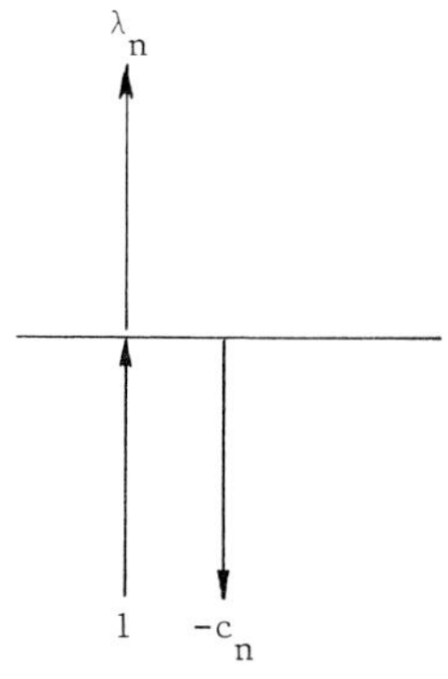

Fig. 3

the pressure must vanish at the lips, so at the lips the downgoing wave in section 1 is the negative of the upgoing wave in section 1.

The velocity of sound waves is constant, and it is convenient to choose the common height of each section such that it takes exactly one-half discrete time unit for a wave to travel from one face of the section to the other. We introduce the unit delay operator z. The delay operator z is defined as that operator which delays a time variable by one time unit. It follows that $z^{\frac{1}{2}}$ is the operator that delays a time variable by one-half time unit, namely the one-way travel time through a section.

For the deep-source model let $x_n(t)$ and $y_n(t)$ be the downgoing and upgoing waves respectively at the top of section n, and let $x'_n(t)$ and $y'_n(t)$ be the corresponding waves at the bottom of section n. See Figure 4.

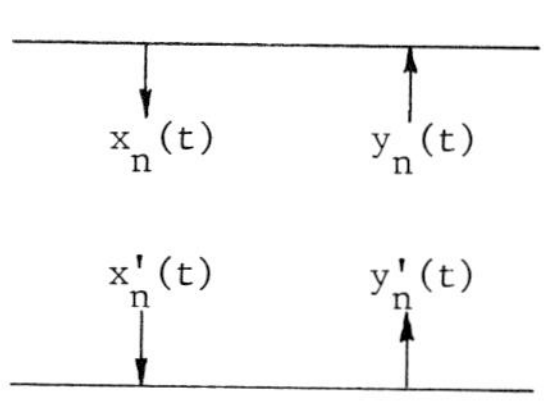

Fig. 4

Then we have

$$\begin{bmatrix} x'_n(t) \\ y'_n(t) \end{bmatrix} = \begin{bmatrix} z^{\frac{1}{2}} & 0 \\ 0 & z^{-\frac{1}{2}} \end{bmatrix} \begin{bmatrix} x_n(t) \\ y_n(t) \end{bmatrix}$$

Here and throughout, the variable t is the discrete time variable.

The waves at interface n are shown in Figure 5. They can be related as follows. See Figure 6. The wave $y'_n(t)$ is made up of a part due to the reflection of $x'_n(t)$ and a part due to the transmission of $y_{n+1}(t)$, that is,

$$y'_n(t) = c_n x'_n(t) + \lambda_n y_{n+1}(t).$$

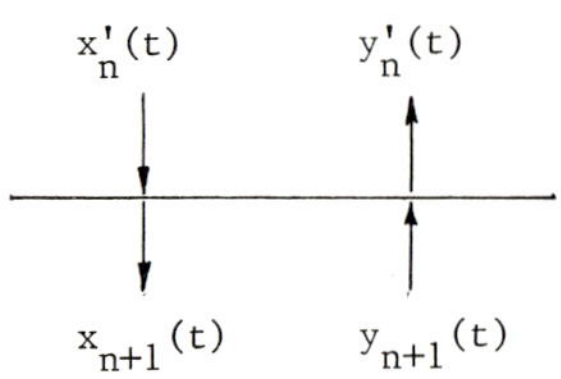

Fig. 5

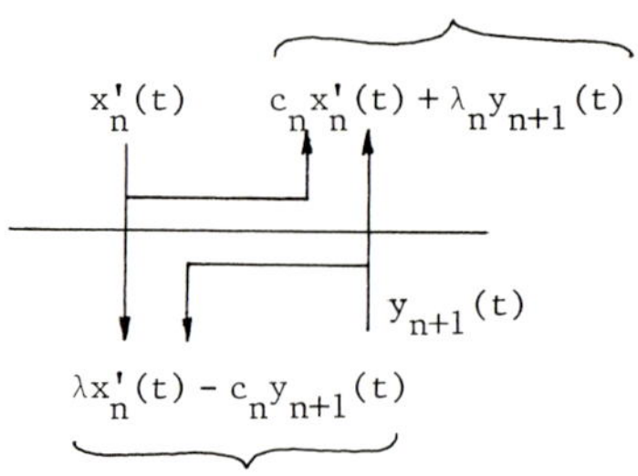

Fig. 6

The wave $x_{n+1}(t)$ is made up of a part due to the reflection of $y_{n+1}(t)$ and a part due to the transmission of $x'_n(t)$, that is

$$x_{n+1}(t) = - c_n y_{n+1}(t) + \lambda_n x'_n(t) .$$

If we solve these two equations for the unprimed quantities we obtain

$$\begin{bmatrix} x_{n+1}(t) \\ y_{n+1}(t) \end{bmatrix} = \frac{1}{\lambda_n} \begin{bmatrix} 1 & -c_n \\ -c_n & 1 \end{bmatrix} \begin{bmatrix} x'_n(t) \\ y'_n(t) \end{bmatrix}$$

Combining this matrix equation with the previous matrix equation, we obtain the *scattering equation:*

$$\begin{bmatrix} x_{n+1}(t) \\ y_{n+1}(t) \end{bmatrix} = \frac{z^{-\frac{1}{2}}}{\lambda_n} \begin{bmatrix} z & -c_n \\ -c_n z & 1 \end{bmatrix} \begin{bmatrix} x_n(t) \\ y_n(t) \end{bmatrix}$$

The z-transform of the scattering equation is

$$\begin{bmatrix} X_{n+1}(z) \\ Y_{n+1}(z) \end{bmatrix} = \frac{z^{-\frac{1}{2}}}{\lambda_n} \begin{bmatrix} z & -c_n \\ -c_n z & 1 \end{bmatrix} \begin{bmatrix} X_n(z) \\ Y_n(z) \end{bmatrix}$$

The z-transform is defined as

$$X_n(z) = \Sigma\, x_n(t) z^t$$

where the summation is over all discrete t values. That is, the z-transform of a time sequence denoted by a lower case letter is denoted by the corresponding capital letter. For simplicity, we often drop the argument z, so $X_n(z)$ would be simply X_n. A bar over a z-transform indicates that each z in the z-transform has been replaced by z^{-1}, that is $\bar{X}_n = X_n(z^{-1})$.

If we assume that the time sequence $x_n(t)$ has finite energy, then the quantity $X_n\bar{X}_n$ represents the energy spectrum of the time sequence $x_n(t)$; that is, $X_n\bar{X}_n$ is the z-transform of the autocorrelation of $x_n(t)$. From the scattering equation we obtain

$$X_{n+1}\bar{X}_{n+1} - Y_{n+1}\bar{Y}_{n+1} = X_n\bar{X}_n - Y_n\bar{Y}_n$$

which says that the net downgoing energy in section n+1 is equal to the net downgoing energy in section n. A recursive application of this result shows that the net downgoing energy in any two sections is the same, provided of course that there are no energy sources or sinks between the two sections in question.

The polynomials P_n and Q_n are defined as

$$\begin{bmatrix} P_n^R & Q_n^R \\ Q_n & P_n \end{bmatrix} = \begin{bmatrix} z & -c_n \\ -c_n z & 1 \end{bmatrix} \begin{bmatrix} z & -c_{n-1} \\ -c_{n-1} z & 1 \end{bmatrix} \cdots \begin{bmatrix} z & -c_1 \\ -c_1 z & 1 \end{bmatrix}$$

where the polynomial P^R is the reverse of P_n and Q_n^R is the reverse of Q_n, that is

$$P_n^R = z^n \bar{P}_n \quad , \qquad Q_n^R = z^n Q_n$$

We also define $\sigma_n = \lambda_n \lambda_{n-1} \cdots \lambda_1$ and $A_n = P_n - Q_n$. We note that in the polynomial A_n the coefficient of z^0 is one and the coefficient of z^n is c_n. The recursive use of the scattering equation gives

$$\begin{bmatrix} X_{n+1} \\ \\ Y_{n+1} \end{bmatrix} = \frac{z^{-n/2}}{\sigma_n} \begin{bmatrix} P_n^R & Q_n^R \\ \\ Q_n & P_n \end{bmatrix} \begin{bmatrix} X_1 \\ \\ Y_1 \end{bmatrix}$$

3. DEEP SOURCE MODEL

We are interested in the impulse response of the acoustic tube in the case of speech, that is, in the case of a deep source. If we let the upgoing wave at the glottis be a unit impulse, that is, if we let $Y_{N+1} = 1$, and if we impose the boundary condition of perfect reflection $Y_1 = -X_1$ at the lips (see Figure 7), we obtain

$$\begin{bmatrix} X_{N+1} \\ \\ 1 \end{bmatrix} = \frac{z^{-N/2}}{\sigma_N} \begin{bmatrix} P_N^R & Q_N^R \\ \\ Q_N & P_N \end{bmatrix} \begin{bmatrix} X_1 \\ \\ -X_1 \end{bmatrix}$$

which gives

$$1 = -\frac{z^{-N/2}}{\sigma_N} (P_N - Q_N) X_1 \quad .$$

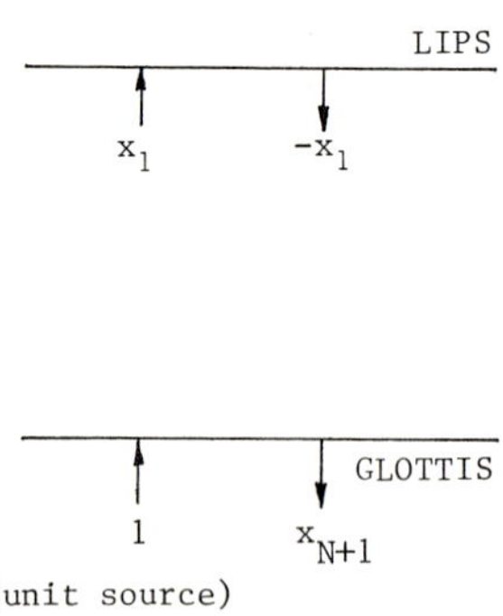

Fig. 7

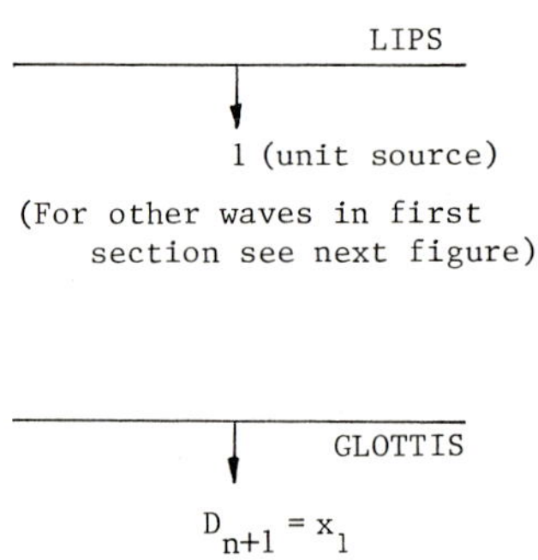

Fig. 8

Thus the impulse response of the vocal tract is

$$X_1 = -\frac{\sigma_N z^{N/2}}{P_N - Q_N} = -\frac{\sigma_N z^{N/2}}{A_N}$$

4. SURFACE-SOURCE MODEL

Speech is produced by an acoustic tube subject to a deep source. Let us now consider the same acoustic tube but subject to a surface source. More specifically, we want to find the downgoing response D_{N+1} at the glottis subject to a downgoing unit impulsive source at the lips. See Figure 8. In the previous section we have found the vocal-tract impulse response, namely the upgoing response X_1 at the lips subject to an upgoing unit impulse source at the glottis. The physical principle of reciprocity states that if we interchange source and receiver we observe the same waveform. Thus the above two responses are the same, i.e. $D_{N+1} = X_1$. We note that no energy is reflected back from the glottis in this surface-source model, so the upgoing waveform at the glottis is zero, i.e. $U_{N+1} = 0$. The net downgoing energy in the glottis is therefore

$$D_{N+1}\bar{D}_{N+1} - U_{N+1}\bar{U}_{N+1} = D_{N+1}\bar{D}_{N+1} = X_1\bar{X}_1 .$$

Let us now look at the first section of the surface-source model (i.e. the section at the lips). The downgoing impulsive source can be represented by the Kronecker delta function δ_t at the top of the first section. This source gives rise to an upgoing wave in the first section as a result of internal reflections from interfaces below; we denote this upcoming wave by $-r_1, -r_2, -r_3, \ldots$ where the subscript represents the time of occurrence. Because the top interface (i.e. the lips) is a perfect reflector, this upgoing wave is reflected at the lips back into the downgoing wave $r_1, r_2, r_3, \ldots$ Thus the entire downgoing wave in section 1 is made up of the downgoing impulse δ_t and the reflected wave $r_1, r_2, r_3, \ldots$; that is $\delta_t + r_t$. The z-transform of this downgoing wave is

$$D_1 = 1 + R$$

The upgoing wave in section 1 is $-r_t$ with z-transform

$$U_1 = -R$$

See Figure 9.

The net downgoing energy in section 1 is

$$D_1\bar{D}_1 - U_1\bar{U}_1 = (1+R)(1+\bar{R}) - R\bar{R} = 1 + R + \bar{R}$$

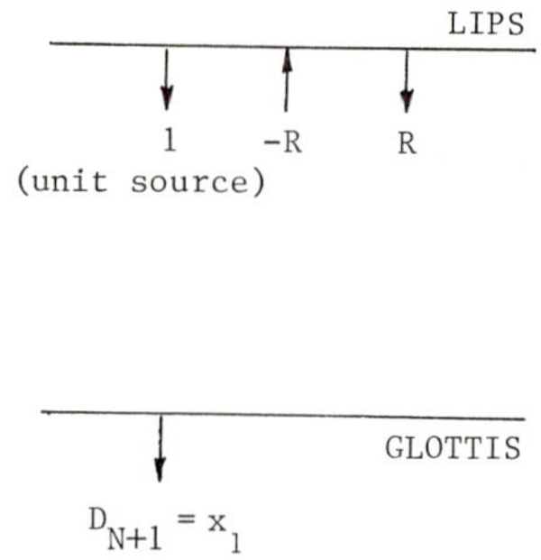

Fig. 9

which is equal to the net downgoing energy in the glottis, that is

$$X_1\bar{X}_1 = 1 + R + \bar{R}$$

Thus we see that $1 + R + \bar{R}$ is the z-transform of the autocorrelation of $x_1(t)$. That is, the autocorrelation of the impulse response function $x_1(t)$ of the vocal tract for speech is

$$\ldots, r_2, r_1, 1, r_1, r_2, \ldots$$

where 1 is the autocorrelation coefficient for lag 0, r_1 for lag -1 and also for lag 1, and so on.

5. NORMAL EQUATIONS

In the surface source model let $d_{n+1}(t)$ and $u_{n+1}(t)$ denote the downgoing and upgoing waves respectively at the top of section n+1, and let their respective z-transforms be D_{n+1} and U_{n+1}. Then the scattering equation gives

$$\begin{bmatrix} D_{n+1} \\ U_{n+1} \end{bmatrix} = \frac{z^{-n/2}}{\sigma_n} \begin{bmatrix} P_n^R & Q_n^R \\ Q_n & P_n \end{bmatrix} \begin{bmatrix} 1 + R \\ - R \end{bmatrix}$$

In this equation we first solve for D_{n+1} and U_{n+1}, from which we find that

$$\bar{D}_{n+1} - U_{n+1} = \frac{z^{-n/2}}{\sigma_n} A_n \, (1 + R + \bar{R})$$

which is

$$A_n (1 + R + \bar{R}) = z^{n/2} \sigma_n (\bar{D}_{n+1} - U_{n+1}) .$$

The function D_{n+1} is the z-transform of the downgoing wave at the top of section n+1 in the surface source model. This downgoing wave is made up of the direct pulse $d_{n+1}(\frac{n}{2})$ together with succeeding pulses $d_{n+1}(\frac{n}{2}+1)$, $d_{n+1}(\frac{n}{2}+2)$, ... Because the direct pulse is the result of transmissions through the first n interfaces, the direct pulse is equal to the product of the first n transmission coefficients, i.e.

$$d_{n+1} (\tfrac{n}{2}) = \lambda_1 \lambda_2 \cdots \lambda_n = \sigma_n$$

and the time of the direct pulse is n/2. Thus

$$D_n = \sigma_n z^{n/2} + \text{terms in higher powers of } z.$$

The function U_{n+1} is the z-transform of the upgoing wave at the top of section n+1. The first pulse in the upgoing wave is the reflection of the direct downgoing pulse at interface n+1. Thus the magnitude of this pulse is $\sigma_n c_{n+1}$; that is, the magnitude is equal to the magnitude of the direct downgoing pulse times the reflection coefficient. Because one time unit elapses for the round trip in section n+1, the first pulse of the upgoing wave occurs at one time unit later than the first pulse of the downgoing wave in section n+1; that is, the first upgoing pulse at the top of section n+1 occurs at $\frac{n}{2} + 1$. Thus

$$U_{n+1} = \sigma_n c_{n+1} z^{\frac{n}{2}+1} + \text{terms in higher powers of } z.$$

We therefore have

$$A_n(1 + R + \bar{R}) = \sigma_n \text{ (terms in negative powers of } z + \sigma_n - \sigma_n c_{n+1} z^{n+1}$$
$$- \text{ terms in higher powers of } z)$$

We note that the powers of z from 1 to n are missing on the right-hand side of the above equation; it is this fact that allows us to extract the normal equations. More specifically, let us now equate coefficients on each side of this equation for the powers of z from 0 to n+1. We obtain

$$a_{no} r_o + a_{n1} r_1 + \ldots + a_{nn} r_n = \sigma_n^2$$
$$a_{no} r_1 + a_{n1} r_o + \ldots + a_{nn} r_{n-1} = 0$$
$$\ldots$$
$$a_{no} r_n + a_{n1} r_{n-1} + \ldots + a_{nn} r_o = 0$$
$$a_{no} r_{n+1} + a_{n1} r_n + \ldots + a_{nn} r_1 = -c_{n+1} \sigma_n^2$$

where $a_{no}, a_{n1}, \ldots, a_{nn}$ are the coefficients of A_n.

We note that $r_o \equiv 1$ and $a_{no} \equiv 1$.

Let us now look at these equations. The time function r_t is the autocorrelation of the vocal-tract impulse response function $x_1(t)$ for speech. Thus the first n+1 of these equations are the normal equations for the prediction-error operator $a_{no}, a_{n1}, \ldots, a_{nn}$. The last of these equations allows us to find c_{n+1}, and we shall make use of this fact in the next section.

By the definition of P_n and Q_n we have

$$\begin{bmatrix} P^R_{n+1} & Q^R_{n+1} \\ Q_{n+1} & P_{n+1} \end{bmatrix} = \begin{bmatrix} z & -c_{n+1} \\ -c_{n+1}z & 1 \end{bmatrix} \begin{bmatrix} P^R_n & Q^R_n \\ Q_n & P_n \end{bmatrix}$$

which gives the recursions

$$P_{n+1} = P_n - c_{n+1} \, z \, Q^R_n$$

$$Q_{n+1} = Q_n - c_{n+1} \, z \, P^R_n$$

Because $A_n = P_n - Q_n$ we can subtract the above two equations to obtain the *polynomial recursion*

$$A_{n+1} = A_n + c_{n+1} \, z \, A^R_n$$

where $A^R_n = P^R_n - Q^R_n = z^n \bar{A}$. Thus if we know A_n and c_{n+1} we can compute the polynomial A_{n+1}.

6. DECONVOLUTION OF SPEECH

As we have seen, each sound in speech is produced by the convolution of excitation function with vocal-tract impulse response function. In case the excitation function is white noise, it follows that the autocorrelation of the vocal-tract impulse response function can be approximated by the autocorrelation of the sound signal. In case the excitation function is a periodic pulse sequence, the autocorrelation of the vocal-tract impulse response function also can be derived approximately from knowledge of the autocorrelation of the sound signal. Thus the first step in the deconvolution of a sound signal measured at the lips is to compute its autocorrelation which is then used to approximate the autocorrelation

$$\ldots, r_2, r_1, 1, r_1, r_2, \ldots \qquad (\text{where } r_o \equiv 1)$$

of the impulse- response function of the vocal tract.

The next step is to compute the sequence of prediction error operators and reflection coefficients for n=1 to N. Initially, we have

$$a_{oo} = 1$$

$$\sigma_o^2 = a_{oo} r_o = 1 \ .$$

Then given the prediction error operator $a_{no}, \ldots, a_{nn}$ and σ_n^2 we compute the reflection coefficient

$$c_{n+1} = - (a_{no} r_{n+1} + a_{n1} r_n + \ldots + a_{nn} r_1) / \sigma_n^2 \ ,$$

the variance

$$\sigma_{n+1}^2 = \lambda_{n+1}^2 \sigma_n^2 = (1 - c_{n+1}^2) \sigma_n^2$$

and the new prediction error operator by the polynomial recursion

$$a_{n+1,o} = a_{no} = 1$$

$$a_{n+1,1} = a_{n1} + c_{n+1} a_{nn}$$

$$\ldots$$

$$a_{n+1,n} = a_{nn} + c_{n+1} a_{n1}$$

$$a_{n+1,n+1} = c_{n+1} a_{no} = c_{n+1}$$

Using this recursive algorithm we can compute the reflection coefficients $c_1, c_2, \ldots, c_N$ of the vocal tract. These reflection coefficients are related to the cross-sectional areas $S_1, S_2, \ldots, S_N$ of the vocal tract by the equation

$$c_n = \frac{S_n - S_{n+1}}{S_n + S_{n+1}}$$

Solving this equation for S_{n+1} we have

$$S_{n+1} = \frac{1 - c_n}{1 + c_n} S_n$$

Thus given the cross-sectional area S_1 at the lips we can compute in a stepwise fashion the areas of all the sections from the lips to the glottis. Thus the deconvolution process has given us the shape of the vocal tract.

The final prediction error operator $a_{N0}, a_{N1}, \ldots, a_{NN}$ is the required deconvolution operator. This statement follows from the fact that the impulse response function of the vocal tract, as we have seen, is

$$X_1 = -\frac{\sigma_N z^{N/2}}{A_N}$$

If E is the z-transform of the excitation function then the z-transform of the sound signal at the lips is $X_1 E$. The z-transform of the deconvolution operator is A_N. The z-transform of the deconvolved sound signal is

$$A_N X_1 E = -\sigma_N z^{N/2} E$$

Since σ_N represents a scale factor and $z^{N/2}$ represents a delay (both of which we know), the deconvolution has yielded E. That is, the deconvolution of the sound signal yields the excitation function.

7. CONCLUSION

The deconvolution of a sound produced in speech yields the reflection coefficients, or equivalently the cross-sectional areas, which make up the numerical parameters necessary to characterize the vocal tract shape for that sound. The deconvolution also yields the excitation function which drives the vocal tract for that sound. In order to establish the results, we have made use of an acoustic tube model with a source at the lips, as this model physically gives us the autocorrelation of the vocal tract impulse response, the normal equations for the prediction error operators, and the reflection coefficients.

REFERENCES

[1] ROBINSON, ENDERS A. (1967). *Multichannel Time Series Analysis with Digital Computer Programs.* Holden-Day Publishing Co., San Francisco.

[2] WAKITA, HISASHI (1973). Direct estimation of the vocal tract shape by inverse filtering of acoustic speech waveforms. *IEEE Transactions on Audio and Electroacoustics*, AU-21, 417-427.

P.R. Krishnaiah, ed., APPLICATIONS OF STATISTICS

STATISTICAL ANALYSIS OF COMPARTMENTAL MODELS WITH APPLICATIONS TO PHARMACOKINETICS AND BIOAVAILABILITY

JAGDISH S. RUSTAGI

Department of Statistics
The Ohio State University
Columbus, Ohio 43210

UMED SINGH

Haryana Agricultural University
Hissar, Haryana, India

1. INTRODUCTION

Large number of scientific problems concerned with the study of a system are mainly devoted to the study of rates of exchange of material between components of the system. In biological contexts, the components are called compartments. The study of the kinetics of absorption, distribution, metabolism and excretion of drugs and their corresponding pharmacologic, therapeutic or toxic responses in animal and man, is described by the term *Pharmacokinetics*. Compartmental models play a central role in pharmacokinetics.

In deciding whether one formulation of a drug is better than the other, there are several measures in current use. The term *availability of the drug for absorption* measures the proportion of the drug administered but not excreted in a given interval of time. *Physiological* availability or *bioavailability* measures the relative efficiency of absorption of a test drug relative to a standard drug. Many measures of bioavailability are available in the literature, Wagner (1971). These methods invariably are based on measurement of plasma, serum, or whole blood concentrations of unchanged drug or measurements of urine concentrations of unchanged drug. They may be based on measurement of the total amount of metabolite drug in urine after the administration of a single or multiple dose. In estimation of these quantities, the use of compartmental models is a necessity. Comprehensive surveys of pharmacokinetics and bioavailability using compartment models have recently been published by Wagner (1971, 1975).

Compartmental models are also commonly applied to many other areas besides those of medicine and biology. Exhaustive surveys of compartmental analysis in biological applications are available, Jacquez (1972) and Atkins (1969).

For basic understanding of the compartmental system, theory of differential equations is generally used. When the exchange rates between compartments are constants, linear differential equations provide the models for most cases. It is the solutions of these differential equations which provide the experimenter suitable models in pharmacokinetic and bioavailability studies.

In this paper, introduction to simple compartmental systems is given. In place of using the solution of the differential equation for experimental verifications and estimations, the differential equation is used directly. The discrete analog of a differential equation is given by a difference equation and we utilize the difference equation for our purpose. In some cases the difference equations reduce to simple time-series models and provide more convenient solutions to experimental results. This method is utilized to analyze data from recent studies.

There is vast literature in the area of Pharmacokinetics and Biopharmaceutics. More than one thousand articles are listed by Wagner (1971). A complete survey to the present time may include more than two thousand papers. A few of these are mentioned in the bibliography.

2. COMPARTMENTAL MODELS

The development of a plausible model for a biological system, requires basic understanding of the biological mechanism. By its nature, a model is a simple and crude approximation of the phenomenon and is dictated, under many circumstances, by the nature of the problem. If one is interested in knowing whether the amount of an atmospheric pollutant is retained in the human body when its uptake and excretion rate are given, only one compartment may be all that is necessary. It seems that compartmental analysis was used for the first time by Hevesey (1923), to describe the uptake and loss of lead ions by roots of *Vicia faba*. Since then its applications to various areas of scientific research continue unabated. Not only compartmental analysis allows interpretations of observations in reference to experiments in biology and medicine but it also provides a mechanism to study many phenomena in chemistry, economics, ecology and engineering. A one compartmental model is given by Figure 1.

Fig. 1

Assume that the amount of a substance in the compartment at any time t is y(t). Further assuming that the rate of output is proportional to the total amount in the compartment, at time t, the following first order differential equation gives its behavior.

$$\frac{dy}{dt} = -k(t) \qquad (2.1)$$

The solution of the equation (2.1) given by

$$y = k_0 e^{-kt}, \qquad (2.2)$$

is generally regarded as the equation of the model. If the observations are available on the compartment at various times, the data provide estimates of the constants involved, thus determining the behavior of the system.

A two-compartment model may have various forms. A few of these are given in Figure 2.

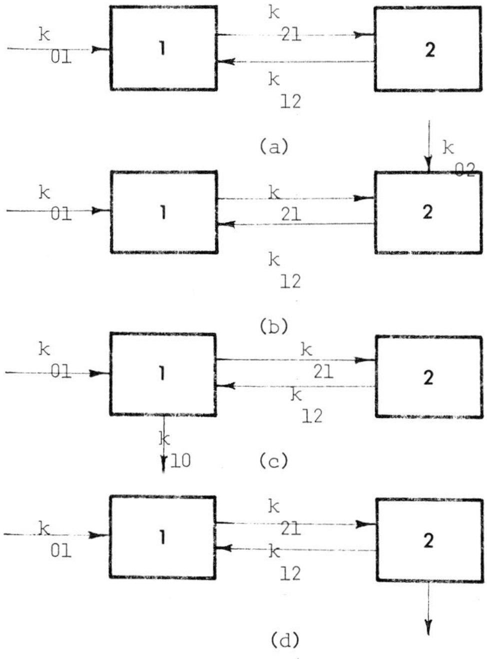

Fig. 2

The above graphical models assume that output may occur from either or both or none of the compartment as well as the intake may be from either one or both of them.

Suppose that $y_1(t)$ is the amount of a substance in compartment 1 and $y_2(t)$ in compartment 2 at time t. Then assuming the rates k_{12}, k_{21}, and k_{20}, and k_{01} as constants, the linear differential equation satisfying the system given by Figure 2 (d) are

$$\frac{dy_1}{dt} = k_{12}y_2 - k_{21}y_1 + k_{01}, \tag{2.3}$$

$$\frac{dy_2}{dt} = k_{21}y_1 - k_{12}y_2 - k_{20}\, y_2. \tag{2.4}$$

The compartmental system may be *closed* or *open*. In a closed system, no material can leave or enter it. This may be the case in the study of the pharmacokinetics of a system after the drug is given and it metabolizes within the system without being excreted. In an open system, material is exchanged with the environment. A *Catenary* System involves p compartments which are arranged in a chain like manner such that each compartment has nonzero rates of exchange (transitions) only with the compartment adjacent to it. A typical system is given in Figure 3.

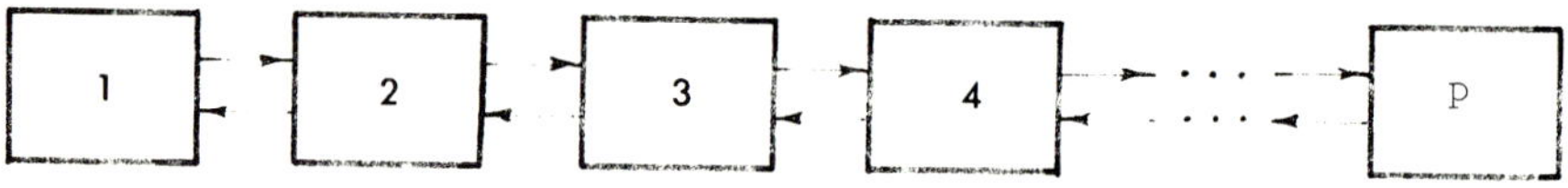

Fig. 3. Catenary System

A *Mammilary* System contains p-1 compartments having transition rates with a central compartment but no transition between (p-1) peripheral compartments. A typical system is given in Figure 4.

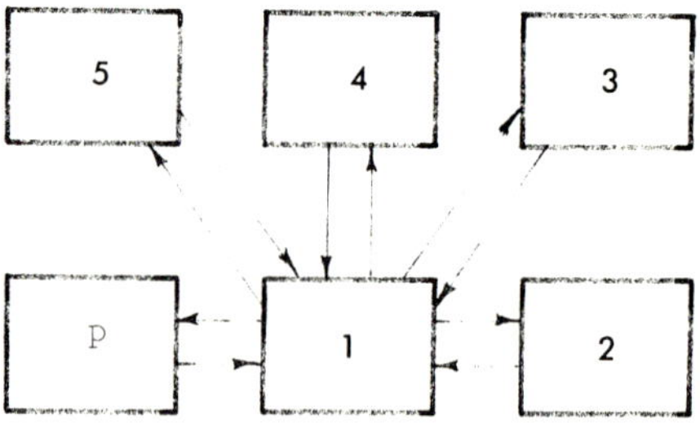

Fig. 4. Mammilary System

Although "The use of a general theory of compartmental analysis has found very little direct application to practical work" Atkins (1968), the general model is necessary to understand the shortcomings of the analysis.

Consider a system having p compartments. Let $y_1(t),\ldots,y_p(t)$ be the amount of material in the system in p compartment.

$$\underset{\sim}{y}(t) = [y_1(t),\ldots,y_p(t)]' .$$

Let ℓ_{ij} be the turnover rates (rate constants) of the material from compartment j to compartment i, $i \neq j$, $i=1,2,\ldots,p$ and let ℓ_{ii} be the rate constants for the ith compartment.

Let $L = (\ell_{ij})$ be the matrix of these constants. In general under the assumptions of linear system the material in the ith compartment satisfies the following system of differential equations

$$\frac{dy_i}{dt} = \ell_{i1}y_1(t) + \ell_{i2}y_2(t) +\ldots+ \ell_{ip}y_p(t), \quad i=1,2,\ldots,p.$$

In the matrix notation, the system has the following form.

$$\frac{d\underset{\sim}{y}}{dt} = \underset{\sim}{L}\,\underset{\sim}{y}. \tag{2.5}$$

Let the concentration of the substance in the ith compartment of Volume V_i be $C_i(t)$, then

$$y_i(t) = V_iC_i(t), \quad i=1,2,\ldots,p. \tag{2.6}$$

or

$$\underset{\sim}{y}(t) = \underset{\sim}{V}\,\underset{\sim}{C}(t) \tag{2.7}$$

where

$$\underset{\sim}{V} = \begin{bmatrix} V_1 & 0 & \cdot & \cdot & \cdot & 0 \\ 0 & V_2 & \cdot & \cdot & \cdot & 0 \\ \cdot & \cdot & \cdot & \cdot & \cdot & \cdot \\ \cdot & \cdot & \cdot & \cdot & \cdot & \cdot \\ \cdot & \cdot & \cdot & \cdot & \cdot & \cdot \\ 0 & 0 & \cdot & \cdot & \cdot & V_p, \end{bmatrix}$$

and $\underset{\sim}{C}(t) = [C_1(t),C_2(t),\ldots,C_p(t)]'$.

The basic compartment equations then reduce to the following.

$$\frac{d\underset{\sim}{C}}{dt}(t) = \underset{\sim}{V}^{-1}\underset{\sim}{L}\ \underset{\sim}{V}\ \underset{\sim}{C}(t) = \underset{\sim}{W}\ \underset{\sim}{C}(t), \tag{2.8}$$

$$\text{where } \underset{\sim}{W} = \underset{\sim}{V}^{-1}\ \underset{\sim}{L}\ \underset{\sim}{V}.$$

The solution of the system (2.8) is well known. It is given by,

$$C(t) = \left[\underset{\sim}{A}\left\{e^{-\lambda_1 t}, e^{-\lambda_2 t}, \ldots, e^{-\lambda_p t}\right\}\right]' , \tag{2.9}$$

where $\underset{\sim}{A}$ is a matrix of unknown constants and $-\lambda_1, -\lambda_2, \ldots, -\lambda_p$ are different characteristic roots of the matrix $\underset{\sim}{W}$.

In applications it is not possible to observe concentrations on all the compartments for determining the rate constants. What is generally done that observations are made on only one compartment and rate constants are estimated if possible. It is not possible to estimate all the rate constants in general. Certain pertinent results are given by Shah (1976). Shah has given arguments to show that in the case of a system having more than 2 compartments, first order rate constants are not uniquely estimable if the drug is introduced in the first compartment and elimination occurs from kth compartment, the system being connected linearly. The situation is shown in the following Figure 5

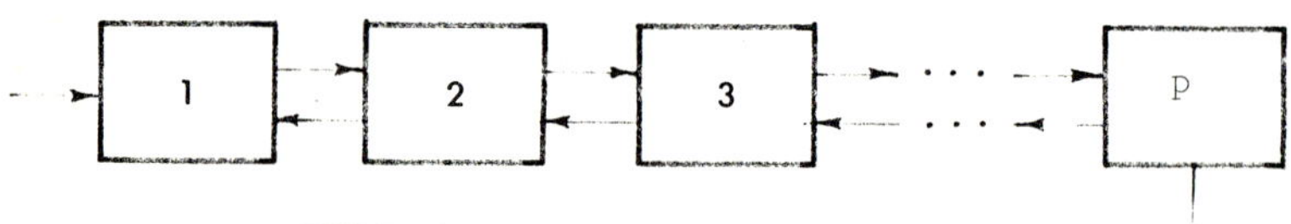

Fig. 5

The elimination procedure provides the pth order linear differential equation equivalent to the system (2.8) as

$$\frac{d^pC}{dt^p} + \sum_{i=1}^{p} \lambda_i \frac{d^{p-1}C}{dt^{p-1}} + \ldots + \lambda_1 \ldots \lambda_p C = 0 \tag{2.10}$$

where C(t) is the concentration of a typical compartment.

3. DIFFERENCE EQUATION APPROACH

Although in most applications of compartmental analysis, the models are in terms of differential equations, it is the solutions of the differential equations which are fitted to data in order to obtain unknown rate constants or in order to determine the number of compartments. The attempt in this paper is to utilize the difference equation analog of the differential equation for this purpose. This approach avoids the estimation of constants of integration which enter the solution of the differential equations and are not an inherent part of the model. The assumptions on errors in the model of difference equations are not the same as those for the solutions of the differential equations but they seem plausible. In addition, for the linear system, the difference equation approach leads to autoregressive time-series models which have been studied in statistics extensively. The time-series solutions are exact as compared to the approximate solutions of nonlinear models especially the most commonly used sum of exponential model. The approach of difference equations has been used earlier by Hartley (1948) as the technique of internal least squares and by Shah (1973) for obtaining initial estimates in pharmacokinetic models.

Since models are generally approximate descriptions of the phenomenon under study, the approximation to difference equations especially when the observations are made at discrete time points, does not introduce serious deviation from the assumptions of the model. The difference equation approach avoids the solution of complicated differential equations. Also sometimes the exact solution of an approximate problem is easier than the approximate solution of an exact problem and should be preferred.

The difference equations result from differential equation using approximations of the derivatives by differences. There are two cases when we approximate the derivatives. First we consider the case when observations are made at equal intervals of time, say at $t=0,1,2,\dots,n$.

Case 1. Equal intervals of time

Let the concentration $C(t)$ of a substance in a compartment be observed at times $t=0,1,2,\dots,n$. Define the following differences.

$$
\begin{aligned}
\Delta^0 C(t) &= 1,\\
\Delta^1 C(t) &= C(t+1) - C(t)\\
\Delta^2 C(t) &= \Delta^1 C(t+1) - \Delta^1 C(t)\\
&\vdots\\
\Delta^p C(t) &= \Delta^{p-1} C(t+1) - \Delta^{p-1} C(t).
\end{aligned}
\qquad (3.1)
$$

Replacing $\frac{d^jC}{dt}$ by $\Delta^j(t)$, the pth order differential equation (2.10) reduces to the following.

$$\Delta^p C(t) + \sum_{i=1}^{p} \lambda_i \, \Delta^{p-1} C(t) + \ldots + \lambda_1, \ldots, \lambda_p C(t) = 0 \tag{3.2}$$

The equation (3.2) reduces to the following.

$$C(t+p) + b_1 C(t+p-1) + b_2 C(t+p-2) \ldots + b_p C(t) = 0 \tag{3.3}$$

The constants $b_1, b_2, \ldots, b_p$ are given by

$$\underset{\sim}{b} = \underset{\sim}{Q} \, \underset{\sim}{\Lambda} \tag{3.4}$$

where $\underset{\sim}{b} = (1, b_1, b_2, \ldots, b_p)'$,

$$\underset{\sim}{\Lambda} = (1, \Sigma\lambda_i, \Sigma\lambda_i\lambda_j, \ldots, \lambda_1\lambda_2 \ldots \lambda_p), \tag{3.5}$$

and

$$\underset{\sim}{Q} = \begin{bmatrix} 1 & 0 & 0 & \cdot & \cdot & \cdot & 0 \\ -\binom{p}{1} & 1 & 0 & \cdot & \cdot & \cdot & 0 \\ \binom{p}{2} & \binom{p-1}{1} & 1 & 0 & \cdot & \cdot & 0 \\ \vdots & \vdots & \vdots & \vdots & & & \vdots \\ (-1)^{p-1}\binom{p}{p-1} & (-1)^{p-2}\binom{p-1}{p-2} & \cdot & \cdot & -2 & 1 & 0 \\ (-1)^p & (-1)^{p-1} & \cdot & \cdot & \cdot & -1 & 1 \end{bmatrix} \tag{3.6}$$

The equation (3.3) can now be written as

$$\underset{\sim}{C}_p(t) \, \underset{\sim}{b} = 0 \tag{3.7}$$

where

$$\underset{\sim}{C}_p(t) = [C(t+p), C(t+p-1), \ldots, C(t)] \tag{3.8}$$

For illustration consider a two compartment model, p=2. The differential equation of the system is given by

$$\frac{d^2C}{dt^2} + (\lambda_1+\lambda_2) \frac{dC}{dt} + \lambda_1\lambda_2 C = 0 \,. \tag{3.9}$$

The difference equation analog of (3.9) is

$$\Delta^2 C(t) + (\lambda_1+\lambda_2) \, \Delta C(t) + \lambda_1\lambda_2 C(t) = 0 \tag{3.10}$$

so that (3.10) gives

$$C(t+2) - 2C(t+1) + C(t) + (\lambda_1+\lambda_2)\,[C(t+1)-C(t)] + \lambda_1\lambda_2 C(t) = 0 \tag{3.11}$$

or

$$C(t+2) + b_1 C(t+1) + b_2 C(t) = 0 \tag{3.12}$$

where

$$b_1 = \lambda_1+\lambda_2-2,\ b_2 = 1+\lambda_1\lambda_2-\lambda_1-\lambda_2$$

It can be verified that b_1, b_2 satisfy the equation (3.4). Similar equations are given by Shah (1973).

Case 2. Unequal intervals of time

We assume now that the observations on the system are being made at unequal intervals of time. Suppose $C(t)$ is given for $t=t_1,t_2,\dots,t_n$ so that $t_2-t_1, t_3-t_2, \dots, t_n-t_{n-1}$ are not equal. The differences in this case are defined in terms of divided differences given as follows.

$$\delta C(t_i) = \frac{C(t_{i+1}) - C(t_i)}{t_{i+1} - t_i} \tag{3.13}$$

$$\delta^2 C(t_i) = \frac{\delta C(t_{i+1}) - \delta C(t_i)}{t_{i+2} - t_i} \tag{3.14}$$

and so on.

Replacing the derivative $\frac{d^j C}{dt^j}$ by $j!\delta^j C(t)$, we have the equations (3.2) now to reduce to

$$p!\delta^p C(t_i) + (p-1)!\delta^{p-1} C(t_i) \sum_{i=1}^{p} \lambda_i + \dots + \lambda_1 \dots \lambda_p C(t_i) = 0 \tag{3.15}$$

The divided differences do not lead to further simplification as was possible in the case of equal time intervals. However, the equations (3.15) can be fitted directly.

The solutions of the difference equations converge to the solutions of the corresponding differential equations in many cases. We do not pursue these questions in this paper. An elaborate discussion is given by Fort (1948).

Examples

Example 1

A one-compartment model with elimination according to Michaelis-Menten scheme is given in Figure 6

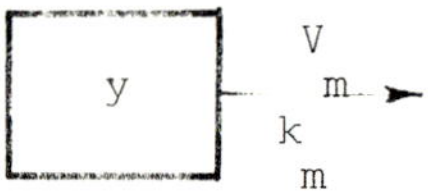

Fig. 6

The mathematical model is given by

$$\frac{dy}{dt} = -\frac{V_m y}{k_m + y} \tag{3.16}$$

when V_m = the maximum reaction rate (mg/m per hr) and k_m = Michaelis constant.

Using difference equation analog of (3.16) for the equal time interval case, we have

$$y_{t+1} - y_t = -\frac{V_m y_t}{k_m + y_t}, \quad t=1,2,\dots,n \tag{3.17}$$

The equation (3.17) simplifies to

$$k_m(y_{t+1}-y_t) + V_m y_t + y_t(y_{t+1}-y_t) = 0 \tag{3.18}$$

The statistical model for estimating constants k_m and V_m is the following.

$$k_m(y_{t+1}-y_t) + V_m y_t + y_t(y_{t+1}-y_t) = u_t. \quad t=1,2,\dots,n \tag{3.19}$$

where u_t are random errors with means zero and variance σ_t^2. The weighted least-square solution for V_m and k_m are obtained by minimizing

$$\sum_{t=1}^{n} w_t[k_m(y_{t+1}-y_t) + V_m y_t + y_t(y_{t+1}-y_t)]^2 \tag{3.20}$$

The following equations provide the unique least -square solutions.

$$k_m \sum_{t=1}^{n} w_t(y_{t+1}-y_t)^2 + V_m \sum_{t=1}^{n} w_t y_t(y_{t+1}-y_t) + \sum_{t=1}^{n} w_t y_t(y_{t+1}-y_t)^2 = 0 \tag{3.21}$$

and

$$k_m \sum w_t(y_{t+1}-y_t)y_t + V_m \sum w_t y_t^2 + \sum w_t y_t^2 (y_{t+1}-y_t) = 0 \qquad (3.22)$$

Usual methods for estimating weights can be applied.

Example 2

Although difference equations are especially useful in linear systems, it is really in the study of non linear systems that their utility is great. Consider the following two compartment models studies by Sedman and Wagner (1975). It is known that iterative methods do not provide reasonable estimates of these constants by well known computer programs such as NONLIN, Wagner (1975). The model is given in Figure 7

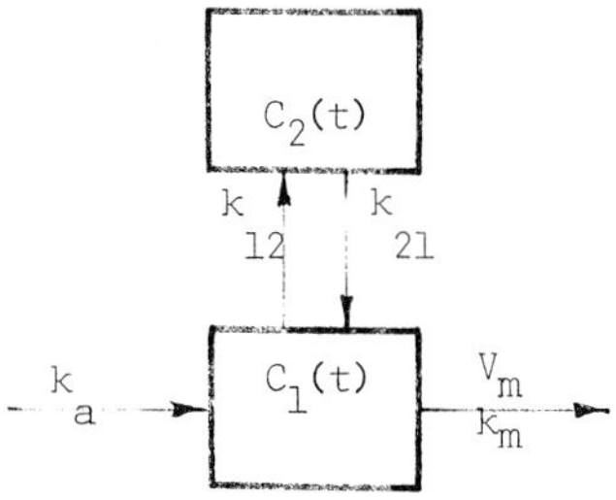

Fig. 7

The mathematical model for concentrations $C_1(t)$ and $C_2(t)$ is given by the following equations.

$$\frac{dC_1}{dt} = k_a C_0 e^{-k_a t} - \left(k_{12} + \frac{Vm}{k_m + C_1} \right) C_1 + k_{21} C_2, \qquad (3.23)$$

$$\frac{dC_2}{dt} = k_{12} C_1 - k_{21} C_2, \qquad (3.24)$$

where the constants have the usual meaning. The corresponding second order differential equation involving C_1 is obtained by eliminating C_2 from equations (3.23), and (3.24). The computation requires substitution of

$$C_2 = e^{-k_{21}t} \int_0^t k_{12} C_1 e^{k_{21}t} dt \qquad (3.25)$$

as a solution of (3.24) in the equation (3.23) and then differentiating once with respect to t. The equation (3.23) reduces to

$$\frac{d^2C}{dt^2}(k_m+C)^2 + \frac{dC}{dt}\{(k_m+C)^2(k_{21}+k_{12}) - V_m k_m\} + C k_{21} V_m (k_m+C)$$
$$+ k_a (k_a-k_{21})e^{-k_a t} C_o(k_m+C) = 0 \qquad (3.26)$$

The corresponding difference equation (assuming equal time intervals) is the following

$$(C_{t+2}-2C_{t+1}+C_t)(k_m+C_t)^2 + (C_{t+1}-C_t)\{(k_m+C_t)^2$$
$$(k_{21}+k_{12}) - V_m k_m\} + C_t k_{21} V_m (k_m+C_t)$$
$$+ k_a(k_a-k_{21})(k_m+C_t)e^{-k_a t} C_o = 0 \qquad (3.27)$$

Least squares methods can be directly applied to the equation (3.27) for estimating k_a, k_{12}, k_{21}, k_m and V_m from data.

4. STATISTICAL CONSIDERATIONS

The difference equation models are generally time-series models where the theory for standard models is well known. In general, we use least square theory to determine the unknown constants. Distribution assumptions are needed to test hypotheses about the unknown parameters. In bioavailability studies, it is the estimates of the rate constants which are pertinent for comparison. We consider the general p compartment model for statistical considerations.

The linear p compartment system is fitted to the equation (3.3) as

$$C(t+p) + b_1 C(t+p-1) + b_2 C(t+p-2) + \ldots + b_p C(t) = u_t$$

where u_t are random errors with means 0 and variances ^{2}.
The least-square solutions are obtained by minimizing

$$\sum_{t=1}^{n} w_t u_t^2 \qquad (4.1)$$

Let

$$\underset{\sim}{C}(t+p-1) = [C(t+p-1), C(t+p-2), \ldots, C(t)]',$$

$$\underset{\sim}{b}_0 = (b_1, b_2, \ldots, b_p)'$$

$$S = \sum_{t=1}^{n-p} w_t \underset{\sim}{C}(t+p-1)\, \underset{\sim}{C}'(t+p-1),$$

$$\underset{\sim}{y} = \sum_{t=1}^{n-p} w\, \underset{\sim}{C}(t+p)\underset{\sim}{C}(t+p-1),$$

then the least-square estimates of $\underset{\sim}{b}_0$ are given by

$$\hat{\underset{\sim}{b}}_0 = -\underset{\sim}{S}^{-1}\, \underset{\sim}{y} \tag{4.2}$$

If we assume further that u_t are normally distributed, we can test hypotheses and provide confidence interval estimates about the parameters $\underset{\sim}{b}_0$.

Suppose we want to determine if the number of compartments is p - 1. It can be verified that in (3.4),

$$1+b_1+b_2+\ldots+b_p = \lambda_1\lambda_2\ldots\lambda_p \tag{4.3}$$

so that the test of the hypothesis that there are p-1 compartments, can be made by

$$H_0:\quad 1+b_1+\ldots+b_p = 0,$$

$$H_1:\quad 1+b_1+\ldots b_p \neq 0.$$

or

$$H_0:\quad \underset{\sim}{1}'\underset{\sim}{b}_0 = -1$$

$$H_1:\quad \underset{\sim}{1}'\underset{\sim}{b}_0 \neq -1 \tag{4.4}$$

where $\underset{\sim}{1} = (1,1,1,\ldots 1)'$.

The test for hypotheses (4.4) is based on the statistic

$$T = \frac{\underset{\sim}{1}'\hat{\underset{\sim}{b}}_0 + 1}{\hat{\sigma}\sqrt{\underset{\sim}{1}'\underset{\sim}{S}^{-1}\underset{\sim}{1}}}, \tag{4.5}$$

where $\hat{\sigma}$ is the unbiased estimate of σ and T has t-distribution with n-p+1 degrees of freedom.

Sequential F tests can be used to determine the number of compartments in case the data is given at unequal intervals. The test is for hypothesis $b_p = 0$ vs. $b_p \neq 0$.

5. ILLUSTRATIONS

(a) The experiment conducted by Kramer *et. al* for the study of pharmacokinetics of Digoxin is first used to illustrate the theory presented here. The authors using traditional methods, showed that a three-compartmental model fits better than the two compartmental model to the Digoxin concentration. However the difference equation model shows otherwise.

Five healthy normal male volunteers were given a dose of 1.0 mg of Digoxin by rapid intravenous injection. Blood samples of 2 ml were withdrawn at various times starting at 2 minutes to 72 hours after administration. Samples were analyzed using ^{125}I radioimmunoassay procedure providing the average of two independent determinations. The data are given in the TABLE I.

TABLE I

Observed Serum Digoxin Concentrations (ng/ml)

Time in Minutes	Subject 1	Subject 2	Subject 3	Subject 4	Subject 5
2	21.00	22.50	26.50	22.00	20.50
4	22.00	21.00	21.00	18.00	17.50
6	18.00	19.00	18.80	20.50	14.50
8	15.00	18.50	16.50	18.50	12.50
10	15.50	16.80	14.30	16.50	13.00
14	14.00	15.80	14.00	16.50	12.00
18	11.30	14.50	13.00	11.50	11.00
22	13.50	14.00	12.30	13.00	9.10
30	12.50	13.30	10.30	9.80	9.60
45	9.10	9.50	7.60	7.30	5.60
60	6.70	7.30	5.80	5.50	4.90
120	3.60	3.60	3.30	2.80	3.20
180	2.60	2.30	2.20	2.00	2.00
240	2.20	1.80	1.90	1.40	1.80
360	1.20	1.20	0.52	0.73	-
480	1.05	1.10	0.52	0.55	0.90
960	0.75	1.03	-	0.50	0.85
1420	1.00	0.93	0.50	-	0.70
2160	0.54	0.53	0.52	0.40	0.45
2880	0.63	0.44	0.33	0.36	0.43
4320	0.53	0.27	0.24	0.16	0.39

Divided difference formulae are used here to analyze the data for Patient 1, it is found that using two-compartment model, we have the statistic

$$F = 9.78$$

which has F-distribution with (1,17) degrees of freedom. It is found significant at 1% level.

However the test that the model has three compartment leads to statistic $F = 2.54$ which is not significant at 5% level of significance. Hence our approach shows that there are only two compartments. The constants estimated are given by

$$\hat{\lambda}_1 = 29.63$$

$$\hat{\lambda}_2 = 2.14$$

and the sample variance is estimated as 209.67.

(b) Consider an experiment which studies, in a proportional counter, the background pulses generated by neutron interaction with walls and gas, as well as pulses generated by circuit noises, Cornell (1962). The logarithm of frequencies y(t) and pulse heights t are given in TABLE II.

TABLE II

Background Pulse Data

t	0	1	2	3	4	5	6	7
y(t)	10.430	4.703	2.327	1.140	0.615	0.325	0.170	0.117

t	8	9	10	11	12	13	14	15
y(t)	0.050	0.040	0.046	0.022	0.036	0.021	0.018	0.016

The model uses the sum of two exponentials. The difference equation model provides the following estimates with the help of equation (3.12).

$$\hat{b}_1 = -0.489,$$

$$\hat{b}_2 = -0.003,$$

$$\hat{\lambda}_1 = 1.007,$$

$$\hat{\lambda}_2 = 0.504,$$

and the residual sum of squares = 0.0059.

Estimates given by Cornell are,

$$\hat{\lambda}_1 = 1.682,$$

$$\hat{\lambda}_2 = 0.655,$$

and the residual sum of squares = 0.0042.

Theses results are quite comparable.

(c) To develop a theory of "Tropicamide" absorption, a study of rabbits was conducted by Smolen and Schoenfeld (1956). Intravenous injection of the drug was given to four rabbits and pupillary diameters were measured A function f (I) of quantity I, called "Mydriatic Response Intensity" was calculated and is reported in the following for a dose equal to 50 mcg per kilogram of weight.

Time in Minutes	0	10	20	30	40	50	60	70
f(I)	48.1	24.7	12.7	6.2	3.1	1.4	0.7	0.3

The difference equation model for two compartments provides the following estimates:

$$\hat{b}_1 = -0.71,$$

$$\hat{b}_2 = 0.10,$$

with the estimate of variance S^2 = 0.000,011

Using the test statistic T, as given in (4.5), we find that there is only one compartment. This fact is confirmed by clinical approach as well where the eignevalues are given by the following.

$$e^{-\hat{\lambda}_1} = 0.5095$$

$$e^{-\hat{\lambda}_2} = 0.5100$$

Note that these values are almost equal, showing that there is only one compartment.

6. DISCUSSION

(i) The proposed method using the difference equations and the resulting time series models make use of different assumptions for the errors of the model. However the numerical examples show that the mean square errors in many cases are considerably reduced.

(ii) In the case of linear differential equations leading to autoregressive time series, the solutions are unique under the difference equation models. Also the variance covariance matrix of the estimates can be determined. In the case of non-linear estimation, only approximate asymptotic results are available. In a recent study Boxenbaum, Riegelman and Elashoff (1974) have emphasized tests of significance for the parameters of compartmental models. The asymptotic theory is utilized to obtain standard deviation of estimates and their estimates are obtained in terms of iterated least squares estimates. In the case of equally spaced time data, we have seen that the variances of the estimates can be obtained exactly by the difference equation approach.

(iii)The approach discussed here provides solutions to problems to which the iterative least squares do not give satisfactory solutions. Also the models can provide initial estimates in many cases for iteration if the actual model to be fitted is non-linear.

(iv) One compartment model has been analyzed using ordered simultaneous estimation procedure by Rodda, Sampson and Smith (1975). This method is sensitive to outliers and performs better than the least squares method when outliers are present. The method is a generalization of a nonparametric estimate of linear regression parameters based on all possible pairs of points. The estimates in the case of linear regression, were given by Sen (1971) using the median of all slopes computed from all possible pairs and using midrange for the intercept. Simulations provided comparison of the nonparametric method to the NONLIN program.

(v) Since total bioavailability of orally administered drugs such as digoxine, is out of reach of the clinician, Karl-Weser (1974), it is important that differences in drugs that are related to patient characteristics, pathologic states and drug interaction should be emphasized. Compartmental models act only as an aid in such clinical studies.

ACKNOWLEDGEMENTS

Our interest in this problem area arose as a result of conversations with Dr. S. D. Dubey, U. S. Food and Drug Administration and we are grateful to him. We are also very grateful to the referee whose comments helped improve the paper considerably.

REFERENCES

[1] ALBERT, K. S., SAKMAR, E., HALLMARK, M. R., WEILDER, D. J. and WAGNER, J. G. (1974). Bioavailability of Diphenylhdantoin. *Clinical Pharmacology and Therapeutics*, 16, 127-135.

[2] ALBERT, K. S., SEDMAN, A. J., WILKINSON, P., STOLL, R. G., MURRAY, W. J. and WAGNER, J. G. (1974). Bioavailability studies of Acetaminophen and Nitrofurantoin. *Journal of Clinical Pharmacology*, 14, 264-270.

[3] ALBERT, K. S., SEDMAN, A. J. and WAGNER, J. G. (1974). Pharmacokinetics of orally administered Acetaminophen in man. *Journal of Pharmacokinetics and Biopharmaceutics*, 2, 381-393.

[4] ALKALAY, D., KHEWANI, L., WAGNER, W. E. and BARTLETT, M. F. (1975). Pharmacokinetics of Phenoformin in man. *J. Clin. Pharm.*, 15, 446-448.

[5] ANDERSON, T. W. (1971). *The Statistical Analysis of Time Series.* John Wiley and Sons, New York.

[6] ATKINS, G. L. (1969). *Multicompartment Models for Biological Systems.* Methuen and Co. Ltd., London.

[7] BELLMAN, RICHARD (1971). Topics in pharmacokinetics - approximation in process space and fitting by sums of exponentials. *Math. Biosc.*, 12, 1-5.

[8] BELLMAN, R. D., JACQUEZ, J. A. and KALABA, R. (1960). Mathematical models of chemotherapy. *Proc. Fourth Berkeley Symp. Math. Statist. Prob.*, p. 57.

[9] BERMAN, M. and SCHOENFELD, R. (1956). Invariants in experimental data on linear kinetics and the formulation of models. *Journal of Applied Physics*, 27, 1360-1371.

[10] BOXENBAUM, H. G., RIEGELMAN, SIDNEY and ELASHOFF, ROBERT M. (1974). Statistical estimations in pharmacokinetics. *J. of Pharm. and Biopharm.*, 2, 123-148.

[11] CORNELL, R. G. (1962). A method for fitting linear combinations of exponentials. *Biometrics*, 18, 104-113.

[12] DARVEY, I. G., SHRAGER and KOHN, L. D. (1975). Integrated steady state rate constants. *Journal of Biological Chemistry*, 250, 4696-4701.

[13] FORT, TOMLINSON (1948). *Finite Differences and Difference Equations in the Real Domain.* Clarendon Press, Oxford.

[14] HARTLEY, H. O. (1948). The estimation of non-linear parameters by internal least-squares. *Biometrika*, 35, 32-45.

[15] HEARON, J. Z. (1972). Residence times in compartmental systems and the moments of a certain distribution. *Mathematical Biosciences*, 15, 69-77.

[16] HEVESEY, G. (1923). The absorption and translocation of lead by plants. *Biochem. J.*, 17, 439.

[17] JACQUEZ, JOHN A. (1972). *Compartmental Analysis in Biology and Medicine.* Elsevier Publishing Co., New York.

[18] JUSKO, W. J., SZEFLER, S. J. and GOLDFARB, A. L. (1974). Pharmacokinetic design of Digoxin dosage regimens in relation to renal functions. *J. Clin. Phar.*, 14, 525-535.

[19] KOCH-WESTER, JAN (1974). Bioavailability of drugs. *New England Journal of Medicine*, 291, 233-237, 503-506.

[20] KRAMER, W. G., LEWIS, R. P., COBB, T. C., FORESTER, W. F., JR., VISCONTI, J. A., WANKE, L. A., BOXENBAUM, H. G. and REUNING, R. H. (1974).Pharmacokinetics of Digoxin: comparison of a two- and three-compartment model in man. *J. of Pharm. and Biopharm.*, 2, 200-312.

[21] LIN, T. K. and DUTT, J. E. (1974). A new method for the analysis of sums of exponential decay curves. *Math. Biosc.*, **20**, 381-391.

[22] METZLER, C. M. (1969). *Nonlin, A Computer Program for Parameter Estimation in Nonlinear Situations*. The Upjohn Company, Kalamazoo, Michigan.

[23] NOTARI, R. D. (1971). *Biopharmaceutics and Pharmacokinetics, An Introduction*. Dekker, New York.

[24] OTTEN, H. O. (1974). Some remarks on the Michaelis-Menten kinetic equations. *Math. Biosc.*, **20**, 155-162.

[25] RODDA, B. E., SAMPSON, C. B. and SMITH, D. W. The one-compartment open model: Some statistical aspects of parameter estimation. *Appl. Statist.*, **24**, 309-318.

[26] RUBINOW, S. I. and WINZER, A. (1971). Compartment analysis: An inverse problem. *Math. Biosc.*, **11**, 203-247.

[27] RUSTAGI, J. S. (1965). Mathematical models of body burden. *Arch. Environ. Health*, **10**, 761-767.

[28] RUSTAGI, J. S. (1976). *Variational Methods in Statistics*. Academic Press, New York.

[29] SEDMAN, A. J. and WAGNER, J. G. (1974). Quantitative pooling of Michaelis-Menten equations in models with parallel metabolite formation paths. *Journal of Pharmacokinetics and Biopharmaceutics*, **2**, 161-173.

[30] SEDMAN, A. J. and WAGNER, J. G. (1974). Importance of the use of the appropriate pharmacokinetic model to analyze *In Vivo* enzyme constants. *Journal of Pharmacokinetics and Biopharmaceutics*, **2**, 161-173.

[31] SEGRE, E. J. (1975). Naproxen metabolism in man. *J. Clin. Pharm.*, **15**, 316-323.

[32] SEN, P. K. (1971). Robust statistical procedures in problems of linear regression with special reference to quantitative bioassay I. *Rev. Int. Statist. Inst.*, **39**, 21-38.

[33] SHAH, B. K. (1973). Obtaining preliminary estimates to fit two exponential model to blood concentration data. *J. Pharmaceutical Sciences*, **62**, 1208-1209.

[34] SHAH, B. K. (1976). Data analysis problems in the area of pharmacokinetics research, *Biometrics*, **32**, 145-157.

[35] SHEPPARD, C. W. and HOUSEHOLDER, A. S. (1951). The mathematical basis of the interpretation of tracer experiments in closed steady state systems, *J. Appl. Phys.*, **22**, 510.

[36] SHEPPARD, C. W. (1962). *Basic Principles of the Tracer Method*. John Wiley and Sons, Inc., New York.

[37] SINGH, UMED (1975). Contributions to statistical studies of compartmental models, Doctoral Dissertation, The Ohio State University, Columbus, Ohio.

[38] SMOLEN, V. F. and SCHOENWELD, R. D. (1971). Drug-absorption analysis from pharmacological data I: Method and confirmation exemplified for the mydriatic drug Tropicamide. *Journal of Pharmaceutical Sciences*, **60**, 96-103.

[39] SOONG, T. T. (1971). Pharmacokinetics with uncertain ties in rate constants. *Math. Biosc.*, **12**, 235-243.

[40] SOONG, T. T. (1972). Pharmacokinetics with uncertain ties in rate constants II. Sensitivity analysis and optimal dosage control. *Math. Biosc.*, **13**, 391-396.

[41] SOONG, T. T. and DOWDEE, J. W. (1974). Pharmacokinetics with uncertainties in rate constants, III. The inverse problem. *Math. Biosc.*, **19**, 343-353.

[42] STAMBAUGH, J. E. and WAINER, I. W. (1975). The bioavailability of Meperidine. *J. Clin. Pharm.*, **15**, 552-559.

[43] TEORELL, T. (1937). Kinetics of distribution of substances administered to the body. The extravascular modes of administration, II. The intravascular modes of administration. *Archs Intern. Pharmacodyn.*, **57**, 205-225, 226-240.

[44] WAGNER, J. G. (1971). *Biopharmaceutics and Relevant Pharmacokinetics*. Drug Intelligence Publications, Hamilton, Illinois.

[45] WAGNER, J. G. (1975). *Fundamentals of Clinical Pharmacokinetics*. Drug Intelligence Publications, Hamilton, Illinois.

[46] WAGNER, J. G. (1975). Personal Communication.

[47] WAGNER, J. G. (1975). Application of the Loo-Riegelman absorption method. *Journal of Pharmacokinetics and Biopharmaceutics*. **3**, 51-67.

[48] WAGNER, J. G. (1973). Properties of the Michaelis-Menten equations and its integrated form which are useful in pharmacokinetics. *Journal of Pharmacokinetics and Biopharmaceutics*, **1**, 103-121.

[49] WAGNER, J. G., SAKMAR, E., and STOLL, R. G. (1973). Statistical analysis of reproducibility or radioimmunoassay of Digoxin. *Research Communications in Chemical Pathology and Pharmacology*, **5**, 883-884.

[50] WHITTAKER, E. T. and ROBINSON, G. (1944). *The Calculus of Observations*. Blackie and Sons, Ltd., London.

[51] WISE, M. E. (1974). Interpreting both short and long term power laws in physiological clearance curves. *Math. Biosc.*, **20**, 327-337.

P.R. Krishnaiah, ed., APPLICATIONS OF STATISTICS
© North-Holland Publishing Company (1977)

THE RELIABILITY OF STRUCTURES SUBJECT TO FATIGUE UNDER INSPECTION AND REPAIR

SAM C. SAUNDERS
Washington State University
Pullman, Washington

1. INTRODUCTION

It is the purpose of this discussion to describe mathematically a structure, each component of which is subject to failure due to the combination of chance overload and weakening due to fatigue. Moreover, it is intended to derive formulae,for its reliability under inspection and repair,which can be easily evaluated and utilized. The structure considered is a series arrangement; every component in order to be operational, must possess enough residual strength to withstand the shocks which occur at intermittant times.

The type of system that is being modeled here is that of the major structural components within a larger, more complex, system which is subjected both to fatigue from its duty cycle as well as random shocks of considerable magnitude. An archtypical example of such a system is an air-frame with its structural components sustaining both ground-air-ground cycles and intermittent gust loading.

The complicated reliability of such a structure is modeled through the joint distribution of shock loads which are apportioned to each component. In addition each component is wearing out at a rate determined, independently of the shock loading, by the GAG cycle.

The problem is to formulate this behavior in mathematical terms and to introduce the effect of inspection and repair in a sufficiently realistic way that the increase in reliability so occasioned can be compared as a function of increased costs and the cost benefit thereby assessed. Such an analysis makes possible comparison between design philosophies and the relationship between strength of design and maintainability.

Of course there have been earlier studies of the reliability and safety of structures which incorporated the two aspects of fatigue and shock. One of the earliest studies was by Freudenthal and Shinozuka [6] who examined stress interaction with fatigue in the determination of structural safety. This work was continued in a more comprehensive analysis of structural safety involving imposed stress and the resistance of the structure, see [5]. One of the most recent studies, by Eggwertz in [4], discussed the specific problem of the interaction between fatigue life and residual strength for wing panels.

2. RESIDUAL STRENGTH FOR A COMPONENT

Due to the continuing dynamic loading of the structure in time there will be a gradual weakening, or decrease of its residual strength which we call fatigue, and we shall regard this weakening as being due to the initiation and growth of a dominant fatigue crack. In order to fix ideas we suppose that a given component has a residual strength $C(x)$ when that component contains a fatigue crack of length $x \geq 0$. It will fail if it encounters a chance overload, such as a shock or gust depending upon the application, with a magnitude exceeding $C(x)$. The concept of strength degradation will certainly depend upon the method of construction, the type of material as well as the loading spectrum and the environment, but residual strength $C(x)$ is always taken to be a known decreasing function of x with a range of values between ultimate and design strength limits.

Based on certain design procedures, such as those presently employed by some airframe manufacturers, there is a design limit load, say δ , and a fail-safe crack length, say σ , which are determined by ratios dictated by design criterion, call them ν_1 and ν_2, as follows:

$$\nu_1 = \frac{C(0)}{\delta} > 1 \ , \ \nu_2 = \frac{C(\sigma)}{\delta} \leq 1 \ . \qquad (2.1)$$

these are respectively the ratios of ultimate strength $C(0)$ and fail-safe strength $C(\sigma)$ with the limit load. Here ν_1 , ν_2 depend upon the components position in the structure and the load expected during usage. However, in the illustration chosen ν_1 is approximately 1.5 and ν_2 approximately .9. The residual strength function must be obtained from engineering tests of the material and a knowledge of the fracture mechanics particular to the geometry of that component.

In general the residual strength C will be a concave decreasing function but an explicit equation is not always easily obtained. In one specific instance involving an aluminum alloy, see [8], the residual strength was taken to be

$$C(x) = \delta[\nu_1 - (\nu_1 - \nu_2)(x/\sigma)^{1/3}] \quad \text{for } x \geq 0. \qquad (2.2)$$

3. THE STRUCTURAL OVERLOAD MODEL FOR A COMPONENT

We assume there is a stressful working environment which contributes to fatigue of the structural component. The actual process, of course, depends upon a variety of specific factors as varied as whether the structure is an igloo subjected to a warm wind or the plenum chamber of a nuclear reactor during operation. The progress of fatigue in each component shall be modeled by a stochastic process determined by the work cycle. The chance overloads will be superimposed upon this duty cycle. In the case of an airframe the shocks are intermittant gust loads which are distributed to various components in the wings and fuselage. This formulation will not lead to a shock model in the sense defined previously, see [1]and references given there, because the shocks themselves do not contribute to the fatigue growth. In many instances fatigue is due to fretting and galling in service while sudden gusts, which relieve local stress concentrations by yielding the material, actually inhibit the growth of fatigue cracks.

We now assume that:

1° the time between gusts of overload magnitude is exponential with a mean of $1/\lambda$;

2° the magnitude of each gust is independently and identically distributed with distribution F.

If N(t) for $t > 0$ is the counting process for the number of gusts on or before time t and Y_1, $Y_,$ ... are the (random) magnitudes of the gust loads encountered then L(t), the maximum gust load over the time period (0,t), has a distribution determined by

$$P[L(t) > y] = \sum_{n=1}^{\infty} P[N(t)=n]P[\max(Y_1,\ldots,Y_n) > y]$$

$$= \sum_{n=1}^{\infty} \frac{e^{-\lambda t}(\lambda t)^n}{n!} [1 - F^n(y)] = 1 - e^{-\lambda t\overline{F}(y)} \qquad (3.1)$$

where $\overline{F} = 1 - F$. Then

$$P[L(t) \leq y] = e^{-\lambda t\overline{F}(y)} \qquad \text{for } y \geq 0. \qquad (3.2)$$

Note that for any time t there is a positive probability, to wit, $e^{-\lambda t}$, that no gust load was encountered during that time.

4. THE CRACK GROWTH MODEL FOR A COMPONENT

The stochastic crack growth, of course, cannot be modeled in exact detail but must be summarized with sufficient accuracy that inferences can be relied upon. Moreover, there are different models that must be fashioned depending upon structural applications. For certain components it is thought appropriate that there be a (random) initial period of crack initiation, say W_0, with a distribution which depends upon the spectrum of service loads. Typically the crack growth process is of the form

$$S(t) = \min\{ Z_0(t - W_0)^+ + Z_1(t - W_0 - W_1)^+, \sigma_2 \}, \; t \geq 0 \qquad (4.1)$$

where $x^+ = \max(x,0)$ and W_0 is the time to crack initiation, W_1 is the time between crack initiation and the time critical crack size is reached and Z_0, Z_1 are the rates of crack advance during the two separate phases of crack growth.

As a graphical illustration of this piecewise linear growth regard Figure 1. Here σ_1 is critical crack size and σ_2 is the ultimate size as determined by the dimensions and geometry of the component.

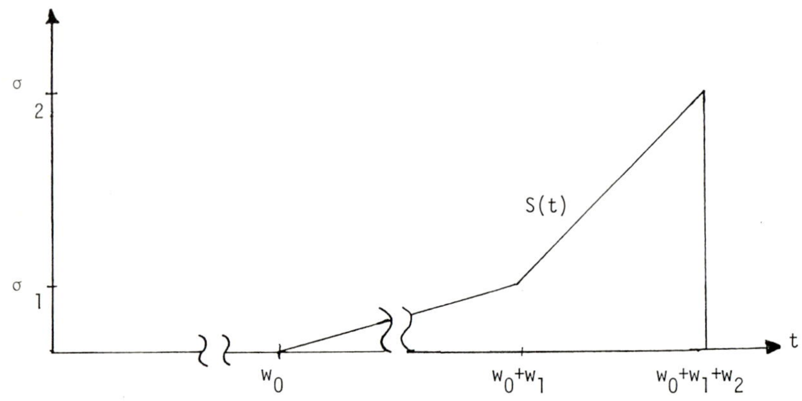

Fig. 1. Typical behavior of crack growth.

More elaborate models are now being studied which include a wider variety of structural components. In particular, those components which contain certain elements (such as fasteners) which can induce initial flaws of significant size, are now being examined, see [9]. There are other studies which avoid flaws entirely.

The model defined in (4.1) can be understood by taking as given the (random) times of crack initiation, w_0, and criticality of crack size, $w_0 + w_1$. Then for any time $t \geq 0$, the crack length is given by

$$s(t) = \begin{cases} 0 & \text{for } t < w_0 \\ z_0(t - w_0) & \text{for } w_0 \leq t < w_0 + w_1 \\ \sigma_1 + z_1(t - w_0 - w_1) & \text{for } w_0 + w_1 \leq t < w_0 + w_1 + w_2 \\ \sigma_2 & \text{for } w_0 + w_1 + w_2 \leq t \end{cases} \tag{4.2}$$

Here, as before, σ_1 and σ_2 are constants determined from the geometry of the component.

It is realized that this assumption of piecewise linearity of crack growth is a simplification but it is thought to be an approximation with sufficient accuracy to the actual crack growth behavior, as verified by full-scale structural tests, that it can be used with confidence for the analysis which follows.

The distribution of the stochastic process $S(t)$, which describes crack growth, is determined by the four random variables W_0, W_1, Z_0, Z_1. The distributions of these variates are chosen from the following considerations. The random variable W_0 is assumed to have a two-parameter Weibull distribution. This means that for some parameters $\alpha, \beta > 0$, governed by quality control during manufacture of the component and determined from historical data, the distribution of time until crack initiation is given by

$$P[W_0 > t] = \exp\{-(t/\beta)^{\alpha}\} \quad \text{for } t > 0. \tag{4.3}$$

This choice of distribution is made because the time until initiation is thought to be an extreme value phenomenon and the Weibull distribution is the asymptotic distribution of extreme smallest values. Consequently, it is posited that this distribution will describe the behavior of time to crack initiation more accurately that any other. At the present time there is no evidence that refutes this assumption.

The variates Z_j for $j = 0,1$ have the ξ-normal distribution defined for parameters κ_j, η_j for $j = 0,1$ as follows

$$P[Z_j \le z] = \Phi[\frac{1}{\kappa_j} \ \xi\ (z/\eta_j)] \quad \text{for } z > 0. \tag{4.4}$$

Here Φ denotes the standard normal distribution and $\xi(t) = \sqrt{t} - 1/\sqrt{t}$ for $t > 0$. This distribution is the limiting distribution of the number of steps (or time, if steps are taken at equal intervals) necessary to cover a fixed distance when the length of each step is random but independently and identically distributed. This distribution and its derivation are discussed in [2] and [3] .

It is thought that the regular advance of the crack due to the work cycle is an instance of such behavior in each of its two postulated phases. Now it follows as a mathematical consequence that: if X has a ξ-normal distribution then $1/X$ has a ξ-normal distribution also, see [7]. Hence, the assumption that the time for the crack to grow to length σ_1, following initiation, has a ξ-normal distribution is equivalent to assuming that the rate of crack growth has a ξ-normal distribution since the time and the rate are inversely related for this simple model. Of course, the choice of a distribution for rate is made because the parameters σ_1 and σ_2 can be determined from the geometry. Moreover there has already been some data collected, see [9], from which the parameters for the distribution of crack growth rate can be determined.

In order to provide additional assurance that this postulated distribution of growth rate agrees with the physical data some simulation with it has been performed and the results displayed in [9]. A random sample so obtained is there seen to be virtually indistinguishable from the data.

From the same considerations W_1 is also assumed to have a ξ-normal distribution and the same consequences obtain.

The value for each parameter was determined from the data, using fracture mechanical principles, and is detailed in [9]. There were also some intuitive conditions imposed involving certain expectations such as

$$ES(W_0+W_1) = \sigma_1$$

and the ratio

$$E(W_0+W_1)/E(W_0+W_1+W_2)$$

having a known value.

5. CRACK DETECTION AND INSPECTION

The purpose of fail-safe design is to allow sufficient time for inspection of fatigue damage in the structure, between the time of crack initiation and subsequent failure of the structure, that detection and repair can occur with high probability.

The optimization problem is one of balancing the needless expense of too frequent and too expensive inspections, in order to detect and repair all cracks before they become structurally critical, with the loss of revenue from the unscheduled down-times necessary for repair, etc. which are consequences of a structural failure in service.

Before any analysis is possible some relationship must be established between the cost and the probability of detection of a crack of given size. Of course, this is determined by the detection technique which is used. A particular formulation, which has been utilized in [8], for the non-detection probability, i.e. the probability of not detecting a crack of size $x \geq 0$, is

$$q(x:\omega) = \begin{cases} \exp\{-\theta(x - \tau_1)^+\} & 0 \leq x \leq \tau_2 \\ 1 - \exp\left\{\frac{\tau_2 - \tau_1}{x - \tau_1} \ln[1 - q(\tau_2:\omega)]\right\} & \tau_2 < x \leq \sigma_2 \end{cases} \tag{5.1}$$

Here $\omega = (\theta, \tau_1, \tau_2)$ is a three dimensional parameter with τ_1 being the inspection threshold i.e. $q(x:\omega) = 1$ for $x \leq \tau_1$. For example, it is much larger if only visual inspection is used than if eddy current techniques are used instead. Also, θ, τ_2 are inspection parameters determined by the thoroughness and type of inspection. Of course, different types of inspections may be scheduled at different times. This could be represented by an alteration of ω in q. The behavior for a representative ω is illustrated in Figure 2.

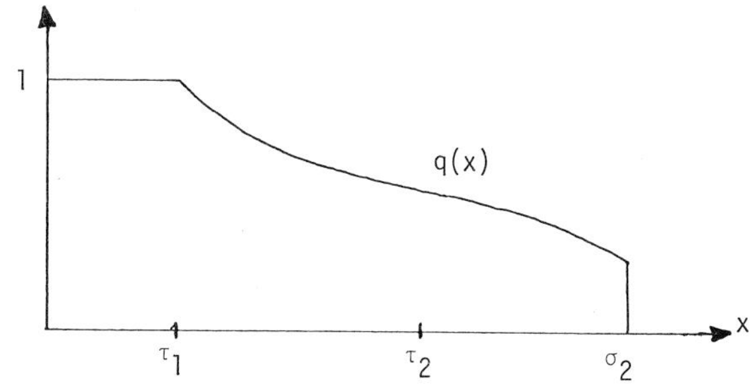

Fig. 2. The probability of not detecting a crack of size x.

Note that there is a positive probability $q(\sigma_2) > 0$ of not detecting the fatigue crack even though the component is cracked to its fail-safe limit σ_2. This assumption is made in accord with instances already observed.

Let $s(t)$ for $t > 0$ be a given crack growth function determined from one observation (w_0, w_1, z_0, z_1) of the four variates. Then by proper scaling of time we can, without loss of generality, take $t = 1, 2, \ldots$ as the predetermined times for inspection of the structure. Thus we need to calculate only $s(t)$ for $t = 1, 2, \ldots$, namely the crack lengths at successive inspection periods, if required.

Let $q_i(s)$ denote the probability of not detecting a crack, which is growing according to s, at the ith inspection. Then if (5.1) were the appropriate non-detection probability with parameter ω_i specified at the ith inspection we would have

$$q_i(s) = q[s(i):\omega_i] \quad \text{for } i = 1, 2, \ldots . \tag{5.2}$$

Further, for convenience set

$$p_i(s) = 1 - q_i(s), \quad s_i = s(i), \quad \rho(s_i) = \lambda\bar{F}[C(s_i)] . \tag{5.3}$$

Define $r_i(s)$ as the probability that the component does not fail during the ith service interval, i.e. between the $(i-1)^{st}$ and i^{th} inspections, given that growth of the crack is determined by s.

Following the assumption 1° that between any two time periods of unit difference, e.g. i-1 and i, the number of gusts has a Poisson distribution with intensity λ and the distribution of the time of the occurence of the gusts is uniform over that interval, we find

$$r_i(s) = \sum_{n=0}^{\infty} \frac{e^{-\lambda}\lambda^n}{n!} \int_{i-1}^{i} \cdots \int_{i-1}^{t_3} \int_{i-1}^{t_2} \prod_{j=1}^{n} FC\, s(t_j)\; n!\; dt_1 \cdots dt_n \tag{5.4}$$

where the juxtaposition of functions indicates composition.

Clearly for each j = 1, ..., n we have $s_{i-1} \le s(t_j) \le s_i$, and so

$$FC(s_{i-1}) \ge FCs(t_j) \ge FC(s_i),$$

since C is an order-reversing and F an order-preserving transformation. Now by (5.3) we have the inequality

$$r_i(s) \ge \sum_{n+0}^{\infty} \frac{e^{-\lambda}\lambda^n}{n!} F^n[C(s_i)] = e^{-\lambda\bar{F}C(s_i)} = e^{-\rho(s_i)}.$$

Thus we obtain the bounds

$$-\rho(s_i) \le \ln r_i(s) \le -\rho(s_{i-1}). \tag{5.5}$$

It is noted that these bounds are not sharp, in general , but in the region of high reliability they are sufficiently close to be quite usable.

The residual strength of any component is a concave decreasing function of the crack length. But when composed with the crack growth as a function of time, which is convex increasing, the residual strength can be smoothed out into a variety of shapes. However the residual strength will always be decreasing. One particular example is given in Figure 3.

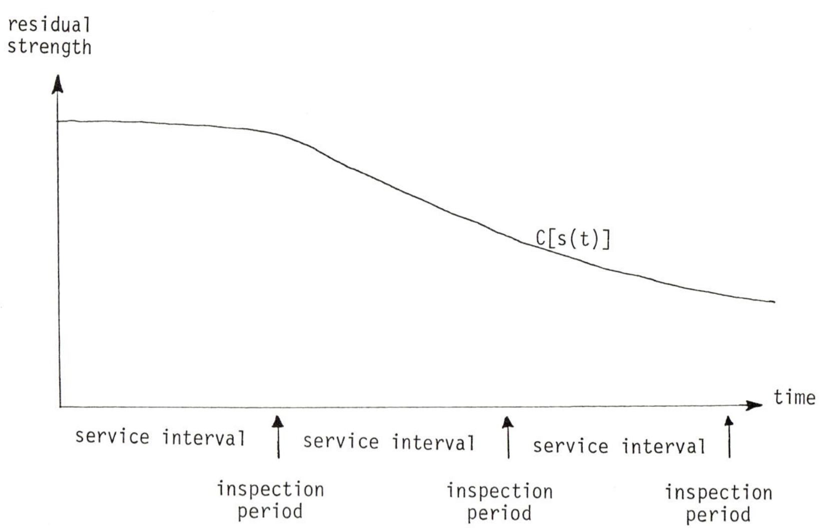

Fig. 3. Generic behavior of residual strength as a function of time.

6. COMPONENT FAILURE UNDER INSPECTION AND REPAIR

Let us presume that the crack growth function $s(t)$, for $t \geq 0$, is given. As a functional of s we set

$A_j(s)$ = probability that a component will not fail during first j service intervals although a crack of length s_i may be present but undetected for $i = 1, \ldots, j$.

Then

$$A_j(s) = \prod_{i=1}^{j} q_i(s) r_i(s) \text{ for } j = 1, 2, \ldots; \; A_0(s) = 1. \tag{6.1}$$

Let $J|s$ be the random number of inspections until detection, and repair, of a cracked structure undergoing crack growth described by s. Its density is given by

$$g(j|s) = P[J=j|s] = A_{j-1}(s) p_j(s) r_j(s) \quad j = 1, 2, \ldots \tag{6.2}$$

and its distribution

$$G(n|s) = \sum_{j=1}^{n} g(j|s) \quad \text{for } n = 1, 2, \ldots .$$

One may write $J|s \sim G(\cdot|s)$ but note that this distribution may be improper, since it can happen that $G(\infty|s) < 1$.

Let $K|s$ be the (random) number of inspection intervals until failure of unrepaired structure given that the structure is sustaining crack growth by the function s. Then

$$h(k|s) = P[K=k|s] = A_{k-1}(s)[1 - r_k(s)] \tag{6.3}$$

$$H(n|s) = \sum_{k=1}^{n} h(k|s) \quad n = 1, 2, \ldots .$$

Again, while for given s, $K|s \sim H(\cdot|s)$, this also may be improper, however one checks that

$$H(\infty|s) + G(\infty|s) = 1.$$

Thus for any given crack growth description s, the structure will either fail in service or the crack will be detected and the component repaired. In fact

$$H(n|s) + G(n|s) = 1 - A(n|s) \tag{6.4}$$

where for consistency we now write

$$A_n(s) \equiv A(n|s).$$

What must be done is to determine the unconditional distribution through conditional expectation. Using the theorem of total probability one obtains

$$H(n) = E\, H(n|S) \quad n = 1, 2, \ldots .$$

The evaluation of this expectation would be a very difficult task analytically but it can be accomplished by Monte Carlo methods. Taking expectations of both sides of equation (6.4), and omitting the arguments of the functions yields the fundamental equation

$$H = 1 - A - G. \tag{6.5}$$

We now rewrite the (unconditional) probability that a structure fails during the nth service interval as

$$u(n) = P[K_1=n] + P[J_1+K_2=n] + \ldots + P[J_1+\ldots+J_{n-1}+K_n=n] \tag{6.6}$$

where successive terms represent, respectively, the probabilities of failing before being repaired, being repaired once and then failing, etc. ... .

Again, if we write the distribution functions

$$U(n) = u(1) + \ldots + u(n) \quad n = 1, 2, \ldots,$$

equation (6.6) can be written as

$$U = H + G*H + G^{2*}*H + \ldots . \tag{6.7}$$

where * denotes the convolution of distribution functions.

The relationship of (6.7) can be written as a renewal equation

$$U = H + U*G.$$

That U is a bona fide probability distribution follows from the fact that $H(\infty) + G(\infty) = 1$.

To obtain the reliability of the renewable component let $V = 1 - U$. The corresponding renewal equation becomes $V = A + V*G$ or

$$V = A + A*G + A*G^{2*} + \ldots \tag{6.8}$$

The numerical evaluation of (6.8) is what must be accomplished. Recall that there are at most (n+1) terms which are non-zero in the expression for V(n).

Thus we must evaluate the two functions,

$$A(k) = E\,A(k|S) = \int A(k|s)d\mu(s) \tag{6.9}$$

$$G(k) = \sum_{j=1}^{k} \int [A(j-1|s)r_j(s) - A(j|s)]d\mu(s), \tag{6.10}$$

where the integration extends across the sample space of the stochastic process S on which the appropriate probability measure is μ. It is more convenient to evaluate (6.9) and

$$D(k) = \int A(k-1|s)r_k(s)d\mu(s) \tag{6.11}$$

from which (6.10) can be easily computed.

If we introduce the notation

$$Q_k(s) = \prod_{i=1}^{k} q_i(s)$$

and utilize (6.1), we have

$$A(k) = \int Q_k(s) \exp\left\{ \sum_{j=1}^{k} \ln r_j(s) \right\} d\mu(s)$$

and from the inequalitites in (5.5), setting

$$P_k(s) = \sum_{i=1}^{k} \rho(s_i),$$

we obtain

$$\int Q_k(s)e^{-P_k(s)}d\mu(s) \leq A(k) \leq \int Q_k(s)e^{-P_{k-1}(s)}d\mu(s). \tag{6.12}$$

Of course a similar inequality obtains for D(k), namely

$$\int Q_{k-1}(s)e^{-P_k(s)}d\mu(s) \leq D(k) \leq \int Q_{k-1}(s)e^{-P_{k-1}(s)}d\mu(s). \tag{6.13}$$

The reason for obtaining these bounds is that $r_j(s)$ is not a simple explicit function of the stochastic process s but $\rho(s_j)$ is. By using (5.5) and computing only bounds for A(k), Monte Carlo methods can be easily used and fortunately in the cases considered in [8] they were virtually identical for nearly all values of k.

Briefly, the method of computation was as follows. Let $s^{(j)}$ for j = 1, ..., M with M large be independent realizations of the stochastic process S defined in (4.1). Then, say $\underset{\sim}{A}(k)$, the lower bound for A(k) defined as the left hand side of (6.12), is approximated by

$$\frac{1}{M}\sum_{j=1}^{M} Q_k[s^{(j)}]e^{-P_k[s^{(j)}]} \to \underset{\sim}{A}(k) \quad \text{as } M \to \infty,$$

from the central limit theorem.

From (6.12) and (6.13) we see $\tilde{D}(k+1) = \underset{\sim}{A}(k)$ and thus all the lower and upper bounds can be computed simultaneously to see which $r_j(s)$ could be determined with use of (5.4). Further details of the computation, as well as the results, can be found in [8].

7. GENERALIZATION TO STRUCTURES WITH JOINTLY DISTRIBUTED LOADS

An application of the preceding concepts to fleets of components, each element being considered in some designated component within a structure, when safety is defined in terms of the first failure within the fleet has already been made in [8]. We now generalize some of the concepts to series structures which are formed from components subject to both fatigue and chance overload.

We now denote the state of the structure with m components by $\underline{x} = (x_1,\ldots, x_m)$ where $x_i \geq 0$ is the length of the crack in the ith component. The joint residual strength of the structure is given by the function

$$\underline{C}(\underline{x}) = \left(C_1(\underline{x}),\ldots,C_m(\underline{x})\right)$$

where $C_j(\underline{x})$ is the residual strength in the jth component when the state of the structure is $\underline{x}$. It is assumed that $\underline{C}$ is a concave decreasing function of $\underline{x}$.

An extensive prior analysis of the structure is presumed which has determined the critical crack region in m-space and the ultimate crack sizes. Moreover each C_j could now depend not only upon several critical crack length parameters but also upon safety ratios involving both ultimate and fail-safe strengths as determined by design. This would require a generalization of equations (2.1) and (2.2) but the exact form is not specified since it would depend upon the material, environment and a multitude of influences beyond the scope of this paper.

Again, as a mathematical simplification which idealizes actual behavior, it is assumed there is no alteration of fatigue crack growth caused by the intermittant shocks other than the fracture which occurs when the residual strength is exceeded. Thus a realization of the m-dimensional stochastic process $\underline{S}(t) = \left(S_1(t),\ldots,S_m(t)\right)$, denoted by $\underline{s}(t)$ and describing crack growth for $t \geq 0$, is generated once and for all in a manner similar to that described previously either with or without dependence between components. Each crack will ultimately grow to its fail-safe length at which time the residual strength of that component will be low. The first component that sustains a shock exceeding its residual strength causes the structure to fail. This is the series assumption.

The probability of detection will not change essentially from the previous model but remains a function of the crack length vector when the inspection occurs. There will now exist a set of possible inspection procedures some of which inspect certain components more, and other components less, carefully than others at certain times. Call the set of possible non-detection probabilities $\mathcal{Q}$. Thus at the ith inspection $q_i \in \mathcal{Q}$ is selected as the appropriate non-detection probability and if $\underline{s}(i)$ is the state of the structure then $q_i(\underline{s}_i)$ is the probability of not detecting any crack. If a crack is detected in any component, the entire structure is repaired and renewed. This is the renewal assumption.

The shocks are now multi-component vectors $\underline{Y}_i$ which are independent and identically distributed:

$$\underline{Y}_i \sim F \quad \text{for } i = 1, 2, \ldots .$$

Let $L_j(t)$ be the maximum shock deterred to the jth component in a time interval of length t for $j = 1, \ldots, m$. Then

$$\underline{L}(t) = \left(L_1(t), \ldots, L_m(t)\right).$$

Again from the nature of the Poisson Process governing shocks and the theorem of total probability

$$P[\underline{L}(t) \leq \underline{y}] = \sum_{n=0}^{\infty} P[N(t)=n] P \left\{ \bigcap_{i=1}^{n} [\underline{Y}_i \leq \underline{y}] \right\}$$

$$= \sum_{n=0}^{\infty} \frac{e^{-\lambda t}(\lambda t)^n}{n!} F^n(\underline{y}) = \exp\{-\lambda t \bar{F}(\underline{y})\}.$$

Let $\underline{s}$ be a realization of the multi-variate stochastic process describing crack growth. Then:

$q_j(\underline{s})$ = probability of not detecting a crack in the structure given $\underline{s}(j)$ are the crack lengths in the component during the jth inspection;

$r_i(\underline{s})$ = probability that the structure does not fail during the ith service interval, i.e. between the $(i-1)^{st}$ and i^{th} inspections.

We now determine the probability that the magnitude of the ith shock vector $\underline{Y}_i$ does not exceed in any component the residual strength available at the time the shock occurs. If the ith shock occurs at time t then the probability of no fracture is

$$P[\underline{Y}_i \leq \underline{C}\, \underline{s}\,(t)] = F\, \underline{C}\, \underline{s}\,(t)$$

where as before juxtaposition of functions indicates composition.

Suppose that $n \geq 0$ shocks are received during the (unit) time interval between i-1 and i. Let $t_1, \ldots, t_n$ be the respective ordered times of their occurrence and

$$i-1 < t_1 < \cdots < t_n \leq i.$$

By the nature of the Poisson Process it follows, as before, that

$$r_1(\underline{s}) = \sum_{n=0}^{\infty} \frac{e^{-\lambda}\lambda^n}{n!} \int_{i-1}^{i} \cdots \int_{i-1}^{t_3} \int_{i-1}^{t_2} \prod_{j=1}^{n} \underline{FC}\ \underline{s}(t_j) \frac{1}{n!}\, dt_1 \cdots dt_n .$$

If we let $A_j(\underline{s})$ equal the probability that the structure will not fail during the first j service intervals, given crack growth $\underline{s}$ and the structure remains unrepaired, then we have in an analogous fashion to (6.1).

$$A_j(\underline{s}) = \prod_{i=1}^{j} q_i(\underline{s}) r_i(\underline{s}) \quad \text{for } j = 1, 2, \ldots; \quad A_0(\underline{s}) = 1.$$

Thus we are in exactly the same position theoretically for this type of structure with repair that we were in sections 5 and 6 for a single component. All of the preceding results will apply with only a change of the interpretation of the symbols involved. However there has yet been no implementation of this theory to actual structures nor the concomitant parameter determinations.

The preceding analysis for a series structure makes it possible when the assumptions apply, to formulate several problems of optimality.

Problem 1: Optimal Inspection for a Given Structure.

Let $\underline{q} = (q_1, q_2, \ldots)$ be the choice of inspection procedures and $(d_1, d_2, \ldots)$ the cost of carrying out such inspections. If k inspection periods are the warranteed life of the structure, it is desired to find $\underline{q}^*$ such that the reliability as a function of $\underline{q}^*$ is

$$V(k:\underline{q}^*) = \sup_{\underline{q}} V(k:\underline{q}) ,$$

subject to the constraint

$$\sum_{1}^{k} d_i \leq \text{constant}.$$

Problem 2: Optimal Design.

Let $\underline{\zeta} = (\underline{\delta}, \underline{\sigma}, \underline{\nu})$ be the disposable parameters of design where $\underline{\delta}$ are the design limit loads, $\underline{\sigma}$ the fail-safe limits and $\underline{\nu}$ the load safety factors for the structure. The inspection procedure $\underline{q}$ and the disposable parameters $\underline{\zeta}$ are determined so that the reliability for k periods satisfies

$$V(k:\underline{q},\underline{\zeta}) \geq 1 - \varepsilon \qquad \text{for preassigned } \varepsilon > 0.$$

The cost of inspection then is $\sum_{1}^{k} d_i$ while the expected cost of maintenance over k periods is M(k) say, where if d is the cost of each renewal then

$$M(k) = d\ H(k) + 2d\ G*H(k) + 3d\ G^{2*}*H(k) + \ldots$$

$$= d\ H* \sum_{i=1}^{\infty} i\ G^{(i-1)*}\ (k).$$

Using Abel's summation formula, after replacing i by $\sum_{j=1}^{i}$, yields

$$M(k) = d\ H* \sum_{j=1}^{\infty} \sum_{i=j}^{\infty} G^{(i-j)*}(k)$$

$$= d \sum_{j=1}^{\infty} G^{(j-1)*} * \sum_{i=j}^{\infty} H*G^{(i-j)*}(k)$$

$$= d\ U* \sum_{j=0}^{\infty} G^{j*}(k)$$

$$= d\ U*[1 + G(k) + G^{2*}(k) + \ldots].$$

The problem is to pick $\underline{q}$, $\underline{\zeta}$ such that the total cost of inspection plus maintenance plus cost of original design is minimized subject to the reliability constraint.

The solution of these problems must await the acquisition of appropriate data but they would allow a mathematical determination of either optimality of structural design for proposed structures or optimality of inspection for previously designed structures based on a given distribution of shocks and rate of crack growth.

REFERENCES

[1] BARLOW, RICHARD E. and PROSCHAN, FRANK (1975). *Statistical Theory of Reliability and Life Testing.* Holt Rinehart and Winston, Inc.

[2] BIRNBAUM, Z. W. and SAUNDERS, S. C. (1969). A new family of life distributions. *J. Appl. Probability,* 6, 319-327.

[3] BIRNBAUM, Z. W. and SAUNDERS, S. C. (1969). Estimations for a new family of life distributions. *J. Appl. Probability,* 6, 328-347.

[4] EGGWERTZ, SIGGE (1972). Investigation of fatigue life and residual strength of wing panel for reliability purposes. *Probabilistic Aspects of Fatigue,* ASTM STP 511, American Society for Testing on Materials, 75-105.

[5] FREUDENTHAL, A. M., GARRELTS, J. M. and SHINOZUKA, M. (1966). The analysis of structural safety. *A.S.C.E. Structural Div. Journ., Proc. Amer. Soc. Civil Engineers,* 92, 1, 267-325.

[6] FREUDENTHAL, A. M. and SHINOZUKA, M. (1961). Structural safety under conditions of ultimate load failure and fatigue. WADD Technical Report 61-177.

[7] SAUNDERS, S. C. (1974). A family of random variables closed under reciprocation. *J. Amer. Statist. Assoc.,* 69, 533-539.

[8] WHITTAKER, I. C. and SAUNDERS, S. C. (1973). Exploratory development on application of reliability analysis to aircraft structures considering interaction of cumulative fatigue damage and ultimate strength. Air Force Materials Laboratory Technical Report AFML-TR-72-283.

[9] WHITTAKER, I. C. and SAUNDERS, S. C. (1973). Application of reliability analysis to aircraft structures subject to fatigue crack growth and periodic structural inspection. Air Force Materials Laboratory Technical Report AFML-TR-73-92.

[10] WHITTAKER, I. C. and SAUNDERS, S. C. (1975). Further development of reliability analysis application to structural fatigue evaluation. Air Force Materials Laboratory Technical Report AFML-TR-75-191.

P.R. Krishnaiah, ed., APPLICATIONS OF STATISTICS

SYSTEM FAILURES AND STOCHASTIC HAZARD RATE

I. N. SHIMI
Air Force Office of Scientific Research
Bolling Air Force Base, Washington, D.C.

The present paper is concerned with the overall pattern of failure for a complex system composed of a large number of components. In particular, as the number of components increase, the asymptotic distribution of a number of system failures is determined when the hazard functions for the components are realizations of stochastic processes. The usefulness of this model is demonstrated by the two examples given at the end of the paper. The first example deals with the situation when the system is operating in a random environment caused by a random load on the system. The second example deals with a case in which the components are being selected from heterogeneous populations due to variations in their manufacturing processes.

1. INTRODUCTION

The requirement for light weight aircraft structures necessitates planning with as small a safety margin as is practical for the anticipated use of the aircraft. Cracks and partial damage in the aircraft structures occur as a result of this small safety margin and the uncertainties in the assumed design loads including the fatigue loads, the loading sustained, due to differences between the actual flight conditions and the test conditions, and the damage due to foreign objects and corrosion. Fail-safe design procedures have been proposed and used to allow the structure to fulfill its role in supporting the service loads during the useful life with the possible existence of cracks or partial damage (see Liebowitz (1974) and Broeck (1974)). The uncertainties causing the cracks and partial failures of structures necessitate the use of a probabilistic approach in the development of a fail-safe design. Due to fatigue loading and stress condition, initial flaws in structural materials, which may be micro-sized, grow to detectable sizes that become more and more critical as factors in the failure of these structures. The time or number of cycles required for the crack growth to attain a detectable size, called crack initiation time, or until the crack size reaches a critical stage, is obviously random. The probability distribution for the crack sizes is continuously changing with time or number of cycles. To the best of our knowledge, in the situations discussed above, changes in the crack lengths associated with the fatigue process, the variability of the flaws in the structural materials, and the continuous changes in the stress loading on the structures, have been only recently considered, Toor (1973), as factors in the development of failure models.

Ignoring these factors and others has led to the use of very simplistic models for failure of these structures, such as the Weibull distribution, to describe the time for fatigue failure. If we consider the hazard function associated with the failure of such structures, it becomes obvious that a deterministic hazard function leads to ignoring the randomness induced by the above mentioned factors. In order to incorporate the effects of the stochastic process representing the fatigue loading, the stochastic process representing the crack lengths and the variability due to the initial flaws in the structural material, it becomes more reasonable and realistic to assume that the hazard function is itself a stochastic process.

The assumption of stochastic hazard rate is not new. This type of reliability model has been considered by Antelman and Savage (1965), Mercer (1961), Harris and Singpurwalla (1967), Gaver (1963), Birnbaum and Saunders (1958), Cozzalino (1968) and McNolty (1964).

The present paper is concerned with the overall pattern of failure for a complex system composed of a large number of components. In particular, as the number of components increases, the asymptotic distribution of the number of system failures is determined when the hazard rates for the components are assumed to be realizations of stochastic processes. The usefulness of this model is demonstrated by two examples given at the end of the paper. The first example deals with the situation when the system is operating in a random environment caused by a random load on the system. The second example considers the case when the components are being selected from heterogeneous populations due to variations in their manufacturing processes.

This type of system was considered by Drenick (1960), when the hazard functions for the different components are deterministic. Under some conditions on the behavior of the hazard functions near zero, Drenick proved that the asymptotic distribution for the initial lifelength of a system is the exponential distribution. His conditions are quite restrictive and not, for example, satisfied by the Weibull distribution. Cox and Smith (1954) considered a similar model for a system made up of a large number of components all alike. They assumed that the common distribution function, $F(x)$, of the interval between component failures is known. In particular, they assumed that $F(x)=0(x^b)$ as $x \to 0$, where $0 \leq b \leq 1$. They proved that for a system consisting of a large number of components, the asymptotic distribution for the number of failures of the system is Poisson.

2. THE MODEL

Consider a system composed of N components each of which is subject to failure. Given the hazard rate, $\phi_{N,i}(\cdot)$ for component i, the probability that component i fails in the time interval $(t, t+\Delta t)$, given that it is in service at time t, is $\phi_{N,i}(t)\,\Delta t + o\,(\Delta t)$. We assume that $\phi_{N,i}(\cdot)$ is a realization of a stochastic process

$$\{\Phi_{N,i}(t), t \geq 0\} \text{ for all } i \ (1 \leq i \leq N).$$

The system has the following properties: (i) different components are not, in general, alike and, given their hazard rates, are stochastically independent; (ii) the system fails with any component failure; (iii) each component is immediately replaced upon its failure by a component of the same type; the hazard rate of the new component is a realization of the stochastic process associated with this type of component.

Throughout this paper we shall assume that $\{\Phi_{N,i}(t), t \geq 0\}$ is a non-negative measurable stochastic process for all N and all i $(1 \leq i \leq N)$, $m_{N,i}(t) = E\ \Phi_{N,i}(t)$ exists for $t \geq 0$ and if A is a Lebesgue measurable parameter set with finite measure, then

$$\int_A m_{N,i}(x)\, d\,x$$

These assumptions insure that all sample functions are Lebesgue integrable over any parameter set A with finite measure; see Doob (1953, p. 62).

We shall need the following lemma to prove the main theorem.

LEMMA. Let $X_{n1}, X_{n2}, \ldots, X_{nk_n}$ be a double sequence of non-negative random variables where $k_n \to \infty$ as $n \to \infty$ and such that:

$$\text{(i) } X_n = \max_{1 \leq j \leq k_n} X_{nj} \xrightarrow{P} 0 \text{ as in}$$

$$\text{(ii) } \sum_{i=1}^{k_n} X_{n,i} \xrightarrow{P} X \text{ as } n \longrightarrow \infty .$$

Then

$$\prod_{j=1}^{k_n} (1+X_{nj}) \xrightarrow{P} \exp X.$$

Proof. From the inequality

$$\left| \sum_{i=1}^{k_n} \log\,(1 + X_{ni}) - \sum_{i=1}^{k_n} X_{ni} \right| \leq \sum_{i=1}^{k_n} \left(\sum_{j=2}^{\infty} \frac{(X_{ni})^j}{j} \right),$$

it follows that

$$P\{|\sum_{i=1}^{k_n} \log(1 + X_{ni}) - \sum_{i=1}^{k_n} X_{ni}| > \varepsilon\}$$

$$\leq P\{\sum_{i=1}^{k_n} \sum_{j=2}^{\infty} \frac{(X_{ni})^j}{j} > \varepsilon\}$$

$$\leq P(\hat{X}_n > \tfrac{1}{2}) + P\{(\sum_{i=1}^{k_n} \sum_{j=2}^{\infty} \frac{(X_{ni})^j}{j} > \varepsilon)(\hat{X}_n \leq \tfrac{1}{2})\}$$

$$\leq P(\hat{X}_n > \tfrac{1}{2}) + P\{(\tfrac{1}{2} \sum_{i=1}^{k_n} \frac{X_{ni}^2}{1 - X_{ni}} > \varepsilon)(\hat{X}_n \leq \tfrac{1}{2})\}$$

$$\leq P(\hat{X}_n > \tfrac{1}{2}) + P\{(\sum_{i=1}^{k_n} X_{ni}^2 > \varepsilon)(\hat{X}_n \leq \tfrac{1}{2})\}$$

$$\leq P(\hat{X}_n > \tfrac{1}{2}) + P(\hat{X}_n \sum_{i=1}^{k_n} X_{ni} > \varepsilon) .$$

In the second and last inequality we used the fact that $P(AB) \leq P(A)$. From (i) and (ii) we can see that

$$\log \prod_{i=1}^{k_n} (1 + X_{ni}) \xrightarrow{P} X \qquad \text{as } n \longrightarrow \infty .$$

Using the well known result that if f is a measurable mapping from the reals to the the reals which is continuous over a Borel set B for which $P(X \in B) = 1$, and if $X_n \xrightarrow{P} X$, then $f(X_n) \xrightarrow{P} f(X)$, the conclusion follows.

Before proving the main result we note the following. Let $Y_N(t)$ denote the number of failures in $[0,t]$ of the original components. Then $Y_N(t) = \sum_{k=1}^{N} V_k(t)$, where $V_k(t)$ equals 0 if the kth component does not fail in $[0,t]$ and equals 1 if the kth component fails in $[0,t]$. But the conditional probability of the event $V_k(t) = 0$, given the hazard rate $\Phi_{N,k}(\cdot)$, is given by

$$Q_{N,k}(t) = \exp - \int_0^t \Phi_{N,k}(x)dx \qquad \text{for } k = 1, 2, \ldots, N,$$

Let $P_{N,k}(t) = 1 - Q_{N,k}(t)$ for $k = 1, 2, \ldots, N$. From this we can show that the conditional probability generating function (p.g.f.) for $V_k(t)$ is equal to $(1 - P_{N,k}(t)(1-s))$. Since the components are conditionally independent, the conditional p.g.f. for $Y_N(t)$, given the hazard rates, is equal to

$$\prod_{i=1}^{N} (1-P_{N,i}(t)(1-s)).$$

Thus the unconditional p.g.f. for $Y_N(t)$ is equal to

$$G_N(s) = E \prod_{i=1}^{N} (1-P_{N,i}(t)(1-s)).$$

3.THE MAIN RESULT

The purpose of our next theorem is to find the limiting p.g.f. for the total number of failures, of original and replacement components, in an arbitrary but fixed time interval $[0, t]$. Let $X_N(x)$ be the number of failures of originals or replacements in $[0,x]$.

Theorem. For an arbitrary but fixed t, if the hazard rates $\Phi_{N,i}(\cdot)$, $1\leq i\leq N$, are such that

$$\text{(i)} \quad \max_{1\leq i\leq N} \Lambda_{N,i}(t) \xrightarrow{P} 0 \quad \text{as } N \longrightarrow \infty$$

and

$$\text{(ii)} \quad \sum_{i=1}^{N} \Lambda_{N,i}(t) \xrightarrow{P} P(t) \quad \text{as } N \longrightarrow \infty$$

where

$$\Lambda_{N,i}(t) = \int_0^t \Phi_{N,i}(x)\, dx ,$$

then

$$\lim_{N\to\infty} G_N(s) = E \exp\{-(1-s)P(t)\} .$$

If (i) is replaced by

$$\text{(iii)} \quad g(N) \max_{1\leq i\leq N} \Lambda_{N,i}(t) \xrightarrow{P} 0, \text{ where } g(N)\longrightarrow\infty \text{ as } N\longrightarrow\infty$$

then

$$P\, \{X_N(t') = Y_N(t') \ ,\ 0 \leq t' \leq t\} \longrightarrow 1 \qquad \text{as } N \longrightarrow \infty .$$

Proof. Condition (i) implies that for all i, $1\leq i\leq N$, $\Lambda_{N,i}(t)\xrightarrow{P} 0$ as $N\longrightarrow\infty$. We use the fact that if a Borel function f has a derivative at a and $X_n \xrightarrow{P} a$, then

$f(X_n)=f(a) + f'(a)\,(X_n-a) + (X_n-a)\,Z_n$, where $Z_n\xrightarrow{P} 0$. Thus we can write, for any k, $1\leq k\leq N$,

$$P_{N,k}(t) = 1 - \exp - \Lambda_{N,k}(t)$$

$$= 1 - \{1 - \Lambda_{N,k}(t) + Z_{N,k} \Lambda_{N,k}(t)\}$$

$$= \Lambda_{N,k}(t) - Z_{N,k} \Lambda_{N,k}(t)$$

where $Z_{N,k} \xrightarrow{p} 0$ as $N \longrightarrow \infty$. We can also show that

$$Z_{N,k} \Lambda_{N,k}(t) \leq \frac{1}{2} (\Lambda_{N,k}(t))^2$$

Thus

$$\sum_{k=1}^{N} Z_{N,k} \Lambda_{N,k}(t) \leq \frac{1}{2} \max_{1 \leq i \leq N} \Lambda_{N,i}(t) \sum_{k=1}^{N} \Lambda_{N,k}(t) .$$

Now, from (i) and (ii), we can show that the right hand side of the last inequality tends to zero in probability and hence

$$\sum_{i=1}^{N} P_{N,i}(t) \longrightarrow P(t) \quad \text{as} \quad N \longrightarrow \infty .$$

Since

$$P_{N,i}(t) = 1 - \exp - \Lambda_{N,i}(t) \leq \Lambda_{N,i}(t)$$

then condition (i) implies that

$$\max_{1 \leq i \leq N} P_{N,i}(t) \xrightarrow{p} 0 \quad \text{as} \quad N \longrightarrow \infty .$$

Thus Lemma 1 can be used to show that, as $N \longrightarrow \infty$.

$$\prod_{j=1}^{N} (1 - (1-s) P_{N,i}(t)) \xrightarrow{p} \exp\{ - (1-s) P(t)\}.$$

Since the above product is bounded above by one for all N, then using the bounded convergence theorem we can show that

$$\lim_{N\to\infty} G_N(s) = \lim_{N\to\infty} E \prod_{i=1}^{N} (1-(1-s)P_{N,i}(t)) = E \exp \{ - (1-s)P(t)\}.$$

To prove the last part of the theorem we notice that if a component of type i is replaced by another component of the same type at time $T(< t)$, then the conditional probability of its survival up to time t, given the hazard rates is equal to $\exp\{-\int_0^{t-T} \Phi_{N,i}(x)dx\}$.

Since $\Phi_{N,i}$ is non-negative, this conditional probability is greater than or equal to $\exp\{-\int_0^t \Phi_{N,i}(x)dx$ a.s.. Using this we can show that

$$q(i_1, i_2, \ldots, i_k) = \Pr\{\text{ no new failures of } i_1, i_2, \ldots, i_k \text{ types in } [0,t] \mid V_{i_j}(t) = 1,\ j = 1, 2, \ldots, k, \text{ and } V_i = 0 \text{ otherwise}\}$$

$$\geq E\exp\{-\sum_{j=1}^{k} \int_0^t \Phi_{N,i_j}(x)dx\} \geq E\exp\{-k \max_{1 \leq i \leq N} \Lambda_{N,i}(t)\} .$$

Thus for an integer $k \leq g(N)$, we can show, using condition (iii) of the theorem that $q(i_1, \ldots, i_k) \to 1$ as $N \to \infty$. Let $q_N = E\exp\{-g(N) \max_{1 \leq i \leq N} \Lambda_{N,i}(t)\}$. We can see that $q(i_1, i_2, \ldots, i_k) \geq q_N$ for any set of integers $\{i_1, i_2, \ldots, i_k\}$ where $k \leq g(N)$, and $q_N \to 1$ as $N \to \infty$. Therefore we can show that

$$P\{X_N(t') = Y_N(t'),\ 0 \leq t' \leq t\} \leq P\{Y_N(t) \leq g(N)\}\ q_N.$$

Since $\{Y_N(t)\}$ converges in distribution, it follows that $P\{Y_N(t) \leq g(N)\} \to 1$ as $N \to \infty$, and thus

$$P\{X_N(t') = Y_N(t'),\ 0 \leq t' \leq t\} \to 1 \qquad \text{as } N \to \infty .$$

It can also be shown that the limiting process is a stochastic process with independent increments.

4. EXAMPLES

We consider next the application of our result to some practical examples.

Example 1. Consider a system consisting of N components and exposed to a total instantaneous load $L(t)$, where $\{L(t), t \geq 0\}$ is a non-negative stochastic process. The load is shared, not necessarily equally, by all the components. We make the assumption, due to Birnbaum and Saunders ((1958) p. 153), that the hazard rate of component i, $\Phi_{N,i}(t)$, can be represented as $\Phi_{N,i}(t)=w_{N,i}(t)D_{N,i}(t)$. In this representation $D_{N,i}(t)$ is the instantaneous damage at time t to component i and $w_{N,i}(t)$ is the cumulative deterioration of component i up to time t. The stochastic process $D_{N,i}(t)$ is a function of $L(t)$ depending on the structure and interconnections of the system. A simple interpretation of the function $w_{N,i}(t)$

is that it represents the hazard rate of component i when exposed to a constant non-random load which gives unit instantaneous damage for all t. This is the standard load for which the component is designed. For further discussion of this assumption see the Birnbaum and Saunders (1958) paper, where an empirical example is presented showing remarkable agreement between the experimental data and the predicted result under the above assumptions. Their results are based on the assumption of a non-random load function.

Now assume that the load function L(x) is such that for all finite t, $D_N = \max_{1 \leq i \leq N} \sup_{0 \leq x \leq t} D_{N,i}(x)$ is finite a.s.. This can be the case when sup L(x) is a.s. within the elastic region for each of the components. Let $D_N = o_p(N^{-a})$ for some $a > 0$, where D_N might depend on t. Also assume that for any finite t, $\int_0^t w_{N,i}(x)dx < A(t)$ for all i, where A(t) is some finite function of t. We notice that this condition on $w_{N,i}(\cdot)$ does not exclude any of the hazard rates that are commonly used. If we assume further that $\sum_{i=1}^{N} \int_0^t w_{N,i}(x)D_{N,i}(x)dx \xrightarrow{P} P(t)$, we can see that the conditions of the theorem are satisfied and the limiting distribution will have a p.g.f. given by $E\{\exp - (1-s)P(t)\}$.

We observe that if we have components sharing the load equally, then $D_{N,i}(t) = \frac{\Delta(t)}{N}$, where $\Delta(t)$ is the total instantaneous damage at time t. Also, if we let $\frac{1}{N} \sum_{i=1}^{N} w_{N,i}(t) \to w(t)$, then

$$\sum_{i=1}^{N} \int_0^t \Phi_{N,i}(x)dx \to \int_0^t w(x)\,\Delta(x)dx.$$

The conditions of the theorem are satisfied and $G_N(s) \to E\exp\{-(1-s)\int w(x)\Delta(x)dx$. We notice that in case $w(x)\Delta(x) = a$, where a is a constant, the limiting p.g.f. is one corresponding to a homogeneous Poisson process.

Example 2. Due to heterogeneity of the populations from which the different components are selected and the interconnection of the components of the system we may assume that

$$\Phi_{N,i}(t) = L_{N,i} + M_{N,i}(\theta_{N,i}+1)t^{\theta_{N,i}}$$

where the shape parameters $\theta_{N,i}$ are non-random and $\theta_{N,i} > -1$ for all i, $L_{N,i}$ and the scale parameters $M_{N,i}$ are non-negative random variables. We are assuming that the failure of component i is due to two factors: one independent of age and the other increasing (for $\theta_{N,i} > 0$) or decreasing (for $\theta_{N,i} < 0$) with increasing age. The assumption of non-random known shape parameters seems to be reasonable for components that have an efficient quality control during their production. We make this assumption to avoid unnecessary complications in our exposition. Let us assume further that $-1 < \theta_{N,i} \leq A < \infty$ for all i and large N. With these assumptions we can see that if $t \leq 1$,

$$\int_0^t \Phi_{N,i}(x)dx \leq L_{N,i}t + M_{N,i}$$

and if $t \geq 1$

$$\int_0^t \Phi_{N,i}(x)dx < L_{N,i}t + M_{N,i}t^{A+1}$$

Thus if $\max_i L_{N,i}$ and $\max_i M_{N,i}$ are both $o_p(N^{-a})$, condition (iii) of the theorem follows. If we further assume that $\sum_{i=1}^N L_{N,i}$ and $\sum_{i=1}^N M_{N,i}t^{\theta_{N,i}}$ both converge in probability, then condition (ii) of the theorem follows and the limiting p.g.f. can be determined.

REFERENCES

[1] ANTELMAN, G. and SAVAGE, I. R. (1965). Characteristic functions of stochastic integrals and reliability theory. *Naval Res. Logist. Quart.*, **12**, 199-222.

[2] BIRNBAUM, Z. W. and SAUNDERS, S. C. (1958). A statistical model for life-length of materials. *J. Amer. Statist. Assoc.*, **53**, 151-160.

[3] COX, D. R. and SMITH, W. L. (1954). On the superposition of renewal processes. *Biometrika* , **41**,91-99.

[4] COZZOLINI, J. M. (1968). Probabilistic models of decreasing failure rate processes. *Naval Res. Logist. Quart.*, **15**, 361-374.

[5] DOOB, J. L. (1953). *Stochastic Processes.* John Wiley and Sons, Inc.

[6] DRENICK, R. F. (1960). The failure law of complex equipment. *J. Soc. Indust. Appl. Math.*, **8**, 680-690.

[7] GAVER, D. P., JR. (1963). Random hazard in reliability problems. *Technometrics* , **5**, 211-226.

[8] HARRIS, C. M. and SINGPURWALLA, N. D. (1967). *Life distributions derived from stochastic hazard functions.* Research Analysis Corporation, Technical Paper RAC-TP-280.

[9] MCNOLTY, F. (1964). Reliability density functions when the failure rate is randomly distributed. *Sankhyā* , **26**, 287-292.

[10] MERCER, A. (1961). Some simple wear-dependent renewal processes. *J. Royal Statist. Assoc. (B)* , **23**, 368-376.

[11] LIEBOWITZ(editor) (1974). Fracture mechanics of aircraft structures. AGARDograph No. 176, (AGARD-AG-176) Jan. 1974.

[12] BROEK, D. (1974). Fail-safe design procedures. AGARDograph No. 176, pp 212-166, Jan. 1974.

[13] TOOR, P. M. (1973) A review of some damage tolerant design approach for aircraft structures. *Engineering Fracture Mechanics* , **5**, 837-880.

P.R. Krishnaiah, ed., APPLICATIONS OF STATISTICS

A SURVEY OF SOME SMOOTHING PROBLEMS AND THE METHOD OF GENERALIZED CROSS-VALIDATION FOR SOLVING THEM*

GRACE WAHBA

Department of Statistics
University of Wisconsin
Madison, Wisconsin

Some applications of the method of generalized cross-validation (GCV) for determining the correct degree of smoothing to minimize mean square error, are surveyed. These are ridge regression, spline smoothing, density estimation, and the approximate solution of linear operator equations when the data are noisy (Tihonov regularization).

1. INTRODUCTION

Consider the model

$$y(t) = f(t) + \varepsilon(t), \qquad t \in [0,1], \tag{1.1}$$

where f is a "smooth" function, and ε is a noise, $E\varepsilon(t)=0$, $E\varepsilon(s)\varepsilon(t)=\sigma^2$, s=t; =0 otherwise. Values of y are observed for $t=t_1,t_2,\ldots,t_n$, and it is desired to recover an estimate of f. Suppose $f \in H_m$: $\{f: f,f',\ldots,f^{(m-1)}$ abs. cont., $f^{(m)} \in L_2[0,1]\}$. Take as an estimate of f the solution, call it $f_{n,\lambda}$, to the problem: Find $f \in H_m$ to minimize

$$\frac{1}{n}\sum_{j=1}^{n} (f(t_j)-y_j)^2 + \lambda \int_0^1 (f^{(m)}(u))^2 du, \quad y_j \equiv y(t_j). \tag{1.2}$$

If $n \geq m$, the solution $f_{n,\lambda}$ is known [20], [22] to be unique, and to be a polynomial spline of degree 2m-1, that is

1) $f_{n,\lambda}$ is a polynomial of degree 2m-1 in each interval $[t_i,t_{i+1}]$, and
2) $f_{n,\lambda}$ has 2m-2 continuous derivatives.

The parameter λ controls the tradeoff between the infidelity of $f_{n,\lambda}$ to the data, as measured by

$$\frac{1}{n}\sum_{j=1}^{n} (f_{n,\lambda}(t_j)-y_j)^2 \tag{1.3}$$

---

*This research was supported by the United States Air Force under Grant No. AFOSR 72-2363-C.

and the "smoothness",

$$\int_0^1 [f_{n,\lambda}^{(m)}(u)]^2 \, du. \tag{1.4}$$

If λ is large, then (1.4) will be small while (1.3) will be large, and vice versa. As $\lambda \to \infty$, $f_{n,\lambda}$ tends (pointwise) to the polynomial of degree m best fitting the data in a least squares sense, and as $\lambda \to 0$, the solution tends to a spline function which interpolates the data. The parameter λ may be thought of as controlling the tradeoff between the squared bias and the variance of the estimate $f_{n,\lambda}(t)$ for $f(t)$. If $\lambda=0$, the bias of $f_{n,0}(t_i)$ equals 0, and the variance is σ^2. Provided f is not a polynomial of degree m or less, as λ increases, the squared bias increases and the variance decreases. For a large class of problems there is a value of λ strictly between 0 and ∞ which minimizes the squared bias plus the variance. (If f is a polynomial of degree m or less the optimum value of λ to minimize mean square error is ∞). The experimenter who wishes to smooth data by this technique must have a valid method of choosing λ. This problem, of controlling the tradeoff between squared bias and variance occurs in a variety of contexts, beginning with the problem of admissible estimates for the mean of a multivariate normal vector with quadratic loss studied by Stein and associates (See [9,17] and references cited there), ridge regression ([8,11,16]), the correct degree of smoothing (equivalently the window width) in density and spectral density estimation ([5,7,19, 25]), the smoothing of surfaces [27], and the approximate solution of linear operator equations when the data are noisy. The operator equation problem is known in approximation theory circles as Tichonov regularization, see [26]).

In 1975 Wahba and Wold [29,30] applied an idea they learned from Stone [23], see also Geisser [10], known as cross-validation, or predictive sample reuse, to choose λ. Monte Carlo results were unbelieveably (at least to the authors!) good. Following this initial success attempts were made to establish theoretical properties of the method of cross-validation and to see if it could be applied in other contexts. It turns out that the method of "ordinary" cross validation must be modified to have desirable properties in general. With this modification, which we call generalized cross validation (GCV), we have been able to show theoretically desirable properties in general, and pleasing Monte Carlo results when the method is used to estimate an optimum smoothing parameter in ridge regression and to estimate the correct degree of smoothing in density estimation.

In this paper we give a general formulation of the GCV method for estimating a good value of the parameter which controls the tradeoff between square bias and variance. The purpose of this paper is to provide a unified overview of some of the various contexts in which the GCV method is applicable, and survey some of the known results, published and unpublished. We will state theorems concerning the properties of the GCV estimate of λ for ridge regression, spline smoothing and density estimation, without proofs. The proofs have appeared or will appear elsewhere. Brief mention will be made of the application to smoothing of surfaces and to Tichonov regularization. In the ridge regression and density estimation cases we will present preliminary, but typical, Monte Carlo results, so that the reader may judge for himself or herself whether he or she would like to use the method on real data.

2. RIDGE REGRESSION

Consider the usual regression model,

$$y_{n\times 1} = X_{n\times p}\beta_{p\times 1}+\varepsilon_{n\times 1} \tag{2.1}$$

where subscripts denote the dimensions of the vector or matrix. Let $\varepsilon \sim N(0,\sigma^2 I_{n\times n})$. The ridge estimate $\hat{\beta}_\lambda$ of β that we consider is

$$\hat{\beta}_\lambda = (X^T X+n\lambda I)^{-1}X^T y \ . \tag{2.2}$$

It is well known that $\hat{\beta}_\lambda$ is the posterior mean of β if β has the prior $\beta \sim N(0,aI_{p\times p})$, where $\lambda=\sigma^2/na$, see for example [11]. It is also not hard to show that $\hat{\beta}_\lambda$ is the solution to the minimization problem:

Find $\hat{\beta}$ in Euclidean p-space to minimize

$$\frac{1}{n}||y-X\beta||_n^2 + \lambda||\beta||_p^2, \tag{2.3}$$

where $||\cdot||_q$ is the Euclidean q norm. It will be convenient to think of $\frac{1}{n}||y-X\beta||_n^2$ as the "infidelity", and $||\beta||_p^2$ as the "energy" or "smoothness". For fixed β, λ controls the tradeoff between squared bias and variance. Thus the reader may take a Bayesian point of view, or minimize the infidelity plus smoothness functional, either approach can lead to estimates of the form (2.2). The "ordinary" cross validation estimate of λ for ridge regression was suggested by Allen (see [1] and the discussion in Stone [23]) and given the name PRESS. It goes as follows: Let $\hat{\beta}^{(k)}$ be the solution to:

Find $\hat{\beta} \varepsilon E_p$ to minimize

$$\frac{1}{n}\ ||y^{(k)}-X^{(k)}\beta||_{n-1}^2 + \lambda\ ||\beta||_p^2\ ,$$

where $y^{(k)}$ and $X^{(k)}$ are obtained from y and X by deleting the kth row in each case. Thus, $\hat{\beta}_\lambda^{(k)}$ is a ridge estimate with the kth data point left out. The idea is, that if a particular λ is a good choice, then $[X\beta_\lambda^{(k)}]_k$, the kth entry of $X\hat{\beta}_\lambda^{(k)}$, should be a good predictor of the missing data point y_k. This predictive ability is measured by $([X\hat{\beta}_\lambda^{(k)}]_k - y_k)^2$. The Allen's PRESS estimate for λ is then obtained by choosing λ to minimize $V_0(\lambda)$ given by

$$V_0(\lambda) = \sum_{k=1}^{n} ([X\hat{\beta}_\lambda^{(k)}]_k - y_k)^2 . \tag{2.4}$$

This idea is intuitively appealing. However, if the components x_{ij} of X satisfy $x_{ij}=0$, $i \neq j$, that is, $y_i = x_{ii}\beta_i + \varepsilon_i$, $i=1,2,\ldots,p$, $y_i = \varepsilon_i$, $i=p+1,\ldots,n$, then $y_1,\ldots,y_{k-1}, y_{k+1},\ldots,y_n$ cannot be expected to provide much information about the kth component β_k of β, and in fact it can be shown that $V_0(\lambda)$ is independent of λ in that case, and insensitive to λ in nearby cases.

The generalized cross validation (GCV) estimate of λ is obtained as follows: Let the singular value decomposition (see [14]) of X be

$$X = U_{n\times n} D_{n\times p} V^T_{p\times p}$$

where U and V are orthogonal, and D is diagonal. Consider the model

$$U^T y = DV^T\beta + U^T\varepsilon \tag{2.5}$$

This model is basically the same as the original model (2.1) with X replaced by a diagonal matrix, since if $\beta \sim N(0, aI_{p\times p})$ and $\varepsilon \sim N(0,\sigma^2 I_{n\times n})$, then $V^T\beta \sim N(0, aI_{p\times p})$ and $U^T\varepsilon \sim N(0,\sigma^2 I_{n\times n})$. Now, (2.5) is just in that form where no information about $\gamma_k = [V^T\beta]_k$ is provided from $\{[U^T y]_j\}$, $j \neq k$. We now want to rotate the coordinate system so that when a "data point" is left out the remaining data points provide maximal information about the missing point. Let W be the n×n unitary matrix which diagonalizes the circulant matrices. (See [4,24]) In complex form, the jk^{th} entry of W is given by

$$[W]_{jk} = \frac{1}{\sqrt{n}} e^{2\pi ijk/n}.$$

(Recall that if $C_{n\times n}$ is any diagonal matrix, then WCW^* is a circulant matrix). Write

$$\tilde{y} \equiv WU^T y = WD(V^T\beta) + WU^T\varepsilon. \tag{2.6}$$

Since we have only effected a rotation, the model is still essentially the same as (2.1), except that it is in complex form. The "new" design matrix, call it $\tilde{X}$ is now

$$\tilde{X} = WDV^T \tag{2.7}$$

and $\tilde{X}\,\tilde{X}^{*T}$ is a circulant matrix. The GCV estimate of λ is defined as the result of doing "ordinary" cross validation, or Allen's PRESS, on the rotated system

$$\begin{aligned}\tilde{y} = WU^T y &= WDV^T \beta + WU^T \varepsilon \,. \\ &= \tilde{X}\beta + WU^T \varepsilon \,. \end{aligned} \tag{2.8}$$

Upon substituting $\tilde{X}$ and $\tilde{y}$ into (2.4) and after some manipulation (see [13]), it can be shown that the GCV estimate defined this way is the minimizer of $V(\lambda)$ given by

$$V(\lambda) = ||(I-A(\lambda))y||^2/[\mathrm{Trace}(I-A(\lambda))]^2 \tag{2.9}$$

where

$$A(\lambda) = X(X^TX+n\lambda I)^{-1}X^T. \qquad A(\lambda)y \equiv X\hat{\beta}_\lambda \,. \tag{2.10}$$

$V(\lambda)$ can also be written

$$V(\lambda) = \sum_{\nu=1}^{n} \left(\frac{n\lambda}{\lambda_\nu^2+n\lambda}\right)^2 z_\nu^2 \Bigg/ \left(\sum_{\nu=1}^{n} \frac{n\lambda}{\lambda_\nu^2+n\lambda}\right)^2 \tag{2.11}$$

where $\lambda_1,\lambda_2,\ldots,\lambda_p$ are the singular values of X, $\lambda_{p+1}=\ldots=\lambda_n=0$, and $z=(z_1,z_2,\ldots,z_n)^T$ is given by $z=U^Ty$.

We will first give some of the theoretical properties relating to the behavior of $\hat{\lambda}$, the minimizer of $V(\lambda)$. Proofs may be found in Golub, Heath, and Wahba [13]. The general idea of Theorem 1 is that the mean square prediction error using $\hat{\lambda}$ is nearly as low as that with the best possible λ.

Theorem 1. Let the mean square prediction error $T(\lambda)$ be defined by

$$T(\lambda)= \frac{1}{n}||X\beta-X\hat{\beta}_\lambda||^2$$

Let $\tilde{\lambda}$ be the minimizer of $EV(\lambda)$.

Case a) $n \to \infty$, p, β fixed. Then

$$[ET(\tilde{\lambda})/\min_\lambda ET(\lambda)] = 1 + 0(\tfrac{p}{n}) \tag{2.12}$$

Case b) $n,\ p \to \infty$, $\sum_\nu \beta_\nu^2 < \infty$, $\lambda_\nu = 0(n\nu^{-m})$, $\nu=1,2,\ldots,n$, for some $m > 1$. Then

$$[ET(\tilde{\lambda})/\min_\lambda ET(\lambda)] = 1 + o(1).$$

The next theorem would appeal to a Bayesian, but, practically speaking, tells us not to be surprised if $T(\hat{\lambda})$ is close to the minimum of $T(\lambda)$ even if n is small.

Theorem 2. Let E_p denote expectation with respect to the prior distribution $\beta \sim N(0,aI_{p\times p})$, (and E denote expectation with respect to the distribution $\epsilon \sim N(0,\sigma^2 I)$). Let Q be any p×p matrix. Then $E_pE||Q(\beta-\beta_\lambda)||^2$ and $E_pEV(\lambda)$ are *both* minimized for $\lambda=\sigma^2/na$.

An unbiased estimate of the negative risk improvement $ET(\lambda)-\sigma^2$, is given by $R(\lambda)$,

$$R(\lambda) = \frac{1}{n} ||(I-A(\lambda))y||^2-2\hat{\sigma}^2(\frac{1}{n} \text{Trace}(I-A(\lambda))), \tag{2.13}$$

where $\hat{\sigma}^2$ is an unbaised estimate of σ^2. The minimizer of $R(\lambda)$ as an estimate of λ has been suggested by Hudson [17]. The GCV estimate may be considered as a form of risk improvement (RI) estimate which does not require knowledge of, or an estimate of σ^2.

For comparison we describe the maximum likelihood estimate of $\lambda=\sigma^2/na$ when $\beta \sim N(0,aI)$. By writing the prior distribution of y as

$$\begin{aligned} y &\sim N(0,aXX^T+\sigma^2 I) \\ &=N(0,a(XX^T+n\lambda I)) \end{aligned} \tag{2.14}$$

one can obtain that the maximum likelihood estimate of λ in the model (2.14) is the minimizer of

$$M(\lambda)=(y^T(I-A(\lambda))y)/[\text{Det}(I-A(\lambda))]^{1/n}\equiv(\sum_{\nu=1}^{n} \frac{n\lambda}{\lambda_\nu^2+n\lambda}z_\nu^2)\Big/[\prod_{\nu=1}^{n} (\frac{n\lambda}{\lambda_\nu^2+n\lambda})]^{1/n}. \tag{2.15}$$

It can be shown that Theorem 2 is also true for the maximum likelihood estimate, that is, $E_pE\ M(\lambda)$ is minimized for $\lambda=\sigma^2/na$. However, Theorem 1 does not hold for all β, for the MLE estimate.

We present a summary of the results from the first run of a Monte Carlo study evaluating the GCV estimate of λ and comparing it with several other estimates. The study will appear in Golub, Heath and Wahba [13]. The values for n and p were 21 and 10, and the design matrix and β come from discretizing a numerical inversion of the Laplace transform. The condition number of X, namely the ratio of the largest to the smallest singular value, was 1.54×10^5. Four values of σ^2, namely $\sigma^2=10^{-8}$, 10^{-6}, 10^{-4} and 10^{-2} were tried and for each value of σ^2 the experiment was replicated 4 times, giving a total of 16 runs. The ϵ_i were generated as pseudo random $N(0,\sigma^2)$ independent r.v.'s, $V(\lambda)$ was computed using the right hand side of (2.11) and the Golub-Reinsch singular value decomposition [14], and the minimizer $\hat{\lambda}$ of $V(\lambda)$ determined by a global search. $T(\lambda)$ was also computed, and the relative inefficiencies I_D and I_R, of $\hat{\lambda}$, defined by

$$I_D = ||\beta-\hat{\beta}_{\hat{\lambda}}||^2/(\min_\lambda||\beta-\hat{\beta}_\lambda||^2);\ I_R = T(\hat{\lambda})/\min_\lambda T(\lambda) \qquad (2.16)$$

were computed. (D = "domain", R = "range"). Four other methods were studied for comparison. They are

1) Hoerl-Kennard-Baldwin (HKB) [16]. $n\hat{\lambda}=\hat{\sigma}^2/\frac{1}{p}||\hat{\beta}_0||^2$, where $\hat{\sigma}^2$ is the "usual" estimator of σ^2, $\hat{\sigma}^2 = \frac{1}{n-p}||y-X\hat{\beta}_0||^2$.

2) Maximum Likelihood. $\hat{\lambda}$ is the minimizer of (2.15).

3) Allen's PRESS. $\hat{\lambda}$ is the minimizer of (2.4).

4) Dempster's RIDGM [8]. $\hat{\lambda}$ is the solution to

$$\frac{1}{p}\sum_{\nu=1}^{p} n\lambda\ \hat{\gamma}^2\ \frac{\lambda_\nu^2}{\lambda_\nu^2+n\lambda} = \hat{\sigma}^2 ,$$

where $\hat{\gamma}=(\hat{\gamma}_1,\hat{\gamma}_2,\ldots,\hat{\gamma}_p)^T=V^T\hat{\beta}_0$.

Comparison with Hudson's risk improvement estimate is in progress.

Dempster's RIDGM was initially chosen because it was reported to be the best of a number of estimates tried in a fairly extensive Monte Carlo study [8]. It can be seen that $\hat{\lambda}$ from RIDGM will be larger than $\hat{\lambda}$ from Hoerl-Kennard and will be nearly the same if $\lambda_\nu^2/(\lambda_\nu^2+n\lambda)\approx 1$. I_D and I_R were determined for each of these four methods as well as GCV and the results are presented in TABLE I. The entries next to "Min Sol'n" and "Min Data" are the inefficiencies (2.15) with $\hat{\lambda}$ taken as the minimizer of $||\beta-\hat{\beta}_\lambda||^2$ and $T(\lambda)$ respectively. S/N, the "signal to noise ratio" is defined by $S/N = [\sigma^2/\frac{1}{n}||X\beta||^2]^{1/2}$. Figure 1 gives a plot of $V(\lambda)$, $M(\lambda)$, $V_0(\lambda)$, $||\beta-\hat{\beta}_\lambda||^2$ and $T(\lambda)$ for Replicate 2 of the $\sigma^2=10^{-6}$ case. The $T(\lambda)$ and $V(\lambda)$ curves tend to follow each other, and this is fairly typical.

We make a few comments on these very preliminary results. The superiority of the GCV estimate of λ is apparent, at least in this example, compared to the other methods listed. The Gauss Markov estimate of λ was not included because it is uniformly worse than all the other estimates in this very ill-conditioned problem. A first run comparing GCV and the RI estimate of (2.13) as this goes to press indicated that the two methods give similar results for n-p=11.

We feel that GCV is generally going to be superior to PRESS, unless XX^T is circulant, in which case they will be the same. It is reasonable to think MLE will look good when β acts "as though" it came from the prior. This really means either p is small or $\{\beta_i^2\}_{i=1}^p$ have about the same spread or less as independent $\chi^2(1)$ random variables times some constant. If p is large it is possible to construct examples where MLE will look bad compared to GCV. In these examples, $\beta_i^2\to 0$ in a manner uncharacteristic of the prior and the MLE estimate is too small. We think the Hoerl-Kennard estimate will look worse compared to GCV the more ill-conditioned the X matrix is, since the Hoerl-Kennard estimate divides by the

TABLE I

Observed Inefficiencies in Sixteen Monte Carlo Runs

	Replication 1		Replication 2		Replication 3		Replication 4	
	I_D	I_R	I_D	I_R	I_D	I_R	I_D	I_R
				$\sigma^2=10^{-8}$, S/N ≃ 4200				
GCV	4.43*	1.06*	1.65*	1.03*	16.71*	1.10*	1.02*	1.01*
HKB	2.18E3	1.31	1.14E2	1.23	3.99E4	1.89	1.03E3	1.52
RIDGM	2.14E3	1.31	1.13E2	1.22	3.95E4	1.88	1.03E3	1.52
MLE	1.67E3	1.31	1.45E2	1.23	2.00E3	1.53	9.12E3	1.51
PRESS	2.31E3	4.8E4	6.31E2	8.6E4	3.84E3	2.1E5	2.87E3	1.2E5
Min Sol'n	1.00	1.02	1.00	1.54	1.00	2.27	1.00	1.00
Min Data	1.20	1.00	2.89	1.00	5.97	1.00	1.00	1.00
				$\sigma^2=10^{-6}$, S/N ≃ 420				
GCV	1.92*	1.05	1.32*	1.00	1.51E2	1.26*	2.20*	1.02*
HKB	1.21E3	1.35	6.24E3	1.58	3.60E5	2.25	5.07E3	1.40
RIDGM	1.21E3	1.35	6.23E3	1.58	3.60E5	2.25	5.05E3	1.40
MLE	1.99E2	1.19	1.70E2	1.45	1.76E2	1.29	1.49E2	1.32
PRESS	5.80	1.01	2.41E2	1.39E4	36.37*	2.43E3	67.00	6.07E2
Min Sol'n	1.00	1.38	1.00	1.02	1.00	1.20	1.00	1.03
Min Data	3.56	1.00	1.28	1.00	7.85	1.00	41.29	1.00
				$\sigma^2=10^{-4}$,S/N ≃ 42				
GCV	1.27*	1.07*	1.50*	2.58*	1.00*	1.11*	1.00*	1.03*
HKB	4.34E3	1.75	1.34E4	4.30	2.03E4	1.98	2.82E6	2.66
RIDGM	4.33E3	1.75	1.31E4	4.30	2.02E4	1.97	2.81E6	2.66
MLE	1.56	1.20	12.16	3.43	1.90	1.49	2.97	1.07
PRESS	3.53	1.57	2.03	3.43	8.66	2.63	2.90	24.34
Min Sol'n	1.00	1.21	1.00	2.05	1.00	1.11	1.00	1.03
Min Data	3.26	1.00	1.16	1.00	2.39	1.00	1.16	1.00
				$\sigma^2=10^{-2}$,S/N ≃ 4.2				
GCV	1.40	2.47	2.01*	1.60*	1.59*	1.01	3.12	1.72
HKB	7.05E7	13.20	3.32E5	3.35	4.58E4	2.20	5.39E5	31.0
RIDGM	7.05E7	13.20	3.31E5	3.35	4.56E4	2.20	5.26E5	31.0
MLE	2.13	3.56	3.81	1.87	2.00	1.00*	2.88	1.69*
PRESS	1.04*	1.01*	2.02	2.68	1.00	1.22	2.16*	2.15
Min Sol'n	1.00	1.31	1.00	1.01	1.00	1.25	1.00	1.98
Min Data	1.02	1.00	1.00	1.00	2.66	1.00	1.21	1.00

"* indicates lowest inefficiency among the estimates

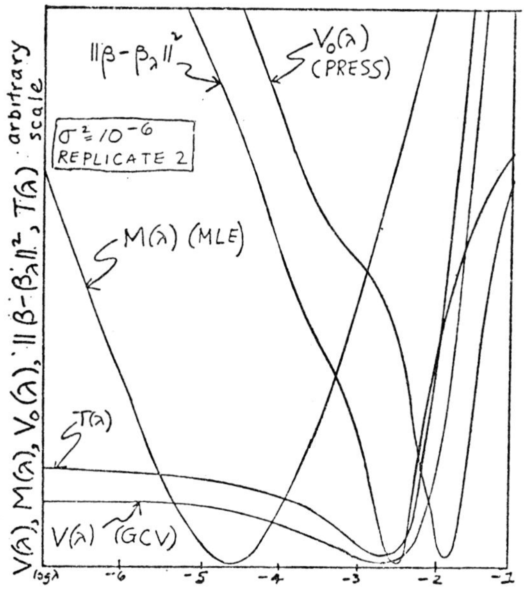

Fig. 1.

eigenvalues λ_i as an intermediate step in the computation of $\hat{\beta}_0$. This division by λ_i introduces a large variance for whatever quantity is being computed. RI should look good if n-p is large.

3. SPLINE SMOOTHING

We return to the model (1.1), $y(t)=f(t)+\varepsilon(t)$, and the estimate $f_{n,\lambda}$ for f. Let $f_{n,\lambda}$ be the solution to the minimization problem (1.2) of Section 1. As in the ridge regression case, there is a prior on the "signal", here f, for which $f_{n,\lambda}(t)$, is the posterior mean. It is given as follows:

Let f(t) be a stochastic process modeled by

$$f(t) = \sum_{j=0}^{m-1} \theta_j t^j + \sqrt{a} \int_0^t \frac{(t-u)_+^{m-1}}{(m-1)!} dW(u) \qquad (3.1)$$

where W(u) is the Weiner process, a is a constant, and $\{\theta_j\}$ are Gaussian random variables with a "flat" prior, i.e. infinite variance. It can be shown (by using the results in Kimeldorf and Wahba [18]) that $f_{n,\lambda}(t)$ is the posterior mean of f(t) given $y(t_i)=y_i$, with $\lambda=\sigma^2/na$.

The vector $\bar{f}_{n,\lambda}=(f_{n,\lambda}(t_1),f_{n,\lambda}(t_2),\ldots,f_{n,\lambda}(t_n))^T$ depends linearly on the data $y=(y_1,y_2,\ldots,y_n)$, and we define the matrix $A(\lambda)$ by

$$\bar{f}_{n,\lambda} = A(\lambda)y \;. \tag{3.2}$$

We omit giving an explicit formula for $A(\lambda)$, it can be obtained for computational purposes in the m=2 case from Reinsch [20]; in general from Kimeldorf and Wahba, [18], and is explicitly given in Craven and Wahba [6].

The GCV estimate of λ is defined as the minimizer of

$$V(\lambda) = ||(I-A(\lambda))y||^2/[\text{Trace }(I-A(\lambda))]^2 \tag{3.3}$$

(Compare (2.9)). The MLE estimate of λ with the prior (3.1) can be shown to be the minimizer of

$$M(\lambda) = (y^T(I-A(\lambda))y)/[\text{Det}(I-A(\lambda))]^{1/n} \tag{3.4}$$

This MLE estimate for λ is essentially that proposed by Anderssen and Bloomfield in their papers [2] and [3], and generalizes their estimate to the non-periodic non-equally spaced data case.

In the equally spaced data case ($t_i=i/n$) for large n, GCV and "ordinary" cross validation are very similar. (In analogy to the ridge regression case, the matrix that plays the role of U is very close to the matrix that plays the role of W). Wahba and Wold [29]reported on Monte Carlo results using "ordinary" cross-validation in connection with spline smoothing. Reinsch's program [20] was used to repeatedly calculate $f_{n,\lambda}^{(k)}$, the solution to the minimization problem of (1.2) with the kth data point omitted. Favorable results were reported with the method. Cheap computational procedures for calculating $V(\lambda)$ from (3.3) are under development and will be reported in [6]. Data sets of 50 can be handled for a few dollars. The following theorem is proved in Craven and Wahba [6].

Theorem 3. Let $f \in H_m = \{f: f^{(\nu)}$ abs. cont., $\nu=),1,\ldots,m-1$, $f^{(m)} \in L_2$ and suppose $\{t_i\}_{i=1}^n = \{t_{in}\}_{i=1}^n$ where $\int_0^{t_{in}} w(u)du = i/n$, $i=1,2,\ldots,n$, $n=1,2,\ldots$, and w is a strictly positive continuous weight function. Let

$$T(\lambda) = \frac{1}{n}\sum_{i=1}^{n}(f_{n,\lambda}(t_{in})-f(t_{in}))^2$$

and let λ^* be the minimizer of $ET(\lambda)$ and $\hat{\lambda}$ be the minimizer of $EV(\lambda)$.

i) If f is a polynomial of degree m-1 or less then

$$\tilde{\lambda} = \lambda^* = \infty.$$

ii) If f is not a polynomial of degree m-1 or less then

$$\tilde{\lambda} = \lambda^*(1+o(1)) \tag{3.5}$$

where $o(1) \to 0$ as $n\to\infty$.

Letting E be expectation with respect to the prior of (3.1), the following can be proved.

Theorem 4. The functions $E_P EV(\lambda)$, $E_P ET(\lambda)$ and $E_P EM(\lambda)$ have the common minimizer $\lambda=\sigma^2/na$.

If f is a very smooth function in H_m, however it can be shown that (3.5) is false for the MLE estimate. It can be shown that as $n\to\infty$ the MLE estimate will tend to 0 more rapidly than the GCV estimate. This is not surprising, since it can be shown that sample functions from the prior (3.1) are, with probability 1, not in H_m.

The smoothing spline $f_{n,\hat{\lambda}}$ may be used to estimate $f^{(\nu)}(t)$ for any $\nu\leq m-1$. The estimate is $f^{(\nu)}_{n,\hat{\lambda}}(t)$. Pleasing numerical results for $\nu=1$, $m=2$ can be found in Wahba and Wold [29]. This problem was in fact the motivation for Anderssen and Bloomfield [23] who also had nice numerical results.

If H_m is constrained to be a space of periodic functions, the computations are vastly simplified. See Section 4.

4. DENSITY ESTIMATION

Every reasonable non-parametric density estimate has (implicitly or explicitly) a parameter or parameters which control the tradeoff between the square bias and the variance. For example, in an estimate of kernel type, the density estimate f is given by

$$\hat{f}(t) = \frac{1}{nh} \sum_{j=1}^{n} K\left(\frac{t-X_j}{h}\right)$$

where $X_1, X_2, \ldots, X_n$ are iid observations from some density f, and $K(\tau)$ is a "hill function" integrating to 1. See Parzen [19]. The parameter h controls the tradeoff between the square bias and the variance and the h which minimizes mean square error depends on the unknown f as well as the sample size n. To our knowledge, no completely automatic, practical method for choosing the smoothing parameter to minimize mean square error has been reported (see, however, Good and Gaskins [15], Woodroofe [31]). However, the method of GCV can be used to do this in conjunction with a smoothing spline density estimate. Detailed results will appear in [28]. We summarize the main idea here, for the purpose of illustrating the wide applicability of the GCV method.

We discuss the method as it applies only to densities that are in H_m of Section 3 and furthermore satisfy the periodic boundary conditions

$$f^{(\nu)}(1)-f^{(\nu)}(0)=0, \quad \nu=0,1,\ldots,m-1 .$$

In practice for small to medium sample sizes m=2 can be chosen, or the m which minimizes $\min_\lambda V(\lambda)$ can be used. The value m=2 corresponds to visual smoothness. We restrict ourselves to the periodic case for *practical* rather than *theoretical* reasons. The computations are easy to program and are "dirt cheap" in the periodic case as can be seen below. To obtain the method we first discuss periodic smoothing splines. It can be shown (See Golomb [12], Kimeldorf and Wahba [18], for the necessary background), that the solution $f_{n,\lambda}$ to the minimization problem:

Find f in H_m satisfying

$$\int_0^1 f(u)du=1, \quad f^{(\nu)}(1)-f^{(\nu)}(0)=0, \quad \nu=0,1,\ldots,m-1$$

and minimizing

$$\frac{1}{n}\sum_{j=1}^{n}(f(\tfrac{j}{n})-y_j)^2 + \lambda\int_0^1 (f^{(m)}(u))^2\,du\ , \quad y_j = y(\tfrac{j}{n})$$

satisfies

$$f_{n,\lambda}(\tfrac{j}{n}) = (1 + \sum_{\substack{\nu=-n/2\\ \nu\neq 0}}^{n/2} \frac{\hat{f}_\nu}{(1+\lambda(2\pi\nu)^{2m})} e^{2\pi i\nu j/n})\ (1+o(1))$$

where

$$\hat{f}_\nu = \frac{1}{n}\sum_{j=1}^{n} y_j\, e^{2\pi i\nu j/n} \tag{4.1}$$

and $o(1)\to 0$ (rapidly) as $n\to\infty$. Thus, the periodic smoothing spline with equally spaced data points can be viewed as the result of passing the data through a low pass filter. *The parameter λ controls the half power point of the filter*, by the relationship that the frequency ν_0 at which f_ν is damped by a factor of 2 satisfies $\lambda(2\pi\nu_0)^{2m}=1$, or $\nu_0=(2\pi\lambda^{1/2m})^{-1}$. The value of λ which minimizes mean square error is estimated by GCV, as the minimizer of

$$V(\lambda) = \frac{\sum_{\nu=1}^{n}\left(\frac{\lambda}{(2\pi\nu)^{-2m}+\lambda}\right)^2 \hat{f}_\nu^2}{\left(\sum_{\nu=1}^{n}\frac{\lambda}{(2\pi\nu)^{-2m}+\lambda}\right)^2}$$

(Compare (2.11)). The eigenvalues of the (formal) operator $[(\frac{d}{dt})^m]^{-1}$ are $(2\pi\nu)^{-m}$ and play the role of the singular values of X. See Wahba [28] for details. In the estimation of a periodic density, (with m=2), the estimate is

$$f_{n,\lambda}(t) = 1 + \sum_{\substack{\nu=-n/2\\ \nu\neq 0}}^{n/2} \frac{\hat{f}_\nu}{(1+\lambda(2\pi\nu)^4)} e^{2\pi i\nu t}$$

where the $\hat{f}_\nu$ of (4.1) are replaced by the *sample fourier coefficients*,

$$\hat{f}_\nu = \frac{1}{n} \sum_{j=1}^{n} e^{2\pi i \nu X_j} .$$

The estimate is related to that of Cogburn and Davis [7] for estimating spectral densities, but the method for choosing λ is new. The GCV method can be used in the Cogburn-Davis spectral density estimate to choose λ, see[30].

The following Theorem is proved in [28].

Theorem 5. Let f be a periodic density in H_m. Let $T(\lambda) = \frac{1}{n}\sum_{j=1}^{n} (f_{n,\lambda}(\frac{j}{n}) - f(\frac{j}{n}))^2$ and let $\hat{\lambda}$ and λ^* be the minimizers of $EV(\lambda)$ and $ET(\lambda)$ respectively. Then

$$\tilde{\lambda} = \lambda^*(1+o(1))$$

where $o(1)\to 0$ as $n\to\infty$.

It can be shown that if, further, $f^{(\nu)}$ abs. cont., $\nu=m, m+1,\dots,2m-1$, $f^{(m)} \in L_2$, and $f^{(\nu)}(1)-f^{(\nu)}(0) = 0$, $\nu = m,m+1,\dots,2m-1$, then

$$ET(\tilde{\lambda}) = 0(n^{4m/(4m+1)}) ,$$

and hence that this density estimate shares the convergence properties of most of the well known estimates. See [25] for a discussion of the best mean square error at a point convergence rates obtainable for densities in H_m.

Figure 2 provides the results of a demonstration run of this method. One hundred and seventy-four observations from a density which is a mixture of two betas were generated by Monte Carlo methods. The density f is

$$f \sim \frac{6}{10}\,\beta(12,7) + \frac{4}{10}\,\beta(3,11) .$$

Figure 2a gives a plot of the true density and a histogram of the 174 Monte Carlo samples. The "bin size", which is the "smoothing" parameter in the histogram estimate, was chosen by eye. Figure 2b gives a plot of $T(\lambda)$ and $V(\lambda)$. It can be seen that the minimizers are quite close. The minimizer of $V(\lambda)$ is $\hat{\lambda}$ and the relative inefficiency $T(\hat{\lambda})/(\min_{\lambda} T(\lambda))$, was 1.04. Figure 2c gives a plot of $f_{n,0}(t)$ and the density estimate $f_{n,\hat{\lambda}}(t)$. $f_{n,\hat{\lambda}}(t)$ is in fact the low-pass filtered version of $f_{n,0}(t)$. Figure 2d gives a plot of f and $f_{n,\hat{\lambda}}$. We remind the reader that the computation of $f_{n,\hat{\lambda}}$ is "completely automatic", with no human intervention required to choose the optimal smoothing.

5. THE APPROXIMATE SOLUTION OF LINEAR OPERATOR EQUATIONS WHEN THE DATA ARE NOISY

Let

$$y(t) = \int_0^1 K(t,s)f(s)ds+\varepsilon(t) \qquad (5.1)$$

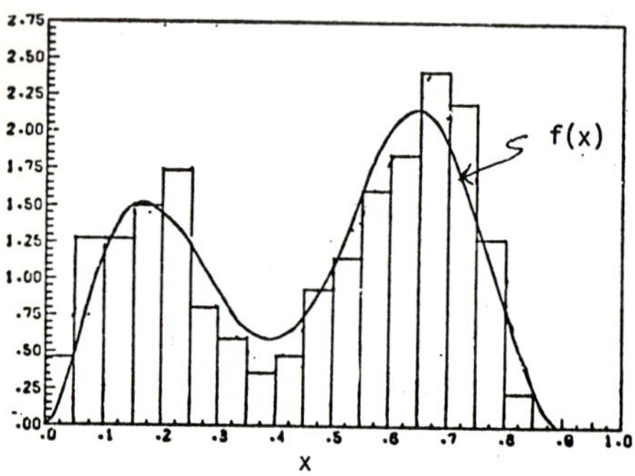

Figure 2a

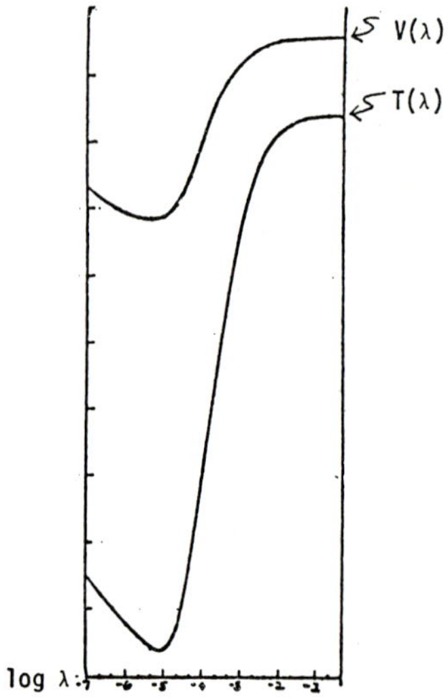

Figure 2b

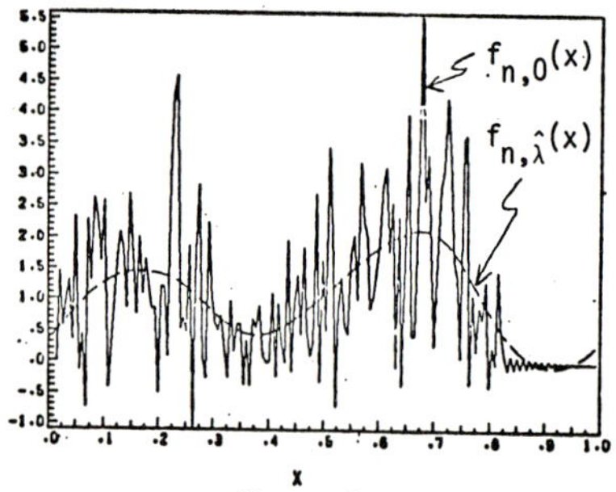

Figure 2c

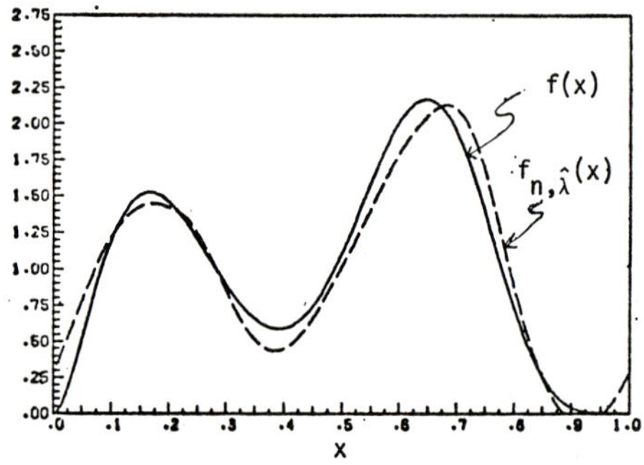

Figure 2d

where K is a continuous kernel and ε is a noise satisfying $E\varepsilon(t)=0$, $E\varepsilon(s)\varepsilon(t)=\sigma^2$, $s=t$, $=0$, otherwise. y is observed for $t=t_1,t_2,\ldots,t_n$, and it is desired to recover f. This problem is a common one in many physical sciences. It is assumed that f is in some Hilbert space H, of smooth functions. The approximate solution $f_{n,\lambda}$ is taken as the solution to the minimization problem: Find $f \varepsilon H$ to minimize

$$\frac{1}{n}\sum_{j=1}^{n}\left[\int_0^1 K(t_j,s)f(s)ds-y_j\right]^2+\lambda||f||_H^2 , \quad y_j = y(t_j). \tag{5.2}$$

This method for solving (5.1) approximately is known in the approximation theory literature as Tichonov regularization, and is a Hilbert space version of ridge regression. See [26].

The GCV estimate of λ is the minimizer of $V(\lambda)$ given by (3.3) where $A(\lambda)$ is defined by (3.2), with $\overline{f}_{n,\lambda}$ replaced by $\overline{g}_{n,\lambda}=(\int_0^1 K(t_1,s)f_{n,\lambda}(s)ds,\ldots,\int_0^1 K(t_n,s).$ $f_{n,\lambda}(s)ds)^T$, where $f_{n,\lambda}$ is the solution to the minimization problem of (5.2). Analogous theorems to those of Sections 2 and 3 are found in [26].

There is nothing special about the index set $t \varepsilon [0,1]$ other than that it is bounded. The results of Section 3 and above can be shown to hold for t in an interval in Euclidean d-space, that is, for the surface smoothing problem. See [27].

NOTES ADDED IN PROOF

The risk improvement estimator of (2.13) is equivalent to C_L, suggested earlier by Mallows (Mallows, C. L. (1973). Some comments on C_p. *Technometrics*, **15**, 661-675.)

The method of Generalized Cross-Validation has numerous other applications including subset selection for regression (instead of ridge regression). In the subset selection case there is a simple relationship between GCV and Akaike's Information Criteria (AIC), discussed elsewhere in this volume. In subset selection, instead of forming a regularized estimate β_λ of β, one selects a subset $\beta_{j_1}, \beta_{j_2},\ldots, \beta_{j_k}$ of the β's to be non-zero, and the remaining β's will be set equal to zero. The non-zero subset will be estimated by Gauss-Markov. Let $X^{(\lambda)}$ be the $n \times k$ design matrix corresponding to the λ-th subset of β_j's, $\lambda=1,2,\ldots,2^p$. Then $A(\lambda) = X^{(\lambda)}(X^{(\lambda)T}X^{(\lambda)})^{-1}X^{(\lambda)}$, and, if the λ^{-th} subset has k members, then

$$n^2\, V(\lambda) = \text{RSS}\,(1-\tfrac{k}{n})^{-2},$$

since $||(I-A(\lambda))y||^2 = \text{RSS}$ and $\text{Tr}\, A(\lambda) = k$. In this context, it can be shown that

$$n\, e^{AIC(\lambda)/n} = RSS e^{-2k/n},$$

so that if p is fixed and n sufficiently large AIC(λ) and V(λ) can be expected to have the same minimizer. We are grateful to E. Parzen for pointing this out.

Gunst and Mason (Generalized mean square error properties of regression estimators, *Commun. Statist. - Theor. Meth.*, A5(15), 1501-1508 (1976)) have posed the problem of deciding, from the data, between (the best) ridge estimate and (the best) principal components estimate. (Parentheses mine). It is clear that this can be done using GCV if the criteria is to minimize the mean square prediction error, T. The minimum V is obtained for the ridge estimate and compared to the minimum V for principal components. The V for principal components is analogous to the V for subset selection above, $n^2 V = RSS\ (1 - \frac{k}{n})^{-2}$, where k is the number of positive principal components.

Details will appear in [13].

REFERENCES

[1] ALLEN, D. M. (1974). The relationship between variable selection and data augmentation and a method for prediction. *Technometrics*, **16**, 1, 125-127.

[2] ANDERSSEN, R. S. and BLOOMFIELD, P. (1974). A time series approach to numerical differentiation. *Technometrics*, **16**, 1, 69-75.

[3] ANDERSSEN, B. and BLOOMFIELD, P. (1974). Numerical differentiation procedures for non-exact data. *Numer. Math.*, **22**, 157-182.

[4] BELLMAN, R. (1960). *Introduction to Matrix Analysis.* McGraw Hill, New York.

[5] BLACKMAN, R. B. and TUKEY, J. W. (1958). *The Measurement of Power Spectra.* Dover, New York.

[6] CRAVEN, P. and WAHBA, G. (1976). Smoothing noisy data with spline functions; Estimating the correct degree of smoothing by the method of generalized cross-validation, in preparation.

[7] COGBURN, R. and DAVIS, H. T. (1974). Periodic splines and spectra estimation. *Ann. Statist.*, **2**, 1108-1126.

[8] DEMPSTER, A. P. (1973). Alternatives to least squares in multiple regression, In *Multivariate Statistical Inference*. Proceedings of the Research Seminar at Dalhousie University, Halifax, March 23-25, 1972. (D. G. Kabe and R. P. Gupta, eds.)

[9] EFRON, B. and MORRIS, C. (1975). Data analysis using Stein's estimator and its generalizations. *J. Amer. Statist. Assoc.*, **70**, 350, 311-319.

[10] GEISSER, S., (1975). The predictive sample reuse method with applications. *J. Amer. Statist. Assoc.*, **70**, 350, 320-328.

[11] GOLDSTEIN, M. and SMITH, A. F. M. (1974). Ridge-type estimators for regression analysis. *J. Roy. Statist. Soc. Ser. B*, **36**, 2, 284-291.

[12] GOLOMB, M. (1968). Approximation by periodic spline interpolants on uniform meshes. *J. Approximation Theory*, **1**, 26-65.

[13] GOLUB, G., HEATH, M. and WAHBA, G. (1975). Cross-validation and optimum ridge regression Abstract in Abstracts of Papers To Be Presented at the SIAM-SIGNUM 1975 Fall Meeting, December 3, 4, 5, 1975, San Francisco. University of Wisconsin-Madison, Dept. of Statistics Tech. Rept. #491 (1977).

[14] GOLUB, G. H. and REINSCH, C. (1970). Singular value decomposition and least squares solutions. *Numer. Math.*, **14**, 403-420.

[15] GOOD, I. J. and GASKINS, R. A. (1971). Non-parametric roughness penalties for probability densities. *Biometrika*, **58**, 255-277.

[16] HOERL, A. E., KENNARD, R. W. and BALDWIN, K. F. (1975). Ridge regression: Some simulations. *Comm. Statist.*, **4**, 2, 105-123.

[17] HUDSON, H. M. (1974). Empirical Bayes estimation. Technical Report #58, Stanford University Department of Statistics, Stanford, California.

[18] KIMELDORF, GOERGE and WAHBA, GRACE. (1971). Some results on Tchebycheffian spline functions. *J. Math. Anal. Appl.*, **33**, 82-95.

[19] PARZEN, E. (1962). On the estimation of a probability density function and mode. *Ann. Math. Statist.*, **33**, 1065-1076.

[20] REINSCH, C. H. (1967). Smoothing by spline functions. *Numer. Math.*, **10**, 177-183.

[21] REINSCH, C. H. (1971). Smoothing by spline functions, II. *Numer. Math.*, **16**, 451-454.

[22] SCHOENBERG, I. J. (1964). Spline functions and the problem of graduation. *Proc. Nat. Acad. Sci. (USA)*, **52**, 947-950.

[23] STONE, M. (1974). Cross-validatory choice and assessment of statistical prediction. *J. Roy. Statist. Soc. Ser. B* , **36**, 2, 111-147.

[24] WAHBA, G. (1968). On the distribution of some statistics useful in the analysis of jointly stationary time series. *Ann. Math. Statist.*, **39**, 6, 1849-1862.

[25] WAHBA, GRACE. (1975). Optimal convergence properties of variable knot kernel, and orthogonal series methods for density estimation. *Ann. Statist.*, **3**, 15-29.

[26] WAHBA, G. (1975). Practical approximate solutions to linear operator equations when the data are noisy. University of Wisconsin-Madison, Department of Statistics, Technical Report #430. To appear, *SIAM J. NUM. ANAL.*

[27] WAHBA, G. (1976). A canonical form for the problem of estimating smooth surfaces. University of Wisconsin-Madison, Department of Statistics, Technical Report #420.

[28] WAHBA, G. (1976). Optimally smoothed density estimates for classification. To appear, *Proceedings of the Advanced Seminar on Classification and Clustering* held May 3-5, 1976 at Madison, Wisconsin, (John Van Ryzin, ed.)

[29] WAHBA, G. and WOLD, S. (1975). A completely automatic Franch curve: Fitting spline functions by cross-validation. *Comm. Statist.*, **4**, 1-7.

[30] WAHBA, G. and WOLD, S. (1975). Periodic splines for spectral density estimation: The use of cross-validation for determining the degree of smoothing. *Comm. Statist.*, **4**, 2, 125-121.

[31] WOODROOFE, M. (1970). On choosing a delta-sequence. *Ann. Math. Statist.*, **41**, 166-171.

P.R. Krishnaiah, ed., APPLICATIONS OF STATISTICS

STATISTICAL PROBLEMS IN WATER RESOURCES

VUJICA YEVJEVICH

Department of Civil Engineering
Colorado State University
Fort Collins, Colorado

Statistical problems in water resources relate to both natural and combined natural and man-affected processes and systems, including water demand. In general, random processes are continuous, four-dimensional, space-time processes.

Line, area and space random processes are mainly trend-stochastic. Annual precipitation series are approximately time independent. Continuous time series of water quantity and quality are among the most complex geophysical processes. Basic statistical problems in water resources are:

(1) Development of appropriate statistical inference techniques for complex processes and their models with relatively large numbers of parameters; (2) Proper use of the experimental statistical method by a generation of samples, for purposes of estimating sampling distributions, extraction of information, support for expressions developed on the basis of physical hypotheses, stochastic optimization in decision making, determination of risks and uncertainties, and similar applied problems; (3) Detection of changes by man and by sudden or evolutive disruptions in nature of water resources stochastic processes, with proper techniques for their removal and extrapolation; (4) Design of experiments in case of complex or intermittent (truncated) processes in nature; (5) Search for exact distributions of special characteristics of stochastic water resources processes, such as the range of partial sums, runs, extremes, and similar variables which are important in planning, design, operation, and maintenance of water resources systems; and (6) Adjustments for multivariate analyses (such as regression, principal components, factor analysis, canonical correlation) in cases of large numbers of involved, random water resources variables.

1. CHARACTER OF STOCHASTIC PROCESSES

Statistical problems in water resources relate to both natural and combined natural and man-affected processes and systems of various types of random phenomena, including the random variables of water demand. Most random phenomena involve the general, continuous, four-dimensional, space-time random processes, as it is the case with most of the other geophysical time-space processes. To estimate characteristics of these processes, the space variations are sampled by a set of points and time variations either continuously recorded or sampled at regular or irregular time intervals. Two types of statistical data are available for analysis of water resources processes, data of space-frozen temporal stochastic processes, and data of time-frozen line, area or space stochastic processes. Accumulation of data is limited by both the time and funds available in the first case, and by economical factors in the second case.

Line, area and space random processes occur mainly in the form of trend-stochastic processes. Their basic parameters may be considered of changing only by trend functions. Time processes are either periodic-stochastic in case of natural man-unaffected processes, or they are trend-periodic-stochastic in case of water demand and water consumption processes. The closest process to a Gaussian time independent process are annual precipitation series, as an extreme simple case. Continuous time processes of water quality variables belong to the most complex geophysical time processes in nature.

No other geophysical science processes have been subject to such significant man-induced changes as were the hydrology and water resources related processes in nature. For example, the entire large rivers have been transformed into the cascades of man-made reservoirs. Changes in nature by man, or by various sudden or evolutive disruptions present the most difficult statistical problems to water resources specialists. Such problems are: the identification of changes, the removal of these changes in case of inferences about the population(s) from which the future samples will be drawn, and the extrapolations into the future of the expected changes.

2. SOURCES OF STOCHASTICITY

The atmosphere is the major source of stochasticity in natural water resources processes. The other environments contribute to or change the level of stochasticity in time periodic-stochastic processes. They usually attenuate the process by eventually changing the variance and by increasing the time dependence. To support these statements, very simple graphs are presented for the evidence. Figure 1 demonstrates what would be the periodic input of total daily solar radiation energy over the 365 days for three experimental river basins [1]. This is equivalent of assuming that there would not be an atmosphere, with the latitude, orientation and slope of river basins accounting for differences between the three curves. For each solar day the variance of daily radiation input is zero. To contrast this zero variance, Figure 2 shows the fitted periodic functions σ_τ of the daily standard deviation (a number of harmonics is fitted to the 365 estimates of daily standard deviation), while Figure 3 gives the fitted periodic functions for the daily means for the 14 radiation measuring stations in the West of the United States [2]. The average coefficient of variations is about 0.25. The opacity of the air for the incoming solar radiation, plus the diffused radiation, account for

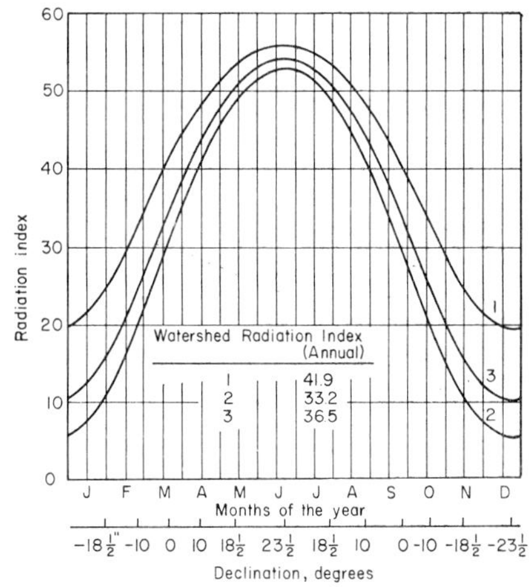

Fig. 1. Radiation indices for three experimental catchments, giving variation with time, Andrews Experimental Forest, Oregon, USA. (after Richard Lee, 1963).

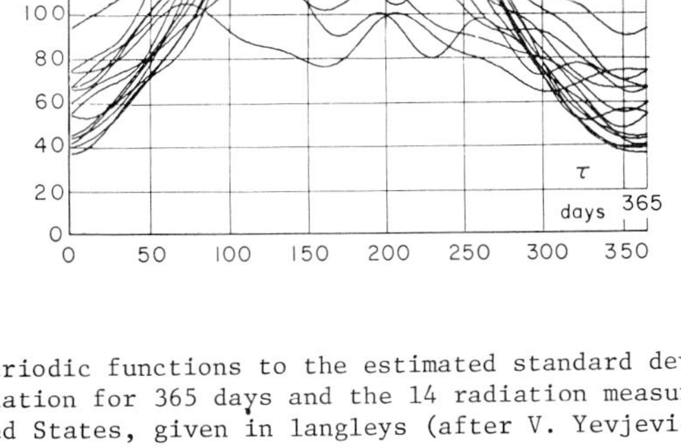

Fig. 2. Fitted periodic functions to the estimated standard deviations of daily incoming solar radiation for 365 days and the 14 radiation measuring stations in the West of the United States, given in langleys (after V. Yevjevich, 1971).

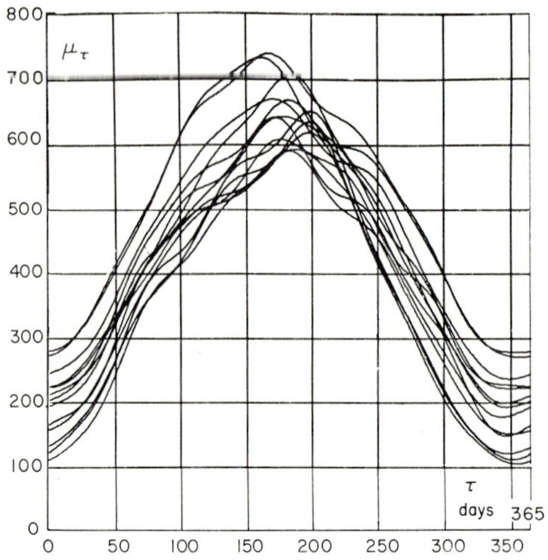

Fig. 3. Fitted periodic functions to the estimated means of daily incoming solar radiation for 365 days and 14 radiation measuring stations in the West of the United States, given in langleys (after V. Yevjevich, 1971).

the variance of each of the 365 days not being zero at the ground. Definitely, the atmosphere, with its opacity for radiation waves (incoming and outgoing), acts as a periodic-stochastic filter, with the periodicity in opacity the result of the established, average climatic patterns at each point of the Earth's surface.

When the radiation variables are standardized by removing the periodicity in the mean and the standard deviation, the stochastic part becomes relatively close to a time independent stochastic component (TISC-variable), with short-lag but significant autocorrelation due to the opacity, and with somewhat longer but small autocorrelation due to the diffused radiation.

Figure 4 shows the correlogram of the mean monthly standardized temperature series, with the removed periodicities of the monthly mean and monthly standard deviation, for the three areas of the Pacific Ocean to the West of the United States [2]. The high, long-range autocorrelation (up to 20-35 months) is due to the heat stored in the upper layers of oceans and seas. It represents a highly attenuated effect of oceans and seas on the stochastic component of incoming solar radiation. Figure 5, however, shows that the standardized monthly precipitation series, by $\varepsilon_i = (x_i - \mu_\tau)/\sigma_\tau$ (with μ_τ = the periodic mean and σ_τ = the periodic standard deviation), and averaged for the seven areas of Western Coast of the United States, are nearly independent stochastic components, regardless of the ocean stored heat as the major factor in the evaporation process. High stochasticity of atmosphere for receiving and transporting the evaporated water is the major reason for the lack of time dependence in the standardized series of monthly precipitation. Again, the atmosphere acts as the stochasticity producing causal factor.

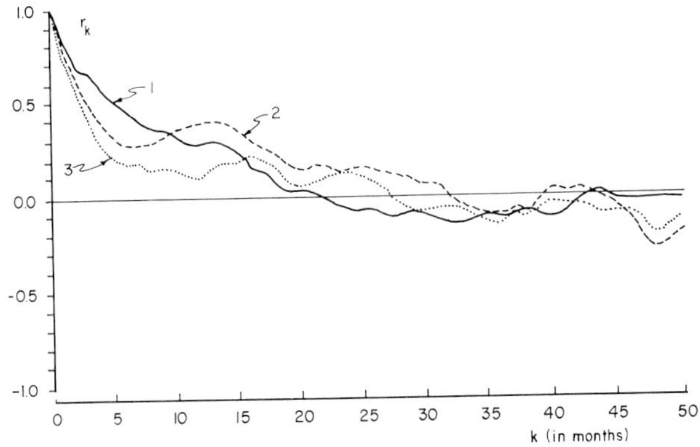

Fig. 4. Correlograms of the standardized series of the monthly ocean surface temperature for areas of the Pacific Ocean West of the United States (after V. Yevjevich, 1971).

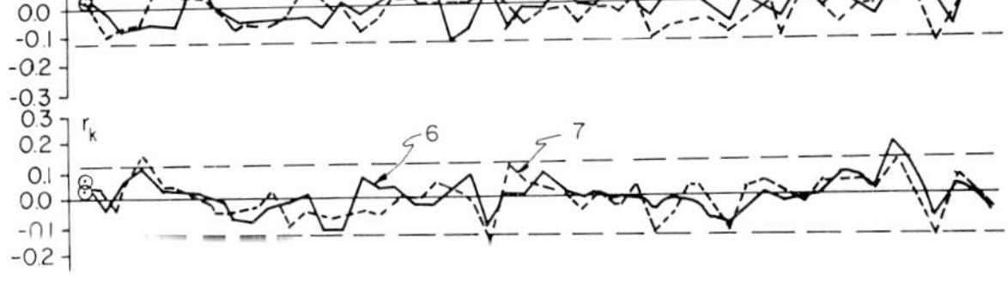

Fig. 5. Correlograms of the standardized series monthly precipitation of seven areas of Western Coast of the United States (after V. Yevjevich, 1971).

3. COMPLEXITIES OF SHORT-INTERVAL RUNOFF TIME SERIES

The basic finding for short-interval hydrologic time series at present is that these series become more complex the more the water crosses various Earth's environments and interacts with the other random or periodic-stochastic processes.

For the case of short-interval time series of precipitation, many river basin factors intervene in disposing of rainfall and in producing the runoff. Such factors are: direct precipitation on streams and lakes, interception by plants, surface water retention, interflow (short-depth, short-length underground flow), infiltration, groundwater flow, evaporation, snow and ice accumulation and melt, and similar factors. Besides these highly varying factors, the effects of inconsistency (random and systematic errors) and of nonhomogeneity in data make the daily runoff series very complex time processes. Figures 6-10 show the estimates of five parameters for each day of the year and for 40 years of data: mean, standard deviation, first autocorrelation coefficient, and skewness and kurtosis coefficients, of daily flows of the Boise River, near Twin Springs, Idaho, for the period 1931 through 1970. All five parameters are periodic; however, some have small and some have large amplitudes of basic harmonics which best fit their periodic variations. While the current analyses of daily hydrologic time series consider the existence of periodicities in the mean and the standard deviation, and sometimes in the autocorrelation coefficients, the treatment of periodicities in higher-order parameters (especially in skewness and kurtosis coefficient) still requires developments of proper approaches.

Besides the effect of periodicities, the linear dependence models of stationary stochastic components (such as the autoregressive, moving average, ARMA and ARIMA models) have been developed and are mainly valid for normal processes or processes approximately normalized by the appropriate transformations. These transformations, therefore, imply a use of non-linear dependence models of the type of the non-linear functions used in transformations. The applications of the above linear models for symmetrically distributed time independent stochastic components (the TISC-processes) produces distortions in mathematical description and in generation of new samples of the complex hydrologic runoff processes.

The deterministic-periodic influx of solar radiation to Earth's environments is subject of the periodic-stochastic filtering by the three environments: atmosphere, oceans (seas) and river basins (continents), in producing the precipitation and river flow time series. At the first glance, one should expect that a clear periodic solar radiation input to the Earth would only produce the periodicity in the means of precipitation and runoff time processes. The fact that the filters operate differently during various seasons of the year, is responsible that nearly all the parameters of these series result as the more or less periodic parameters.

The mathematical abstraction of the physical runoff process most often stops at taking into account the periodicity in the second-order parameters. Furthermore, approximations are made in transforming the asymmetric distributions into the normal processes and in mathematical modeling of the real process of time dependent second-order stationary stochastic component. Therefore, a discrepancy always

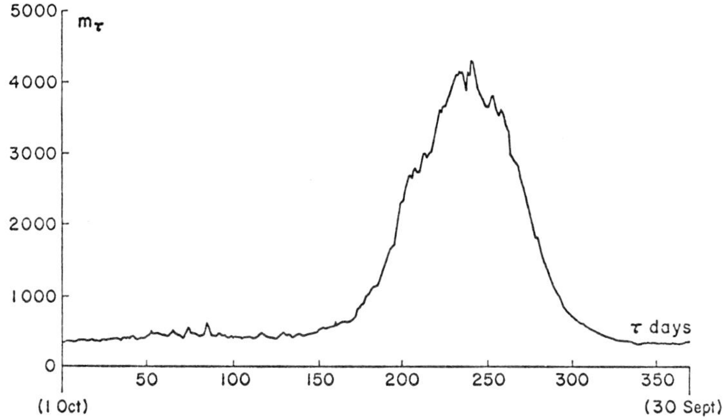

Fig. 6. The sequence of 365 means of daily flows (estimated from 40 values of the 40 year long sample) of the boise River near Twin Springs, Idaho.

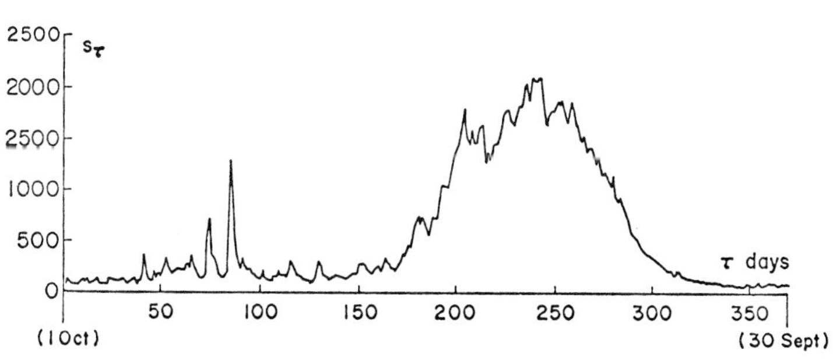

Fig. 7. The sequence of 365 standard deviations of daily flows of the Boise River, Idaho.

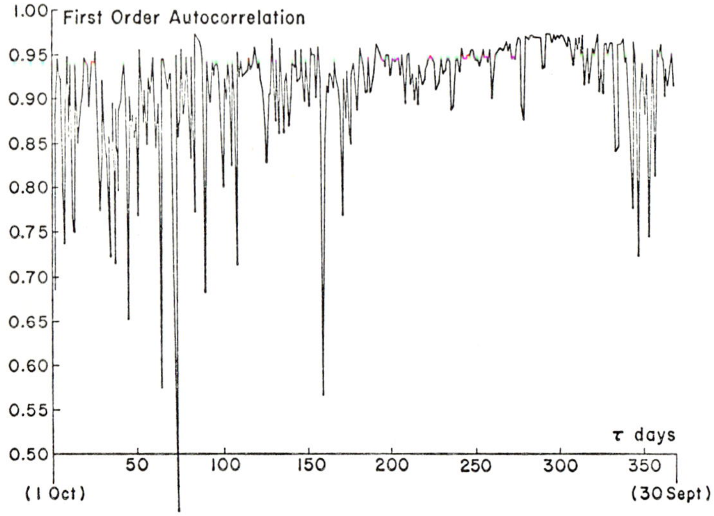

Fig. 8. The sequence of 365 first serial correlation coefficients of daily flows of the Boise River, Idaho.

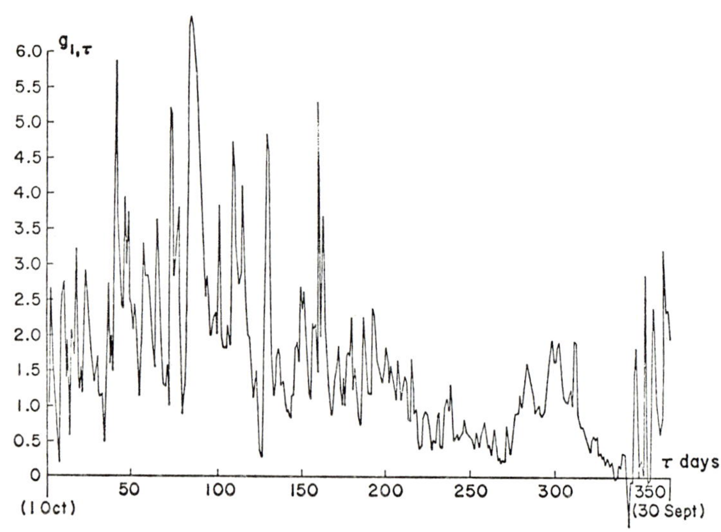

Fig. 9. The sequence of 365 coefficients of skewness of daily flows of the Boise River, Idaho.

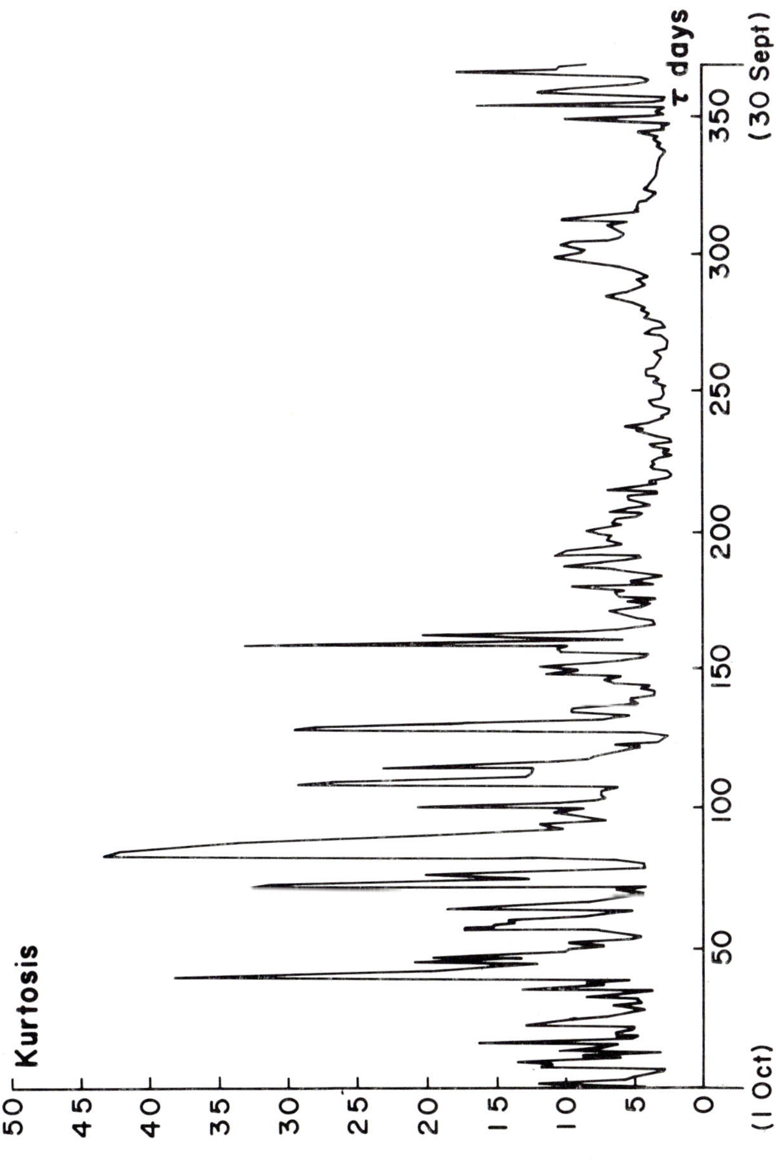

Fig. 10. The sequence of 365 coefficients of Kurtosis of daily flows of the Boise River, Idaho.

exists between the mathematical model and the real population process of periodic-stochastic, short-interval (daily in particular), river flow time series. The present approaches by hydrologists in modeling the complex precipitation, runoff and water quality, short interval, time series require by necessity an optimization. Processes are modeled in order to generate the new samples by Monte Carlo method, among several objectives. The optimization is, therefore, between the work involved in deriving models in complex forms (say by taking into account the higher-order non-stationarity, the higher-order time dependence, and the mixed probability distribution functions to fit the frequency distributions of time independent stochastic components), and in using significant computer time in generating new samples of good preservation of basic population characteristics, on one side, or in using the opposite approach in simplifying the models, on the other side. Trends have been toward using this second approach in the past, with the simple and approximate models, an easy generation of new samples, but with a constant complaint that the generated samples did not preserve the properties of historic time series.

Another dilemma facing hydrologists in generating new samples, both for the short-interval and the long-interval time series, is whether to generate first the short-interval samples, and by summation to produce also the long-interval samples, on one side, or to generate first the long-interval samples and then to add the generated short-interval differences of series to the generated long-interval samples.

4. PROBLEM OF STATIONARITY OF ANNUAL HYDROLOGIC AND CLIMATOLOGIC TIME SERIES

One of the central problems in water resources concerns the stationarity of annual runoff and annual precipitation series. In studies of longest time series of best quality data on annual precipitation and annual runoff, available for the last 125-175 years, no periodicity could be detected. Annual precipitation series are independent stationary stochastic processes. Annual runoff series are dependent stationary stochastic processes. Climatic changes based on the long-range fluctuation in surface temperature as the result of changes in distribution of solar radiation over the Earth's surface with time, according to the Milankovich's theory of the three almost-periodic astronomical processes (eccentricity, tilt or obliquity, and precession), were not significant over the last 150 years [3]. Series could be considered temporary stationary stochastic processes, at least for a couple of centuries, in the case of annual precipitation and annual runoff. The concept of temporary stationarity is necessary for hydrology. Otherwise, the undisputable long-range but slow climatic changes over the Earth may be indiscriminately used for infer the sudden changes for the short periods. The long-range climatic and hydrologic persistences are nebulous concepts at present. This persistence is often advocated without the plausible, physical or statistical proofs for either periodic, or trend or stochastic changes in basic population parameters of annual series of precipitation and runoff. The statistical rigor in discriminating changes is very much needed in both climatology and hydrology. What was the astrology to astronomy in the past, the superficial inferences on climatologic changes based on a limited or one-sided information are the corresponding pseudo-scientific approaches to climatology and hydrology at present.

5. COMBINATION OF STATISTICAL AND PHYSICAL INFORMATION

In studies of hydrologic stochastic processes the statistical approach of investigation is often used by neglecting the physical background or physical interrelationships. The physics of hydrologic processes contains information that can well supplement the statistical analysis of data. Most hydrologic studies have been carried out either only by a deterministic or only by a statistical approach. Bridges between the physical and statistical methods of analysis have not yet been well established in investigation of water resources processes. It is also too early to ascertain whether the statistical analysis only will be able to infer on the best models to be used and to estimate their parameters. The generation of new samples promises to shed some insight on the character of hydrologic processes. The better the analysis and inference on mathematical models and the more efficient the estimation of their parameters, the more reliable becomes the use of generated samples for investigations. The experience shows that reducing the time and space dependent hydrologic historic samples to time independent and normally distributed random variables, linearly correlated in space, are needed in order to apply the principal component analysis in generating the new multivariate samples. The information on physical cause-effect relationships should basically contribute to the selection of the best mathematical models, while the statistical data analyses should give the best estimates of model parameters. Methods are needed for an appropriate combination of the hydrologic, statistical and physical, informations.

6. DISCRIMINATION OF CHANGES IN CLIMATOLOGIC AND HYDROLOGIC TIME PROCESSES

The conclusions to be derived from the analysis of changes in climatologic and hydrologic time processes are so important to economic and social problems, that the analysis should not be left alone to climatologists and hydrologists. The basic difficulties which are inherent in the detection and analysis of changes in these processes are schematically presented in Figure 11. The modern data collection by using the advanced instrumentation is approximately 175-200 years old. The gathering of historic data of climatologic, hydrologic, and other geophysical

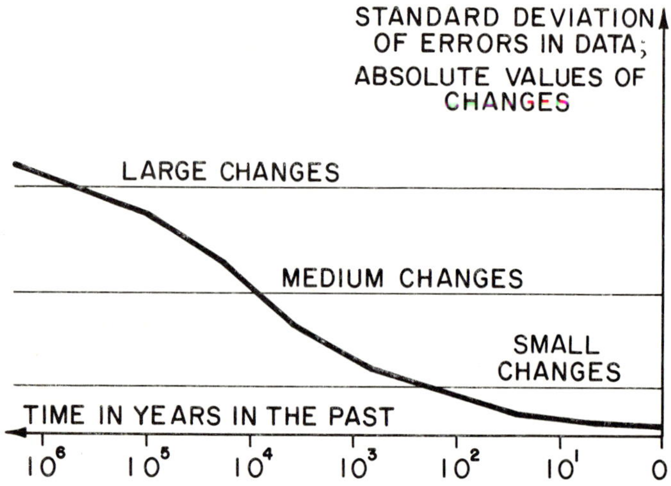

Fig. 11. Difficulty in discrimination of changes in chronologic, hydrologic and climatologic series, because of errors in data.

processes is referred to as the chronology. Tree rings, varves (sediment layers on the bottom of old lakes and other depositions), properties of accumulated ice in glaciers, pollen depositions in sediments, various approaches in determining the surface water temperatures of oceans and seas from the sediment at the bottom of oceans and seas, and similar data, are some of the basic chronologic materials that have been studied for the derivation of conclusions on climatologic changes. The most basic characteristic of the reconstructed chronologic series is that the further one goes into the past, the larger the standard deviation of errors in data (errors defined as differences between the true values at the time of formation of the chronologic records and the estimates of these values from the present data). Therefore, only the large changes can be detected in the remote past, the medium changes in the past couple of thousand years, and the somewhat smaller changes in the modern collection of data in the last 200 years. Because of the high dependence in all of the chronologic time series, (with some of them not resulting from the nature but from the way how the data have been processed), any discrimination technique used should take into account the complexity of processes beyond the accounting for various errors in data. Therefore, the need exists for the very advanced, rigorous, statistical techniques for discriminating the changes in climatology and hydrology. Because the national welfare requires a development of large water resources projects, it is important for hydrologists to be able to infer what kind of climatologic and hydrologic conditions are expected to prevail during the physical and particularly during the economical life of these large water resources systems.

7. NEW METHODS FOR STATISTICAL INFERENCE

Because of complexity of most time processes in water resources, the need exists for the new methods of making statistical inferences for both the selection of mathematical models and the estimation of their parameters. This is important in cases of complex processes, and models with the relatively large numbers of parameters. It is often difficult to find the exact sampling distributions of parameters and derived variables of these complex models. A sharpening of techniques for the use of the experimental statistical method (Monte Carlo, or samples generation method) is very much in need. Statisticians often concentrate only to developing analytically the exact, asymptotic or approximate sampling distributions of various parameters and derived variables. The sample generation method, however, promises to produce often a more accurate estimate of sampling distributions, provided that the inferred complex models and the estimated parameters permit a good reproduction of the physical process.

In improving the statistical inference techniques for the case of complex water resources processes, and in using the data generation method and computers, as the practical approach in developing the approximate sampling distributions or in testing the accuracy of the approximate or asymptotic sampling distributions developed analytically, several other useful purposes can be served by the experimental (Monte Carlo) statistical method. Some of these uses are: (1) To enable a full extraction of information from historic series for highly varying derived variables of time series (such variables as the range of partial sums, the run variables, the extreme high or low values or their volumes, etc.); (2) To use the method for the approximate proofs of analytical expressions which have been developed on the basis of hypotheses advanced from the physics of a process; (3) To use the generated samples for the stochastic optimization and decision making in planning and operating the water resources systems, especially the large reserviors; (4) To determine more accurately the risks and uncertainties which are involved in planning, design and operation of water resources systems; and (5) To use it for several other applied problems in water resources. Therefore, the experimental statistical method (Monte Carlo, or samples generation method) has a large potential for application to stochastic processes in hydrology and to input and output time processes of water resources systems.

8. DESIGN OF EXPERIMENTS

Statistical design of experiments has been developed for various disciplines, especially in agricultural and biological experimentation. An extreme complexity of line, area, space and time water resources processes require additional research and development of statistical design of experiments to fit the specific characteristics of these processes. Hydrologic services have been developed on the basis of inferences by hydrologists of what should be observed many decades ago. The intermittent (truncated) processes in nature, such as precipitation, sediment transport, and other important variables, require such analyses of these processes that would lead to conclusions of how best to observe processes in order to optimize information for the available funds. Some hydrologic services around the world have attempted to find the worth of their data to various users, as well as to determine how long to observe variables in order to sufficiently sample their time processes. Because of a continuous influence of man on the environment, the question has been raised not only of how long the natural processes should continue to be observed but also in order to detect what effects the various interventions and/or conservation measures produce on these processes. Proper statistical design of experiments, including the design of long-range observational services, may significantly improve the economy and efficiency of various field experiments or long-range observations.

9. PROPERTIES OF DERIVED VARIABLES

Developments of exact distributions of specific characteristics of water resources stochastic processes are very important whenever it is feasible to derive them by using the analytical methods of probability theory, stochastic processes and mathematical statistics. The derived variables are such as the maximum, minimum and range of partial sums, characteristics of runs such as the duration, deficit and intensity of runs, and similar random variables which are important in planning, design and operation of water resources systems. Furthermore, the extremes such as floods and droughts are represented by different variables such as the maximum or minimum instantaneous values, large excess or deficit volumes, and similar random variables. Because of periodicities in parameters of most hydrologic processes, it is difficult to derive analytically some of these characteristics in form of distributions of derived random variables. Often, it is difficult even to define which are the appropriate variables that are related to particular problems in planning, design, operation and maintenance of water resources. Therefore, analytical methods of finding statistical properties of derived variables of particular importance to water resources are needed. It is not surprising that in the absence of statistical solutions many hydrologists have been induced to analyze the hydrologic extremes, floods and droughts, water storage problems and similar hydrologic characteristics by theoretical approaches. Likely the maximum contributions can be obtained by works of teams of hydrologists and statisticians, with the former ones expected to be acquainted with physical intricacies of processes and the quality of hydrologic data, and the latter with the potentials and limitations of present-day theories in stochastic processes.

10. MULTIVARIATE ANALYSIS

Nearly all the contributions to and the state-of-the-art of multivariate statistical analyses can be applied to water resources and hydrology. Besides the extensive use of the classical regression and correlation analysis, the principal component analysis, the factor analysis, and the canonical correlation analysis promise large applications to water resources. The principal component analysis is the only available method for generating samples of the four-dimensional hydrologic processes at present, which is able to preserve both the time and space characteristics of these processes in generating the new multivariate samples. The factor analysis can be very useful in finding the new hydrologic quantities that should be observed and better used in explaining the complex processes, than is the case with the classical, observation variables that are used and analyzed at present. Because the large sets of variables must often be correlated in hydrology, the canonical correlation has a definite promise of application. A set of time series of snowmelt runoff may be well correlated for a large mountain chain with a set of time series of accumulated snow on the ground and of time series which affect the snow evaporation and melt. Similarly, a set of ocean temperature time series along the coastal areas may be well related to a set of the areal precipitation time series along those coastal areas. There is a number of problems with the similar sets of variables, that can be studied by the canonical correlation. The multivariate techniques that can be applied to a variety of problems in water resources, involving a large number of set of variables of random processes, need various adjustments in order for them to best fit the physical reality.

It is, maybe, beneficial to state at this point, that blind, non-critical applications of various statistical methods, analytically derived for given conditions and hypotheses, have made a lot of harm in the application of statistics in hydrology and water resources. The reason is that in transferring these methods to water resources the reality is neglected in order not to question the hypotheses underlying the development of these methods. What the applications of stochastic processes in water resources need are the proper adjustments and changes in the developed statistical techniques, with the hypotheses in development to be in agreement with the basic physical properties of hydrologic and water resources processes.

REFERENCES

[1] LEE, RICHARD (1963). Evaluation of solar beam irradiation as a climatic parameter of mountain watersheds. Colorado State University, Hydrology Paper Series, No. 2, August.

[2] YEVJEVICH, VUJICA (1971). Stochasticity in geophysical and hydrological time series. *Nordic Hydrology*, II, 4, 217-242.

[3] CHIN, Q. WILLIAM and YEVJEVICH, VUJICA (1974). Almost-periodic, stochastic process of long-term climatic changes. Colorado State University, Hydrology Papers Series, No. 65, March.

TITLES OF CONTRIBUTED PAPERS

E. Billikam (Air Force Avionics Laboratory) *and A. H. Moore* (Air Force Institute of Technology): Estimation of Mission Reliability from Multiple Independent Grouped Censored Data

E. Bock (Purdue University) *and M. Mintz* (University of Pennsylvania): On Statistical Games with Noncompact Strategy Spaces and Loss Functions That Exhibit a Product Structure

N. K. Bodenstein (Aeronautical Systems Division, Wright-Patterson Air Force Base): The Use of Computers and Adaptive Statistical Filtering to Monitor Electrocardiograms and Detect Premature Venctricular Contractions in Heart Beats

P. Coughlin (State University of New York at Albany): Optimal Statistical Voting in n-Person and Infinite Person Binary Election Games

J. A. Craig and R. A. Huffman (Vought Corporation, Dallas, Texas): An Application of D-Optimal Regression Analysis Design to Optimal Aircraft Design

T. Dillon (Youngstown State University): An Application of Statistics to Mathematics

A. L. Goel (Syracuse University): Reliability and Acceptance Sampling Plans for the Exponential Distribution with Random Parameter

A. K. Gupta (University of Michigan) *and A. P. Basu* (University of Missouri): Estimation of the Restricted Scale Parameter of the Exponential Distribution

R. M. Hashway (Laboratory for Statistical and Policy Research, Boston College): Can Tests Which Maximally Discriminate between Individuals also Detect Between Treatment Differences?

J. J. Higgins and C. P. Tsokos (University of South Florida): The Effect of Loss Function on Bayes Estimation

S. K. Katti (University of Missouri, Columbia): Use of Subjective Information in Objective Technological Research

F. Kozin and R. Gran (Polytechnic Institute of New York): An Approach to Characterizing Modeling and Analyzing Earthquake Excitation Records

M. Kunt (Massachusetts Institute of Technology): Statistical Models and Information Measurements for Two-Level Facsimile

W. Lai (Wright State University): Statistical Analysis and Evaluation of Data Base for Simulation Studies

J. I. McCool (SKF Industries, King of Prussia, Pennsylvania): Random Process Characterization of Rough Surfaces

A. K. Mukherjee (Texas Department of Mental Health and Mental Retardation): A Comparative Study of Multiple Correlation Coefficients Leased on Predictor Variables and on Factor Scores

K. E. Peace and R. E. Flora (Medical College of Virginia, Richmond): Some Results in the Estimation of Survival Parameters

J. Petaku (Massachusetts Institute of Technology): Sequential Medical Trials for Comparing an Experiment with a Standard Treatment

J. Richardson (Eglin Air Force Base, Florida): Delineation of Multivariate Data in Two Dimensions

P. H. Schonemann and J. H. Steiger (Department of Psychological Sciences, Purdue University): A Note on Factor Score Validity

D. G. Shankland (Air Force Institute of Technology): Use of Pseudo-Inverses to Obtain Estimates of System Transfer Functions

N. Shklov (University of Windsor, Canada): Statistical Estimates of Pipeline Model Parameters

P. Smith (Texas A & M University): The Selection of Variables by Groups

R. Strong (Wright-Patterson Air Force Base, Ohio): Life Expectancy Methodology

S. Suthasinekul and S. J. Dwyer, III. (University of Missouri-Columbia): Profile of Level Crossing for Certain Class of Stochastic Point Processes

F. Symons (Katholieke Universiteit de Leuven, Belgium): Testing Whether at least Two out of a Group of Multivariate Swarms do Differ in Position: Some Heuristic Methods

V. M. Rao Tummala (University of Detroit): Minimum Expected Loss (MELO) Estimators of Reliability and Parameters of Certain Life Time Distributions

E. J. Wegman (University of North Carolina): A Monte Carlo Study of Robust Estimators of Location

L. J. Wei (University of South Carolina): Sequential Searching Schemes for an Optimal Dosage - A Bayesian Approach

N. K. Womer (Air Force Institute of Technology): Aircraft Cost Estimation: An Application of Joint Generalized Least Squares

TITLES OF PRESENTATIONS IN CLINICAL SESSIONS

R. Albanese (USAF School of Aerospace Medicine): Problems in Practical Systems Analysis Involving Human Operators: Identification and Level Crossings

A. R. Boehm (Hq. Air Weather Service): Problems of Significance in Verifying Weather Forecast

P. Crump (USAF School of Aerospace Medicine): Missing Data

I. I. Gringorten (Air Force Geophysics Laboratory): Computer Algorithm for Models of the Extent of Weather Conditions in Time and Space

M. Kabrisky (Air Force Institute of Technology): Pattern Recognition

J. Klion (Rome Air Development Center): Reliability

R. G. Merkle (Air Force Flight Dynamics Laboratory): Statistical Treatment of Vibration Measurements

K. A. Myers (Air Force Avionics Laboratory): Applications of Kalman Filtering in Integrated GPS/Inertial Navigation System

M. Obal and J. Lee (Air Force Flight Dynamics Laboratory): Angular Vibration

W. C. Summers (Aerospace Medical Research Laboratories): Time Varying Human Tracking Models

E. W. Turner (Air Force Flight Dynamics Laboratory): Multichannel Flight Data Analysis